# Fundamentals
# of Classical
# Thermodynamics

SERIES IN THERMAL AND TRANSPORT SCIENCES
Gordon J. Van Wylen, Coordinator

FUNDAMENTALS OF CLASSICAL THERMODYNAMICS
*Gordon J. Van Wylen and Richard E. Sonntag*

FUNDAMENTALS OF STATISTICAL THERMODYNAMICS
*Richard E. Sonntag and Gordon J. Van Wylen*

FLUID MECHANICS
*Arthur G. Hansen*

INTRODUCTION TO THERMODYNAMICS: CLASSICAL AND STATISTICAL
*Richard E. Sonntag and Gordon J. Van Wylen*

# Fundamentals of Classical Thermodynamics

Second Edition,

**SI Version**

## Gordon J. Van Wylen

*President*
*Hope College, Holland, Michigan*

## and Richard E. Sonntag

*Professor of Mechanical Engineering*
*University of Michigan, Ann Arbor, Michigan*

JOHN WILEY AND SONS, INC.
New York · London · Sydney · Toronto

*Library of Congress Cataloging in Publication Data:*

Van Wylen, Gordon John.
    Fundamentals of classical thermodynamics.

    (Series in thermal and transport sciences)
    Bibliography: p.
    1. Thermodynamics.  I.  Sonntag, Richard Edwin, joint author.  II.  Title.
TJ265.V23      1976      536'.7      76-2405
ISBN 0-471-90229-2

Printed in the United States of America

10-9 8 7 6 5

# Preface

In this second edition we have retained the basic objective of the first edition: to present a comprehensive and rigorous treatment of classical thermodynamics while retaining an engineering perspective and, in doing so, to both lay the groundwork for subsequent studies in such fields as fluid mechanics, heat transfer, and statistical thermodynamics, and to prepare the student to effectively use thermodynamics in the practice of engineering.

We have deliberately directed our presentation to students. New concepts and definitions are presented in the context where they are first relevant. The first thermodynamic properties to be defined (Chapter 2) are those that can be readily measured: pressure, specific volume, and temperature. In Chapter 3, tables of thermodynamic properties are introduced, but only in regard to these measurable properties. Internal energy and enthalpy are introduced in connection with the first law, entropy with the second law, and the Helmholtz and Gibbs functions in the chapter on availability. Many examples have been included in the book to assist the student in gaining an understanding of thermodynamics, and the problems at the end of each chapter have been carefully sequenced to correlate with the subject matter and provide some progression in difficulty.

We have attempted to cover fairly comprehensively the basic subject matter of classical thermodynamics, and believe that the book provides adequate preparation for study of the application of thermodynamics to the various professional fields as well as for study of more advanced topics in thermodynamics, such as those related to materials, surface phenomena, plasmas, and cryogenics. We also recognize that a number of colleges offer a single introductory course in thermodynamics for all departments, and we have tried to cover those topics that the various departments might wish to have included in such a course. However, since specific courses vary considerably in prerequisites, specific objectives, duration, and background of the students, we have arranged the material, particularly in the later chapters, so that there is considerable flexibility in the amount of material that may be covered.

Throughout the book we have attempted to maintain an engineering perspective, particularly through the choice of examples and problems. One of the major changes from the first edition is in the problems. Although some of the original problems have been retained, most are new or have been significantly revised. A number of applications of thermodynamics have been introduced in the first chapter, and these serve as a basis for many of the problems. We have retained a chapter on cycles (Chapter 9) because we find that many students enjoy this subject, and it can serve to effectively strengthen the student's understanding of the first and second laws of thermodynamics, and introduce him to engineering design and practice. We have also included several comments and problems that relate thermodynamics to environmental concerns.

Chapter 14, which includes an introduction to compressible flow and flow through blade passages, has been retained for those students who would not otherwise undertake extensive studies in fluid flow. We recognize that in many cases the material in this chapter will be covered in other courses.

Several significant changes have been made in the text. The principal change is the extensive revision and simplification of the control volume development of the first and second laws. The revised version has been used in class several times, and the students have found it much easier to comprehend. There are also substantial changes in Chapters 10 through 13, which deal with thermodynamic relations, mixtures, chemical reactions and equilibrium. The material on ideal-real gas behavior and equations of state has been extensively revised, as has that dealing with partial properties and the general behavior of mixtures. New material on the third law, absolute entropy and fuel cells, and a simplified presentation of the material on phase equilibrium are also included. There are numerous places throughout the book where explanations or discussions have been improved. The book has also been generally updated in regard to definitions, applications and thermodynamic data. The new Steam Tables, which include values of internal energy, have been used and a summary included in the appendix.

In regard to the symbols used in this text, we have tried to maintain consistency throughout the book. In a limited number of cases, we have used a given symbol for more than one purpose. We believe, however, that the context will clarify the meaning of the symbol in these cases.

All units in this version of the Second Edition are SI units (*Le Systeme International d'Unites*). Throughout the text material and tables of thermodynamic properties and in most of the examples and problems, we have utilized the basic units for pressure (pascal) and volume (cubic metre). We recognize that some persons will wish to use the bar more extensively as the pressure unit, and that some will wish to use the litre more extensively as the volume unit. We trust that this will present no particular difficulties. Concerning the extensive properties, a lower-case letter ($u$, $h$, $s$) designates the property per unit mass, an uppercase letter ($U$, $H$, $S$) the property for the entire system; a lowercase letter with a bar ($\bar{u}$, $\bar{h}$, $\bar{s}$), the property per unit mole; and an uppercase letter with a bar ($\bar{U}$, $\bar{H}$, $\bar{S}$), the partial molal property. Following this pattern, we have found it convenient to designate the total heat transfer as $Q$, the heat transfer per unit mass of the system as $q$, the total work as $W$, and the work per unit mass of the system as $w$.

Further, we represent the rate of flow across a system boundary or control surface by a dot over a given quantity. Thus, $\dot{Q}$ represents a rate of heat transfer across the system boundary; $\dot{W}$, the rate at which work crosses the system boundary (that is, the power); and $\dot{m}$, the mass rate of flow across a control surface ($\dot{n}$ is used when the mass rate of flow is expressed in moles per unit time). The rate of heat transfer across a control surface is designated $\dot{Q}_{c.v.}$. We realize that we have departed from the usual mathematical use of a dotted symbol since, in mathematics, the dotted symbol typically refers to a derivative with respect to time. However, we have used the dotted symbol only to indicate a flow of heat and work across a system boundary, and heat, work, and mass across a control surface, and we believe that it has contributed to a simple and consistent use of symbols for this book.

We acknowledge with appreciation the suggestions, counsel, and encouragement from many colleagues, both at the University of Michigan and elsewhere. This assistance has been most helpful to us during the writing of both the first and second editions. Several secretaries have aided us immeasurably during this period, and we acknowledge this with many thanks. Students have also aided us. Their perceptive questions have often caused us to rewrite or rethink a given portion of the text. Finally, for each of us, the encouragement and patience of our wives and families have been indispensable, and have made this time of writing pleasant and enjoyable in spite of the pressures of the project.

Our hope is that this book will contribute to the effective teaching of thermodynamics to students who face very significant challenges and opportunities during their professional careers. Your comments, criticism, and suggestions will also be appreciated.

GORDON J. VAN WYLEN

*Ann Arbor, April, 1976*                    RICHARD E. SONNTAG

# Contents

# English Engineering System Data Conversion to SI Units

The basic units in the English Engineering System are the pound mass (lbm), pound force (lbf), foot (ft) or inch (in), and second (s or sec). The force/mass conversion constant is 32.174 $\frac{lbm\ ft}{lbf\ sec^2}$.

Temperature scales are Fahrenheit (F) and Rankine (R). The energy unit in thermodynamics is the British Thermal Unit (1 International Table Btu = 778.17 ft lbf), and the unit of power is the horsepower (hp).

## Conversions

| To convert from | to | Multiply by |
|---|---|---|
| Btu (IT) | kJ | 1.055 056 |
| Btu/lbm | kJ/kg | 2.326* |
| Btu/lbm R | kJ/kg K | 4.1868* |
| ft | m | 0.3048* |
| ft³ | m³ | 0.028 317 |
| in² | mm² | 645.16* |
| lbf | N | 4.448 222 |
| lbm | kg | 0.453 592 |
| lbf/in² | kPa | 6.894 757 |
| ft³/lbm | m³/kg | 0.062 428 |
| hp | kW | 0.7457 |
| R | K | 1/1.8* |

\* Exact relation

# Symbols

| | |
|---|---|
| $a$ | acceleration |
| $a$ | activity |
| $a, A$ | specific Helmholtz function and total Helmholtz function |
| $AF$ | air-fuel ratio |
| $c$ | velocity of sound |
| $C_D$ | coefficient of discharge |
| $C_p$ | constant-pressure specific heat |
| $C_v$ | constant-volume specific heat |
| $C_{po}$ | zero-pressure constant-pressure specific heat |
| $C_{vo}$ | zero-pressure constant-volume specific heat |
| $e, E$ | specific energy and total energy |
| $f$ | fugacity |
| $\bar{f}_i$ | fugacity of component $i$ in a mixture |
| $F$ | force |
| $FA$ | fuel-air ratio |
| $g$ | acceleration due to gravity |
| $g, G$ | specific Gibbs function and total Gibbs function |
| $h, H$ | specific enthalpy and total enthalpy |
| $i$ | electrical current |
| $I$ | irreversibility |
| $J$ | proportionality factor to relate units of work to units of heat |
| $k$ | specific heat ratio: $C_p/C_v$ |

| | |
|---|---|
| $K$ | equilibrium constant |
| KE | kinetic energy |
| $L$ | length |
| $lw, LW$ | lost work per unit mass and total lost work |
| $m$ | mass |
| $\dot{m}$ | mass rate of flow |
| $M$ | molecular weight |
| M | Mach number |
| $mf$ | mass fraction |
| $n$ | number of moles |
| $n$ | polytropic exponent |
| $P$ | pressure |
| $P_i$ | partial pressure of component $i$ in a mixture |
| PE | potential energy |
| $P_r$ | relative pressure as used in gas tables |
| $q, Q$ | heat transfer per unit mass and total heat transfer |
| $\dot{Q}$ | rate of heat transfer |
| $Q_H, Q_L$ | heat transfer with high-temperature body and heat transfer with low-temperature body; sign determined from context |
| $R$ | gas constant |
| $\bar{R}$ | universal gas constant |
| $s, S$ | specific entropy and total entropy |
| $t$ | time |
| $T$ | temperature |
| $u, U$ | specific internal energy and total internal energy |
| $v, V$ | specific volume and total volume |
| $v_r$ | relative specific volume as used in gas tables |
| $vf$ | volume fraction |
| $\mathsf{V}$ | velocity |
| $\mathsf{V}_r$ | relative velocity |
| $w, W$ | work per unit mass and total work |
| $\dot{W}$ | rate of work, or power |
| $w_{\mathrm{rev}}$ | reversible work between two states assuming heat transfer with surroundings |
| $x$ | quality |
| $x$ | liquid-phase or solid-phase mole fraction |
| $y$ | vapor-phase mole fraction |
| $Z$ | elevation |
| $Z$ | compressibility factor |
| $Z$ | electrical charge |

## SCRIPT LETTERS

| | |
|---|---|
| $\mathscr{A}$ | area |
| $\mathscr{C}$ | number of components |
| $\mathscr{E}$ | electrical potential |
| $\mathscr{H}$ | magnetic field intensity |
| $\mathscr{M}$ | magnetization |
| $\mathscr{P}$ | number of phases |
| $\mathscr{S}$ | surface tension |
| $\mathscr{T}$ | tension |
| $\mathscr{V}$ | variance |

## GREEK LETTERS

| | |
|---|---|
| $\alpha$ | residual volume |
| $\alpha_p$ | volume expansivity |
| $\beta$ | coefficient of performance for a refrigerator |
| $\beta'$ | coefficient of performance for a heat pump |
| $\beta_S$ | adiabatic compressibility |
| $\beta_T$ | isothermal compressibility |
| $\gamma$ | activity coefficient |
| $\eta$ | efficiency |
| $\mu$ | chemical potential |
| $\mu_J$ | Joule-Thomson coefficient |
| $\nu$ | stoichiometric coefficient |
| $\rho$ | density |
| $\phi$ | relative humidity |
| $\phi$ | availability for a system |
| $\psi$ | availability associated with a steady-state, steady-flow process |
| $\omega$ | humidity ratio or specific humidity |

## SUBSCRIPTS

| | |
|---|---|
| $c$ | property at the critical point |
| c.v. | control volume |
| $e$ | state of a substance leaving a control volume |
| $f$ | formation |
| $f$ | property of saturated liquid |
| $fg$ | difference in property for saturated vapor and saturated liquid |
| $g$ | property of saturated vapor |
| $i$ | state of a substance entering a control volume |

| | |
|---|---|
| $i$ | property of saturated solid |
| $ig$ | difference in property for saturated vapor and saturated solid |
| $r$ | reduced property |
| $s$ | isentropic process |
| 0 | property of the surroundings |
| 0 | stagnation property |

## SUPERSCRIPTS

| | |
|---|---|
| — | bar over symbol denotes property on a molal basis (over $V$, $H$, $S$, $U$, $A$, $G$, the bar denotes partial molal property) |
| ° | property at standard-state condition |
| * | ideal gas |
| * | property at the throat of a nozzle |
| $L$ | liquid-phase |
| $S$ | solid-phase |
| $V$ | vapor-phase |

# Fundamentals
# of Classical
# Thermodynamics

# 1

# Some Introductory Comments

In the course of our study of thermodynamics, a number of the examples and problems presented refer to processes that occur in such equipment as a steam power plant, a fuel cell, a vapor compression refrigerator, a thermoelectric cooler, a rocket engine, and an air separation plant. In this introductory chapter a brief description of this equipment is given. There are at least two reasons for including such a chapter. First, many students have had limited contact with such equipment, and the solution of problems will be more significant and relevant when they have some familiarity with the actual processes and the equipment involved. Second, this chapter will provide an introduction to thermodynamics, including the use of certain terms (which will be more formally defined in later chapters), some of the problems for which thermodynamics is relevant, and some accomplishments that have resulted, at least in part, from the application of thermodynamics.

It should be emphasized that thermodynamics is relevant to many other processes than those cited in this chapter. It is basic to the study of materials, chemical reactions, and plasmas. The student should bear in mind that this chapter is only a brief and necessarily very incomplete introduction to the subject of thermodynamics.

## 1.1  The Simple Steam Power Plant

A schematic diagram of a simple steam power plant is shown in Fig. 1.1. High-pressure superheated steam leaves the boiler, which is also referred to as a steam generator, and enters the turbine. The steam expands in the turbine and in doing so, does work, which enables the turbine to drive the electric generator. The low-pressure steam leaves

1

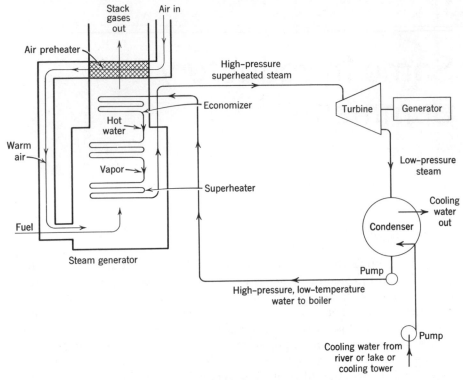

**Fig. 1.1**   Schematic diagram of a steam power plant.

the turbine and enters the condenser, where heat is transferred from the steam (causing it to condense) to the cooling water. Since large quantities of cooling water are required, power plants are frequently located near rivers or lakes. It is this transfer of heat to the water in lakes and rivers that leads to the thermal pollution problem, which is now being extensively studied. During our study of thermodynamics we shall gain an understanding of why this heat transfer is necessary and ways in which it can be minimized. When the supply of cooling water is limited, a cooling tower may be used. In the cooling tower some of the cooling water evaporates in such a way as to lower the temperature of the water that remains as a liquid.

The pressure of the condensate leaving the condenser is increased in the pump, thus enabling the condensate to flow into the steam generator. In many steam generators an economizer is used. An economizer is simply a heat exchanger in which heat is transferred from the products of combustion (just before they leave the steam generator) to the condensate, with the result that the temperature of the condensate is

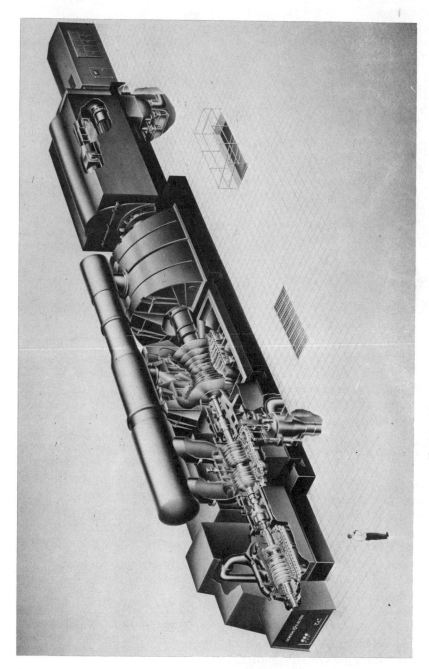

**Fig. 1.2**    A large steam turbine (Courtesy General Electric Co.).

**Fig. 1.3** A condenser used in a large power plant (Courtesy Westinghouse Electric Corp.).

increased, but no evaporation takes place. In other sections of the steam generator, heat is transferred from the products of combustion to the water, causing it to evaporate. The temperature at which evaporation occurs is called the saturation temperature. The steam then flows through another heat exchanger known as a superheater, where the temperature of the steam is increased well above the saturation temperature.

In many power plants the air that is used for combustion is preheated in the air preheater by transferring heat from the stack gases as they are leaving the furnace. This air is then mixed with fuel—which might be coal, fuel oil, natural gas, or other combustible material—and combustion takes place in the furnace. As the products of combustion pass through the furnace, heat is transferred to the water in the superheater, the boiler, the economizer, and to the air in the air preheater. The products of combustion from power plants are discharged to the atmosphere, and this constitutes one of the facets of the air pollution problem we now face.

A large power plant will have many other pieces of equipment, some of which will be considered in later chapters.

Figure 1.2 shows a steam turbine and the generator that it drives.

Steam turbines vary in capacity from less than 10 kilowatts to 1 000 000 kilowatts.

Figure 1.3 shows a cutaway view of a condenser. The steam enters at the top and the condensate is collected in the hot well at the bottom while the cooling water flows through the tubes. A large condenser has a tremendous number of tubes, as shown in Fig. 1.3.

Figure 1.4 shows a large steam generator. The flow of air and products of combustion are indicated. The condensate, also called the boiler feedwater, enters at the economizer inlet, and the superheated steam leaves at the superheater outlet.

The number of nuclear power plants in operation is rapidly increas-

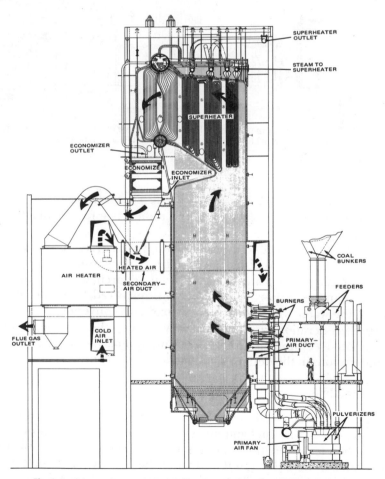

**Fig. 1.4**   A large steam generator (Courtesy Babcock and Wilcox Co.).

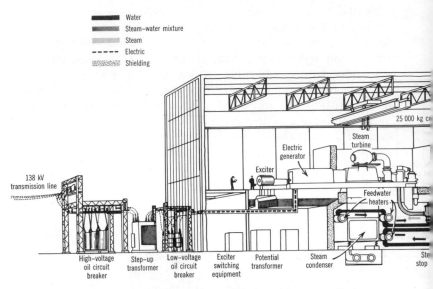

**Fig. 1.5**   Schematic diagram of the Big Rock Point nuclear plant of Consumers Power Company at Charlevoix, Michigan (Courtesy Consumers Power Company).

ing. In these power plants the reactor replaces the steam generator of the conventional power plant, and the radioactive fuel elements replace the coal, oil, or natural gas.

There are several different reactor designs in current use. One of these is the boiling water reactor, such as the system shown in Fig. 1.5. In other nuclear power plants a secondary fluid circulates from the reactor to the steam generator, where heat is transferred from the secondary fluid to the water which in turn goes through a conventional steam cycle. Safety considerations and the necessity to keep the turbine, condenser, and related equipment from becoming radioactive are always major considerations in the design of a nuclear power plant.

## 1.2   Fuel Cells

When a conventional power plant is viewed as a whole, as shown in Fig. 1.6, we see that fuel and air enter the power plant and products of combustion leave the unit. There is also a transfer of heat to the cooling water, and work is done in the form of the electrical energy leaving the power plant. The overall objective of a power plant is to convert the

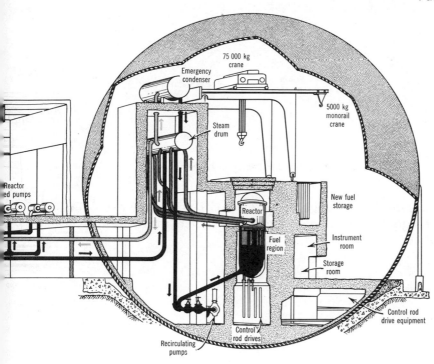

availability (to do work) of the fuel into work (in the form of electrical energy) in the most efficient manner, consistent with such considerations as cost, space, and safety.

We might well ask if all of the equipment in the power plant, such as the steam generator, the turbine, the condenser, and the pump, is necessary. Is it not possible to produce electrical energy from the fuel in a more direct manner?

The fuel cell is a device in which this objective is accomplished. Figure 1.7 shows a schematic arrangement of a fuel cell of the ion-exchange membrane type. In this fuel cell hydrogen and oxygen react to form water. Let us consider the general features of the operation of this type of fuel cell.

The flow of electrons in the external circuit is from anode to cathode. Hydrogen enters at the anode side and oxygen enters at the cathode side. At the surface of the ion-exchange membrane the hydrogen is ionized according to the reaction

$$2H_2 \rightarrow 4H^+ + 4e^-$$

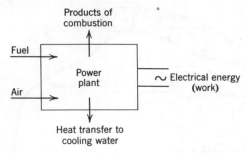

**Fig. 1.6**   Schematic diagram of a power plant.

The electrons flow through the external circuit and the hydrogen ions flow through the membrane to the cathode, where the following reaction takes place.

$$4H^+ + 4e^- + O_2 \rightarrow 2H_2O$$

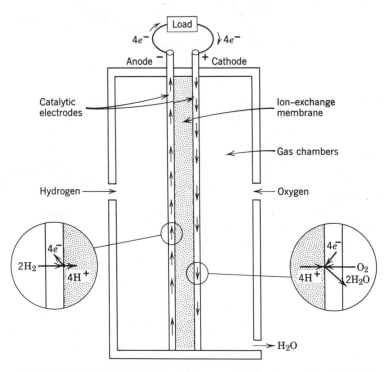

**Fig. 1.7**   Schematic arrangement of an ion-exchange membrane type of fuel cell.

There is a potential difference between the anode and cathode, and thus there is a flow of electricity through a potential difference which, in thermodynamic terms, is called work. There may also be a transfer of heat between the fuel cell and the surroundings.

At the present time the fuel used in fuel cells is usually either hydrogen or a mixture of gaseous hydrocarbons and hydrogen. The oxidizer is usually oxygen. However, current research is directed toward the development of fuel cells that use hydrocarbon fuels and air. Although the conventional (or nuclear) steam power plant is still used in large-scale power generating systems, and conventional piston engines and gas turbines are still used in most transportation power systems, the fuel cell may eventually become a serious competitor. The fuel cell is already being used to produce power for certain space applications.

Thermodynamics plays a vital role in the analysis, development, and design of all power-producing systems, including reciprocating internal combustion engines and gas turbines. Such considerations as the increase of efficiency, improved design, optimum operating conditions, and alternate methods of power generation involve, among other factors, the careful application of the fundamentals of thermodynamics.

## 1.3   The Vapor-Compression Refrigeration Cycle

A simple vapor-compression refrigeration cycle is shown schematically in Fig. 1.8. The refrigerant enters the compressor as a slightly super-

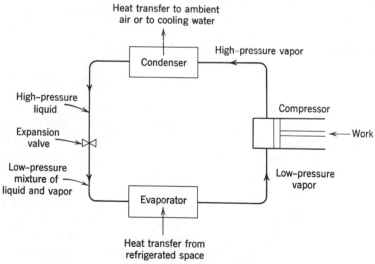

**Fig. 1.8**   Schematic diagram of a simple refrigeration cycle.

heated vapor at a low pressure. It then leaves the compressor and enters the condenser as a vapor at some elevated pressure, where the refrigerant is condensed as a result of heat transfer to cooling water or to the surroundings. The refrigerant then leaves the condenser as a high-pressure liquid. The pressure of the liquid is decreased as it flows through the expansion valve and, as a result, some of the liquid flashes into vapor. The remaining liquid, now at a low pressure, is vaporized in the evaporator as a result of heat transfer from the refrigerated space. This vapor then enters the compressor.

In a typical home refrigerator the compressor is located in the rear near the bottom of the unit. The compressors are usually hermetically sealed; that is, the motor and compressor are mounted in a sealed housing, and the electric leads for the motor pass through this housing. This is done to prevent leakage of the refrigerant. The condenser is also located at the back of the refrigerator and is so arranged that the air in the room flows past the condenser by natural convection. The expansion valve takes the form of a long capillary tube and the evaporator is located around the outside of the freezing compartment inside the refrigerator.

Figure 1.9 shows a large centrifugal unit that is used to provide refrigeration for an air-conditioning unit. In this unit, water is cooled and then circulated to provide cooling where needed.

## 1.4  The Thermoelectric Refrigerator

We may well ask the same question about the vapor compression refrigerator that we asked about the steam power plant, namely, isn't it possible to accomplish our objective in a more direct manner? Isn't it possible, in the case of a refrigerator, to use the electrical energy (which goes to the electric motor that drives the compressor) to produce cooling in a more direct manner, and avoid the cost of the compressor, condenser, evaporator, and all the related piping.

The thermoelectric refrigerator is such a device. This is shown schematically in Fig. 1.10. The thermoelectric device, like the conventional thermocouple, utilizes two dissimilar materials. There are two junctions between these two materials in a thermoelectric refrigerator. One is located in the refrigerated space, and the other in ambient surroundings. When a potential difference is applied, as indicated, the temperature of the junction located in the refrigerated space will decrease and the temperature of the other junction will increase. Under steady-state operating conditions heat will be transferred from the refrigerated space to the cold junction. The other junction will be at a temperature above the ambient, and heat will be transferred from the junction to the surroundings.

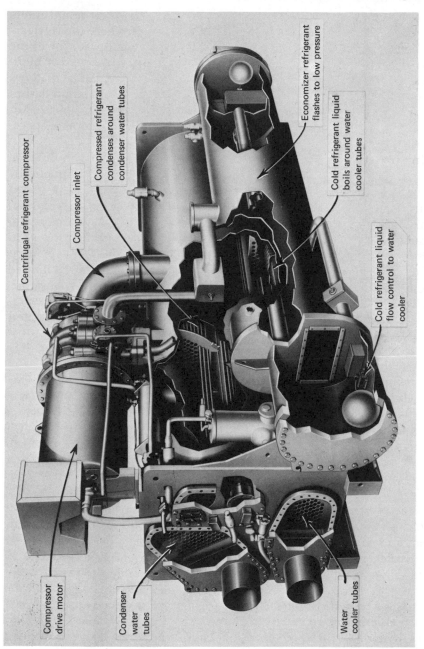

Centrifugal refrigerant compressor

Compressor inlet

Compressed refrigerant condenses around condenser water tubes

Economizer refrigerant flashes to low pressure

Cold refrigerant liquid boils around water cooler tubes

Cold refrigerant liquid flow control to water cooler

Compressor drive motor

Condenser water tubes

Water cooler tubes

**Fig. 1.9** A refrigeration unit for an air-conditioning system (Courtesy Carrier Air Conditioning Co.).

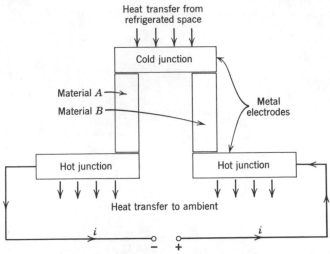

**Fig. 1.10**  A thermoelectric refrigerator.

It should be emphasized that a thermoelectric device can also be used to generate power by replacing the refrigerated space with a body which is at a temperature above the ambient. Such a system is shown in Fig. 1.11.

The thermoelectric refrigerator cannot yet compete economically with the conventional vapor-compression units. However, in certain special applications the thermoelectric refrigerator is already in use, and in view of the major research and development efforts underway in this field, it is quite possible that the use of thermoelectric refrigerators will be much more extensive in the future.

## 1.5   The Air Separation Plant

One process of great industrial significance is the air separation plant, in which air is separated into its various components. The oxygen, nitrogen, argon, and rare gases so produced are used extensively in various industrial, research, space, and consumer goods applications. The air separation plant may be considered as an example from two major fields, the chemical process industry and the field of cryogenics. Cryogenics is a term that refers to technology, processes, and research at very low temperatures (in general, below 150 K). In both chemical processing and cryogenics, thermodynamics is basic to an understanding of many phenomena that occur and to the design and development of processes and equipment.

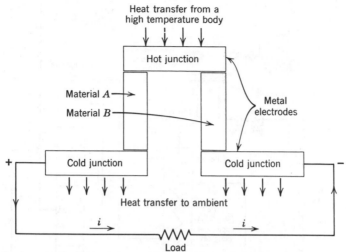

**Fig. 1.11**   A thermoelectric power generation device.

A number of different designs of air separation plants have been developed. Consider Fig. 1.12, which shows a somewhat simplified sketch of a type of plant that is frequently used. Air from the atmosphere is compressed to a pressure of 2 to 3 megapascals. It is then purified, particularly to remove carbon dioxide (which would plug the flow passages as it solidifies when the air is cooled to its liquefaction temperature). The air is then compressed to a pressure of 15 to 20 megapascals, cooled to the ambient temperature in the aftercooler, and dried to remove the water vapor (which would also plug the flow passages as it freezes).

The basic refrigeration in the liquefaction process is provided by two different processes. One involves expansion of the air in the expansion engine. During this process the air does work and as a result the temperature of the air is reduced. The other refrigeration process involves passing the air through a throttle valve that is so designed and so located that there is a substantial drop in the pressure of the air and, associated with this, a substantial drop in the temperature of the air.

As shown in Fig. 1.12, the dry, high pressure air enters a heat exchanger. The air temperature drops as it flows through the heat exchanger. At some intermediate point in the heat exchanger, part of the air is bled off and flows through the expansion engine. The remaining air flows through the rest of the heat exchanger and through the throttle valve. The two streams join, both being at the pressure of 0.5-1 mega-

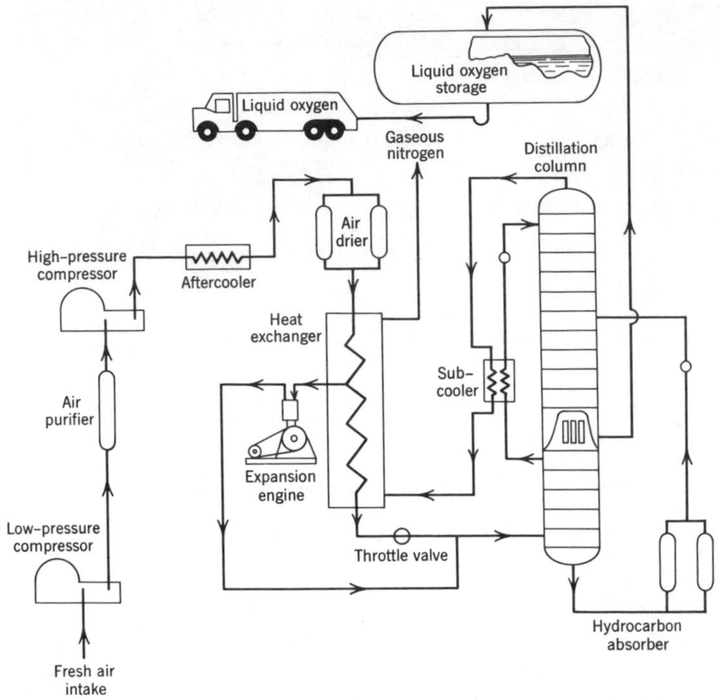

**Fig. 1.12**   A simplified diagram of a liquid oxygen plant (Courtesy Air Products and Chemicals, Inc.).

pascals, and enter the bottom of the distillation column, which is referred to as the high pressure column. The function of the distillation column is to separate the air into its various components, principally oxygen and nitrogen. Two streams of different composition flow from the high pressure column through throttle valves to the upper column (also called the low-pressure column). One of these is an oxygen-rich liquid that flows from the bottom of the lower column, and the other is a nitrogen-rich stream that flows through the subcooler. The separation is completed in the upper column, with liquid oxygen leaving from the bottom of the upper column and gaseous nitrogen from the top of the column. The nitrogen gas flows through the subcooler and the main heat exchanger. It is the heat transfer to this cold nitrogen gas that causes the cooling of the high pressure air entering the heat exchanger.

Not only is a thermodynamic analysis essential to the design of the system as a whole, but essentially every component of such a system, including the compressors, the expansion engine, the purifiers and driers, and the distillation column, involves thermodynamics. In this separation process we are also concerned with the thermodynamic properties of mixtures and the principles and procedures by which these mixtures can be separated. This is the type of problem encountered in the refining of petroleum and many other chemical processes. It should also be noted that cryogenics is particularly relevant to many aspects of the space program, and a thorough knowledge of thermodynamics is essential for creative and effective work in cryogenics.

## 1.6   The Chemical Rocket Engine

The advent of missiles and satellites has brought to prominence the use of the rocket engine as a propulsion power plant. Chemical rocket engines may be classified as either liquid propellant or solid propellant, according to the fuel used.

Figure 1.13 shows a simplified schematic diagram of a liquid-propellant

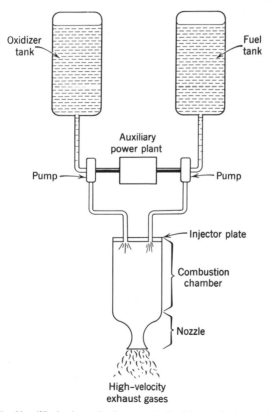

**Fig. 1.13**   Simplified schematic diagram of a liquid-propellant rocket engine.

external to the system is the surroundings, and the system is separated from the surroundings by the system boundaries. These boundaries may be either movable or fixed.

In Fig. 2.1 the gas in the cylinder is considered the system. If a Bunsen

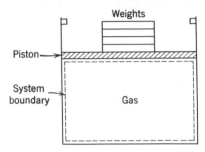

**Fig. 2.1**   Example of a system.

burner is placed under the cylinder, the temperature of the gas will increase and the piston will rise. As the piston rises, the boundary of the system moves. As we shall see later, heat and work cross the boundary of the system during this process, but the matter that comprises the system can always be identified.

An isolated system is one that is not influenced in any way by the surroundings. This means that no heat or work cross the boundary of the system.

In many cases a thermodynamic analysis must be made of a device, such as an air compressor, that involves a flow of mass into and/or out of the device, as shown schematically in Fig. 2.2. The procedure that is

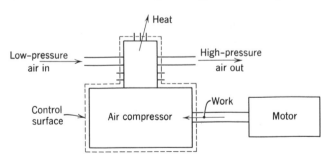

**Fig. 2.2**   Example of a control volume.

followed in such an analysis is to specify a control volume that surrounds the device under consideration. The surface of this control volume is referred to as a control surface. Mass, as well as heat and work (and momentum) can flow across the control surface.

Thus, a system is defined when dealing with a fixed quantity of mass, and a control volume is specified when an analysis is to be made that involves a flow of mass. The difference in these two approaches is considered in detail in Chapter 5. It should be noted that the terms closed system and open system are sometimes used as the equivalent of the terms system (fixed mass) and control volume (involving a flow of mass). The procedure that will be followed in the presentation of the first and second laws of thermodynamics is first to present these laws for a system, and then to make the necessary transformations to apply them to a control volume.

## 2.2    Macroscopic vs. Microscopic Point of View

An investigation into the behavior of a system may be undertaken from either a microscopic or macroscopic point of view. Let us briefly consider the problem we would have if we describe a system from a microscopic point of view. Consider a system consisting of a cube 25 mm on a side and containing a monatomic gas at atmospheric pressure and temperature. This volume contains approximately $10^{20}$ atoms. To describe the position of each atom, three coordinates must be specified; to describe the velocity of each atom, three velocity components must be specified.

Thus, to completely describe the behavior of this system from a microscopic point of view, it would be necessary to deal with at least $6 \times 10^{20}$ equations. Even with a large digital computer, this is a quite hopeless computational task. However, there are two approaches to this problem that reduce the number of equations and variables to a few that can be handled relatively easily in performing computations. One of these approaches is the statistical approach in which, on the basis of statistical considerations and probability theory, we deal with "average" values for all particles in the system. This is usually done in connection with a model of the atom under consideration. This is the approach used in the disciplines known as kinetic theory and statistical mechanics.

The other approach that reduces the number of variables to a few that can be handled is the macroscopic point of view of classical thermodynamics. As the word macroscopic implies, we are concerned with the gross or average effects of many molecules. Furthermore, these effects can be perceived by our senses and measured by instruments. In so doing, however, what we really perceive and measure is the time-averaged influence of many molecules. For example, consider the pressure a gas exerts on the walls of its container. This pressure results from the change in momentum of the molecules as they collide with the wall. However, from a macroscopic point of view, we are not concerned with the action of the individual molecules but with the time-averaged force on a given area, which can be measured by a pressure gage. In fact, these macroscopic observations are completely independent of our assumptions regarding the nature of matter.

Although the theory and development in this book will be presented from a macroscopic point of view, a few supplementary remarks regarding the significance of the microscopic perspective, are included as an aid to the understanding of the physical processes involved. Another book in this series on Thermal and Transport Sciences, namely, *Fundamentals of Statistical Thermodynamics*, by R. E. Sonntag and G. J. Van Wylen deals with thermodynamics from the microscopic and statistical point of view.

A few remarks should be made regarding the continuum. From the macroscopic point of view, we are always concerned with volumes that are very large compared to molecular dimensions, and, therefore, with systems that contain many molecules. Since we are not concerned with the behavior of individual molecules, we can treat the substance as being continuous, disregarding the action of individual molecules, and this is called a continuum. The concept of a continuum, of course, is only a convenient assumption that loses validity when the mean free path of the molecules approaches the order of magnitude of the dimensions of the vessel, as, for example, in high-vacuum technology. In much engineering work the assumption of a continuum is valid and convenient, and goes hand in hand with the macroscopic point of view.

## 2.3   Properties and State of a Substance

If we consider a given mass of water, we recognize that this water can exist in various forms. If it is a liquid initially, it may become a vapor when it is heated, or a solid when it is cooled. Thus we speak of the different phases of a substance. A phase is defined as a quantity of matter that is homogeneous throughout. When more than one phase is present the phases are separated from each other by the phase boundaries. In each phase the substance may exist at various pressures and temperatures or, to use the thermodynamic term, in various states. The state may be identified or described by certain observable, macroscopic properties; some familiar ones are temperature, pressure, and density. In later chapters other properties will be introduced. Each of the properties of a substance in a given state has only one definite value, and these properties always have the same value for a given state, regardless of how the substance arrived at that state. In fact, a property can be defined as any quantity that depends on the state of the system and is independent of the path (i.e., the prior history) by which the system arrived at the given state. Conversely, the state is specified or described by the properties, and later we shall consider the number of independent properties a substance can have, i.e., the minimum number of properties that must be specified in order to fix the state of the substance.

Thermodynamic properties can be divided into two general classes, intensive and extensive properties. An intensive property is independent of the mass; the value of an extensive property varies directly with the mass. Thus, if a quantity of matter in a given state is divided into two equal parts, each part will have the same value of intensive properties as the original, and half the value of the extensive properties. Pressure, temperature, and density are examples of intensive properties. Mass and total volume are examples of extensive properties. Extensive properties per unit mass, such as specific volume, are intensive properties.

Frequently we will refer not only to the properties of a substance but to the properties of a system. When we do so we necessarily imply that the value of the property has significance for the entire system, and this implies what is called equilibrium. For example, if the gas that comprises the system in Fig. 2.1 is in thermal equilibrium, the temperature will be the same throughout the entire system, and we may speak of the temperature as a property of the system. We may also consider mechanical equilibrium and this is related to pressure. If a system is in mechanical equilibrium, there is no tendency for the pressure at any point to change with time as long as the system is isolated from the surroundings. There will be a variation in pressure with elevation, due to the influence of gravitational forces, although under equilibrium conditions there will be no tendency for the pressure at any location to change. However, in many thermodynamic problems this variation in pressure with elevation is so small that it can be neglected. Chemical equilibrium is also important and will be considered in Chapter 13.

When a system is in equilibrium as regards all possible changes of state, we say that the system is in thermodynamic equilibrium.

## 2.4   Processes and Cycles

Whenever one or more of the properties of a system change we say that a change in state has occurred. For example, when one of the weights on the piston in Fig. 2.3 is removed, the piston rises and a change in state occurs, for the pressure decreases and the specific volume increases. The path of the succession of states through which the system passes is called the process.

Let us consider the equilibrium of a system as it undergoes a change in state. The moment the weight is removed from the piston in Fig. 2.3,

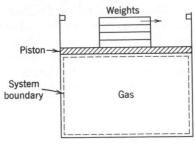

**Fig. 2.3** Example of a system that may undergo a quasiequilibrium process.

mechanical equilibrium does not exist and as a result the piston is moved upward until mechanical equilibrium is again restored. The question that arises is this: since the properties describe the state of a system only when it is in equilibrium, how can we describe the states of a system during a process if the actual process occurs only when equilibrium does not exist. One step in the answer to this question concerns the definition of an ideal process, which we call a quasiequilibrium process. A quasiequilibrium process is one in which the deviation from thermodynamic equilibrium is infinitesimal, and all the states the system passes through during a quasiequilibrium process may be considered as equilibrium states. Many actual processes closely approach a quasiequilibrium process, and may be so treated with essentially no error. If the weights on the piston in Fig. 2.3 are small and are taken off one by one, the process could be considered quasiequilibrium. On the other hand, if all the weights were removed at once, the piston would rise rapidly until it hit the stops. This would be a nonequilibrium process, and the system would not be in equilibrium at any time during this change of state.

For nonequilibrium processes, we are limited to a description of the system before the process occurs and after the process is completed and equilibrium is restored. We are not able to specify each state through which the system passes, nor the rate at which the process occurs. However, as we shall see later, we are able to describe certain over-all effects which occur during the process.

Several processes are described by the fact that one property remains constant. The prefix iso- is used to describe this. An isothermal process is a constant-temperature process, an isobaric (sometimes called iso-piestic) process is a constant-pressure process, and an isochoric process is a constant-volume process.

When a system in a given initial state goes through a number of different changes of state or processes and finally returns to its initial state, the system has undergone a cycle. Therefore, at the conclusion of a cycle all the properties have the same value they had at the beginning. Steam (water) that circulates through a steam power plant undergoes a cycle.

A distinction should be made between a thermodynamic cycle, which has just been described, and a mechanical cycle. A four-stroke cycle internal-combustion engine goes through a mechanical cycle once every two revolutions. However, the working fluid does not go through a thermodynamic cycle in the engine, since air and fuel are burned and changed to products of combustion which are exhausted to the atmosphere. In this text the term cycle will refer to a thermodynamic 'cycle' unless otherwise designated.

## 2.5   Units for Mass, Length, Time, and Force

Since we are considering thermodynamic properties from a macroscopic perspective, we are dealing with quantities which can, either directly or indirectly, be measured and counted. Therefore, the matter of units becomes an important consideration. In the remaining sections of this chapter we will define certain thermodynamic properties and the basic units involved. The system of units used throughout this text is the International System, commonly referred to as SI units (from Le Systeme International d'Unites). The basic units of mass, length and time in this system are described in the following paragraphs. The concept of force then follows directly from Newton's second law of motion, which states that the force acting on a body is proportional to the product of the mass and the acceleration in the direction of the force.

$$F \propto ma$$

The concept of time is well established. The basic unit of time is the second(s), which in the past has been defined in terms of the solar day, the time interval for one complete revolution of the earth relative to the sun. Since this period will vary with the season of the year, an average value over a one-year period is called the mean solar day, and the mean solar second is $1/86\,400$ of the mean solar day. (The measurement of the earth's rotation is sometimes made relative to a fixed star, in which case the period is called a sidereal day.) In 1967, the General Conference of Weights and Measures (CGPM) adopted a definition of the second in terms of a resonator using a beam of cesium-133 atoms. The time required for $9\,192\,631\,770$ cycles of the cesium resonator is now accepted as the definition of the second.

Other units for time that are often used are the hour (hr) and the day (day), although neither one is a part of the SI units.

The concept of length is also well established. The basic unit of length is the metre, and for many years the accepted standard was the International Prototype Metre, the distance between two marks on a platinum-iridium bar under certain prescribed conditions. This bar is maintained at the International Bureau of Weights and Measures, Sevres, France. In 1960, the CGPM adopted a definition of the metre in terms of the wavelength of the orange-red line of krypton 86. The definition of the length of the metre is

$$1 \text{ metre} = 1\ 650\ 763.73 \text{ wavelengths in vacuo}$$
$$\text{of the orange-red line of Kr-86}$$

In SI units, the unit of mass is the kilogram (kg). As adopted by the first CGPM in 1889, and restated in 1901, it is the mass of a certain platinum-irridium cylinder maintained under prescribed conditions at the International Bureau of Weights and Measures. A related unit that is used frequently in thermodynamics is the mole (mol), defined as an amount of substance containing as many elementary entities as there are atoms in 0.012 kg of carbon-12. These elementary entities must be specified, and may be atoms, molecules, electrons, ions, or other particles or specific groups. For example, one mole of diatomic oxygen, having a molecular weight of 32 (compared to 12 for carbon) is a mass of 0.032 kg. The mole is often termed a gram mole, since it is an amount of substance in grams numerically equal to the molecular weight. In this text, it will often be convenient to use the kilomole (kmol), the amount of substance in kilograms numerically equal to the molecular weight.

In the International System, the unit of force is defined from Newton's second law, and is not an independent concept, as is the case in some systems of units. Therefore, a proportionality constant is unnecessary, and we may write that law as an equality.

$$F = ma$$

The unit of force is the newton (N), which by definition is the force required to accelerate a mass of one kilogram at the rate of one metre per second per second.

$$1 \text{ N} = 1 \text{ kg m/s}^2$$

(It is worth noting that SI units derived from proper nouns utilize capital letters for symbols; others utilize the lower-case letters.)

The term "weight" is often used with respect to a body, and is some-times confused with mass. Weight is really correctly used only as a force. When we say a body weighs so much we mean that this is the force with which it is attracted to the earth (or some other body), that is, the product of its mass and the local gravitational acceleration. The mass of a substance remains constant with elevation, but its weight varies with elevation.

Often, it is convenient and desirable to use multiples of the various units defined in this section. These prefixes are standard for all units, and are listed below in Table 2.1.

### Table 2.1
### SI Unit Prefixes

| Factor | Prefix | Symbol | Factor | Prefix | Symbol |
|--------|--------|--------|--------|--------|--------|
| $10^{12}$ | tera | T | $10^{-3}$ | milli | m |
| $10^{9}$ | giga | G | $10^{-6}$ | micro | $\mu$ |
| $10^{6}$ | mega | M | $10^{-9}$ | nano | n |
| $10^{3}$ | kilo | k | $10^{-12}$ | pico | p |

## 2.6   Specific Volume

The specific volume of a substance is defined as the volume per unit mass, and is given the symbol $v$. The density of a substance is defined as the mass per unit volume, and is therefore the reciprocal of the specific volume. Density is designated by the symbol $\rho$. Specific volume and density are intensive properties.

The specific volume of a system in a gravitational field may vary from point to point. For example, considering the atmosphere as a system, the specific volume increases as the elevation increases. Therefore the definition of specific volume involves the specific volume of a substance at a point in a system.

Consider a small volume $\delta V$ of a system, and let the mass be designated $\delta m$. The specific volume is defined by the relation

$$v = \lim_{\delta V \to \delta V'} \frac{\delta V}{\delta m}$$

where $\delta V'$ is the smallest volume for which the system can be considered a continuum.

Thus in a given system we should speak of the specific volume or density at a point in the system, and recognize that this may vary with elevation. However, most of the systems that we consider are relatively small, and the change in specific volume with elevation is not significant. In this case, we can speak of one value of specific volume or density for the entire system.

In this text, the specific volume and density will be given on either a mass or on a mole basis. A bar over the symbol (lower case) will be used to designate the property on a mole basis. Thus $\bar{v}$ will designate molal specific volume and $\bar{\rho}$ will designate the molal density. In SI units, the units for specific volume are $m^3/kg$ and $m^3/mol$ (or $m^3/kmol$); for density the corresponding units are $kg/m^3$ and $mol/m^3$ (or $kmol/m^3$).

While the SI unit for volume is the cubic metre, a commonly used volume unit is the litre ($\ell$), which is a special name given to a volume of 0.001 cubic metres, that is, $1\ell = 10^{-3}m^3$.

## 2.7   Pressure

When dealing with liquids and gases we ordinarily speak of pressure; in solids we speak of stresses. The pressure in a fluid at rest at a given point is the same in all directions, and we define pressure as the normal component of force per unit area. More specifically, if $\delta\mathscr{A}$ is a small area, and $\delta\mathscr{A}'$ is the smallest area over which we can consider the fluid a continuum, and $\delta F_n$ is the component of force normal to $\delta\mathscr{A}$, we define pressure, $P$, as

$$P = \lim_{\delta\mathscr{A}\to\delta\mathscr{A}'} \frac{\delta F_n}{\delta\mathscr{A}}$$

The pressure $P$ at a point in a fluid in equilibrium is the same in all directions. In a viscous fluid in motion the variation in the state of stress with orientation becomes an important consideration. These considerations are beyond the scope of this book, and we will consider pressure only in terms of a fluid in equilibrium.

The unit for pressure in the International System is the force of one newton acting on a square metre area, which is called the pascal (Pa). That is,

$$1 \text{ Pa} = 1 \text{ N/m}^2$$

Two other units, not within the International System, continue to be widely used, and should be noted here. These are the bar, where

$$1 \text{ bar} = 10^5 \text{ Pa} = 0.1 \text{ MPa}$$

and the standard atmosphere, where

$$1 \text{ atm} = 101\ 325 \text{ Pa}$$

which is slightly larger than the bar. In this text, we will normally use the SI unit, the Pa, and especially the multiples of kPa and MPa. The bar will be utilized often in the examples and problems, but the atmosphere will not be used, except in specifying certain reference points.

In most thermodynamic investigations we are concerned with absolute pressure. Most pressure and vacuum gages, however, read the difference between the absolute pressure and the atmospheric pressure existing at the gage, and this is referred to as gage pressure. This is shown graphically in Fig. 2.4, and the following examples illustrate the principles involved. Pressures below atmospheric and slightly above atmospheric, and pressure differences (for example, across an orifice in a pipe) are frequently measured with a manometer, which contains water, mercury, alcohol, oil, or other fluids. From the principles of hydrostatics one concludes that for a difference in level of $L$ metres, the pressure difference in pascals is calculated by the relation

$$\Delta P = \rho L g$$

where $\rho$ is the density of the fluid, and $g$ is the local acceleration due to gravity. The accepted standard value for gravitational acceleration is

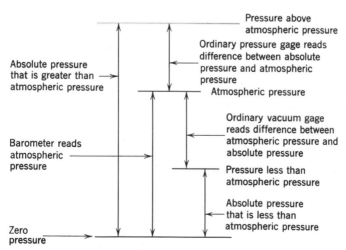

**Fig. 2.4**    Illustration of terms used in pressure measurement.

$$g = 9.80665 \text{ m/s}^2$$

but the value varies with location and elevation. The use of a manometer is illustrated in Figure 2.5.

In order to distinguish between absolute and gage pressure in this text, the term pascal will always refer to absolute pressure. Any gage pressure will then be indicated as such.

## 2.8   Equality of Temperature

Although temperature is a property with which we are all familiar, an exact definition of it is difficult. We are aware of "temperature" first of all as a sense of hotness or coldness when we touch an object. We also learn early in our experience that when a hot body and a cold body are brought into contact, the hot body becomes cooler and the cold body becomes warmer. If these bodies remain in contact for some time, they usually appear to have the same hotness or coldness. However, we also realize that our sense of hotness or coldness is very unreliable. Sometimes very cold bodies may seem hot, and bodies of different materials that are at the same temperature appear to be at different temperatures.

Because of these difficulties in defining temperature, we define equality of temperature. Consider two blocks of copper, one hot and the other cold, each of which is in contact with a mercury-in-glass thermometer. If these two blocks of copper are brought into thermal communication, we then observe that the electrical resistance of the hot block decreases with time and for the cold block it increases with time. After a period of time has elapsed, however, no further changes in resistance are observed. Similarly, when the blocks are first brought in thermal communication, the length of a side of the hot block decreases with time, whereas for the

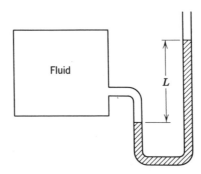

**Fig. 2.5**   Example of pressure measurement using a column of fluid.

cold block it increases with time. After a period of time, no further change in length of either of the blocks is perceived. Also, the mercury column of the thermometer in the hot block drops at first and in the cold block it rises, but after a period of time no further changes in height are observed. We may say, therefore, that two bodies have equality of temperature when no change in any observable property occurs when they are in thermal communication.

## 2.9    The Zeroth Law of Thermodynamics

Now consider the same two blocks of copper, and also another thermometer. Let one block of copper be brought into contact with the thermometer until equality of temperature is established, and then removed. Then let the second block of copper be brought into contact with the thermometer, and suppose that no change in the mercury level of the thermometer occurs during this operation with the second block. Then we can say that both blocks are in thermal equilibrium with the given thermometer.

The zeroth law of thermodynamics states that when two bodies have equality of temperature with a third body, they in turn have equality of temperature with each other. This seems very obvious to us because we are so familiar with this experiment. However, since this fact is not derivable from other laws, and since in the logical presentation of thermodynamics it precedes the first and second laws of thermodynamics, it has been called the zeroth law of thermodynamics. This law is really the basis of temperature measurement, for numbers can be placed on the mercury thermometer, and every time a body has equality of temperature with the thermometer, we can say that the body has the temperature we read on the thermometer. The problem remains, however, of relating temperatures that we might read on different mercury thermometers, or that we obtain when using different temperature-measuring devices, such as thermocouples and resistance thermometers. This suggests the need for a standard scale for temperature measurements.

## 2.10    Temperature Scales

The scale used for measuring temperature is the Celsius scale, the symbol for which is °C. This was formerly called the Centigrade scale, but is now designated after Anders Celsius (1701–1744), the Swedish astronomer who devised this scale.

Until 1954, the Celsius scale was based on two fixed, easily duplicated points, the ice point and the steam point. The temperature of the ice point is defined as the temperature of a mixture of ice and water which is in equilibrium with saturated air at a pressure of 1 atm (0.101 325 MPa). The temperature of the steam point is the temperature of water and steam which are in equilibrium at a pressure of 1 atm. These two points are numbered 0 and 100 on the Celsius scale.

At the tenth CGPM in 1954, the Celsius scale was redefined in terms of a single fixed point and the ideal-gas temperature scale. The single fixed point is the triple point of water (the state in which the solid, liquid, and vapor phases of water exist together in equilibrium). The magnitude of the degree is defined in terms of the ideal-gas temperature scale, which is discussed in Chapter 6. The essential features of this new scale are a single fixed point and a definition of the magnitude of the degree. The triple point of water is assigned the value 0.01°C. On this scale the steam point is experimentally found to be 100.00°C. Thus, there is essential agreement between the old and new temperature scales.

It should be noted that we have not yet considered an absolute scale of temperature. The possibility of such a scale arises from the second law of thermodynamics and is discussed in Chapter 6. On the basis of the second law of thermodynamics a temperature scale which is independent of any thermometric substance can be defined. This absolute scale is usually referred to as the thermodynamic scale of temperature. However, it is very complicated to use this scale directly, and therefore a more practical scale, the International Practical Temperature Scale, which closely represents the thermodynamics scale, has been adopted.

The absolute scale related to the Celsius scale is referred to as the Kelvin scale (after William Thomson, 1824–1907, who is also known as Lord Kelvin), and is designated K (without the degree symbol). The relation between these scales is

$$K = °C + 273.15$$

In 1967, the CGPM defined the Kelvin as 1/273.16 of the temperature at the triple point of water. The Celsius scale is then defined by the equation above, instead of by its earlier definition.

### 2.11   The International Practical Temperature Scale

In 1968 the International Committee on Weights and Measures adopted a revised International Practical Temperature Scale, IPTS-68, which is described below. This scale, similar to earlier ones of 1927 and 1948, has been extended in range and made to conform more closely to the thermodynamic temperature scale. It is based on a number of fixed and easily reproducible points that are assigned definite numerical values of temperature, and on specified formulas relating temperature to the readings on certain temperature-measuring instruments for the purpose of interpolation between the fixed points. The primary fixed points and a summary of the interpolation techniques are listed here for the sake of completeness, although the student will have limited need for it at this time.

The primary fixed-point temperatures in degrees Celsius are as follows:

1. Triple point (equilibrium between solid, liquid and vapor phases) of equilibrium-hydrogen.                                                    −259.34
2. Boiling point (equilibrium between liquid and vapor phases) of equilibrium-hydrogen at 25/76 atm (33.33 kPa) pressure. −256.108
3. Normal boiling point (1 atm  pressure) of equilibrium-hydrogen.
                                                                                           −252.87
4. Normal boiling point of neon.                                         −246.048
5. Triple point of oxygen.                                                   −218.789
6. Normal boiling point of oxygen.                                      −182.962
7. Triple point of water.                                                             0.01
8. Normal boiling point of water.                                               100
9. Normal freezing point (equilibrium between solid and liquid phases at 1 atm. pressure) of zinc.                                          419.58
10. Normal freezing point of silver.                                        961.93
11. Normal freezing point of gold.                                         1064.43

The means available for measurement and interpolation lead to a division of the temperature scale into four general ranges.

1. The range from −259.34°C to 0°C is based on measurements on a platinum resistance thermometer, with temperature expressed in terms of a 20th degree reference function equation. This range is subdivided into four parts. In each, the difference between measured resistance ratios of a particular thermometer and the reference function at the fixed points are used to determine the constants in a specified polynomial interpolation equation.
2. The range from 0°C to 630.74°C (the normal freezing point of antimony, a secondary fixed point) is also based on a platinum resistance thermometer, with constants in a polynomial interpolating equation determined by calibration at the three fixed points in this range.

**3.** The range from 630.74°C to 1064.43°C is based on measurements on a standard platinum vs. rhodium-platinum thermocouple, and a three-term equation expressing EMF as a function of temperature. The constants are determined by a platinum resistance thermometer measurement at the antimony point and by calibration at the two primary fixed points in this range.

**4.** The range above 1064.43°C is based on measurements of the intensity of visible-spectrum radiation compared with that of the same wavelength at the gold point, and on Planck's equation for black body radiation.

## PROBLEMS

2.1 A mass of 10 kg is accelerated with a force of 1 kN. Calculate the acceleration.

2.2 The "standard" acceleration due to gravity is 9.80665 m/s². Calculate the force due to "standard" gravity acting on a mass of 5 kg.

2.3 With what force is a mass of 100 kg attracted to the earth at a point where the gravitational acceleration is 9.60 m/s².

2.4 A kilogram mass is "weighed" with a beam balance at a point where $g = 9.7$ m/s². What reading would be expected? If it is weighed with a spring scale that reads correctly for standard gravity, what reading would be obtained?

2.5 A body of fixed mass is "weighed" at an elevation of 10 000 m. ($g = 9.778$ m/s²) by a spring balance which was calibrated at sea level. The reading on the spring balance is 10 kg. What is the mass of the body?

2.6 A piston has an area of 500 mm². What mass must the piston have if it exerts a pressure of 50 kPa above atmospheric pressure on the gas enclosed in the cylinder? Assume standard gravitational acceleration.

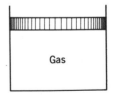

Gas

2.7 Suppose that in an orbiting space station, an artificial gravity of 2 m/s² is induced by rotating the station. How much would a 75 kg man weigh inside?

**Fig. 2.6** Sketch for Problem 2.6.

2.8 A pressure gage reads 2.10 MPa, and the barometer reads 98 kPa. Calculate the absolute pressure.

2.9    A manometer contains a fluid having a density of 800 kg/m³. The difference in height of the two columns is 300 mm. What pressure difference is
indicated? What would the height difference be if this same pressure
difference had been measured by a manometer containing mercury
(density of 13 600 kg/m³)?

2.10   The height of a mercury manometer column which is used to measure a
vacuum is 700 mm, and the barometer reads 97 kPa. Determine the
pressure.

2.11   In an experimental airplane flying at 20 000 m ($g = 9.75$ m/s²), the air
flow in a piece of apparatus is measured by using a mercury manometer.
The difference in the level is 300 mm. At sea level and the same temperature, mercury has a density of 13 600 kg/m³. Determine the pressure drop
across the orifice.

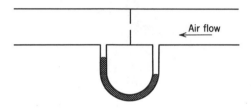

**Fig. 2.7**   Sketch for Problem 2.11.

2.12   A mercury column is used to measure the pressure difference of 200 kPa
in a piece of apparatus which is located out of doors. The minimum temperature in the winter is −15°C and the maximum temperature in the
summer is 35°C. What will be the difference in the height of the mercury
column in the summer as compared to the winter when measuring this
pressure difference of 200 kPa? Assume standard gravitational acceleration. The following data are given for the density of mercury:

| T°C | Density |
|---|---|
| −10 | 13 620 kg/m³ |
| 0 | 13 595 |
| 10 | 13 570 |
| 20 | 13 546 |
| 30 | 13 521 |

2.13   A vertical cylinder containing argon gas is fitted with a piston of 40 kg
mass and cross-sectional area of 0.025 m². The atmospheric pressure outside the cylinder is 95 kPa and the local gravitational acceleration is
9.79 m/s². What is the argon pressure inside the cylinder?

2.14 The level of the water in an enclosed water tank is 40 m above the ground. The pressure in the air space above the water is 120 kPa. The average density of the water is 1000 kg/m³. What is the pressure of the water at the ground level?

2.15 A gas is contained in two cylinders $A$ and $B$, connected by a piston of two different diameters, as shown in Fig. 2.8. The mass of the piston is 10 kg and the gas pressure inside cylinder $A$ is 200 kPa. Calculate the pressure in cylinder $B$.

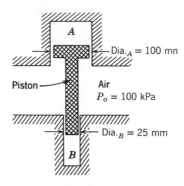

**Fig. 2.8** Sketch for Problem 2.15.

# 3

---

# Properties of a Pure Substance

In the previous chapter we considered three familiar properties of a substance, namely, specific volume, pressure, and temperature. We now turn our attention to pure substances and consider some of the phases in which a pure substance may exist, the number of independent properties a pure substance may have, and methods of presenting thermodynamic properties.

## 3.1 The Pure Substance

A pure substance is one that has a homogeneous and invariable chemical composition. It may exist in more than one phase, but the chemical composition is the same in all phases. Thus, liquid water, a mixture of liquid water and water vapor (steam), or a mixture of ice and liquid water are all pure substances, for every phase has the same chemical composition. On the other hand, a mixture of liquid air and gaseous air is not a pure substance, since the composition of the liquid phase is different from that of the vapor phase.

Sometimes a mixture of gases, such as air, is considered a pure substance as long as there is no change of phase. Strictly speaking, this is not true, but rather, as we shall see later, we should say that a mixture of gases such as air exhibits some of the characteristics of a pure substance as long as there is no change of phase.

In this text the emphasis will be on those substances which may be called simple compressible substances. By this we understand that

35

surface effects, magnetic effects, and electrical effects are not significant when dealing with these substances. On the other hand changes in volume, such as those associated with the expansion of a gas in a cylinder, are most important. However, reference will be made to other substances in which surface, magnetic, or electrical effects are important. We will refer to a system consisting of a simple compressible substance as a simple compressible system.

## 3.2   Vapor-Liquid-Solid Phase Equilibrium in a Pure Substance

Consider as a system 1 kg of water contained in the piston-cylinder arrangement of Fig. 3.1a. Suppose that the piston and weight maintain a

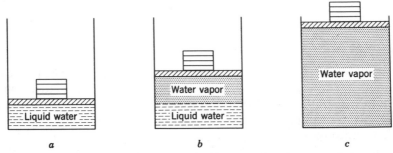

**Fig. 3.1**   Constant-pressure change from liquid to vapor phase for a pure substance.

pressure of 0.1 MPa in the cylinder, and that the initial temperature is 20°C. As heat is transferred to the water the temperature increases appreciably, the specific volume increases slightly, and the pressure remains constant. When the temperature reaches 99.6°C, additional heat transfer results in a change of phase, as indicated in Fig. 3.1b. That is, some of the liquid becomes vapor, and during this process both the temperature and pressure remain constant, but the specific volume increases considerably. When the last drop of liquid has vaporized, further transfer of heat results in an increase in both temperature and specific volume of the vapor, Fig. 3.1c.

The term saturation temperature designates the temperature at which vaporization takes place at a given pressure, and this pressure is called the saturation pressure for the given temperature. Thus for water at 99.6°C the saturation pressure is 0.1 MPa, and for water at 0.1 MPa the saturation temperature is 99.6°C. For a pure substance there is a definite relation between saturation pressure and saturation temperature, a typical curve being shown in Fig. 3.2. This is called the vapor-pressure curve.

If a substance exists as liquid at the saturation temperature and pressure, it is called saturated liquid. If the temperature of the liquid is lower than the saturation temperature for the existing pressure, it is called either a subcooled liquid (implying that the temperature is lower than the saturation temperature for the given pressure) or a compressed liquid (implying that the pressure

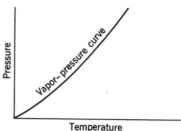

**Fig. 3.2** Vapor-pressure curve of a pure substance.

is greater than the saturation pressure for the given temperature). Either term may be used, but the latter term will be used in this text.

When a substance exists as part liquid and part vapor at the saturation temperature, its quality is defined as the ratio of the mass of vapor to the total mass. Thus, in Fig. 3.1$b$, if the mass of the vapor is 0.2 kg and the mass of the liquid is 0.8 kg, the quality is 0.2 or 20 per cent. The quality may be considered as an intensive property, and it has the symbol $x$. Quality has meaning only when the substance is in a saturated state, i.e., at saturation pressure and temperature.

If a substance exists as vapor at the saturation temperature, it is called saturated vapor. (Sometimes the term dry saturated vapor is used to emphasize that the quality is 100 per cent.) When the vapor is at a temperature greater than the saturation temperature, it is said to exist as superheated vapor. The pressure and temperature of superheated vapor are independent properties, since the temperature may increase while the pressure remains constant. Actually, the substances we call gases are highly superheated vapors.

Consider Fig. 3.1 again, and let us plot on the temperature-volume diagram of Fig. 3.3 the constant-pressure line that represents the states through which the water passes as it is heated from the initial state of 0.1 MPa and 20°C. Let state $A$ represent the initial state, $B$ the saturated-liquid state (99.6°C), and line $AB$ the process in which the liquid is heated from the initial temperature to the saturation temperature. Point $C$ is the saturated vapor state, and line $BC$ is the constant-temperature process in which the change of phase from liquid to vapor occurs. Line $CD$ represents the process in which the steam is superheated at constant pressure. Temperature and volume both increase during this process.

Now let the process take place at a constant pressure of 1 MPa,

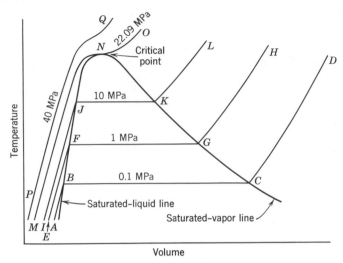

**Fig. 3.3**   Temperature-volume diagram for water showing liquid and vapor phases (not to scale).

beginning from an initial temperature of 20°C. Point *E* represents the initial state, the specific volume being slightly less than at 0.1 MPa and 20°C. Vaporization now begins at point *F,* where the temperature is 179.9°C. Point *G* is the saturated-vapor state, and line *GH* the constant-pressure process in which the steam is superheated.

In a similar manner, a constant pressure of 10 MPa is represented by line *IJKL,* the saturation temperature being 311.1°C.

At a pressure of 22.09 MPa, represented by line *MNO,* we find, however, that there is no constant-temperature vaporization process. Rather, point *N* is a point of inflection with a zero slope. This point is called the critical point, and at the critical point the saturated-liquid and saturated-vapor states are identical. The temperature, pressure, and specific volume at the critical point are called the critical temperature, critical pressure, and critical volume. The critical-point data for some substances are given in Table 3.1, and more extensive data are given in Table A.7 in the Appendix.

A constant-pressure process at a pressure greater than the critical pressure is represented by line *PQ.* If water at 40MPa, 20°C is heated in a constant-pressure process in a cylinder such as shown in Fig. 3.1, there will never be two phases present, and the state shown in Fig. 3.1*b,* will never exist. Rather, there will be a continuous change in density and at all times there will be only one phase present. The question then arises as to when do we have a liquid and when do we have a vapor? The

**Table 3.1**
Some Critical Point Data

|  | Critical Temperature °C | Critical Pressure MPa | Critical Volume m³/kg |
|---|---|---|---|
| Water | 374.14 | 22.09 | 0.003 155 |
| Carbon dioxide | 31.05 | 7.39 | 0.002 143 |
| Oxygen | −118.35 | 5.08 | 0.002 438 |
| Hydrogen | −239.85 | 1.30 | 0.032 192 |

answer is that this is not a valid question at supercritical pressures. Instead, we simply term the substance a fluid. However, rather arbitrarily at temperatures below the critical temperature we usually refer to it as a compressed liquid and at temperatures above the critical temperature as a superheated vapor. It should be emphasized, however, that at pressures above the critical pressure we never have a liquid and vapor phase of a pure substance existing in equilibrium.

In Fig. 3.3 line *NJFB* represents the saturated-liquid line and line *NKGC* represents the saturated-vapor line.

Let us consider another experiment with the piston-cylinder arrangement. Suppose that the cylinder contains 1 kg of ice at −20°C, 100 kPa. When heat is transferred to the ice, the pressure remains constant, the specific volume increases slightly, and the temperature increases until it reaches 0°C, at which point the ice melts while the temperature remains constant. In this state the ice is called saturated solid. For most substances the specific volume increases during this melting process but for water the specific volume of the liquid is less than the specific volume of the solid. When all of the ice has melted, a further heat transfer causes an increase in temperature of the liquid.

If the initial pressure of the ice at −20°C is 0.260 kPa, heat transfer to the ice first results in an increase in temperature to −10°C. At this point, however, the ice would pass directly from the solid phase to the vapor phase in the process known as sublimation. Further heat transfer would result in superheating of the vapor.

Finally consider an initial pressure of the ice of 0.6113 kPa and a temperature of −20°C. As a result of heat transfer let the temperature increase until it reaches 0.01°C. At this point, however, further heat transfer may result in some of the ice becoming vapor and some becoming liquid, for at this point it is possible to have the three phases in equilibrium. This is called the triple point, which is defined as the state in which three phases may all be present in equilibrium. The pressure and temperature at the triple point for a number of substances is given in Table 3.2.

**Table 3.2**
Some Solid-Liquid-Vapor Triple Point Data

|  | Temperature °C | Pressure kPa |
|---|---|---|
| Hydrogen (normal) | −259 | 7.194 |
| Oxygen | −219 | 0.15 |
| Nitrogen | −210 | 12.53 |
| Mercury | −39 | 0.000 000 13 |
| Water | 0.01 | 0.6113 |
| Zinc | 419 | 5.066 |
| Silver | 961 | 0.01 |
| Copper | 1083 | 0.000 079 |

This whole matter is best summarized by the diagram of Fig. 3.4, which shows how the solid, liquid and vapor phases may exist together in equilibrium. Along the sublimation line the solid and vapor phases are in equilibrium, along the fusion line the solid and liquid phases are in equilibrium, and along the vaporization line the liquid and vapor phases are in equilibrium. The only point at which all three phases may exist in

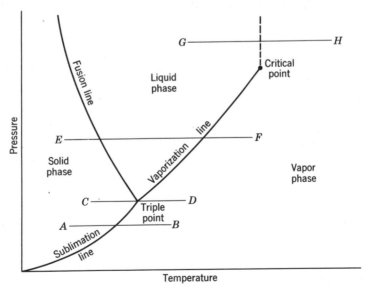

**Fig. 3.4**   Pressure-temperature diagram for a substance such as water.

equilibrium is the triple point. The vaporization line ends at the critical point because there is no distinct change from the liquid phase to the vapor phase above the critical point.

Consider a solid in state $A$, Fig. 3.4. When the temperature is increased while the pressure (which is less than the triple point pressure) is constant, the substance passes directly from the solid to the vapor phase. Along the constant-pressure line $EF$, the substance first passes from the solid to the liquid phase at one temperature, and then from the liquid to the vapor phase at a higher temperature. Constant-pressure line $CD$ passes through the triple point, and it is only at the triple point that the three phases may exist together in equilibrium. At a pressure above the critical pressure, such as $GH$, there is no sharp distinction between the liquid and vapor phases.

Although we have made these comments with rather specific reference to water (only because of our familiarity with water) all pure substances exhibit the same general behavior. However, the triple point temperature and critical temperature vary greatly from one substance to another. For example, the critical temperature of helium, as given in Table A.7 is 5.3 K. Therefore, the absolute temperature of helium at ambient conditions is over 50 times greater than the critical temperature. On the other hand, water has a critical temperature of 374.14°C (647.29K) and at ambient conditions the temperature of water is less than half the critical temperature. Most metals have a much higher critical temperature than water. In considering the behavior of a substance in a given state, it is often helpful to think of this state in relation to the critical state or triple point. For example, if the pressure is greater than the critical pressure, it is impossible to have a liquid and a vapor phase in equilibrium. Or, to consider another example, the states at which vacuum melting a given metal is possible can be ascertained by a consideration of the properties at the triple point. In the case of iron at a pressure just above 0.5 kPa (the triple point pressure), iron would melt at a temperature of about 1535°C (the triple point temperature).

It should also be pointed out that a pure substance can exist in a number of different solid phases. A transition from one solid phase to another is called an allotropic transformation. Figure 3.5 is a pressure-temperature diagram for iron that shows three solid phases, the liquid phase, and the vapor phase. Figure 3.6 shows a number of solid phases for water. It is evident that a pure substance can have a number of triple points, but only one triple point involves solid, liquid, and vapor equilibrium. Other triple points for a pure substance can involve two solid phases and a liquid phase, two solid phases and a vapor phase, or three solid phases.

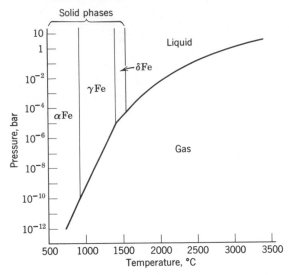

**Fig. 3.5** Estimated pressure-temperature diagram for iron (from *Phase Diagrams in Metallurgy*, by F. N. Rhines, copyright 1956, McGraw-Hill Book Company; used by permission).

### 3.3 Independent Properties of a Pure Substance

One important reason for introducing the concept of a pure substance is that the state of a simple compressible pure substance (i.e., a pure substance in the absence of motion, gravity, and surface, magnetic or electrical effects) is defined by two independent properties. This means, for example, that if the specific volume and temperature of superheated steam are specified, the state of the steam is determined.

To understand the significance of the term independent property, consider the saturated-liquid and saturated-vapor states of a pure substance. These two states have the same pressure and the same temperature, but are definitely not the same state. In a saturation state, therefore, pressure and temperature are not independent properties. Two independent properties such as pressure and specific volume, or pressure and quality, are required to specify a saturation state of a pure substance.

The reason for mentioning previously that a mixture of gases, such as air, has the same characteristics as a pure substance as long as only one phase is present, concerns precisely this point. The state of air, which is a mixture of gases of definite composition, is determined by specifying two properties as long as it remains in the gaseous phase, and in this regard air can be treated as a pure substance.

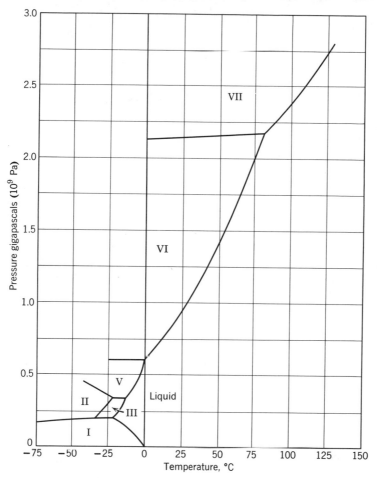

**Fig. 3.6** Phase diagram of water (Adapted from the *American Institute of Physics Handbook*, 2nd Ed., 1963, McGraw-Hill).

## 3.4  Equations of State for the Vapor Phase of a Simple Compressible Substance

From experimental observations it has been established that the *P-v-T* behavior of gases at low density is closely given by the following equation of state.

$$P\bar{v} = \bar{R}T \tag{3.1}$$

where $\bar{R}$ is the universal gas constant, the value of which is

$$\bar{R} = 8.31434 \frac{N\ m}{mol\ K}$$

$$= 8.31434 \frac{J}{mol\ K}$$

Dividing Eq. 3.1 by $M$, the molecular weight, we have the equation of state on a unit mass basis.

or

$$\frac{P\bar{v}}{M} = \frac{\bar{R}T}{M} \tag{3.2}$$

$$Pv = RT$$

where

$$R = \frac{\bar{R}}{M} \tag{3.3}$$

$R$ is a constant for a particular gas. The value of $R$ for a number of substances is given in Table A.8 of the Appendix. It follows from Eqs. 3.1 and 3.2 that this equation of state can be as written in terms of the total volume.

$$PV = n\bar{R}T \tag{3.4}$$

$$PV = mRT$$

It should also be noted that Eq. 3.4 can alternately be written in the form

$$\frac{P_1 V_1}{T_1} = \frac{P_2 V_2}{T_2} \tag{3.5}$$

That is, gases at low density closely follow the well-known Boyle's and Charles' laws. Boyle and Charles, of course, based their statements on experimental observations. (Strictly speaking, neither of these statements should be called a law, since they are only approximately true and even then only under conditions of low density.)

The equation of state given by Eq. 3.1 (or Eq. 3.2) is referred to as the ideal gas equation of state. At very low density all gases and vapors approach ideal gas behavior, with the $P$-$v$-$T$ relationship being given by the ideal gas equation of state. At higher densities the behavior may deviate substantially from the ideal gas equation of state.

Because of its simplicity, the ideal gas equation of state is very convenient to use in thermodynamic calculations. However, two questions can appropriately be raised. The ideal gas equation of state is a good approximation at low density. But what constitutes low density? Or, expressed in other words, over what range of density will the ideal gas equation of state hold with accuracy? The second question is, how much does an actual gas at a given pressure and temperature deviate from ideal gas behavior?

To answer both questions we introduce the concept of the compressibility factor, $Z$, which is defined by the relation

$$Z = \frac{P\bar{v}}{\bar{R}T}$$

or $$P\bar{v} = Z\bar{R}T \tag{3.6}$$

Note that for an ideal gas, $Z = 1$, and the deviation of $Z$ from unity is a measure of the deviation of the actual relation from the ideal gas equation of state.

Figure 3.7 shows a skeleton compressibility chart for nitrogen. From this chart we make three observations. The first is that at all temperatures, $Z \to 1$ as $P \to 0$. That is, as the pressure approaches zero, the

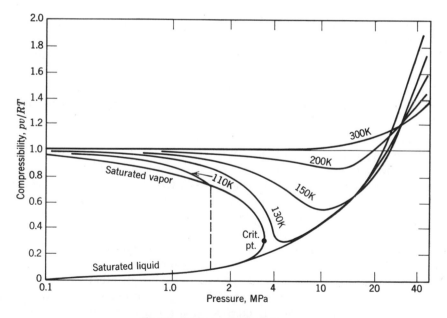

**Fig. 3.7** Compressibility of nitrogen.

$P$-$v$-$T$ behavior closely approaches that predicted by the ideal gas equation of state. Note also that at temperatures of 300 K and above (i.e., room temperature and above) the compressibility factor is near unity up to pressures of about 10 MPa. This means that the ideal gas equation of state can be used for nitrogen (and, as it happens, also air) over this range with considerable accuracy.

Now suppose we reduce the temperature from 300 K while keeping the pressure constant at 4 MPa. The density will increase and we note a sharp decrease below unity in the value of the compressibility factor. Values of $Z < 1$ mean that the actual density is greater than would be predicted by ideal gas behavior. The physical explanation of this is as follows: As the temperature is reduced from 300 K while pressure remains constant at 4 MPa, the molecules are brought closer together. In this range of intermolecular distances, and at this pressure and temperature, there is an attractive force between the molecules. The lower the temperature the greater is this intermolecular attractive force. This attractive force between the molecules means that the density is greater than would be predicted by the ideal gas behavior, which assumes no intermolecular forces. Note also from the compressibility chart that at very high densities, for pressures above 30 MPa, the compressibility factor is always greater than unity. In this range the intermolecular distances are very small, and there is a repulsive force between the molecules. This tends to make the density less than would otherwise be expected.

The precise nature of intermolecular forces is a rather complex matter. These forces are a function of the temperature as well as the density. The preceding discussion should be considered as a qualitative analysis to assist in gaining some understanding of the ideal gas equation of state and how the $P$-$v$-$T$ behavior of actual gases deviate from this equation.

From a practical point of view in the solution of problems, two things should be noted. First, at very low pressures, ideal gas behavior can be assumed with good accuracy, regardless of the temperature. Second, at temperatures that are double the critical temperature or above (the critical temperature of nitrogen is 126 K) ideal gas behavior can be assumed with good accuracy to pressures of around 10 MPa. When the temperature is less than twice the critical temperature, and the pressure above a very low value, say ambient pressure, we are in the superheated vapor region, and the deviation from ideal gas behavior may be considerable. In this region it is preferable to use tables of thermodynamic properties or charts for a particular substance. These tables are considered in the following section. The concept of the generalized compressibility chart is introduced in Chapter 10.

In order to have an equation of state that accurately represents the $P$-$v$-$T$ behavior for a particular gas over the entire superheated vapor range, more complicated equations of state have been developed. Several different forms of these have been used. To illustrate the nature and complexity of these equations we refer to one of the best known, namely, the Beattie-Bridgeman equation of state. This equation is

$$P = \frac{\bar{R}T(1-\epsilon)}{\bar{v}^2}(\bar{v}+B) - \frac{A}{\bar{v}^2} \tag{3.7}$$

where $A = A_0(1-a/\bar{v})$, $B = B_0(1-b/\bar{v})$, $\epsilon = c/\bar{v}T^3$, and $A_0$, $a$, $B_0$, $b$, and $c$ are constants for different gases. The values of these constants for various substances are given in Table 3.3.

**Table 3.3**
Constants of the Beattie-Bridgeman Equation of State

(Pressure in kilopascals; specific volume in m³/kmol; temperature in Kelvin; $\bar{R} = 8.31434$ N m/mol K).

| Gas | $A_0$ | $a$ | $B_0$ | $b$ | $10^{-4}c$ |
|---|---|---|---|---|---|
| Helium | 2.1886 | 0.05984 | 0.01400 | 0.0 | 0.0040 |
| Argon | 130.7802 | 0.02328 | 0.03931 | 0.0 | 5.99 |
| Hydrogen | 20.0117 | −0.00506 | 0.02096 | −0.04359 | 0.0504 |
| Nitrogen | 136.2315 | 0.02617 | 0.05046 | −0.00691 | 4.20 |
| Oxygen | 151.0857 | 0.02562 | 0.04624 | 0.004208 | 4.80 |
| Air | 131.8441 | 0.01931 | 0.04611 | −0.001101 | 4.34 |
| Carbon dioxide | 507.2836 | 0.07132 | 0.10476 | 0.07235 | 66.00 |

The matter of equations of state will be discussed further in Chapter 10. The observation to be made here in particular is that an equation of state that accurately describes the relation between pressure, temperature, and specific volume is rather cumbersome and the solution requires considerable time. When using a large digital computer, it is often most convenient to determine the thermodynamic properties in a given state from such equations. However, in hand calculations it is much more convenient to tabulate values of pressure, temperature, specific volume, and other thermodynamic properties for various substances. The Appendix includes summary tables and graphs of the thermodynamic properties of water, ammonia, Freon-12, oxygen, nitrogen, mercury, and Freon-13. The tables of the properties of water are usually referred to as the "steam tables" and are extracted from "Steam Tables" by Keenan, Keyes, Hill and Moore. The method for compiling the *P-v-T* data for such a table is to find an equation of state that accurately fits the experimental data, and then to solve the equation of state for the values listed in the table.

**Example 3.1**

What is the mass of air contained in a room 6 m × 10 m × 4 m if the pressure is 100 kPa and the temperature is 25°C? Assume air to be an ideal gas.

By using Eq. 3.4, and the value of *R* from Table A.8

$$m = \frac{PV}{RT} = \frac{100 \text{ kN/m}^2 \times 240 \text{ m}^3}{0.287 \text{ kN m/kg K} \times 298.15 \text{ K}} = 280.5 \text{ kg}$$

**Example 3.2**

A tank has a volume of 0.5 m³ and contains 10 kg of an ideal gas having a molecular weight of 24. The temperature is 25°C. What is the pressure?

The gas constant is determined first:

$$R = \frac{\overline{R}}{M} = \frac{8.31434 \text{ kN m/kmol K}}{24 \text{ kg/kmol}}$$

$$= 0.346\,43 \text{ kN m/kg K}$$

We now solve for *P*.

$$P = \frac{mRT}{V} = \frac{10 \text{ kg} \times 0.346\,43 \text{ kN m/kg K} \times 298.15 \text{ K}}{0.5 \text{ m}^3}$$

$$= 2065.8 \text{ kPa}$$

## 3.5   Tables of Thermodynamic Properties

Tables of thermodynamic properties of many substances are available, and in general all these have the same form. In this section we shall refer to the steam tables. The steam tables are selected both as a vehicle for presenting thermodynamic tables and because steam is used extensively in power plants and industrial processes. Once the steam tables are understood, other thermodynamic tables can be readily used.

Several different versions of steam tables have been published. Two new tables have recently been published in the United States. In 1967 The American Society of Mechanical Engineers published a volume entitled "Thermodynamic and Transport Properties of Steam," commonly referred to as the 1967 ASME steam tables. "Steam Tables," by Keenan, Keyes, Hill, and Moore was published in 1969. This is a revision of a very extensively used volume by Keenan and Keyes, which was published in 1936. The Appendix includes a summary of the 1969 edition of "Steam Tables," and reference to these tables is made throughout this text.

In Table A.1.1, the first column after the temperature gives the corresponding saturation pressure in kilopascals or megapascals. The next two columns give specific volume in cubic metres per kilogram. The first of these gives the specific volume of the saturated liquid, $v_f$; the second column gives the specific volume of saturated vapor, $v_g$. The difference between these two, $v_g - v_f$, represents the increase in specific volume when the state changes from staurated liquid to saturated vapor, and is designated $v_{fg}$.

The specific volume of a substance having a given quality can be found by utilizing the definition of quality. Quality has already been defined as the ratio of the mass of vapor to total mass of liquid plus vapor when a substance is in a saturation state. Let us consider a mass of 1 kg having a quality $x$. The specific volume is the sum of the volume of the liquid and the volume of the vapor. The volume of the liquid is $(1-x)v_f$, and the volume of the vapor is $xv_g$. Therefore the specific volume $v$ is

$$v = xv_g + (1-x)v_f \tag{3.8}$$

Since $v_f + v_{fg} = v_g$, Eq. 3.8 can also be written in the following forms

$$v = v_f + xv_{fg} \tag{3.9}$$

$$v = v_g - (1-x)v_{fg} \tag{3.10}$$

As an example, let us calculate the specific volume of saturated steam at 200°C having a quality of 70%. Using Eq. 3.8,

$$v = 0.3 \ (0.001 \ 157) + 0.7 \ (0.127 \ 36)$$

$$= 0.0895 \ \text{m}^3/\text{kg}$$

In Table A.1.2, the first column after the pressure lists the saturation temperature for each pressure. The next columns list specific volume in a manner similar to Table A.1.1. When necessary, $v_{fg}$ can readily be found by subtracting $v_f$ from $v_g$.

Table 3 of the steam tables, which is summarized in Table A.1.3 in the Appendix, gives the properties of superheated vapor. In the superheat region, pressure and temperature are independent properties, and therefore, for each pressure a large number of temperatures is given, and for each temperature four thermodynamic properties are listed,

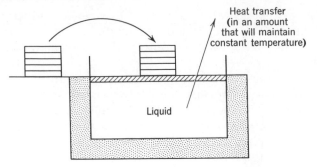

Fig. 3.8    Illustration of compressed liquid state.

the first one being specific volume. Thus, the specific volume of steam at a pressure of 0.5 MPa and 200°C is 0.4249 m³/kg.

Table 4 of the steam tables, summarized in Table A.1.4 in the Appendix, gives the properties of the compressed liquid. To demonstrate the use of this table, consider a piston and a cylinder (as shown in Fig. 3.8) that contains 1 kg of saturated liquid water at 100°C. Its properties are given in Table A.1.1, and we note that the pressure is 0.1013 MPa and the specific volume is 0.001 044 m³/kg. Suppose the pressure is increased to 10 MPa while the temperature is held constant at 100°C by the necessary transfer of heat, $Q$. Since water is slightly compressible, we would expect a slight decrease in specific volume during this process. Table A.1.4 gives this specific volume as 0.001 039 m³/kg. Note that this is only a slight decrease, and only a small error would be made if one assumed that the volume of a compressed liquid was equal to the specific volume of the saturated liquid at the same temperature. In many cases this is the most convenient procedure, particularly in those cases when compressed liquid data are not available.

Furthermore, since specific volume does change rapidly with temperature, care should be exercised in interpolation over the wide temperature ranges in Table A.1.4. (In some cases it may be more accurate to use the saturated liquid data from Table A.1.1, and interpolate differences between Table A.1.1, the saturated liquid data, and Table A.1.4, the compressed liquid data.)

Table 6 of the steam tables, which is summarized in Table A.1.5 in the Appendix, gives the properties of saturated solid and saturated vapor that are in equilibrium. The first column gives the temperature, and the second column gives the corresponding saturation pressure. As would be expected, all these pressures are less than the triple-point pressure. The next two columns give the specific volume of the saturated solid and saturated vapor (note that the tabulated value is $v \times 10^3$).

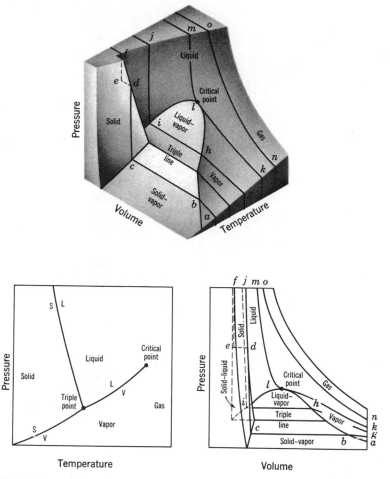

**Fig. 3.9** Pressure-volume-temperature surface for a substance that expands on freezing.

## 3.6  Thermodynamic Surfaces

The matter discussed in this chapter can be well summarized by a consideration of a pressure-specific volume-temperature surface. Two such surfaces are shown in Figs. 3.9 and 3.10. Figure 3.9 shows a substance such as water in which the specific volume increases during freezing, and Fig. 3.10 shows a substance in which the specific volume decreases during freezing.

In these diagrams the pressure, specific volume, and temperature are

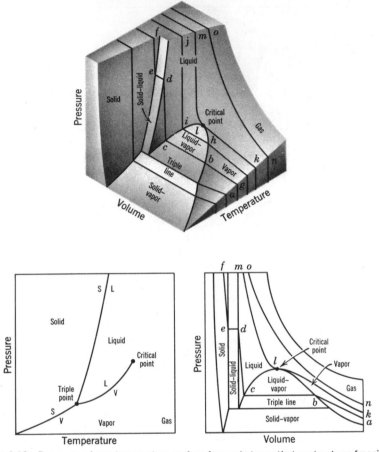

**Fig. 3.10**   Pressure-volume-temperature surface for a substance that contracts on freezing.

plotted on mutually perpendicular coordinates, and each possible equilibrium state is thus represented by a point on the surface. This follows directly from the fact that a pure substance has only two independent intensive properties. All points along a quasiequilibrium process lie on the $P$-$v$-$T$ surface, since such a process always passes through equilibrium states.

The regions of the surface that represent a single phase, namely, the solid, liquid, and vapor phases, are indicated, these surfaces being curved. The two-phase regions, namely, the solid-liquid, solid-vapor, and liquid-vapor regions, are ruled surfaces. By this we understand that they are made up of straight lines parallel to the specific volume axis.

This, of course, follows from the fact that in the two-phase region, lines of constant pressure are also lines of constant temperature, though the specific volume may change. The triple point actually appears as the triple line on the $P$-$v$-$T$ surface, since the pressure and temperature of the triple point are fixed, but the specific volume may vary, depending on the proportion of each phase.

It is also of interest to note the pressure-temperature and pressure-volume projections of these surfaces. We have already considered the pressure-temperature diagram for a substance such as water. It is on this diagram that we observe the triple point. Various lines of constant temperature are shown on the pressure-volume diagram, and the corresponding constant-temperature sections are lettered identically on the $P$-$v$-$T$ surface. The critical isotherm has a point of inflection at the critical point.

One notices that with a substance such as water, which expands on freezing, the freezing temperature decreases with an increase in pressure. With a substance that contracts on freezing, the freezing temperature increases as the pressure increases. Thus, as the pressure of vapor is increased along the constant-temperature line *abcdef* in Fig. 3.9, a substance that expands on freezing first becomes solid and then liquid. For the substance that contracts on freezing, the corresponding constant-temperature line, Fig. 3.10, indicates that as the pressure on the vapor is increased, it first becomes liquid and then solid.

## Example 3.3

A vessel having a volume of 0.4 m³ contains 2.0 kg of a liquid water and water vapor mixture in equilibrium at a pressure of 6 bar (0.6 MPa).

Calculate

1. The volume and mass of liquid.
2. The volume and mass of vapor.

The specific volume is calculated first:

$$v = \frac{0.4}{2.0} = 0.20 \text{ m}^3/\text{kg}$$

From the steam tables (Appendix Table A.1.2)

$$v_{fg} = 0.3157\text{-}0.001\ 101 = 0.3146$$

The quality can now be calculated, using Eq. 3.10

$$0.20 = 0.3157 - (1 - x)\,0.3146$$
$$1 - x = \frac{0.1157}{0.3146} = 0.3678$$

$$x = 0.6322$$

Therefore the mass of liquid is

$$2.0\,(0.3678) = 0.7356 \text{ kg}$$

The mass of vapor is

$$2.0\,(0.6322) = 1.2644 \text{ kg}$$

The volume of liquid is

$$V_{\text{liq}} = m_{\text{liq}}\,v_f = 0.7356\,(0.001\ 101) = 0.0008 \text{ m}^3$$

The volume of the vapor is

$$V_{\text{vap}} = m_{\text{vap}}v_g = 1.2644\,(0.3157) = 0.3992 \text{ m}^3$$

**Example 3.4**

A rigid vessel contains saturated ammonia vapor at 20°C. Heat is transferred to the system until the temperature reaches 40°C. What is the final pressure?

Since the volume does not change during this process, the specific volume also remains constant. From the ammonia tables, Table A.2.

$$v_1 = v_2 = 0.1494 \text{ m}^3/\text{kg}$$

Since $v_g$ at 40°C is less than 0.1494 m³/kg, it is evident that in the final state the ammonia is superheated vapor. By interpolating between the 900 and 1000 kPa columns of Table A.2.2, we find that

$$P_2 = 938 \text{ kPa}$$

## PROBLEMS

3.1   A spherical balloon has a radius of 5 m. The atmospheric pressure is 100 kPa and the temperature is 20°C.

(a) Calculate the mass and the number of moles of air this balloon displaces.

(b) If the balloon is filled with helium at 100 kPa, 20°C, what is the mass and the number of moles of helium?

3.2   Air is contained in a vertical cylinder fitted with a frictionless piston and a set of stops, as shown in Fig. 3.11. The cross-sectional area of the piston is

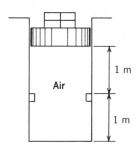

**Fig. 3.11**   Sketch for Problem 3.2.

0.2 m², and the air is initially at 200 kPa, 500°C. The air is then cooled as a result of heat transfer to the surroundings.

(a) What is the temperature of the air inside when the piston reaches the stops?

(b) If the cooling is continued until the temperature reaches 20°C, what is the pressure inside the cylinder at this state?

3.3   A vacuum pump is used to pump a vacuum over a bath of liquid helium. The volume rate of flow into the vacuum pump is 1.5 m³/s. The pressure at the vacuum pump inlet is 15 Pa and the temperature is −25°C. What mass of helium enters the pump per minute?

3.4   A metal sphere of 150 mm inside diameter is weighed on a precision beam balance when evacuated and again when filled to 875 kPa with an unknown gas. The difference in weight is 0.0025 kg. The room temperature is 25°C. What is the gas, assuming it to be a pure substance?

3.5   A rigid vessel $A$ is connected to a spherical elastic balloon $B$ as shown in Fig. 3.12. Both contain air at the ambient temperature 25°C. The volume of vessel $A$ is 0.1 m³ and the initial pressure is 300 kPa. The initial diameter of the balloon is 0.5 m and the pressure inside is 100 kPa. The valve connecting $A$ and $B$ is now opened, and remains open. It may be assumed that the pressure inside the balloon is directly proportional to its diameter, and also that the final temperature of the air is uniform throughout at

25°C. Determine the final pressure in the system and the final volume of the balloon.

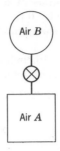

**Fig. 3.12**   Sketch for Problem 3.5.

3.6   Is it reasonable to assume that at the given states the substance behaves as an ideal gas:

(a) Nitrogen at 30°C, 3 MPa.
(b) Carbon dioxide at 30°C, 3 MPa.
(c) Water at 1300°C, 3 MPa.
(d) Water at 50°C, 10 kPa.
(e) Water at 30°C, 10 kPa.

3.7   Determine whether water at each of the following states is a compressed liquid, a superheated vapor, or a mixture of saturated liquid and vapor: 120°C, 150 kPa; 0.35 MPa, 0.4 m³/kg; 160°C, 0.4 m³/kg; 200 kPa, 110°C; 300°C, 0.01 m³/kg; 5 kPa, 10°C.

3.8   Plot the following vapor-pressure curves (saturation pressure vs. saturation temperature):

(a) Water on Cartesian coordinates, −40°C to 20°C.
(b) Water on Cartesian coordinates, 0 to 23 MPa.
(c) Water, Freon-12, and ammonia, on semilog paper (pressure on log scale), 10 kPa to 10 MPa, −50°C to 150°C.

3.9   Calculate the following specific volumes:

(a) Ammonia, 30°C, 80% quality.
(b) Freon-12, 50°C, 15% quality.
(c) Water, 8 MPa, 98% quality.
(d) Nitrogen, 90 K, 40% quality.

3.10 Determine the quality (if saturated) or temperature (if superheated) of the following substances in the given states:

(a) Ammonia, 20°C, 0.1 m³/kg; 800 kPa, 0.2 m³/kg.
(b) Freon-12, 400 kPa, 0.04 m³/kg; 400 kPa, 0.045 m³/kg.
(c) Water, 20°C, 1 m³/kg; 8 MPa, 0.01 m³/kg.
(d) Nitrogen, 0.5 MPa, 0.08 m³/kg; 80 K, 0.14 m³/kg.

3.11 Plot a pressure-specific volume diagram on log log paper (3 × 5 cycles) for water showing the following lines:

    (*a*) Saturated liquid.

    (*b*) Saturated vapor.

    (*c*) The following constant-temperature lines (including the compressed-liquid region): 150°C, 250°C, 350°C, 400°C, 500°C.

    (*d*) The following lines of constant quality: 10%, 50%, 90%.

3.12 Plot a pressure-specific volume diagram on log log paper (2 × 3 cycles) for Freon-12, showing the following lines:

    (*a*) Saturated liquid.

    (*b*) Saturated vapor.

    (*c*) The following constant-temperature lines: −10°C, 50°C, 110°C, 150°C.

    (*d*) The following constant-quality lines: 10%, 50%, 90%.

3.13 A closed tank contains vapor and liquid $H_2O$ in equilibrium at 250°C. The distance from the bottom of the tank to the liquid level is 10 m. What is the pressure reading at the bottom of the tank as compared to the pressure reading at the top of the tank?

3.14 A spacecraft storage vessel of 0.1 $m^3$ capacity contains 100 kg of saturated oxygen at 90 K. Determine the percentages of liquid and vapor in the vessel on a mass basis and also a volume basis.

3.15 A rigid vessel contains saturated water at 100 kPa. Find the percentage of liquid (on a volume basis) at this state such that the water will pass through the critical point when heated.

3.16 A vessel fitted with a sight glass contains Freon-12 at 25°C. Liquid is withdrawn from the bottom at a slow rate, and the temperature remains constant during the process. If the area of the vessel is 0.05 $m^2$ and the level drops 150 mm, determine the mass of Freon-12 withdrawn.

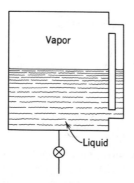

**Fig. 3.13**   Sketch for Problem 3.16.

3.17 For a certain experiment, Freon-12 vapor is contained in a sealed glass tube at 30°C. It is desired to know the pressure at this condition, but there is no means for measuring it, since the tube is sealed. However, if the tube is cooled to 10°C, small droplets of liquid are observed on the glass walls. What is the pressure inside at 30°C?

3.18 A tank contains Freon-12 at 35°C. The volume of the tank is 0.1 m³, and initially the volume of the liquid in the tank is equal to the volume of the vapor. Additional Freon-12 is forced into the tank until the mass of Freon-12 in the tank reaches 80 kg. What is the final volume of liquid in the tank, assuming that the temperature is maintained at 35°C? How much mass enters the tank?

3.19 A rigid vessel of 0.015 m³ volume contains 10 kg of water (liquid plus vapor) at 30°C. The vessel is then slowly heated. Will the liquid level inside eventually rise to the top or drop to the bottom of the vessel? What if the vessel contains 1 kg instead of 10 kg?

3.20 A refrigeration unit of 0.05 m³ volume is evacuated and then slowly charged with Freon-12. During this process, the temperature of the Freon-12 remains constant at the ambient temperature of 25°C.

(a) What will be the mass of Freon-12 in the system when the pressure reaches 250 kPa?

(b) What will be the mass of Freon-12 in the system when the system is filled with saturated vapor?

(c) What fraction of the Freon-12 will exist as a liquid when 5 kg of Freon-12 have been placed in the system?

3.21 Tank A (Fig. 3.14) has a volume of 0.1 m³ and contains Freon-12 at 25°C, 10% liquid and 90% vapor by volume, while tank B is evacuated. The valve is then opened, and the tanks eventually come to the same pressure, which is found to be 200 kPa. During this process, heat is transferred such that the Freon remains at 25°C. What is the volume of tank B?

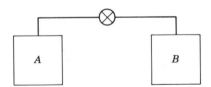

**Fig. 3.14**   Sketch for Problem 3.21.

3.22 A container of liquid nitrogen at 229 kPa pressure has a cross-sectional area of 0.04 m² (Fig. 3.15). As the result of heat transfer to the liquid nitrogen, some of the nitrogen evaporates and in one hour, the level drops 20 mm. The vapor that leaves the insulated container passes through a

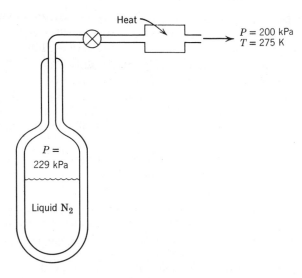

**Fig. 3.15**   Sketch for Problem 3.22.

heater and leaves at 200 kPa, 275 K. Calculate the volume rate of flow out of the heater, assuming ideal gas behavior, and compare this with the result obtained when using the nitrogen tables, Table A.5.

3.23  Water is contained in a cylinder fitted with a frictionless piston, as shown in Fig. 3.16. The mass of water is 1 kg and the area of the piston is 0.5 m². At the initial state the water is at 110°C, with a quality of 90%, and the spring just touches the piston, but exerts no force on it. Now, heat is transferred to the water, and the piston begins to rise. During this process, the resisting force of the spring is proportional to the distance moved, with a force of 10 N/mm. Calculate the pressure in the cylinder when the temperature reaches 200°C.

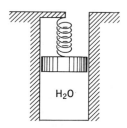

**Fig. 3.16**   Sketch for Problem 3.23.

3.24 A boiler feed pump delivers 50 kg/s of water at 300°C, 20 MPa. What is the volume rate of flow in m³/s? What would be the per cent error if the properties of saturated liquid water at 300°C were used in the calculation?

3.25 Consider compressed liquid water at 40°C. What pressure is required to decrease the specific volume by 1% from its saturated liquid value?

3.26 One kg of $H_2O$ exists at the triple point. The volume of the liquid phase is equal to the volume of the solid phase, and the volume of the vapor phase is equal to $10^4$ times the volume of the liquid phase. What is the mass of $H_2O$ in each phase?

3.27 A 0.1 m³ tank containing saturated vapor water at 200 kPa is cooled to −20°C. What percentage, on a volume basis, of the water is solid at this temperature?

3.28 Compare the specific volume of nitrogen at 6 MPa, 150 K as given in the nitrogen tables, A.5, with the value calculated from the Beattie–Bridgeman equation of state and with the value calculated from the ideal gas equation of state? Which of the three values is the best?

3.29 Write a computer program to solve the following problem. For any specified substance, it is desired to calculate the pressure according to the Beattie–Bridgeman equation of state for any given set (or sets) of temperature and specific volume, and to compare the result with the ideal gas equation of state.

3.30 Write a computer program to solve the following problem. For any specified substance, it is desired to solve the Beattie–Bridgeman equation of state for specific volume at any given set (or sets) of pressure and temperature, and to compare the result with the ideal gas equation of state.

# 4

---

# Work and Heat

In this chapter we shall consider work and heat. It is essential for the student of thermodynamics to understand clearly the definitions of both work and heat, because the correct analysis of many thermodynamic problems depends upon distinguishing between them.

## 4.1  Definition of Work

Work is usually defined as a force $F$ acting through a displacement $x$, the displacement being in the direction of the force. That is,

$$W = \int_1^2 F \cdot dx \qquad (4.1)$$

This is a very useful relationship because it enables us to find the work required to raise a weight, to stretch a wire, or to move a charged particle through a magnetic field.

However, when treating thermodynamics from a macroscopic point of view, it is advantageous to tie in the definition of work with the concepts of systems, properties, and processes. We therefore define work as follows: work is done by a system if the sole effect on the surroundings (everything external to the system) could be the raising of a weight. Notice that the raising of a weight is in effect a force acting through a distance. Notice, also, that our definition does not state that a weight was actually raised, or that a force actually acted through a given distance but that the sole effect external to the system could be the raising of a weight. Work done *by* a system is considered positive and

**61**

work done *on* a system is considered negative. The symbol $W$ designates the work done by a system.

In general, we will speak of work as a form of energy. No attempt will be made to give a rigorous definition of energy. Rather, since the concept is familiar, the term energy will be used as appropriate, and various forms of energy will be identified. Work is the form of energy that fulfills the definition given above.

Let us illustrate this definition of work with a few examples. Consider as a system the battery and motor of Fig. 4.1a and let the motor drive a

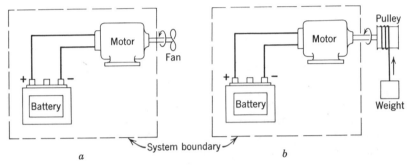

**Fig. 4.1** Example of work done at the boundary of a system.

fan. Does work cross the boundary of the system? To answer this question using the definition of work given above, let the fan be replaced with a pulley and weight arrangement shown in Fig.4.1b. As the motor turns, the weight is raised, and the sole effect external to the system is the raising of a weight. Thus, for our original system of Fig. 4.1a, we conclude that work is crossing the boundary of the system since the sole effect external to the system could be the raising of a weight.

Let the boundaries of the system be changed now to include only the battery shown in Fig. 4.2. Again we ask the question, does work cross the boundary of the system? In answering this question, we will be answering a more general question; namely, does the flow of electrical energy across the boundary of a system constitute work?

The only limiting factor in having the sole external effect the raising of a weight is the inefficiency of the motor. However, as we design a more efficient motor, with lower bearing and electrical losses, we recognize that we can approach a certain limit, which does meet the requirement of having the only external effect the raising of a weight. Therefore, we can conclude that when there is a flow of electricity across the boundary of a system, as in Fig.4.2, it is work with which we are concerned.

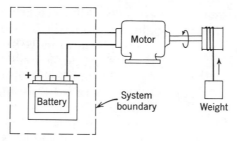

**Fig. 4.2**  Example of work crossing the boundary of a system because of a flow of an electric current across the system boundary.

## 4.2   Units for Work

As already noted, we consider work done *by* a system, such as that done by a gas expanding against a piston as positive, and work done *on* a system, such as that done by a piston compressing a gas, as negative. Thus, positive work means that energy leaves the system and negative work means that energy is added to the system.

Our definition of work involves the raising of a weight, that is, the product of a unit force (one newton) acting through a unit distance (one metre). This unit for work in SI units is called the joule (J).

$$1 \text{ J} = 1 \text{ N m}$$

Power is the time rate of doing work, and is designated by the symbol $\dot{W}$.

$$\dot{W} \equiv \frac{\delta W}{dt}$$

The unit for power is a rate of work of one joule per second, which is a watt (W),

$$1 \text{ W} = 1 \text{ J/s}$$

## 4.3   Work Done at the Moving Boundary of a Simple Compressible System in a Quasiequilibrium Process

We have already noted that there are a variety of ways in which work can be done on or by a system. These include work done by a rotating shaft, electrical work, and the work done by the movement of the system boundary, such as the work done in moving the piston in a cylinder. In this section we will consider in some detail the work done at the moving boundary of a simple compressible system during a quasiequilibrium process.

Consider as a system the gas contained in a cylinder and piston, as in Fig.4.3. Let one of the small weights be removed from the piston, causing

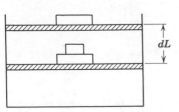

the piston to move upward a distance $dL$. We can consider this a quasi-equilibrium process and calculate the amount of work $W$ done by the system during this process. The total force on the piston is $P\mathscr{A}$, where $P$ is the pressure of the gas and $\mathscr{A}$ is the area of the piston. Therefore, the work $\delta W$ is

**Fig. 4.3**  Example of work done at the moving boundary of a system in a quasi-equilibrium process.

$$\delta W = P\mathscr{A}\,dL$$

But $\mathscr{A}\,dL = dV$, the change in volume of the gas. Therefore,

$$\delta W = P\,dV \tag{4.2}$$

The work done at the moving boundary during a given quasiequilibrium process can be found by integrating Eq.4.2. However, this integration can be performed only if we know the relationship between $P$ and $V$ during this process. This relationship might be expressed in the form of an equation, or it might be shown in the form of a graph.

Let us consider a graphical solution first, using as an example a compression process such as that which occurs during the compression of air in a cylinder, Fig. 4.4. At the beginning of the process the piston is at

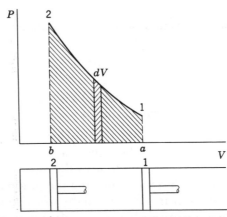

**Fig. 4.4**  Use of pressure-volume diagram to show work done at the moving boundary of a system in a quasiequilibrium process.

position 1, the pressure being relatively low. This state is represented on a pressure-volume diagram (usually referred to as a *P-V* diagram) as shown. At the conclusion of the process the piston is in position 2, and the corresponding state of the gas is shown at point 2 on the *P-V* diagram. Let us assume that this compression was a quasiequilibrium process, and that during the process the system passed through the states shown by the line connecting states 1 and 2 on the *P-V* diagram. The assumption of a quasiequilibrium process is essential here because each point on line 1-2 represents a definite state, and these states will correspond to the actual state of the system only if the deviation from equilibrium is infinitesimal. The work done on the air during this compression process can be found by integrating Eq.4.2.

$$_1W_2 = \int_1^2 \delta W = \int_1^2 P \, dV \tag{4.3}$$

The symbol $_1W_2$ is to be interpreted as the work done during the process from state 1 to state 2. It is clear from examining the *P-V* diagram that the work done during this process, namely,

$$\int_1^2 P \, dV$$

is represented by the area under the curve 1-2, area *a-1-2-b-a*. In this example the volume decreased, and the area *a-1-2-b-a* represents work done on the system. If the process had proceeded from state 2 to state 1 along the same path, the same area would represent work done by the system.

Further consideration of a *P-V* diagram, Fig.4.5, leads to another important conclusion. It is possible to go from state 1 to state 2 along

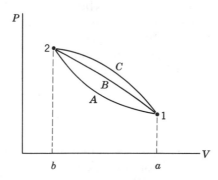

**Fig. 4.5** Various quasiequilibrium processes between two given states, indicating that work is a path function.

many different quasiequilibrium paths, such as $A$, $B$, or $C$. Since the area underneath each curve represents the work for each process, it is evident that the amount of work involved in each case is a function not only of the end states of the process, but in addition is dependent on the path that is followed in going from one state to another. For this reason work is called a path function, or in mathematical parlance, $\delta W$ is an inexact differential.

This leads to a brief consideration of point and path functions or, to use another term, exact and inexact differentials. Thermodynamic properties are point functions, a name that arises from the fact that for a given point on a diagram (such as Fig.4.5) or surface (such as Fig. 3.9), the state is fixed, and thus there is a definite value of each property corresponding to this point. The differentials of point functions are exact differentials, and the integration is simply

$$\int_1^2 dV = V_2 - V_1$$

Thus, we can speak of the volume in state 2 and the volume in state 1, and the change in volume depends only on the initial and final states.

Work on the other hand, is a path function, for, as has been indicated, the work done in a quasiequilibrium process between two given states depends on the path followed. The differentials of path functions are inexact differentials, and the symbol $\delta$ will be used in this text to designate inexact differentials (in contrast to $d$ for exact differentials). Thus, for work we would write

$$\int_1^2 \delta W = {}_1W_2$$

It would be more precise to use the notation $({}_1W_{2,A})$ which would indicate the work done during the change from state 1 to 2 along path $A$. However, implied in the notation ${}_1W_2$ is that the process between states 1 and 2 has been specified. It should be noted, we never speak about the work in the system in state 1 or state 2, and thus we would never write $W_2 - W_1$.

**Example 4.1**

Consider as a system the gas contained in the cylinder shown in Fig. 4.6, which is fitted with a piston on which a number of small weights are placed. The initial pressure is 200 kPa and the initial volume of the gas is 0.04 m$^3$.

(a) Let a Bunsen burner be placed under the cylinder, and let the volume of the gas increase to 0.1 m³ while the pressure remains constant. Calculate the work done by the system during this process.

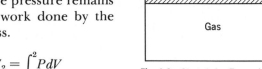

$${}_1W_2 = \int_1^2 P\,dV$$

**Fig. 4.6**   Sketch for Example 4.1.

Since the pressure is constant, we conclude from Eq. 4.3,

$${}_1W_2 = P\int_1^2 dV = P(V_2 - V_1)$$

$${}_1W_2 = 200 \text{ kPa} \times (0.1 - 0.04) \text{ m}^3 = 12.0 \text{ kJ}$$

(b) Consider the same system and initial conditions, but at the same time that the Bunsen burner is under the cylinder and the piston is rising, let weights be removed from the piston at such a rate that, during the process, the relation between pressure and volume is given by the expression $PV = \text{constant} = P_1V_1 = P_2V_2$. Let the final volume again be 0.1 m³.

We first determine the final pressure.

$$P_2 = \frac{P_1V_1}{V_2} = 200 \times \frac{0.04}{0.10} = 80 \text{ kPa}$$

Again we use Eq. 4.3 to calculate the work.

$${}_1W_2 = \int_1^2 P\,dV$$

We can substitute $P = \text{constant}/V = P_1V_1/V$ into this equation.

$${}_1W_2 = P_1V_1 \int_1^2 \frac{dV}{V} = P_1V_1 \ln \frac{V_2}{V_1}$$

$${}_1W_2 = 200 \text{ kPa} \times 0.04 \text{ m}^3 \times \ln \frac{0.10}{0.04} = 7.33 \text{ kJ}$$

(c) Consider the same system, but during the heat transfer let the weights be removed at such a rate that the expression $PV^{1.3} = \text{constant}$ describes the relation between pressure and volume during the process. Again the final volume is 0.1 m³. Calculate the work.

Let us first solve this problem for the general case of $PV^n = $ constant:

$$PV^n = \text{constant} = P_1V_1{}^n = P_2V_2{}^n$$

$$P = \frac{\text{constant}}{V^n} = \frac{P_1V_1{}^n}{V^n} = \frac{P_2V_2{}^n}{V^n}$$

$$_1W_2 = \int_1^2 P\,dV = \text{constant}\int_1^2 \frac{dV}{V^n} = \text{constant}\left[\frac{V^{-n+1}}{-n+1}\right]_1^2$$

$$= \frac{\text{constant}}{1-n}(V_2^{1-n} - V_1^{1-n}) = \frac{P_2V_2{}^nV_2^{1-n} - P_1V_1{}^nV_1^{1-n}}{1-n}$$

$$_1W_2 = \frac{P_2V_2 - P_1V_1}{1-n}$$

For our problem

$$P_2 = 200\left(\frac{0.04}{0.10}\right)^{1.3} = 60.77 \text{ kPa}$$

$$_1W_2 = \frac{P_2V_2 - P_1V_1}{1-1.3} = \frac{60.77 \times 0.1 - 200 \times 0.04}{1-1.3}$$

$$= 6.41 \text{ kJ}$$

(d) Consider the system and initial state given in the first three examples, but let the piston be held by a pin so that the volume remains constant. In addition, let heat be transferred from the system until the pressure drops to 100 kPa. Calculate the work.

Since $\delta W = P\,dV$ for a quasiequilibrium process, the work is zero, because in this case there is no change in volume.

The process for each of four examples is shown on the $P$-$V$ diagram of Fig. 4.7. Process 1–2a is a constant-pressure process, and area 1–2a–f–e–1 represents the work. Similarly, line 1–2b represents the process in which $PV = $ constant, line 1–2c the process in which $PV^{1.3} = $ constant, and line 1–2d represents the constant-volume process. The student should compare the relative areas under each curve with the numerical results obtained above.

## 4.4   Some Other Systems Involving Work at a Moving Boundary

In the preceding section we considered the work done at the moving boundary of a simple compressible system during a quasiequilibrium

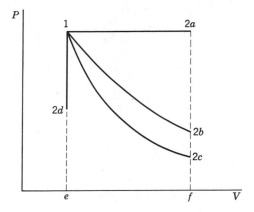

**Fig. 4.7**   Pressure-volume diagram showing work done in the various processes of Example 4.1.

process. There are other types of systems that involve work at a moving boundary, and in this section we shall briefly consider two such systems, a stretched wire and a surface film.

Consider as a system a stretched wire that is under a given tension $\mathcal{T}$. When the length of the wire changes by the amount $dL$, the work done by the system is

$$\delta W = -\mathcal{T}\,dL \tag{4.4}$$

The minus sign is necessary because work is done by the system when $dL$ is negative. This can be integrated to give

$$_1W_2 = -\int_1^2 \mathcal{T}\,dL \tag{4.5}$$

The integration can be performed either graphically or analytically if the relation between $\mathcal{T}$ and $L$ is known. The stretched wire is a simple example of the type of problem in solid body mechanics that involves the calculation of work.

## Example 4.2

A metallic wire of initial length $L_0$ is stretched. Assuming elastic behavior, determine the work done in terms of the modulus of elasticity and the strain.

Let $\sigma =$ stress, $e =$ strain, and $E =$ modulus of elasticity.

$$\sigma = \frac{\mathcal{T}}{\mathcal{A}} = Ee$$

Therefore

$$\mathcal{T} = \mathcal{A}Ee$$

From the definition of strain,

$$de = \frac{dL}{L_0}$$

Therefore,

$$\delta W = -\mathcal{T}\,dL = -\mathcal{A}EeL_0\,de$$

$$W = -\mathcal{A}EL_0 \int_{e=0}^{e} e\,de = \frac{\mathcal{A}EL_0}{2}(e)^2$$

Now consider a system that consists of a liquid film having a surface tension $\mathcal{S}$. A schematic arrangement of such a film is shown in Fig. 4.8,

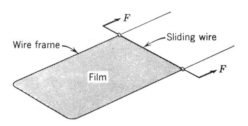

**Fig. 4.8** Schematic arrangement showing work done on a surface film.

where a film is maintained on a wire frame, one side of which can be moved. When the area of the film is changed, for example by sliding the movable wire along the frame, work is done on or by the film. When the area changes by an amount $d\mathcal{A}$, the work done by the system is

$$\delta W = -\mathcal{S}\,d\mathcal{A} \tag{4.6}$$

For finite changes

$$_1W_2 = -\int_1^2 \mathcal{S}\,d\mathcal{A} \tag{4.7}$$

## 4.5 Systems That Involve Other Modes of Work

There are systems that involve other modes of work, and in this section we shall consider two of these, namely, systems involving magnetic and systems involving electrical modes of work. We shall consider a quasi-equilibrium process for these systems, and present expressions for the work done during such a process.

In order to visualize how work can be accomplished by magnetic effects, let us briefly describe magnetic cooling, or adiabatic demagnetization, which is a process used to produce temperatures well below 1 K. A temperature of 1.0 K can be produced by pumping a vacuum over a bath of liquid helium (helium has the lowest normal boiling point of any substance, namely 4.2 K at one atmosphere pressure). An apparatus in which the magnetic cooling is accomplished is shown schematically in Fig. 4.9. The paramagnetic salt is the magnetic substance in which

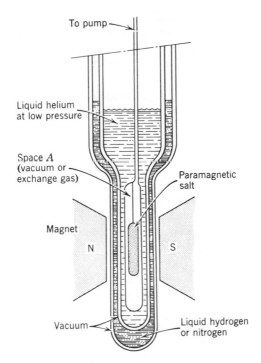

**Fig. 4.9**  Schematic arrangement for magnetic cooling.

temperatures well below 1 K are achieved. When the magnetic field is slowly increased, work is done on the paramagnetic salt. From a microscopic point of view this work is associated with the fact that in the presence of the magnetic field the ions in the salt tend to align themselves with their magnetic axes in the direction of the field. As a result of this work done on the salt, the temperature of the salt tends to increase. However, at this point in the experiment space $A$ is filled with low pressure helium gas, and heat is transferred from the paramagnetic salt to the liquid helium, which is maintained at about 1 K. When the magnetic field is at full strength, and the paramagnetic salt is at the temperature of the liquid helium, space $A$ is evacuated, thus insulating the paramagnetic salt. The magnetic field is now reduced to zero, and in this process work is done by the paramagnetic salt, and its temperature drops sharply. This entire process may be compared to the compression of a gas that is initially at ambient pressure and temperature. As the result of the compression, the temperature of the gas tends to increase. However, the high pressure gas can be cooled to the ambient temperature. If this gas is now isolated from the surroundings and allowed to expand and do work (against a piston for example) the temperature of the gas will decrease below the ambient temperature during the expansion process.

The basic parameters in this process are the intensity of the magnetic field and the magnetization. It may be shown that in a reversible quasiequilibrium process, the work done on a simple magnetic substance is

$$\delta W = -\mu_0 \mathcal{H} d(V \mathcal{M}) \tag{4.8}$$

where:

$\mu_0$ = permeability of free space
$V$ = volume
$\mathcal{H}$ = intensity of the magnetic field
$\mathcal{M}$ = magnetization

The minus sign indicates that as the magnetization increases, work is done on the simple magnetic substance.

We have already noted that electrical energy flowing across the boundary of a system is work. However, we can gain further insight into such a process by considering a system in which the only work mode is electrical. As an example of such a system we can think of a charged condenser, an electrolytic cell, or the type of fuel cell described in Chapter 1. Consider a quasiequilibrium process for such a system, and during this process let the potential difference be $\mathcal{E}$ and the amount of electrical energy that flows into the system be $dZ$. For this quasiequilibrium process the work is given by the relation

$$\delta W = -\mathcal{E} dZ \tag{4.9}$$

Since the current, $i$, equals $dZ/dt$ (where $t$ = time) we can also write

$$\delta W = -\mathscr{E}idt$$

$$_1W_2 = -\int_1^2 \mathscr{E}idt \tag{4.10}$$

Equation 4.10 may also be written as a rate equation for work (the power).

$$\frac{\delta W}{dt} = -\mathscr{E}i \tag{4.11}$$

Since the ampere (electric current) is one of the fundamental units in the International System, and the watt has been defined previously, this relation serves as the definition of the unit for electric potential, the volt (V), which is one watt divided by one ampere.

## 4.6   Some Concluding Remarks Regarding Work

The similarity between the expressions for work in the two processes mentioned in Section 4.5 and the three processes involving a moving boundary should be noted. In each of these quasiequilibrium processes the work is given by the integral of the product of an intensive property and the change of an extensive property. These are summarized below:

Simple compressible system    $_1W_2 = \int_1^2 PdV$

Stretched wire    $_1W_2 = -\int_1^2 \mathscr{T}\, dL$

Surface film    $_1W_2 = -\int_1^2 \mathscr{S}\, d\mathscr{A}$    (4.12)

System involving magnetic work only    $_1W_2 = -\int_1^2 \mu_0 \mathscr{H} d(V\mathscr{M})$

System involving electrical work only    $_1W_2 = -\int_1^2 \mathscr{E}dZ$

Although we will deal primarily with systems involving one mode of work, it is quite possible to have more than one work mode involved in a given process. Thus we could write

$$\delta W = PdV - \mathscr{T}\, dL - \mathscr{S}\, d\mathscr{A} - \mu_0 \mathscr{H}\, d(V\mathscr{M}) - \mathscr{E}\, dZ + \cdots \quad (4.13)$$

where the dotted lines represent other products of an intensive property and the derivative of a related extensive property.

It should also be noted that there are many other forms of work which can be identified in processes that are not quasiequilibrium processes. An example of these is the work done by shearing forces in a process involving friction in a viscous fluid or the work done by a rotating shaft that crosses the system boundary.

The identification of work is an important aspect of many thermodynamic problems. We have already noted that work can be identified only at the boundaries of the system. For example, consider Fig. 4.10,

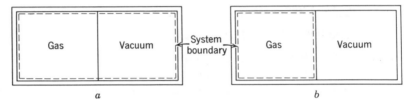

**Fig. 4.10**   Example of a process involving a change of volume for which the work is zero.

which shows a gas separated from the vacuum by a membrane. Let the membrane rupture and the gas fill the entire volume. Neglecting any work associated with the rupturing of the membrane, we can ask if there is work involved in the process. If we take as our system the gas and the vacuum space, we readily conclude that there is no work involved, since no work can be identified at the system boundary. If we take the gas as a system we do have a change of volume, and we might be tempted to calculate the work from the integral

$$\int_1^2 PdV$$

However, this is not a quasiequilibrium process, and therefore the work cannot be calculated from this relation. Rather, since there is no resistance at the system boundary as the volume increases we conclude that for this system there is no work involved in this process.

Another example can be cited with the aid of Fig. 4.11. In Fig. 4.11a the system consists of the container plus the gas. Work crosses the boundary of the system at the point where the system boundary intersects the shaft, and can be associated with the shearing forces in the rotating shaft. In Fig. 4.11$b$ the system includes shaft and weight as well as the

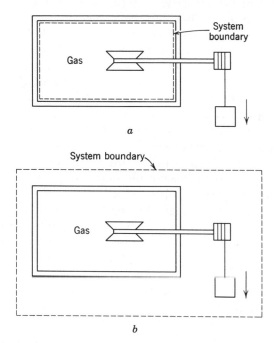

$a$

$b$

**Fig. 4.11**    Example showing how selection of the system determines whether work is involved in a process.

gas and the container. In this case there is no work crossing the system boundary as the weight moves downward. As we will see in the next chapter, we can identify a change of potential energy within the system, but this should not be confused with work crossing the system boundary.

## 4.7    Definition of Heat

The thermodynamic definition of heat is somewhat different from the everyday understanding of the word. Therefore, it is essential to understand clearly the definition of heat given here, because it is involved in so many thermodynamic problems.

If a block of hot copper is placed in a beaker of cold water, we know from experience that the block of copper cools down and the water warms up until the copper and water reach the same temperature. What causes this decrease in the temperature of the copper and the increase in the temperature of the water? We say that it is the result of the transfer of energy from the copper block to the water. It is out of such a transfer of energy that we arrive at a definition of heat.

Heat is defined as the form of energy that is transferred across the boundary of a system at a given temperature to another system (or the surroundings) at a lower temperature by virtue of the temperature difference between the two systems. That is, heat is transferred from the system at the higher to the system at the lower temperature, and the heat transfer occurs solely because of the temperature difference between the two systems. Another aspect of this definition of heat is that a body never contains heat. Rather heat can be identified only as it crosses the boundary. Thus, heat is a transient phenomenon. If we consider the hot block of copper as one system and the cold water in the beaker as another system, we recognize that originally neither system contains any heat (they do contain energy, of course). When the copper is placed in the water and the two are in thermal communication, heat is transferred from the copper to the water, until equilibrium of temperature is established. At that point we no longer have heat transfer, since there is no temperature difference. Neither of the systems contain heat at the conclusion of the process. It also follows that heat is identified at the boundary of the system, for heat is defined as energy being transferred across the system boundary.

## 4.8   Units of Heat

As discussed above, heat, like work, is a form of energy transfer to or from a system. Therefore, the units for heat, and to be more general for any other form of energy as well, are the same as the units for work, which in the International System has been defined as the joule.

Further, heat transferred *to* a system is considered to be positive, and heat transferred *from* a system, negative. Thus, positive heat represents energy transferred to a system, and negative heat represents energy transferred from a system. The symbol $Q$ is used to represent heat.

A process in which there is no heat transfer ($Q = 0$) is called an adiabatic process.

From a mathematical perspective, heat, like work, is a path function and is recognized as an inexact differential. That is, the amount of heat transferred when a system undergoes a change of state from state 1 to state 2 depends on the path that the system follows during the change of state. Since heat is an inexact differential, the differential is written $\delta Q$. On integrating, we write

$$\int_1^2 \delta Q = {}_1Q_2$$

In words, ${}_1Q_2$ is the heat transferred during the given process between state 1 and state 2.

The rate at which heat is transferred to a system is designated by the symbol $\dot{Q}$.

$$\dot{Q} \equiv \frac{\delta Q}{dt}$$

It is also convenient to speak of the heat transfer per unit mass of the system, $q$, which is defined as

$$q \equiv \frac{Q}{m}$$

## 4.9  Comparison of Heat and Work

At this point it is evident that there are many similarities between heat and work, and these are summarized here.

1. Heat and work are both transient phenomena. Systems never possess heat or work, but either or both cross the system boundary when a system undergoes a change of state.
2. Both heat and work are boundary phenomena. Both are observed only at the boundaries of the system, and both represent energy crossing the boundary of the system.
3. Both heat and work are path functions and inexact differentials.

It should also be noted that in our sign convention, $+Q$ represents heat transferred *to* the system, and thus is energy added to the system, and $+W$ represents work done *by* the system and thus represents energy leaving the system.

A final illustration may be helpful to indicate the difference between heat and work. Figure 4.12 shows a gas contained in a rigid vessel. Resistance coils are wound around the outside of the vessel. When current flows through the resistance coils, the temperature of the gas increases. Which crosses the boundary of the system, heat or work?

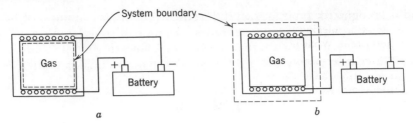

**Fig. 4.12** An example showing the difference between heat and work.

In Fig. 4.12*a* we consider only the gas as the system. In this case the energy crosses the boundary of the system because the temperature of the walls is higher than the temperature of the gas. Therefore, we recognize that heat crosses the boundary of the system.

In Fig. 4.12*b* the system includes the vessel and the resistance heater. Electricity crosses the boundary of the system, and as indicated earlier, this is work.

## PROBLEMS

4.1  Five kilograms of saturated vapor water at 1 MPa is contained in a cylinder fitted with a movable piston. This system is now heated at constant pressure until the temperature of the steam is 300°C. Calculate the work done by the steam during the process.

4.2  One-tenth kilogram of oxygen is contained in a cylinder fitted with a piston. The initial conditions are 150 kPa, 20°C. Weights are then added to the piston, and the $O_2$ is slowly compressed isothermally until the final pressure is 500 kPa. Calculate the work done during this process.

4.3  Consider the system shown in Fig. 4.13. The initial volume inside the cylinder is 0.1 m³. At this state the pressure inside is 100 kPa, which just balances the atmospheric pressure outside plus the piston weight; the spring is touching but exerts no force on the piston at this state. The gas is now heated until the volume is doubled. The final pressure of the gas is 300 kPa, and during the process the spring force is proportional to the displacement of the piston from the initial position.

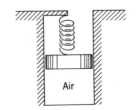

**Fig. 4.13**  Sketch for Problem 4.3.

(a) Show the process on a $P$–$V$ diagram.

(b) Considering the gas inside as the system, calculate the work done by the system. What percentage of this work is done against the spring?

4.4   The cylinder-piston arrangement shown in Fig. 4.14 contains carbon dioxide at 300 kPa, 200°C, at which point the volume is 0.2 m³. Weights are then removed at such a rate that the gas expands according to the relation

$$PV^{1.2} = \text{constant}$$

until the final temperature is 100°C. Determine the work done during this process.

**Fig. 4.14**   Sketch for Problem 4.4.

4.5   A baloon which is initially flat is inflated by filling it with air from a tank of compressed air. The final volume of the balloon is 5 m³. The barometer reads 95 kPa. Consider the tank, the balloon, and the connecting pipe as a system. Determine the work for this process.

4.6   The cylinder shown in Fig. 4.15 contains 1 kg of saturated water at 30°C. The piston has a cross-sectional area of 0.065 m², a mass of 40 kg, and is resting on the stops as shown. The volume at this point is 0.1 m³. Atmospheric pressure outside is 94 kPa, and the local gravitational acceleration is 9.75 m/s². Heat is now transferred to the system until the cylinder contains saturated vapor.

(a) What is the temperature of the water when the piston first rises from the stops?

(b) Calculate the work done by the water during the overall process.

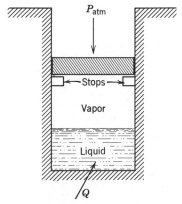

**Fig. 4.15**   Sketch for Problem 4.6.

4.7 The gas space above the water in a closed tank contains nitrogen at 25°C, 100 kPa. The tank has a total volume of 4 m³ and contains 500 kg of water at 25°C. An additional 500 kg of water is now slowly forced into the tank. Assuming that the temperature remains constant, calculate the final pressure of the $N_2$, and the work done on the $N_2$ during the process.

4.8 A spherical balloon has a diameter of 0.3 m, and contains air at a pressure of 150 kPa. The diameter of the balloon increases to 0.4 m due to heating, and during this process the pressure is proportional to the diameter. Calculate the work done by the air during this process.

4.9 A spherical balloon having a radius of 10 m is to be filled with helium from a bank of high pressure gas cylinders that contain helium at 15 MPa, 25°C. The balloon is initially flat, and the atmospheric pressure is 101 kPa.

(a) How much work is done against the atmosphere as the balloon is inflated? Assume no stretching of the material from which the balloon is made and that the pressure in the balloon is essentially equal to the atmospheric pressure.

(b) What is the required volume of the high pressure cylinders, if the final pressure in the cylinders is the same as that in the balloon?

4.10 Ammonia is compressed in a cylinder by a piston. The initial temperature is 30°C, the initial pressure is 500 kPa, and the final pressure is 1400 kPa The following data are available for this process:

| Pressure, kPa | Volume, litres |
|---------------|----------------|
| 500 | 1.25 |
| 650 | 1.08 |
| 800 | 0.96 |
| 950 | 0.84 |
| 1100 | 0.72 |
| 1250 | 0.60 |
| 1400 | 0.50 |

(a. Determine the work for the process considering the ammonia as the system.

(b) What is the final temperature of the ammonia?

4.11 In a rocket, a bottle of helium gas is used to maintain a constant pressure of 0.1 MPa on liquid oxygen as it is fed to the engine pump. The initial conditions are shown in Fig. 4.16. When the tank is drained, the helium bottle and oxygen tank are both filled with helium gas at −75°C, 0.1 MPa pressure.

(a) Calculate the work done by the helium during this process.

(b) Determine the required volume of the helium bottle.

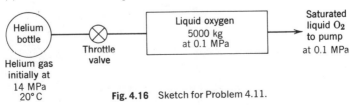

Helium bottle

Throttle valve

Helium gas initially at 14 MPa 20° C

Liquid oxygen 5000 kg at 0.1 MPa

Saturated liquid $O_2$ to pump at 0.1 MPa

**Fig. 4.16** Sketch for Problem 4.11.

4.12 Saturated water vapor at 200°C is contained in a cylinder fitted with a piston. The initial volume of the steam is 0.01 m³. The steam then expands in a quasiequilibrium isothermal process until the final pressure is 200 kPa and in so doing does work against the piston.

(a) Determine the work done during this process.

(b) How much error would be made by assuming the steam to behave as an ideal gas?

4.13 Tank A has a volume of 0.4 m³ and contains argon at 250 kPa, 30°C. Cylinder B contains a frictionless piston of a mass such that a pressure of 150 kPa inside the cylinder is required to raise the piston. The valve connecting the two is now opened, allowing argon to flow into the cylinder. Eventually, the argon is at a uniform state of 30°C, 150 kPa throughout. Calculate the work done by the argon during the process.

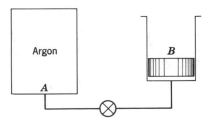

**Fig. 4.17**    Sketch for Problem 4.13.

4.14 Consider the piston and cylinder arrangement indicated in Fig. 4.18, in which a frictionless piston with a cross-sectional area of 0.06 m² rests on stops on the cylinder walls such that the contained volume is 0.03 m³. The mass of the piston is such that 300 kPa pressure is required to raise the piston against atmospheric pressure. When the piston has moved to a point where the contained volume is 0.075 m³, the piston encounters a linear spring that requires 360 kN to deflect it 1 m. Initially the cylinder contains 4 kg of saturated (two-phase) water at 35°C. The final pressure is 7 MPa. Determine the final state of the $H_2O$ and the work done during the process.

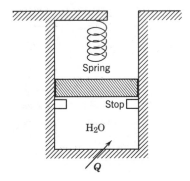

**Fig. 4.18**    Sketch for Problem 4.14.

4.15 A vertical cylinder fitted with a piston contains $0.03 \text{ m}^3$ of Freon-12 at $25°C$ and a quality of 90%. The piston has a mass of 90 kg, cross-sectional area of $0.006 \text{ m}^2$, and is held in place by a pin, as shown in Fig. 4.19. The ambient pressure is 100 kPa.

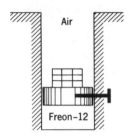

**Fig. 4.19** Sketch for Problem 4.15.

The pin is now removed, allowing the piston to move. After a period of time, the system comes to equilibrium, with the final temperature being $25°C$.

(a) Determine the final pressure and volume of the Freon-12.

(b) Calculate the work done by the Freon-12 during this process, and explain what this work is done against.

4.16 The cylinder indicated in Fig. 4.20 is fitted with a piston which is restrained by a spring so arranged that for zero volume in the cylinder the spring is fully extended. The spring force is proportional to the spring displacement and the weight of the piston is negligible. The enclosed volume in the cylinder is $0.12 \text{ m}^3$ when the piston encounters the stops. The cylinder contains 4 kg of water initially at 350 kPa, 1% quality and the water is then heated until it exists as saturated vapor. Show the process on a $P\text{-}V$ diagram

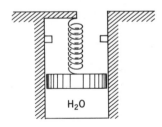

**Fig. 4.20** Sketch for Problem 4.16.

which includes the saturation region and determine:

(a) The final pressure.

(b) The work.

4.17 A steel bar 10 mm in diameter and 0.5 m long is pulled in tension in a tensile testing machine. How much work is required to produce a strain of of 0.1%? The modulus of elasticity of steel is 200 GN/m².

4.18 A sealed vessel having the shape of a rectangular prism with the area of the base $\mathscr{A}$ and height $L_2$, and of negligible mass, is initially floating on a liquid of density $\rho$, Fig. 4.21. Derive an expression in terms of the given

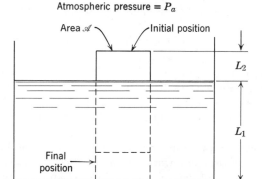

Fig. 4.21 Sketch for Problem 4.18.

variables for the work required to move the vessel to the bottom of a very large tank in which depth of the liquid is $L_1$.

4.19 Repeat Problem 4.18 assuming that the sealed tank has a mass $m$.

4.20 At 20°C ethanol has a surface tension of 22.3 mN/m. Suppose that a film of ethanol is maintained on the wire frame as shown in Fig. 4.22, one side of which can be moved. The original dimensions of the wire frame are as shown. Determine the work done (consider the film to be the system), when the wire is moved 10 mm in the direction indicated.

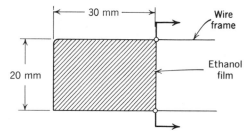

Fig. 4.22 Sketch for Problem 4.20.

4.21 A storage battery is well insulated (thermally) while it is being charged. The charging voltage is 12.3 V (volts) and the current is 24.0 A (amperes). Considering the storage battery as the system, what are the heat transfer and work in a 15-minute period?

4.22  A room is heated with steam radiators on a winter day. Examine the following systems regarding heat transfer (including sign):

(*a*)  The radiator.

(*b*)  The room.

(*c*)  The radiator and the room.

4.23  Consider a hot-air heating system for a home and examine the following systems for heat transfer:

(*a*)  The combustion chamber and combustion gas side of the heat transfer area.

(*b*)  The furnace as a whole including the hot and cold air ducts and chimney.

4.24  Write a computer program to solve the following problem. Determine the boundary movement work for a specified gas undergoing an isothermal process at temperature $T$ from $v_1$ to $v_2$ using the Beattie–Bridgeman equation of state, and compare the result with that found assuming ideal gas behavior.

4.25  Write a computer program to solve the following problem. Determine the boundary movement work for a specified substance undergoing a process, given a set of data (values of pressure and corresponding volume during the process).

# 5

## The First Law
## of Thermodynamics

Having completed our consideration of basic definitions and concepts we are ready to proceed to a discussion of the first law of thermodynamics. Often this law is called the law of the conservation of energy, and as we shall see later, this is essentially true. Our procedure will be to state this law first for a system undergoing a cycle, and then for a change of state of a system. Finally the first law of thermodynamics will be applied to a control volume. The law of the conservation of matter will also be considered in this chapter.

### 5.1 The First Law of Thermodynamics for a System Undergoing a Cycle

The first law of thermodynamics states that during any cycle a system undergoes, the cyclic integral of the heat is proportional to the cyclic integral of the work.

To illustrate this law, consider as a system the gas in the container shown in Fig. 5.1. Let this system go through a cycle that is comprised of two processes. In the first process work is done on the system by the paddle that turns as the weight is lowered. Let the system then be returned to its initial state by transferring heat from the system until the cycle has been completed.

Historically, work was measured in mechanical units of force times distance, as for example foot-pounds force or joules, while measurements of heat were made in thermal units, such as the British Thermal Unit or the calorie. Measurements of work and heat were made during a cycle for a wide variety of systems and for various amounts of work and heat. When the amounts of work and heat were compared, it was found that these two are always proportional. Observations such as this have led to the formulation of the first law of thermodynamics, which in equation form is written

$$J \oint \delta Q = \oint \delta W \qquad (5.1)$$

The symbol $\oint \delta Q$, which is called the cyclic integral of the heat transfer, represents the net heat transfer during the cycle, and $\oint \delta W$, the cyclic integral of the work, represents the net work during the cycle. $J$ is a proportionality factor, which depends on the units used for work and heat.

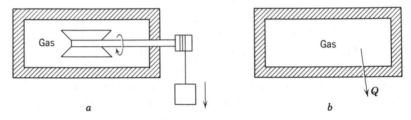

**Fig. 5.1**   Example of a system undergoing a cycle.

The basis of every law of nature is experimental evidence, and this is true also of the first law of thermodynamics. Every experiment that has been conducted thus far has verified the first law either directly or indirectly. It has never been disproved, and has been verified by many different experiments.

As was discussed in Chapter 4, the unit for work and heat, and for any other form of energy as well, in the International System of units is the joule. Thus, in this system we do not need the proportionality factor J, and we may write Eq. 5.1 as

$$\oint \delta Q = \oint \delta W \qquad (5.2)$$

which can be considered as the basic statement of the first law of thermodynamics.

## 5.2 The First Law of Thermodynamics for a Change in State of a System

Equation 5.2 states the first law of thermodynamics for a system during a cycle. Many times, however, we are concerned with a process rather than a cycle, and we now consider the first law of thermodynamics for a system that undergoes a change of state. This can be done by introducing a new property, the energy, which is given the symbol $E$. Consider a system that undergoes a cycle, changing from state 1 to state 2 by process $A$, and returning from state 2 to state 1 by process $B$. This cycle is shown in Fig. 5.2 on a pressure (or other intensive property)-volume (or other extensive property) diagram. From the first law of thermodynamics, Eq. 5.2,

$$\oint \delta Q = \oint \delta W$$

Considering the two separate processes we have

$$\int_{1A}^{2A} \delta Q + \int_{2B}^{1B} \delta Q = \int_{1A}^{2A} \delta W + \int_{2B}^{1B} \delta W$$

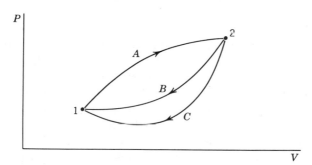

**Fig. 5.2**    Demonstration of the existence of thermodynamic property $E$.

Now consider another cycle, the system changing from state 1 to state 2 by process $A$, as before, and returning to state 1 by process $C$. For this cycle we can write

$$\int_{1A}^{2A} \delta Q + \int_{2C}^{1C} \delta Q = \int_{1A}^{2A} \delta W + \int_{2C}^{1C} \delta W$$

Subtracting the second of these equations from the first, we have

$$\int_{2B}^{1B} \delta Q - \int_{2C}^{1C} \delta Q = \int_{2B}^{1B} \delta W - \int_{2C}^{1C} \delta W$$

or, by rearranging

$$\int_{2B}^{1B}(\delta Q - \delta W) = \int_{2C}^{1C}(\delta Q - \delta W) \qquad (5.3)$$

Since $B$ and $C$ represent arbitrary processes between states 1 and 2, we conclude that the quantity $(\delta Q - \delta W)$ is the same for all processes between state 1 and state 2. Therefore, $(\delta Q - \delta W)$ depends only on the initial and final states and not on the path followed between the two states. We conclude that this is a point function, and therefore is the differential of a property of the system. This property is the energy of the system and is given the symbol $E$. Thus, we can write

$$\delta Q - \delta W = dE$$
$$\delta Q = dE + \delta W \qquad (5.4)$$

Note that since $E$ is a property, its derivative is written $dE$. When Eq. 5.4 is integrated from an initial state 1 to the final state 2, we have

$$_1Q_2 = E_2 - E_1 + {}_1W_2 \qquad (5.5)$$

where $_1Q_2$ is the heat transferred to the system during the process from state 1 to state 2, $E_1$ and $E_2$ are the initial and final values of the energy $E$ of the system, and $_1W_2$ is the work done by the system during the process.

The physical significance of the property $E$ is that it represents all the energy of the system in the given state. This energy might be present in a variety of forms, such as the kinetic or potential energy of the system as a whole with respect to the chosen coordinate frame, energy associated with the motion and position of the molecules, energy associated with the structure of the atom, chemical energy such as is present in a storage battery, energy present in a charged condenser, or in any of a number of other forms.

In the study of thermodynamics it is convenient to consider the bulk kinetic and potential energy separately and then to consider all the other energy of the system in a single property which we call the internal energy, and to which we give the symbol $U$. Thus, we would write

$$E = \text{Internal energy} + \text{Kinetic energy} + \text{Potential energy}$$

or

$$E = U + \text{KE} + \text{PE}$$

The reason for doing this is that the kinetic and potential energy of the system are associated with the coordinate frame that we select and can be specified by the macroscopic parameters of mass, velocity, and elevation. The internal energy $U$ includes all other forms of energy of the system and is associated with the thermodynamic state of the system.

Since the terms comprising $E$ are point functions, we can write

$$dE = dU + d(KE) + d(PE) \qquad (5.6)$$

The first law of thermodynamics for a change of state of a system may therefore be written

$$\delta Q = dU + d(KE) + d(PE) + \delta W \qquad (5.7)$$

In words this equation states that as a system undergoes a change of state, energy may cross the boundary as either heat or work, and each may be positive or negative. The net change in the energy of the system will be exactly equal to the net energy that crosses the boundary of the system. The energy of the system may change in any of three ways, namely, by a change in internal energy, kinetic energy, or potential energy. Since the mass of a system is fixed, we can also say that a quantity of matter may have energy in three forms, internal energy, kinetic energy, or potential energy.

This section will be concluded by deriving an expression for the kinetic and potential energy of a system. Consider first a system that is initially at rest relative to the earth, which is taken as the coordinate frame, and let this system be acted upon by an external horizontal force $F$ which moves the system a distance $dx$ in the direction of the force. Thus there is no change in potential energy. Let there be no heat transfer and no change in internal energy. Then, from the first law, Eq. 5.7,

$$\delta W = -F\,dx = -d\text{KE}$$

But

$$F = ma = m\frac{dV}{dt} = m\frac{dx}{dt}\frac{dV}{dx} = mV\frac{dV}{dx}$$

Then

$$d\text{KE} = F\,dx = mV\,dV$$

Integrating, we obtain

$$\int_{\text{KE}=0}^{\text{KE}} d\text{KE} = \int_{V=0}^{V} mV\,dV$$

$$\text{KE} = \frac{1}{2}mV^2 \qquad (5.8)$$

An expression for potential energy can be found in a similar manner. Consider a system that is initially at rest and at the elevation of some reference level. Let this system be acted upon by a vertical force $F$ which is of such magnitude that it raises (in elevation) the system with constant velocity an amount $dZ$. Let the acceleration due to gravity at this point be $g$. From the first law, Eq. 5.7

$$\delta W = -FdZ = -d\text{PE}$$

$$F = ma = mg$$

Then

$$d\text{PE} = FdZ = mgdZ$$

Integrating

$$\int_{\text{PE}_1}^{\text{PE}_2} d\text{PE} = m\int_{Z_1}^{Z_2} gdZ$$

Assuming that $g$ does not vary with $Z$ (which is a very reasonable assumption for moderate changes in elevation)

$$\text{PE}_2 - \text{PE}_1 = mg(Z_2 - Z_1) \tag{5.9}$$

Substituting these expressions for kinetic and potential energy into Eq. 5.6 we have

$$dE = dU + m\mathsf{V}d\mathsf{V} + mgdZ$$

Integrating for a change of state from state 1 to state 2 with constant $g$ we have

$$E_2 - E_1 = U_2 - U_1 + \frac{m\mathsf{V}_2{}^2}{2} - \frac{m\mathsf{V}_1{}^2}{2} + mgZ_2 - mgZ_1$$

Similarly, substituting these expressions for kinetic and potential energy into Eq. 5.7 we have

$$\delta Q = dU + \frac{d(m\mathsf{V}^2)}{2} + d\left(mgZ\right) + \delta W \tag{5.10}$$

In the integrated form this equation is, assuming $g$ is a constant,

$$_1Q_2 = U_2 - U_1 + \frac{m(\mathbf{V}_2{}^2 - \mathbf{V}_1{}^2)}{2} + mg(Z_2 - Z_1) + _1W_2 \qquad (5.11)$$

Three observations should be made regarding this equation. The first is that the property $E$, the energy of the system, was found to exist, and we were able to write the first law for a change of state using Eq. 5.5. However, rather than deal with this property $E$, we find it more convenient to consider the internal energy and the kinetic and potential energies of the system. In general, this will be the procedure followed in the rest of this book.

The second observation is that Eqs. 5.10 and 5.11 are in effect a statement of the conservation of energy. The net change of the energy of the system is always equal to the net transfer of energy across the system boundary as heat and work. This is somewhat analogous to a joint checking account which a man might have with his wife. There are two ways in which deposits and withdrawals can be made, either by the man or by his wife, and the balance will always reflect the net amount of the transaction. Similarly, there are two ways in which energy can cross the boundary of a system, either as heat or work, and the energy of the system will change by the exact amount of the net energy crossing the system boundary. The concept of energy and the law of the conservation of energy are basic to thermodynamics.

The third observation is that Eqs. 5.10 and 5.11 can give only changes in internal energy, kinetic energy, and potential energy. We can learn nothing about absolute values of these quantities from these equations. If we wish to assign values to internal energy, kinetic energy, and potential energy we must assume reference states and assign a value to the quantity in this reference state. The kinetic energy of a body with zero velocity relative to the earth is assumed to be zero. Similarly, the value of the potential energy is assumed to be zero when the body is at some reference elevation. With internal energy, therefore, we must also have a reference state if we wish to assign values of this property. This matter is considered in the following section.

## 5.3  Internal Energy—A Thermodynamic Property

Internal energy is an extensive property, since it depends upon the mass of the system. Similarly, the kinetic and potential energies are extensive properties.

The symbol $U$ designates the internal energy of a given mass of a substance. Following the convention used with other extensive properties the symbol $u$ designates the internal energy per unit mass. We could speak of $u$ as the specific internal energy, as we do in the case of specific volume. However, since the context will usually make it clear whether $u$ or $U$ is referred to, we will simply use the term internal energy to refer to both internal energy per unit mass and the total internal energy.

In Chapter 3 it was noted that in the absence of the effects of motion, gravity, surface effects, electricity, and magnetism, the state of a pure substance is specified by two independent properties. It is very significant that with these restrictions, the internal energy may be one of the independent properties of a pure substance. This means, for example, that if we specify the pressure and internal energy (with reference to an arbitrary base) of superheated steam, the temperature is also specified.

Thus, in a table of thermodynamic properties such as the steam tables, the value of internal energy can be tabulated along with other thermodynamic properties. Tables 1 and 2 of the steam tables by Keenan et al (Appendix Tables A.1.1 and A.1.2) list the internal energy for saturated states. Included are the internal energy of saturated liquid $u_f$, the internal energy of saturated vapor $u_g$, and the difference between the internal energy of saturated liquid and saturated vapor $u_{fg}$. The values are given in relation to an arbitrarily assumed reference state, which will be discussed later. The internal energy of saturated steam of a given quality is calculated in the same way specific volume is calculated. The relations are

$$u = (1-x)u_f + xu_g$$
$$u = u_f + xu_{fg}$$
$$u = u_g - (1-x)u_{fg}$$

As an example, the specific internal energy of saturated steam having a pressure of 0.6 MPa and a quality of 95% can be calculated as follows:

$$u = u_g - (1-x)\,u_{fg}$$

$$= 2567.4 - 0.05\,(1897.5) = 2472.5 \text{ kJ/kg}$$

**Example 5.1**

A tank containing a fluid is stirred by a paddle wheel. The work input to the paddle wheel is 5090 kJ. The heat transfer from the tank is 1500 kJ. Considering the tank and the fluid as the system, determine the change in the internal energy of the system. The first law of thermodynamics is (Eq. 5.11)

$$_1Q_2 = U_2 - U_1 + m\frac{(V_2{}^2 - V_1{}^2)}{2} + mg\,(Z_2 - Z_1) + {}_1W_2$$

Since there is no change in kinetic and potential energy, this reduces to

$$_1Q_2 = U_2 - U_1 + {}_1W_2$$
$$-1500 = U_2 - U_1 - 5090$$
$$U_2 - U_1 = 3590 \text{ kJ}$$

## Example 5.2

Consider a system composed of a stone having a mass of 10 kg and a bucket containing 100 kg of water. Initially the stone is 10.2 m above the water and the stone and water are at the same temperature. The stone then falls into the water.

Determine $\Delta U$, $\Delta KE$, $\Delta PE$, $Q$, and $W$ for the following changes of state, assuming standard gravitational acceleration, 9.80665 m/s².

1. The stone is about to enter the water.
2. The stone has just come to rest in the bucket.
3. Heat has been transferred to the surroundings in such an amount that the stone and water are at the same temperature they were initially.

The first law of thermodynamics is

$$_1Q_2 = U_2 - U_1 + m\frac{(V_2{}^2 - V_1{}^2)}{2} + mg(Z_2 - Z_1) + {}_1W_2$$

**1.** The stone is about to enter the water: assuming no heat transfer to or from the stone as it falls, we conclude that during the change from the initial state to the state that exists at the moment the stone enters the water,

$$_1Q_2 = 0 \qquad _1W_2 = 0 \qquad \Delta U = 0$$

Therefore the first law reduces to

$$-\Delta KE = \Delta PE = mg \, (Z_2 - Z_1)$$

$$= 10 \text{ kg} \times 9.80665 \, \frac{m}{s^2} \times (-10.2\text{m})$$

$$= -1000 \text{ J} = -1 \text{ kJ}$$

That is, $\Delta Ke = 1$ kJ and $\Delta PE = -1$ kJ

**2.** Just after the stone comes to rest in the bucket:

$$_1Q_2 = 0 \qquad _1W_2 = 0 \qquad \Delta KE = 0$$

Then

$$\Delta PE = -\Delta U = mg(Z_2 - Z_1) = -1 \text{ kJ}$$

$$\Delta U = 1 \text{ kJ} \qquad \Delta PE = -1 \text{ kJ}$$

**3.** After enough heat has been transferred so that the stone and water are at the same temperature they were initially, we conclude that $\Delta U = 0$. Therefore in this case

$$\Delta U = 0 \qquad \Delta KE = 0 \qquad {}_1W_2 = 0$$

$${}_1Q_2 = \Delta PE = mg(Z_2 - Z_1) = -1 \text{ kJ}$$

**Example 5.3**

A vessel having a volume of 5 m³ contains 0.05 m³ of saturated liquid water and 4.95 m³ of saturated water vapor at 0.1 MPa. Heat is transferred until the vessel is filled with saturated vapor. Determine the heat transfer for this process.

Consider the total mass within the vessel as our system. Therefore the first law for this process is

$$Q = U_2 - U_1 + m\frac{(V_2{}^2 - V_1{}^2)}{2} + mg(Z_2 - Z_1) + {}_1W_2$$

Since changes in kinetic and potential energy are not involved, this reduces to

$${}_1Q_2 = U_2 - U_1 + {}_1W_2$$

Further, the work for this process is zero, and therefore

$${}_1Q_2 = U_2 - U_1$$

The thermodynamic properties can be found in Table 2 of the steam tables. The initial internal energy $U_1$ is the sum of the initial internal energy of the liquid and the vapor.

$$U_1 = m_{1\text{liq}}u_{1\text{liq}} + m_{1\text{vap}}u_{1\text{vap}}$$

$$m_{1\text{liq}} = \frac{V_{\text{liq}}}{v_f} = \frac{0.05}{0.001\,043} = 47.94 \text{ kg}$$

$$m_{1\text{vap}} = \frac{V_{\text{vap}}}{v_g} = \frac{4.95}{1.6940} = 2.92 \text{ kg}$$

$$U_1 = 47.94 \, (417.36) + 2.92 \, (2506.1) = 27 \, 326 \text{ kJ}$$

To determine $u_2$ we need to know two thermodynamic properties, since this determines the final state. The properties we know are the quality, $x = 100\%$, and $v_2$, the final specific volume, which can readily be determined.

$$m = m_{1\text{liq}} + m_{1\text{vap}} = 47.94 + 2.92 = 50.86 \text{ kg}$$

$$v_2 = \frac{V}{m} = \frac{5.0}{50.86} = 0.098 \, 31 \text{ m}^3/\text{kg}$$

In Table 2 of the steam tables we find, by interpolation, that at a pressure of 2.03 MPa, $v_g = 0.098 \, 31$ m³/kg. The final pressure of the steam is therefore 2.03 MPa. Then,

$$u_2 = 2600.5 \text{ kJ/kg}$$

$$U_2 = mu_2 = 50.86 \, (2600.5) = 132 \, 261 \text{ kJ}$$

$$_1Q_2 = U_2 - U_1 = 132 \, 261 - 27 \, 326 = 104 \, 935 \text{ kJ}$$

The internal energy of superheated vapor steam is listed as a function of temperature and pressure in Table A.1.3. Similarly, Tables A.1.4 and A.1.5 list values in the compressed liquid and the saturated solid-vapor regions, respectively. In summary, all the values of internal energy are listed in the tables in the same manner as are values of specific volume.

## 5.4  The Thermodynamic Property Enthalpy

In analyzing specific types of processes, we frequently encounter certain combinations of thermodynamic properties, which are therefore also properties of the substance undergoing the change of state. To demonstrate one such situation in which this occurs, let us consider a system undergoing a quasiequilibrium constant-pressure process, as shown in Fig. 5.3. We further assume that there are no changes in kinetic or potential energy and also that the only work done during the process is that associated with the boundary movement. Taking the gas as our system, and applying the first law, Eq. 5.11,

$$_1Q_2 = U_2 - U_1 + {_1W_2}$$

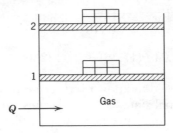

**Fig. 5.3**    The constant-pressure quasiequilibrium process.

The work can be calculated from the relation

$$_1W_2 = \int_1^2 P\, dV$$

Since the pressure is constant

$$_1W_2 = P \int_1^2 dV = P(V_2 - V_1)$$

Therefore

$$_1Q_2 = U_2 - U_1 + P_2 V_2 - P_1 V_1$$
$$= (U_2 + P_2 V_2) - (U_1 + P_1 V_1)$$

We find that in this very restricted case, the heat transfer during the process is given in terms of the change in the quantity $U + PV$ between the initial and final states. Inasmuch as all of these quantities are thermodynamic properties, functions only of the state of the system, their combination must also have these same characteristics. Therefore, we find it convenient to define a new extensive property, called the enthalpy,

$$H = U + PV \qquad (5.12)$$

or, per unit mass,

$$h = u + Pv \qquad (5.13)$$

As in the case of internal energy, we could speak of specific enthalpy, $h$, and total enthalpy, $H$. However, we will simply refer to both as enthalpy, since the context will make it clear which is referred to.

As a result of our definition, the heat transfer in a constant-pressure quasiequilibrium process is equal to the change in enthalpy, which includes both the change in internal energy and the work for this particular process. This is by no means a general result. It is valid for this special case only because the work done during the process is equal to the difference in the *PV* product for the final and initial states. This would not be true if the pressure had not remained constant during the process.

The significance and use of enthalpy is not restricted to the special process described above. Other cases in which this same combination of properties $u + Pv$ appear will be developed later, notably in Section 5.10, in which we discuss control volume analyses. Our reason for introducing enthalpy at this time is that while the steam tables list values for internal energy, most other tables and charts of thermodynamic properties give values for enthalpy but not for the internal energy. In these cases it is necessary to calculate the internal energy at a state using the tabulated values and Eq. 5.13,

$$u = h - Pv$$

Students often become confused about the validity of this calculation when analyzing system processes that do not occur at constant pressure, for which enthalpy has no physical significance. We must keep in mind that enthalpy, being a property, is a state or point function, and its use in calculating internal energy at the same state is not related to, or dependent on any process that may be taking place.

Tabular values of enthalpy, such as those included in Appendix Tables A.1 through A.6, are all relative to some arbitrarily selected base. In the steam tables, the internal energy of saturated liquid at 0.01°C is the reference state and is given a value of zero. For refrigerants, such as ammonia and Freon-12, the reference state is saturated liquid at $-40°C$, the enthalpy in this reference state being assigned the value of zero. Thus, it is possible to have negative values of enthalpy, as is the case for saturated solid $H_2O$ in Table A.1.5 of the Appendix. It should be pointed out that when enthalpy and internal energy are given values relative to the same reference state, as is the case in essentially all thermodynamic tables, the difference between internal energy and enthalpy at the reference state is equal to $Pv$. Since the specific volume of the liquid is very small, this product is negligible as far as the significant figures of the tables are concerned, but the principle should be kept in mind, as in certain cases it is significant.

In the superheat region of most thermodynamic tables, values of the specific internal energy $u$ are not given. As mentioned above, these can be readily calculated from the relation $u = h - Pv$, though it is important to keep the units in mind. As an example, let us calculate the internal energy $u$ of superheated Freon-12 at 0.4 MPa, 70°C.

$$u = h - Pv$$

$$= 232.230 - 400 \times 0.056\ 014$$

$$= 209.824 \text{ kJ/kg}$$

The enthalpy of a substance in a saturation state and having a given quality is found in the same way that the specific volume and internal energy were found. The enthalpy of saturated liquid has the symbol $h_f$, saturated vapor $h_g$, and the increase in enthalpy during vaporization, $h_{fg}$. For a saturation state, the enthalpy can be calculated by one of the following relations:

$$h = (1 - x)h_f + xh_g$$
$$h = h_f + xh_{fg}$$
$$h = h_g - (1 - x)h_{fg}$$

The enthalpy of compressed water may be found from Table A.1.4, while for other substances for which compressed liquid tables are not available, the enthalpy is taken as that of saturated liquid at the same temperature.

## Example 5.4

A cylinder fitted with a piston has a volume of 0.1 m³ and contains 0.5 kg of steam at 0.4 MPa. Heat is transferred to the steam until the temperature is 300°C, while the pressure remains constant.

Determine the heat transfer and the work for this process.

For this system changes in kinetic and potential energy are not significant. Therefore

$$_1Q_2 = m(u_2 - u_1) + {}_1W_2$$

$$_1W_2 = \int_1^2 P\,dV = P\int_1^2 dV = P(V_2 - V_1) = m(P_2 v_2 - P_1 v_1)$$

Therefore

$$_1Q_2 = m(u_2 - u_1) + m(P_2 v_2 - P_1 v_1) = m(h_2 - h_1)$$

$$v_1 = \frac{V_1}{m} = \frac{0.1}{0.5} = 0.2 = 0.001\ 084 + x_1\ 0.4614$$

$$x_1 = \frac{0.1989}{0.4614} = 0.4311$$

$$h_1 = h_f + x_1 h_{fg}$$

$$= 604.74 + 0.4311 \times 2133.8 = 1524.6$$

$$h_2 = 3066.8$$

$$_1Q_2 = 0.5 \,(3066.8 - 1524.6) = 771.1 \text{ kJ}$$

$$_1W_2 = mP \,(v_2 - v_1) = 0.5 \times 400 \,(0.6548 - 0.2)$$

$$= 91.0 \text{ kJ}$$

Therefore

$$U_2 - U_1 = {}_1Q_2 - {}_1W_2 = 771.1 - 91.0 = 680.1 \text{ kJ}$$

The heat transfer could also have been found from $u_1$ and $u_2$.

$$u_1 = u_f + x_1 \, u_{fg}$$

$$= 604.31 + 0.4311 \times 1949.3 = 1444.6$$

$$u_2 = 2804.8$$

and

$$_1Q_2 = 0.5 \,(2804.8 - 1444.6) + 91.0 = 771.1 \text{ kJ}$$

## 5.5   The Constant-Volume and Constant-Pressure Specific Heats

The constant-volume specific heat and the constant-pressure specific heat are useful functions for thermodynamic calculations, and particularly for gases, as will be discussed in the next section.

The constant-volume specific heat $C_v$ is defined by the relation

$$C_v \equiv \left(\frac{\partial u}{\partial T}\right)_v \qquad (5.14)$$

The constant-pressure specific heat $C_p$ is defined by the relation

$$C_p \equiv \left(\frac{\partial h}{\partial T}\right)_P \qquad (5.15)$$

Note that each of these quantities is defined in terms of properties,

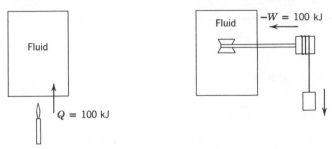

**Fig. 5.4**    Sketch showing two ways in which a given $\Delta U$ may be achieved.

and therefore the constant-volume and constant-pressure specific heats are thermodynamic properties of a substance. These definitions assume constant composition, and also that there are no surface, electrical, or magnetic effects.

The term heat capacity is often used instead of specific heat. However, neither term adequately conveys the nature of the thermodynamic property involved. For example, consider the two identical systems of Fig. 5.4. In the first, 100 kJ of heat is transferred to the system, and in the second, 100 kJ of work is done on the system. Thus, the change of internal energy is the same in each, and therefore the final state and the final temperature are the same in each. In accordance with Eq. 5.14, therefore, exactly the same value for the average constant-volume specific heat would be found for the substance for the two processes. However, if the specific heat is defined in terms of the heat transfer, different values will be obtained for the specific heat, since in the first case the heat transfer is finite and in the second case it is zero. Thus, although the terms "specific heat" and "heat capacity" are not the most appropriate, it is always important to keep in mind that these are thermodynamic properties, and are defined by Eqs. 5.14 and 5.15. When other effects are involved, additional specific heats can be defined, such as specific heat at constant magnetization for a system involving magnetic effects.

**Example 5.5**

Estimate the constant-pressure specific heat of steam at 0.5 MPa, 375°C.

If we consider a change of state at constant pressure, Eq. 5.15 may be written

$$C_p \approx \left(\frac{\Delta h}{\Delta T}\right)_P$$

From the steam tables

$$\text{at } 0.5 \text{ MPa, } 350°C, \, h = 3167.7$$

$$\text{at } 0.5 \text{ MPa, } 400°C, \, h = 3271.9$$

Since we are interested in $C_p$ at 0.5 MPa, 375°C,

$$C_p \simeq \frac{104.2}{50} = 2.084 \text{ kJ/kg K}$$

## 5.6   The Internal Energy, Enthalpy, and Specific Heat of Ideal Gases

At this point certain comments about the internal energy, enthalpy, and the constant-pressure and constant-volume specific heats of an ideal gas should be made. An ideal gas has been defined in Chapter 3 as a gas at sufficiently low density so that intermolecular forces and the associated energy are negligibly small. As a result, an ideal gas has the equation of state

$$Pv = RT$$

It can be shown that for an ideal gas, the internal energy is a function of the temperature only. That is, for an ideal gas

$$u = f(T) \tag{5.16}$$

This means that an ideal gas at a given temperature has a certain definite specific internal energy $u$, regardless of the pressure. This will be demonstrated mathematically in Chapter 10.

In 1843 Joule demonstrated this fact when he conducted the following experiment, which is one of the classical experiments in thermodynamics. Two pressure vessels (Fig. 5.5), connected by a pipe and valve, were

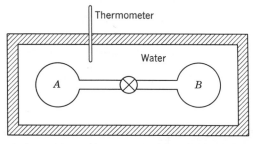

**Fig. 5.5**   Apparatus for conducting Joule's experiment.

immersed in a bath of water. Initially vessel $A$ contained air at 22 atm pressure and vessel $B$ was highly evacuated. When thermal equilibrium was attained the valve was opened, allowing the pressures in $A$ and $B$ to equalize. No change in the temperature of the bath was detected during or after this process. Because there was no change in the temperature of the bath, Joule concluded that there had been no heat transferred to the air. Since the work was also zero, he concluded from the first law of thermodynamics that there was no change in the internal energy of the gas. Since the pressure and volume changed during this process, one concludes that internal energy is not a function of pressure and volume. Because air does not conform exactly to the definition of an ideal gas, a small change in temperature will be detected when very accurate measurements are made in Joule's experiment.

The relation between the internal energy $u$ and the temperature can be established by using the definition of constant-volume specific heat given by Eq. 5.14.

$$C_v = \left(\frac{\partial u}{\partial T}\right)_v$$

Since the internal energy of an ideal gas is not a function of volume, for an ideal gas we can write

$$C_{vo} = \frac{du}{dT}$$

$$du = C_{vo}\, dT \tag{5.17}$$

where the subscript $o$ denotes the specific heat of an ideal gas. For a given mass $m$

$$dU = mC_{vo}\, dT \tag{5.18}$$

From the definition of enthalpy and the equation of state of an ideal gas, it follows that

$$h = u + Pv = u + RT \tag{5.19}$$

Since $R$ is a constant and $u$ is a function of temperature only, it follows that the enthalpy, $h$, of an ideal gas is also a function of temperature only. That is,

$$h = f(T) \tag{5.20}$$

The relation between enthalpy and temperature is found from the constant-pressure specific heat as defined by Eq. 5.15.

$$C_p = \left(\frac{\partial h}{\partial T}\right)_P$$

Since the enthalpy of an ideal gas is a function of the temperature only, and is independent of the pressure, it follows that

$$C_{po} = \frac{dh}{dT}$$

$$dh = C_{po}\, dT \tag{5.21}$$

For a given mass $m$,

$$dH = mC_{po}\, dT \tag{5.22}$$

The consequences of Eqs. 5.17 and 5.21 are demonstrated by Fig. 5.6, which shows two lines of constant temperature. Since internal energy and enthalpy are functions of temperature only, these lines of constant temperature are also lines of constant internal energy and constant enthalpy. From state 1 the high temperature can be reached by a variety of paths, and in each case the final state is different. However, regardless of the path, the change in internal energy is the same, as is the change in enthalpy, for lines of constant temperature are also lines of constant $u$ and constant $h$.

Because the internal energy and enthalpy of an ideal gas are functions of temperature only, it also follows that the constant-volume and constant-pressure specific heats are also functions of temperature only. That is

$$C_{vo} = f(T); \qquad C_{po} = f(T) \tag{5.23}$$

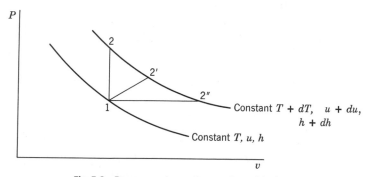

Fig. 5.6   Pressure-volume diagram for an ideal gas.

Since all gases approach ideal gas behavior as the pressure approaches zero, the ideal-gas specific heat for a given substance is often called the zero-pressure specific heat, and the zero-pressure constant-pressure specific heat is given the symbol $C_{po}$. The zero-pressure constant-volume specific heat is given the symbol $C_{vo}$. $\overline{C}_{po}$ as a function of temperature for a number of different substances is shown in Fig. 5.7. These values are determined by using the techniques of statistical thermodynamics, and such calculations will not be discussed in this text. However, a brief qualitative discussion regarding the microscopic viewpoint should provide some insight into this behavior, and is also beneficial in determining under what conditions an assumption of constant specific heat is justified.

From a molecular point of view, the internal energy of a system may be stored in various forms by the molecules comprising the system. One of these is the energy associated with vibration of the atoms within each molecule. The amount of this energy increases as the temperature increases, and in such a manner that it constitutes the major contribution to temperature dependency of the specific heat. A monatomic gas, comprised of individual atoms (such as helium, argon or neon) has no vibrational energy, and thus shows little or no variation in specific heat over a wide range of temperature. A diatomic gas (such as $H_2$, CO, $O_2$) has one vibrational mode, which results in an increase of specific heat with temperature as seen in Fig. 5.7. A polyatomic gas (such as $CO_2$, $H_2O$) shows a much greater increase in specific heat as the temperature increases, and this is due to the additional vibrational modes of a polyatomic molecule.

The results of the specific heat calculations from statistical thermodynamics do not lend themselves to convenient mathematical forms. As a result, empirical equations expressing $\overline{C}_{po}$ as a function of temperature have been developed for many substances, and Table A.9 in the Appendix gives a number of these.

A very important relation between the constant-pressure and constant-volume specific heats of an ideal gas may be developed from the definition of enthalpy.

$$h = u + Pv = u + RT$$

Differentiating, and substituting Eqs. 5.17 and 5.21, we have

$$dh = du + RdT$$

$$C_{po}dT = C_{vo}dT + RdT$$

Therefore

$$C_{po} - C_{vo} = R \tag{5.24}$$

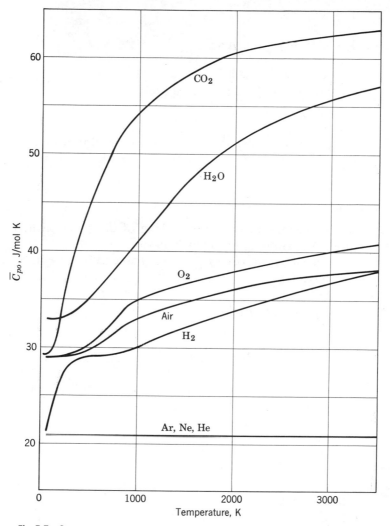

**Fig. 5.7**  Constant-pressure specific heats for a number of gases at zero pressure.

On a mole basis this equation would be written

$$\bar{C}_{po} - \bar{C}_{vo} = \bar{R} \tag{5.25}$$

This tells us that the difference between the constant-pressure and constant-volume specific heats of an ideal gas is always constant, though both are a function of temperature. Thus, using the value of $C_{po}$ from Table A.9, the constant-volume specific heat of oxygen, $C_{vo}$, would be written

$$\bar{C}_{vo} = \bar{C}_{po} - \bar{R}$$

$$\bar{C}_{vo} = 37.432 + 0.020102 \left(\frac{T}{100}\right)^{1.5} - 178.57 \left(\frac{T}{100}\right)^{-1.5}$$

$$+ 236.88 \left(\frac{T}{100}\right)^{-2} - 8.31434$$

where $T$ is in degrees K.

The equations from Table A.9 can be substituted into Eqs. 5.17 and 5.21, and the change in internal energy and enthalpy then found by integration.

**Example 5.6**

Calculate the change of enthalpy as 1 kg of oxygen is heated from 300 K to 1500 K, assuming ideal gas behavior.

Using the specific-heat equation from Table A.9, and integrating Eq. 5.21, we have

$$\bar{h}_2 - \bar{h}_1 = \int_{T_1}^{T_2} \bar{C}_{po} dT = \int_{\theta_1}^{\theta_2} \bar{C}_{po}(\theta) \times 100 \, d\theta$$

$$\bar{h}_{1500} - \bar{h}_{300} = 100 \left( 37.432\theta + \frac{0.020102}{2.5} \theta^{2.5} \right.$$
$$\left. + \frac{178.57}{0.5} \theta^{-0.5} - 236.88\theta^{-1} \right) \quad \begin{vmatrix} \theta_2 = 15 \\ \theta_1 = 3 \end{vmatrix}$$

$$= 40\ 525 \text{ kJ/kmol}$$

$$h_{1500} - h_{300} = \frac{\bar{h}_{1500} - \bar{h}_{300}}{M} = \frac{40\ 525}{32}$$

$$= 1266 \text{ kJ/kg}$$

The average specific heat for any process is defined by the relation

$$C_{p(av)} = \frac{\int_{T_1}^{T_2} C_p dT}{T_2 - T_1} \qquad (5.26)$$

Thus, the average specific heat for Example 5.6 is

$$C_{p(av)} = \frac{1266 \text{ kJ/kg}}{(1500 - 300) \text{ K}} = 1.055 \text{ kJ/kg K}$$

The integration of specific heat equations can be very conveniently done on a digital computer. For this reason it is important to have accurate specific heat equations available. On the other hand, when a digitial computer is not used, there is great value in having tables that give the zero-pressure enthalpy and internal energy of ideal gases as a function of temperature. Values of enthalpy as a function of temperature for a number of ideal gases are given in Table A.11, and for air in Table A.10.

Often, sufficiently accurate results are obtained by assuming that the specific heat is a constant. In this case it follows that

$$h_2 - h_1 = C_{po}(T_2 - T_1); \quad u_2 - u_1 = C_{vo}(T_2 - T_1)$$

**Example 5.7**

A cylinder fitted with a piston has an initial volume of 0.1 m³ and contains nitrogen at 150 kPa, 25°C. The piston is moved, compressing the nitrogen until the pressure is 1 MPa and the temperature is 150°C. During this compression process heat is transferred from the nitrogen, and the work done on the nitrogen is 20 kJ. Determine the amount of this heat transfer.

We consider the nitrogen as our system, and assume that the nitrogen is an ideal gas. Changes in the kinetic and potential energy are negligible.

First law: $_1Q_2 = m(u_2 - u_1) + _1W_2$
Property relation: $Pv = RT$; $u = f(T)$

The mass of nitrogen is found from the equation of state with the value of $R$ from Table A.8.

$$m = \frac{PV}{RT} = \frac{150 \times 0.1}{0.2968 \times 298.15} = 0.1695 \text{ kg}$$

Assuming constant specific heat as given in Table A.8,

$$\begin{aligned}
_1Q_2 &= mC_{vo}(T_2 - T_1) + _1W_2 \\
&= 0.1695(0.7448)(150 - 25) - 20.0 \\
&= 15.8 - 20.0 = -4.2 \text{ kJ}
\end{aligned}$$

## 5.7   The First Law as a Rate Equation

We frequently find it desirable to use the first law as a rate equation that expresses either the instantaneous or average rate at which energy crosses the system boundary as heat and work and the rate at which the energy of the system changes. In so doing we are departing from a strictly classical point of view, because basically classical thermodynamics deals with systems that are in equilibrium, and time is not a relevant parameter for systems that are in equilibrium. However, since these rate equations are developed from the concepts of classical thermodynamics, and are used in many applications of thermodynamics, they are included in this book. This rate form of the first law will be used in the development of the first law for the control volume in Section 5.10, and in this form the first law finds extensive applications in thermodynamics, fluid mechanics and heat transfer.

Consider a time interval $\delta t$ during which an amount of heat $\delta Q$ crosses the system boundary, an amount of work $\delta W$ is done by the system, the internal energy change is $\Delta U$, the kinetic energy change is $\Delta \text{KE}$, and the potential energy change is $\Delta \text{PE}$. From the first law we can write

$$\delta Q = \Delta U + \Delta \text{KE} + \Delta \text{PE} + \delta W$$

Dividing by $\delta t$ we have the average rate of energy transfer as heat and work and increase of the energy of the system.

$$\frac{\delta Q}{\delta t} = \frac{\Delta U}{\delta t} + \frac{\Delta \text{KE}}{\delta t} + \frac{\Delta \text{PE}}{\delta t} + \frac{\delta W}{\delta t}$$

Taking the limit for each of these quantities as $\delta t$ approaches zero we have

$$\lim_{\delta t \to 0} \frac{\delta Q}{\delta t} = \dot{Q}, \quad \text{the heat transfer rate}$$

$$\lim_{\delta t \to 0} \frac{\delta W}{\delta t} = \dot{W}, \quad \text{the power}$$

$$\lim_{\delta t \to 0} \frac{\Delta U}{\delta t} = \frac{dU}{dt}; \qquad \lim_{\delta t \to 0} \frac{\Delta(\text{KE})}{\delta t} = \frac{d(\text{KE})}{dt}; \qquad \lim_{\delta t \to 0} \frac{\Delta(\text{PE})}{\delta t} = \frac{d(\text{PE})}{dt}$$

Therefore, the rate equation form of the first law is

$$\dot{Q} = \frac{dU}{dt} + \frac{d(\text{KE})}{dt} + \frac{d(\text{PE})}{dt} + \dot{W} \tag{5.27}$$

We could also write this in the form

$$\dot{Q} = \frac{dE}{dt} + \dot{W} \tag{5.28}$$

**Example 5.8**

During this charging of a storage battery the current in 20A (amperes) and the voltage is 12.8 V (volts). The rate of heat transfer from the battery is 10 W. At what rate is the internal energy increasing?

Since changes in kinetic and potential energy are not significant, the first law can be written as a rate equation in the form, Eq. 5.27,

$$\dot{Q} = \frac{dU}{dt} + \dot{W}$$

$$\dot{W} = -\mathscr{E}i = -20 \times 12.8 = -256 \text{ W}$$

Therefore,

$$\frac{dU}{dt} = \dot{Q} - \dot{W} = -10\text{W} - (-256) = 246 \text{ J/s}$$

## 5.8   Conservation of Mass

In the previous section we have considered the first law of thermodynamics for a system undergoing a change of state. A system is defined as a a fixed quantity of mass. The question now arises, does the mass of the system change when the energy of a system changes? If it does, then our definition of a system as a fixed quantity of mass is no longer valid when the energy of the system changes.

We know from relativistic considerations that mass and energy are related by the well-known equation

$$E = mc^2 \qquad\qquad (5.29)$$

where $c = $ velocity of light, and $E = $ energy.
We conclude from this equation that the mass of a system does change when its energy changes. Let us calculate the magnitude of this change of mass for a typical problem, and determine whether this change in mass is significant.

Consider as a system a rigid vessel that contains a 1 kg stoichiometric mixture of a hydrocarbon fuel (such as gasoline) and air. From our knowledge of combustion, we know that after combustion takes place it will be necessary to transfer about 2900 kJ from the system in order to restore the system to its initial temperature. From the first law

$$_1Q_2 = U_2 - U_1 + _1W_2$$

we conclude, since $_1W_2 = 0$ and $_1Q_2 = -2900$ kJ, that the internal energy of the system decreases by 2900 kJ during the heat transfer process. Let us now calculate the decrease in mass during this process using Eq. 5.29.

The velocity of light, $c$, is $2.9979 \times 10^8$ m/s. Therefore,

$$2900 \text{ kJ} = 2\,900\,000 \text{ J} = m \text{ (kg)} \times (2.9979 \times 10^8 \text{ m/s})^2$$

$$m = 3.23 \times 10^{-11} \text{ kg}$$

Therefore, when the energy of the system decreases by 2900 kJ, the decrease in mass is $3.23 \times 10^{-11}$ kg.

A change in mass of this magnitude cannot be detected by even our most accurate chemical balance. And, certainly, a fractional change in mass of this magnitude is beyond the accuracy required in essentially all engineering calculations. Therefore, if we use the laws of conservation of mass and conservation of energy as separate laws, we will not introduce significant error into most thermodynamic problems, and our definition of a system as having a fixed mass can be used even though the energy of the system changes.

## 5.9 Conservation of Mass and the Control Volume

A control volume is a volume in space in which one has interest for a particular study or analysis. The surface of this control volume is referred to as a control surface and always consists of a closed surface. The size and shape of the control volume are completely arbitrary, and are so defined as to best suit the analysis which is to be made. The surface may be fixed, or it may move or expand. However, the surface must be defined relative to some coordinate system. In some analyses it may be desirable to consider a rotating or moving coordinate system, and to describe the position of the control surface relative to such a coordinate system.

Mass as well as heat and work can cross the control surface and the mass in the control volume, as well as the properties of this mass, can change with time. Figure 5.8 shows a schematic diagram of a control volume, with heat transfer, shaft work, accumulation of mass within the control volume and a moving boundary.

Let us first consider the law of the conservation of mass as it relates to the control volume. In doing so we consider the mass flow into and out of the control volume and the net increase of mass within the control volume. During a time interval $\delta t$ let the mass $\delta m_i$, as shown in Fig. 5.9 enter the control volume, and the mass $\delta m_e$ leave the control volume.

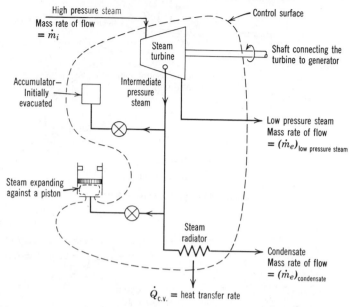

**Fig. 5.8**   Schematic diagram of a control volume showing mass and energy transfers and accumulation.

Further, let us designate the mass within the control volume at the beginning of this time interval as $m_t$ and the mass within the control volume after this interval as $m_{t+\delta t}$. Then, from the law of the conservation of mass we can write

$$m_t + \delta m_i = m_{t+\delta t} + \delta m_e$$

We can also look at this from the point of view of the net flow across the control surface and the change of mass within the control volume.

Net flow into the control volume during $\delta t =$ increase of mass within the control volume during $\delta t$

$$(\delta m_i - \delta m_e) = m_{t+\delta t} - m_t$$

or

$$(m_{t+\delta t} - m_t) + (\delta m_e - \delta m_i) = 0 \tag{5.30}$$

As written, this equation simply states that the change of mass within the control volume during $\delta t$, namely, $(m_{t+\delta t} - m_t)$, and the net mass flow into the control volume during $\delta t$, namely, $(\delta m_i - \delta m_e)$ are equal. However, in many problems requiring a thermodynamic analysis we find it very convenient to have the law of the conservation of mass (as well as

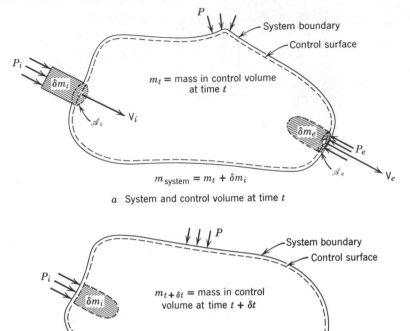

*a  System and control volume at time $t$*

*b  System and control volume at time $t + \delta t$*

**Fig. 5.9** Schematic diagram of a control volume for analysis of the continuity equation as applied to a control volume.

the first and second laws of thermodynamics and the momentum equation) expressed as a rate equation for the control volume. This involves the instantaneous rate of mass flow across the control surface and the instantaneous rate of change of mass within the control volume. Let us therefore write an expression for the average rate of change of mass within the control volume during $\delta t$ and the average mass rates of flow across the control surface during $\delta t$ by dividing Eq. 5.30 by $\delta t$.

$$\left(\frac{m_{t+\delta t} - m_t}{\delta t}\right) + \frac{\delta m_e}{\delta t} - \frac{\delta m_i}{\delta t} = 0 \tag{5.31}$$

To obtain the rate equation for the control volume, we find the limit of each term as $\delta t$ is made to approach zero, at which point the system and control volume coincide.

$$\lim_{\delta t \to 0} \left( \frac{m_{t+\delta t} - m_t}{\delta t} \right) = \frac{dm_{\text{c.v.}}}{dt}$$

$$\lim_{\delta t \to 0} \left( \frac{\delta m_e}{\delta t} \right) = \dot{m}_e$$

$$\lim_{\delta t \to 0} \left( \frac{\delta m_i}{\delta t} \right) = \dot{m}_i$$

The symbol $m_{\text{c.v.}}$ is used to denote the instantaneous mass inside the control volume; $\dot{m}_i$ is the instantaneous rate of flow entering the control volume across area $\mathscr{A}_i$; and $\dot{m}_e$ is that leaving across area $\mathscr{A}_e$. We recognize that, in practice, there may well be several areas on the control surface across which flow occurs (in mixing processes or processes involving chemical reactions, for example). We account for this possibility by including summation signs, implying a summation of flows over the various discrete flow areas. Thus, the instantaneous rate form becomes

$$\frac{dm_{\text{c.v.}}}{dt} + \Sigma \dot{m}_e - \Sigma \dot{m}_i = 0 \tag{5.32}$$

Equation 5.32 for the conservation of mass is commonly termed the continuity equation. While this form of the equation is sufficient for the majority of our applications in this text, it is often convenient (and instructive) to express each of these mass quantities in terms of local fluid properties.

The state of the matter inside the control volume is not necessarily uniform at any given instant of time. However, we may divide the control volume into arbitrarily small volume elements $\delta V$ containing mass $\rho \delta V$, where $\rho$ is the local density. Thus, the total mass inside the control volume, $m_{\text{c.v.}}$, at any instant of time is

$$m_{\text{c.v.}} = \int_V \rho \, dV \tag{5.33}$$

The mass flow rate terms are more difficult to express in this manner, since in general neither the thermodynamic state nor the velocity are uniform across a finite area of flow. To further complicate the problem, the flow may not be normal to the control surface, and the control surface may itself be moving.

In order to demonstrate these possibilities, let us consider a fluid stream moving with some nonuniform velocity, as shown in Fig. 5.10. The control surface is moving with velocity $V_{cs}$ and the flow area on the

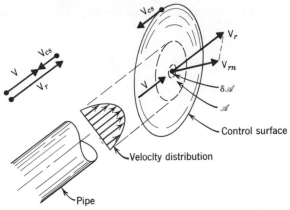

**Fig. 5.10**    Non-uniform flow across a moving control surface.

surface is $\mathscr{A}$. Now consider an elemental area $\delta\mathscr{A}$ sufficiently small that the local state and fluid velocity $V$ are uniform. The velocity relative to the control surface is $V_r$, with a normal component $V_{rn}$. Thus, the instantaneous rate of flow $\delta\dot{m}$ across this element may be expressed in terms of the local density as

$$\delta\dot{m} = \rho V_{rn}\delta\mathscr{A}$$

The total flow rate across area $\mathscr{A}$, then, is

$$\dot{m} = \int_{\mathscr{A}} \delta\dot{m} = \int_{\mathscr{A}} \rho V_{rn}d\mathscr{A} \tag{5.34}$$

We are now in a position to rewrite the continuity equation in terms of local fluid properties. In regard to the flow terms, if we require that $V_{rn}$ be the outward-directed normal to the control volume, then a single term of the form of Eq. 5.34 integrated over the entire control surface includes all flow terms ($V_{rn}$ is then negative for inward-directed flow and zero at locations on the control surface where no flow occurs.) The result is

$$\frac{d}{dt}\int_V \rho dV + \int_{\mathscr{A}} \rho V_{rn}d\mathscr{A} = 0 \tag{5.35}$$

This form of the continuity equation is basic to subsequent developments in the fields of fluid mechanics and heat transfer, which are concerned with state and velocity distributions. In thermodynamics, however, we are commonly concerned with bulk average values for these quantities, and the continuity equation in the form of Eq. 5.32 is sufficient.

As a result, let us reconsider Eq. 5.34 and make the following assumptions:

1. The control surface is stationary.
2. The flow is normal to the control surface.
3. The thermodynamic state and fluid velocity are uniform (bulk average values) over the flow area at any instant of time.

As a result of assumptions **1** and **2**, $V_{rn} = V_n = V$. As a result of assumption **3**, $\rho$ and $V$ are constant over the area $\mathscr{A}$, such that

$$\dot{m} = \int_{\mathscr{A}} \rho V d\mathscr{A} = \rho \mathscr{A} V \qquad (5.36)$$

This result applies to any of the various flow streams entering or leaving the control volume, subject to the conditions assumed. The significance of each of these assumptions should be well understood and borne in mind, as they will be considered again in the development of the first and second laws of thermodynamics for the control volume.

**Example 5.9**

Air is flowing in a 0.2 m diameter pipe at a uniform velocity of 0.1 m/s. The temperature and pressure are 25°C and 150 kPa. Determine the mass flow rate.

From Eq. 5.36,

$$\dot{m} = \rho \mathscr{A} V = \frac{\mathscr{A} V}{v}$$

For air, using $R$ from Table A.8,

$$v = \frac{RT}{P} = \frac{0.287 \times 298.15}{150} = 0.5705 \text{ m}^3/\text{kg}$$

The cross-section area is

$$\mathscr{A} = \frac{\pi}{4}(0.2)^2 = 0.0314 \text{ m}^2$$

Therefore,

$$\dot{m} = \frac{0.0314 \times 0.1}{0.5705} = 0.0055 \text{ kg/s}$$

## 5.10    The First Law of Thermodynamics for a Control Volume

We have already considered the first law of thermodynamics for a system, which consists of a fixed quantity of mass, and noted, Eq. 5.5, that it may be written

$$_1Q_2 = E_2 - E_1 + {_1}W_2$$

We have also noted that this may be written as an average rate equation, over the time interval $\delta t$ by dividing by $\delta t$

$$\frac{\delta Q}{\delta t} = \frac{E_2 - E_1}{\delta t} + \frac{\delta W}{\delta t} \tag{5.37}$$

In order to write the first law as a rate equation for a control volume we proceed in a manner analogous to that used in developing a rate equation for the law of the conservation of mass. A system and control volume are shown in Fig. 5.11. The system consists of all the mass initially in the control volume plus the mass $\delta m_i$.

Consider the changes that take place in the system and control volume during the time interval $\delta t$. During this time $\delta t$ the mass $\delta m_i$ enters the control volume through the discrete area $\mathscr{A}_i$, and the mass $\delta m_e$ leaves the control volume through the area $\mathscr{A}_e$.

In our analysis we will assume that the increment of mass, $\delta m_i$ has uniform properties, and similarly that $\delta m_e$ has uniform properties. The total work done by the system during this process, $\delta W$, is that associated with the masses $\delta m_i$ and $\delta m_e$ crossing the control surface, usually referred to as flow work, and the work $\delta W_{c.v.}$ which includes all other forms of work, such as work associated with a rotating shaft that crosses the system boundary, shear forces, electrical, magnetic or surface effects, or expansion or contraction of the control volume. An amount of heat $\delta Q$ crosses the boundary during $\delta t$.

Let us now consider each of the terms of the first law as it is written for the system, and transform each term into an equivalent form that applies to the control volume. Consider first the term $E_2 - E_1$.

Let $E_t =$ the energy in the control volume at time $t$

$E_{t+\delta t} =$ the energy in the control volume at time $t + \delta t$.

Then

$$E_1 = E_t + e_i \, \delta m_i = \text{energy of the system at time } t$$

$$E_2 = E_{t+\delta t} + e_e \delta m_e = \text{energy of the system at time } t + \delta t.$$

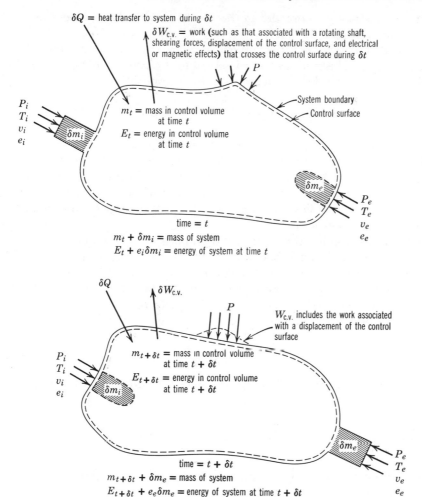

**Fig. 5.11**   Schematic diagram for a first law analysis of a control volume, showing heat and work as well as mass crossing the control surface.

Therefore

$$E_2 - E_1 = E_{t+\delta t} + e_e\,\delta m_e - E_t - e_i\,\delta m_i$$

$$= (E_{t+\delta t} - E_t) + (e_e\,\delta m_e - e_i\,\delta m_i) \tag{5.38}$$

The term $(e_e\,\delta m_e - e_i\,\delta m_i)$ represents the net flow of energy that crosses the control surface during $\delta t$ as the result of the masses $\delta m_e$ and $\delta m_i$ crossing the control surface.

Let us consider in more detail the work associated with the masses $\delta m_i$ and $\delta m_e$ crossing the control surface. Work is done by the normal force (normal to area $\mathscr{A}$) acting on $\delta m_i$ and $\delta m_e$ as these masses cross the control surface. This normal force is equal to the product of the normal tensile stress, $-\sigma_n$, and the area, $\mathscr{A}$. The work done is

$$-\sigma_n \mathscr{A} \, dl = -\sigma_n \delta V = -\sigma_n v \delta m \tag{5.39}$$

A complete analysis of the nature of the normal stress, $\sigma_n$, for an actual fluid involves both static pressure and viscous effects, and is beyond the scope of this book. We will assume in this book that the normal stress, $\sigma_n$, at a point is equal to the static pressure at this point, $P$, and simply note that in many applications this is a very reasonable assumption and yields results of good accuracy.

With this assumption the work done on mass $\delta m_i$ as it enters the control volume is $P_i v_i \, \delta m_i$, and the work done by mass $\delta m_e$ as it leaves the control volume is $P_e v_e \, \delta m_e$. We shall refer to these terms as the flow work. Various other terms are encountered in the literature such as flow energy and work of introduction and work of expulsion.

Thus the total work done by the system during time $\delta t$ is.

$$\delta W = \delta W_{\text{c.v.}} + [P_e v_e \, \delta m_e - P_i v_i \, \delta m_i] \tag{5.40}$$

Let us now divide Eqs. 5.38 and 5.40 by $\delta t$ and substitute into the first law, Eq. 5.37. Combining terms and rearranging,

$$\frac{\delta Q}{\delta t} + \frac{\delta m_i}{\delta t}\,(e_i + P_i v_i) = \left(\frac{E_{t+\delta t} - E_t}{\delta t}\right) + \frac{\delta m_e}{\delta t}\,(e_e + P_e v_e) + \frac{\delta W_{\text{c.v.}}}{\delta t} \tag{5.41}$$

We recognize that each of the flow terms in this expression can be rewritten in the form

$$e + Pv = u + Pv + \frac{\mathbf{V}^2}{2} + gZ$$

$$= h + \frac{\mathbf{V}^2}{2} + gZ \tag{5.42}$$

using the definition of the thermodynamic property enthalpy, Eq. 5.13. It should be apparent that the appearance of the combination $(u + Pv)$ whenever mass flows across a control surface is the principal reason for defining the property enthalpy. Its introduction earlier in conjunction with the constant-pressure process was to facilitate use of the tables of thermodynamic properties at that time.

Utilizing Eq. 5.42 for the masses entering and leaving the control volume, Eq. 5.41 becomes

$$\frac{\delta Q}{\delta t} + \frac{\delta m_i}{\delta t}\left(h_i + \frac{V_i^2}{2} + gZ_i\right) = \left(\frac{E_{t+\delta t} - E_t}{\delta t}\right)$$

$$+ \frac{\delta m_e}{\delta t}\left(h_e + \frac{V_e^2}{2} + gZ_e\right) + \frac{\delta W_{\text{c.v.}}}{\delta t} \qquad (5.43)$$

In order to reduce this expression to a rate equation, let us now examine each of the terms and establish the limits as $\delta t$ is made to approach zero.

As regards the heat transfer, as $\delta t$ approaches zero the system and the control volume coincide, and the heat transfer rate to the system is also the heat transfer rate to the control volume. Therefore,

$$\lim_{\delta t \to 0} \frac{\delta Q}{\delta t} = \dot{Q}_{\text{c.v.}}$$

the rate of heat transfer to the control volume. Further,

$$\lim_{\delta t \to 0}\left(\frac{E_{t+\delta t} - E_t}{\delta t}\right) = \frac{dE_{\text{c.v.}}}{dt}$$

$$\lim_{\delta t \to 0}\left(\frac{\delta W_{\text{c.v.}}}{\delta t}\right) = \dot{W}_{\text{c.v.}}$$

$$\lim_{\delta t \to 0}\left(\frac{\delta m_i}{\delta t}\left(h_i + \frac{V_i^2}{2} + gZ_i\right)\right) = \dot{m}_i\left(h_i + \frac{V_i^2}{2} + gZ_i\right)$$

$$\lim_{\delta t \to 0}\left(\frac{\delta m_e}{\delta t}\left(h_e + \frac{V_e^2}{2} + gZ_e\right)\right) = \dot{m}_e\left(h_e + \frac{V_e^2}{2} + gZ_e\right)$$

We originally assumed uniform properties for the incremental mass $\delta m_i$ entering the control volume across area $\mathcal{A}_i$, and made a similar assumption for $\delta m_e$ leaving across area $\mathcal{A}_e$. Consequently, in taking the limits above, this reduces to the restriction of uniform properties across area $\mathcal{A}_i$ and also across $\mathcal{A}_e$ at an instant of time. These may of course be time dependent.

In utilizing these limiting values to express the rate equation of the first law for a control volume, we again include summation signs on the flow terms to account for the possibility of additional flow streams entering or leaving the control volume. Therefore, the result is

$$\dot{Q}_{\text{c.v.}} + \sum \dot{m}_i \left( h_i + \frac{\mathsf{V}_i^2}{2} + gZ_i \right) = \frac{dE_{\text{c.v.}}}{dt}$$

$$+ \sum \dot{m}_e \left( h_e + \frac{\mathsf{V}_e^2}{2} + gZ_e \right) + \dot{W}_{\text{c.v.}} \qquad (5.44)$$

which is, for our purposes, the general expression of the first law of thermodynamics. In words, this equation says that the rate of heat transfer into the control volume plus rate of energy flowing in as a result of mass transfer is equal to the rate of change of energy inside the control volume plus rate of energy flowing out as a result of mass transfer plus the power output associated with shaft, shear, and electrical effects and other factors that have already been mentioned.

Equation 5.44 can be integrated over the total time of a process to give the total energy changes that occur during that period. However, to do so requires knowledge of the time dependency of the various mass flow rates and the states at which mass enters and leaves the control volume. An example of this type of process will be considered in Section 5.13.

Another point to be noted is that if there is no mass flow into or out of the control volume, those terms simply drop out of Eq. 5.44, which then reduces to the rate equation form of the first law for a system,

$$\dot{Q} = \frac{dE}{dt} + \dot{W}$$

discussed in Section 5.7.

Since the control volume approach is more general, and reduces to the usual statement of the first law for a system when there is no mass flow across the control surface we will, when making a general statement of the first law, utilize Eq. 5.44, the rate form of the first law for a control volume.

The term flux should also be introduced at this point. Flux is defined as a rate of flow of any quantity per unit area across a control surface. Thus the mass flux is the rate of mass flow per unit area and heat flux is the heat flow rate per unit area across the control surface.

Finally, we should consider the question of uniformity of states and representation of the first law, Eq. 5.44, in terms of local fluid properties, as was done for the continuity equation in Section 5.9. As discussed in that section, the state of the matter inside the control volume is not necessarily uniform at any given instant of time. Thus, the quantity $E_{\text{c.v.}}$, representing the total integrated energy inside the control volume, may be expressed as

$$E_{\text{c.v.}} = \int_V e\rho \, dV \qquad (5.45)$$

Analysis of the flow terms for nonuniform states of mass crossing the control surface is the same as that depicted in Fig. 5.9 and leading to Eq. 5.34. For each area of flow $\mathscr{A}$, the resulting first law term is

$$\int_{\mathscr{A}} \left(h + \frac{V^2}{2} + gZ\right)\delta\dot{m} = \int_{\mathscr{A}} \left(h + \frac{V^2}{2} + gZ\right)\rho V_{rn} d\mathscr{A} \tag{5.46}$$

in terms of the local values $h$, $V$, $Z$, $\rho$ and $V_{rn}$. Substitution of this expression for flow terms (with $V_{rn}$ the outward-directed normal, as discussed above, Eq. 5.35) and Eq. 5.45 for the control volume term into Eq. 5.44 results in

$$\dot{Q}_{\text{c.v.}} = \frac{d}{dt}\int_V e\rho\, dV + \int_{\mathscr{A}} \left(h + \frac{V^2}{2} + gZ\right)\rho V_{rn} d\mathscr{A} + \dot{W}_{\text{c.v.}} \tag{5.47}$$

which is the first law for a control volume stated in terms of local properties. This form of the first law is not required for the majority of applications in the remainder of this text, and in general the simplified form of the first law as given by Eq. 5.44 will be adequate. However, we should always bear in mind the assumptions implicit in Eq. 5.44 and the significance of the various terms in that equation.

## 5.11   The Steady-State, Steady-Flow Process

Our first application of the control volume equations will be to develop a suitable analytical model for the long-term steady operation of devices such as turbines, compressors, nozzles, boilers, condensers—a very large class of problems of interest in thermodynamic analysis. This model will not include the short-term transient start-up or shutdown of such devices, but only the steady operating period of time.

Let us consider a certain set of assumptions (beyond those leading to Eqs. 5.32 and 5.44) that lead to a reasonable model for this type of process, which we refer to as the steady-state, steady-flow process. For convenience, we shall often refer to this process as the SSSF process.

1. The control volume does not move relative to the coordinate frame.
2. As for the mass in the control volume, the state of the mass at each point in the control volume does not vary with time.
3. As for the mass that flows across the control surface, the mass flux and the state of this mass at each discrete area of flow on the control surface do not vary with time. The rates at which heat and work cross the control surface remain constant.

As an example of a steady-state, steady-flow process consider a centrifugal air compressor that operates with constant mass rate of flow into and out of the compressor, constant properties at each point across the inlet and exit ducts, a constant rate of heat transfer to the surroundings, and a constant power input. At each point in the compressor the properties are constant with time, even though the properties of a given elemental mass of air vary as it flows through the compressor. Usually such a process is referred to simply as a steady flow process, since we are concerned primarily with the properties of the fluid entering and leaving the control volume. On the other hand, in the analysis of certain heat transfer problems in which the same assumptions apply, we are primarily interested in the spatial distribution of properties, particularly temperature, and such a process is often referred to as a steady state process. Since this is an introductory book we will use the term steady-state, steady-flow process in order to emphasize the basic assumptions involved. The student should realize that the terms steady-state process and steady-flow process are both used extensively in the literature.

Let us now consider the significance of each of these assumptions for the SSSF process.

**1.** The assumption that the control volume does not move relative to the coordinate frame means that all velocities measured relative to the coordinate frame are also velocities relative to the control surface, and there is no work associated with the acceleration of the control volume.

**2.** The assumption that the state of the mass at each point in the control volume does not vary with time requires that

$$\frac{dm_{c.v.}}{dt} = \frac{d}{dt} \int_V \rho dV = 0$$

and also

$$\frac{dE_{c.v.}}{dt} = \frac{d}{dt} \int_V e\rho dV = 0$$

Therefore, we conclude that for the SSSF process, we can write, from Eqs. 5.32 and 5.44

Continuity Equation:

$$\sum \dot{m}_i = \sum \dot{m}_e \tag{5.48}$$

First Law:

$$\dot{Q}_{c.v.} + \sum \dot{m}_i \left( h_i + \frac{V_i^2}{2} + gZ_i \right) = \sum \dot{m}_e \left( h_e + \frac{V_e^2}{2} + gZ_e \right) + \dot{W}_{c.v.} \tag{5.49}$$

**3.** The assumption that the various mass flows, states, and rates at which heat and work cross the control surface remain constant requires that every quantity in Eqs. 5.48 and 5.49 is steady with time. This means that application of Eqs. 5.48 and 5.49 to the operation of some device is independent of time.

Many of the applications of the SSSF model are such that there is only one flow stream entering and one leaving the control volume. For this type of process, we can write

Continuity Equation:

$$\dot{m}_i = \dot{m}_e = \dot{m} \tag{5.50}$$

First Law:

$$\dot{Q}_{\text{c.v.}} + \dot{m}\left(h_i + \frac{V_i^2}{2} + gZ_i\right) = \dot{m}\left(h_e + \frac{V_e^2}{2} + gZ_e\right) + \dot{W}_{\text{c.v.}} \tag{5.51}$$

Rearranging this equation we have:

$$q + h_i + \frac{V_i^2}{2} + gZ_i = h_e + \frac{V_e^2}{2} + gZ_e + w \tag{5.52}$$

where, by definition,

$$q = \frac{\dot{Q}_{\text{c.v.}}}{\dot{m}} \quad \text{and} \quad w = \frac{\dot{W}_{\text{c.v.}}}{\dot{m}} \tag{5.53}$$

Note that the units for $q$ and $w$ are kJ/kg. From their definition $q$ and $w$ can be thought of as the heat transfer and work (other than flow work) per unit mass flowing into and out of the control volume for this particular SSSF process.

The symbols $q$ and $w$ are also used for the heat transfer and work per unit mass of a system. However, since it is always evident from the context whether it is a system (fixed mass) or control volume (involving a flow of mass) with which we are concerned, the significance of the symbols $q$ and $w$ will also be readily evident in each situation.

The SSSF process is often used in the analysis of reciprocating machines, such as reciprocating compressors or engines. In this case the rate of flow, which may actually be pulsating, is considered to be the average rate of flow for an integral number of cycles. A similar assumption is made as regards the properties of the fluid flowing across the control surface, and the heat transfer and work crossing the control surface. It is also assumed that for an integral number of cycles the reciprocating device undergoes, the energy and mass within the control volume do not change.

A number of examples are now given to illustrate the analysis of SSSF processes.

**Example 5.10**

The mass rate of flow into a steam turbine is 1.5 kg/s, and the heat transfer from the turbine is 8.5 kW. The following data are known for the steam entering and leaving the turbine.

|  | Inlet Conditions | Exit Conditions |
|---|---|---|
| Pressure | 2.0 MPa | 0.1 MPa |
| Temperature | 350°C |  |
| Quality |  | 100% |
| Velocity | 50 m/s | 200 m/s |
| Elevation above reference plane | 6 m | 3 m |
| $g = 9.8066$ m/s² |  |  |

Determine the power output of the turbine.

Consider a control surface around the turbine, as shown in Fig. 5.12. From the data available it is evident that we can assume an SSSF process. Since flow enters at only one point and leaves at only one point the equation for the first law is given by Eq. 5.51.

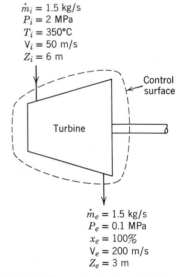

$\dot{m}_i = 1.5$ kg/s
$P_i = 2$ MPa
$T_i = 350°C$
$V_i = 50$ m/s
$Z_i = 6$ m

Control surface

Turbine

$\dot{m}_e = 1.5$ kg/s
$P_e = 0.1$ MPa
$x_e = 100\%$
$V_e = 200$ m/s
$Z_e = 3$ m

**Fig. 5.12**  Illustration for Example 5.10.

$$\dot{Q}_{c.v.} + \dot{m}\left(h_i + \frac{V_i^2}{2} + gZ_i\right)$$

$$= \dot{m}\left(h_e + \frac{V_e^2}{2} + gZ_e\right) + \dot{W}_{c.v.}$$

$$\dot{Q}_{c.v.} = -8.5 \text{ kW}$$

$h_i = 3137.0$ kJ/kg (from the steam tables)

$$\frac{V_i^2}{2} = \frac{50 \times 50}{2 \times 1000} = 1.25 \text{ kJ/kg}$$

$$gZ_i = \frac{6 \times 9.8066}{1000} = 0.059 \text{ kJ/kg}$$

Similarly, $h_e = 2675.5$ kJ/kg

$$\frac{V_e^2}{2} = \frac{200 \times 200}{2 \times 1000} = 20.0 \text{ kJ/kg}$$

$$gZ_e = \frac{3 \times 9.8066}{1000} = 0.029 \text{ kJ/kg}$$

Therefore, substituting into Eq. 5.51,

$$-8.5 + 1.5(3137 + 1.25 + 0.059)$$

$$= 1.5(2675.5 + 20.0 + 0.029) + \dot{W}_{\text{c.v.}}$$

$$\dot{W}_{\text{c.v.}} = -8.5 + 4707.5 - 4043.3 = 655.7 \text{ kW}$$

If Eq. 5.52 is used, the work per kilogram of fluid flowing is found first.

$$q + h_i + \frac{V_i^2}{2} + gZ_i = h_e + \frac{V_e^2}{2} + gZ_e + w$$

$$q = \frac{-8.5}{1.5} = -5.667 \text{ kJ/kg}$$

Therefore, substituting into Eq. 5.52,

$$-5.667 + 3137 + 1.25 + 0.059$$

$$= 2675.5 + 20.0 + 0.029 + w$$

$$w = 437.11 \text{ kJ/kg}$$

$$W_{\text{c.v.}} = 1.5 \text{ kg/s} \times 437.11 \text{ kJ/kg} = 655.7 \text{ kW}$$

Two further observations can be made by referring to this example. First, in many engineering problems potential energy changes are insignificant when compared with the other energy quantities. In the above example the potential energy change did not affect any of the significant figures. In most problems where the change in elevation is small the potential energy terms may be neglected.

Second, if velocities are small, say under 30 m/s, in many cases the kinetic energy is insignificant compared with other energy quantities.

Further, when the velocities entering and leaving the system are essentially the same, the change in kinetic energy is small. Since it is the change in kinetic energy that is important in the SSSF energy equation, the kinetic energy terms can usually be neglected when there is no significant difference between the velocity of the fluid entering and leaving the control volume. Thus in many thermodynamic problems one must make judgments as to which quantities may be negligible for a given analysis.

### Example 5.11

Steam at 0.6 MPa, 200°C, enters an insulated nozzle with a velocity of 50 m/s. It leaves at a pressure of 0.15 MPa and a velocity of 600 m/s. Determine the final temperature if the steam is superheated in the final state, and the quality if it is saturated.

Considering a control surface around the nozzle (Fig. 5.13) it is evident that a SSSF process is a reasonable assumption. Further, $\dot{Q}_{c.v.}$ and $\dot{W}_{c.v.}$

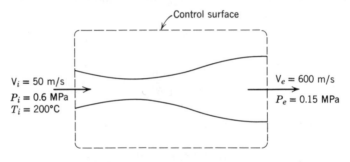

**Fig. 5.13**  Illustration for Example 5.11.

are zero in this problem. Also, since there is no significant change in elevation between the inlet and exit of the nozzle, the change in potential energy is negligible. Therefore, for this process the first law, Eq. 5.52 reduces to

$$h_i + \frac{V_i^2}{2} = h_e + \frac{V_e^2}{2}$$

$$h_e = 2850.1 + \frac{(50)^2}{2 \times 1000} - \frac{(600)^2}{2 \times 1000} = 2671.4 \text{ kJ/kg}$$

The two properties of the fluid leaving which we now know are pressure and enthalpy, and therefore the state of this fluid is determined. Since $h_e$ is less than $h_g$ at 0.15 MPa, the quality is calculated.

$$h = h_g - (1-x)h_{fg}$$

$$2671.4 = 2693.6 - (1 - x_e)\ 2226.5$$

$$(1 - x_e) = \frac{22.2}{2226.5} = 0.010$$

$$x_e = 0.99$$

**Example 5.12**

In a refrigeration system, in which Freon-12 is the refrigerant, the Freon-12 enters the compressor at 150 kPa, −10°C, and leaves at 1.0 MPa, 90°C. The mass rate of flow is 0.016 kg/s, and the power input to the compressor is 1 kW.

On leaving the compressor the Freon-12 enters a water-cooled condenser at 1.0 MPa, 80°C, and leaves as a liquid at 0.95 MPa, 35°C. Water enters the condenser at 10°C and leaves at 20°C. Determine

**1.** The heat transfer rate from the compressor.
**2.** The rate at which cooling water flows through the condenser.

Consider first a control volume analysis of the compressor. A steady-state, steady-flow process is a reasonable assumption.

$$w = \frac{-1}{0.016} = -62.5 \text{ kJ/kg}$$

From the Freon-12 tables

$$h_i = 184.62 \text{ kJ/kg} \qquad h_e = 240.10 \text{ kJ/kg}$$

Since the velocity of the vapor entering a refrigeration compressor is low and not greatly different from the velocity leaving, kinetic energy changes, as well as potential energy changes, can be neglected. Therefore, the first law for this process, Eq. 5.52 reduces to

$$q + h_i = h_e + w$$

$$q = 240.10 - 62.5 - 184.62 = -7.02 \text{ kJ/kg}$$

$$\dot{Q}_{c.v.} = 0.016 \text{ kg/s} \times (-7.02 \text{ kJ/kg}) = -0.1123 \text{ kW}$$

Next consider a control volume analysis of the condenser and a steady-

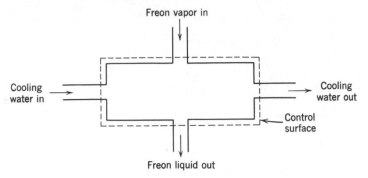

**Fig. 5.14**   Schematic diagram of a Freon condenser.

state, steady-flow process. The schematic diagram for this condenser is shown in Fig. 5.14.

With this control volume we have two fluid streams, the Freon and the water, entering and leaving the control volume. It is reasonable to assume that the kinetic and potential energy changes are negligible. We note that the work is zero, and we make the other reasonable assumption that there is no heat transfer across the control surface. Therefore, the first law, Eq. 5.49, reduces to

$$\sum \dot{m}_i h_i = \sum \dot{m}_e h_e$$

In this case we have two streams entering and two streams leaving. Using the subscript $r$ for refrigerant and $w$ for water

$$\dot{m}_r (h_i)_r + \dot{m}_w (h_i)_w = \dot{m}_r (h_e)_r + \dot{m}_w (h_e)_w$$

From the Freon-12 and steam tables we have

$$(h_i)_r = 232.74 \text{ kJ/kg} \qquad (h_i)_w = 42.01 \text{ kJ/kg}$$

$$(h_e)_r = 69.49 \text{ kJ/kg} \qquad (h_e)_w = 83.96 \text{ kJ/kg}$$

Solving the above equation for $\dot{m}_w$, the rate of flow of water, we have

$$\dot{m}_w = \dot{m}_r \frac{(h_i - h_e)_r}{(h_e - h_i)_w} = 0.016 \text{ kg/s} \frac{(232.74 - 69.49)\text{kJ/kg}}{(83.96 - 42.01)\text{kJ/kg}} = 0.0623 \text{ kg/s}$$

This problem can also be solved by considering two separate control volumes, one of which has the flow of Freon-12 across its control surface, and the other having the flow of water across its control surface. Further there is heat transfer from one control volume to the other. This is shown schematically in Fig. 5.15.

The heat transfer for the control volume involving Freon is calculated first. In this case the SSSF energy equation, Eq. 5.51, reduces to

$$\dot{Q}_{c.v.} = \dot{m}_r (h_e - h_i)_r$$

$$\dot{Q}_{c.v.} = 0.016 \text{ kg/s } (69.49 - 232.74) \text{ kJ/kg} = -2.612 \text{ kW}$$

This is also the heat transfer to the other control volume, for which $\dot{Q}_{c.v.} = +2.612$ kW

$$\dot{Q}_{c.v.} = \dot{m}_w (h_e - h_i)_w$$

$$\dot{m}_w = \frac{2.612 \text{ kW}}{(83.96 - 42.01)\text{kJ/kg}} = 0.0623 \text{ kg/s}$$

**Example 5.13**

Consider the simple steam power plant, as shown in Fig. 5.16. The following data are for such a power plant.

| Location | Pressure | Temperature or quality |
|---|---|---|
| Leaving boiler | 2.0 MPa | 300°C |
| Entering turbine | 1.9 MPa | 290°C |
| Leaving turbine, entering condenser | 15 kPa | 90% |
| Leaving condenser, entering pump | 14 kPa | 45°C |
| Pump work = 4 kJ/kg | | |

Determine the following quantities per kilogram flowing through the unit.

**1.** Heat transfer in line between boiler and turbine.

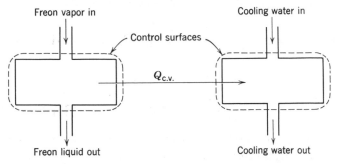

Freon vapor in                     Cooling water in

Control surfaces

$Q_{c.v.}$

Freon liquid out                   Cooling water out

**Fig. 5.15** Schematic diagram of Freon condenser considering two control surfaces.

**2.** Turbine work.
**3.** Heat transfer in condenser.
**4.** Heat transfer in boiler.

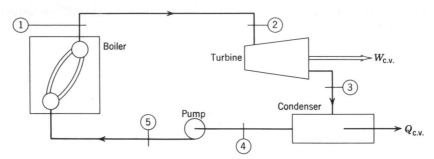

**Fig. 5.16**   Simple steam power plant.

There is a certain advantage in assigning a number to various points in the cycle. For this reason the subscripts $i$ and $e$ in the steady state, steady-flow energy equation are often replaced by appropriate numbers. Thus for a steady-state, steady-flow process for a control volume around the turbine one can write

$$q_{\text{turb}} + h_2 + \frac{V_2{}^2}{2} + gZ_2 = h_3 + \frac{V_3{}^2}{2} + gZ_3 + w_{\text{turb}}$$

Similarly, for a control volume around the boiler

$$q_{\text{boiler}} + h_5 + \frac{V_5{}^2}{2} + gZ_5 = h_1 + \frac{V_1{}^2}{2} + gZ_1$$

We might also use the alternate but equally acceptable notation

$$q_{\text{turb}} = {}_2q_3; \qquad q_{\text{boiler}} = {}_5q_1$$

In this problem we shall neglect changes in kinetic energy and potential energy.

The following property values are obtained from the steam tables, where the subscripts refer to Fig. 5.16.

$$h_1 = 3023.5 \text{ kJ/kg}$$

$$h_2 = 3002.5 \text{ kJ/kg}$$

$$h_3 = 2599.1 - 0.1(2373.1) = 2361.8 \text{ kJ/kg}$$

$$h_4 = 188.5 \text{ kJ/kg}$$

**1.** Considering a control volume that includes the pipe line between boiler and turbine,

$$_1q_2 + h_1 = h_2$$
$$_1q_2 = h_2 - h_1 = 3002.5 - 3023.5 = -21.0 \text{ kJ/kg}$$

**2.** A turbine is essentially an adiabatic machine, and therefore, considering a control volume around the turbine we can write

$$h_2 = h_3 + _2w_3$$
$$_2w_3 = 3002.5 - 2361.8 = 640.7 \text{ kJ/kg}$$

**3.** For the condenser the work is zero, and therefore

$$_3q_4 + h_3 = h_4$$
$$_3q_4 = 188.5 - 2361.8 = -2173.3 \text{ kJ/kg}$$

**4.** The enthalpy at point 5 may be found by considering a control volume around the pump

$$h_4 = h_5 + _4w_5$$
$$h_5 = 188.5 - (-4) = 192.5 \text{ kJ/kg}$$

For the boiler we can write

$$_5q_1 + h_5 = h_1$$
$$_5q_1 = 3023.5 - 192.5 = 2831 \text{ kJ/kg}$$

**Example 5.14**

The centrifugal air compressor of a gas turbine receives air from the ambient atmosphere where the pressure is 1 bar and the temperature is 300 K. At the discharge of the compressor the pressure is 4 bar, the temperature is 480 K, and the velocity is 100 m/s. The mass rate of flow into the compressor is 15 kg/s. Determine the power required to drive the compressor.

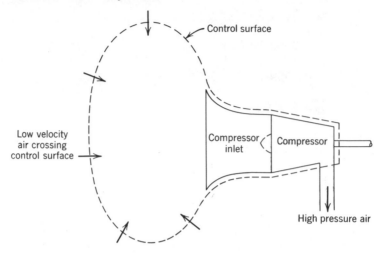

**Fig. 5.17**   Sketch for Example 5.14.

We consider a control volume around the compressor, and locate the control volume at some distance from the compressor so that the air crossing the control surface has a very low velocity and is essentially at ambient conditions. If we located our control volume directly across the inlet section it would be necessary to know the temperature and velocity at the compressor inlet. We assume an adiabatic steady-state, steady-flow process in which the change in potential energy and the inlet kinetic energy are zero. Therefore, the first law for this process, Eq. 5.52 reduces to

$$h_i = h_e + \frac{\mathbf{V}_e^2}{2} + w$$

If we assume constant specific heat, from Table A.8 we have

$$-w = h_e - h_i + \frac{\mathbf{V}_e^2}{2} = C_{po}(T_e - T_i) + \frac{\mathbf{V}_e^2}{2}$$

$$= 1.0035(480 - 300) + \frac{100 \times 100}{2 \times 1000}$$

$$= 180.6 + 5.0 = 185.6 \text{ kJ/kg}$$

$$-\dot{W}_{\text{c.v.}} = 15 \times 185.6 = 2784 \text{ kW}$$

If we use the values for the enthalpy of air that are given in the Air Tables (Table A.10), we have

$$h_i = 300.19 \text{ kJ/kg} \qquad h_e = 482.48 \text{ kJ/kg}$$

$$-w = h_e - h_i + \frac{V_e^2}{2}$$

$$= 482.48 - 300.19 + \frac{100 \times 100}{2 \times 1000}$$

$$= 182.3 + 5.0 = 187.3 \text{ kJ/kg}$$

$$-\dot{W}_{c.v.} = 15 \times 187.3 = 2810 \text{ kW}$$

## 5.12   The Joule-Thomson Coefficient and the Throttling Process

The Joule-Thomson coefficient $\mu_J$ is defined by the relation

$$\mu_J \equiv \left(\frac{\partial T}{\partial P}\right)_h \tag{5.54}$$

As in the definition of specific heats in Section 5.5, this quantity is defined in terms of thermodynamic properties, and therefore is itself a property of a substance.

The significance of the Joule-Thomson coefficient may be demonstrated by considering a throttling process that is a steady-state, steady-flow process across a restriction, with a resulting drop in pressure. A typical example is the flow through a partially opened valve or a restriction in the line. In most cases this occurs so rapidly and in such a small space, that there is neither sufficient time nor a large enough area for much heat transfer. Therefore, we usually may assume such processes to be adiabatic.

If we consider the control surface shown in Fig. 5.18 we can write the SSSF energy equation for this process. There is no work, no change in potential energy, and we make the reasonable assumption that there is

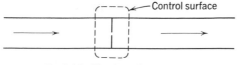

**Fig. 5.18**   The throttling process.

no heat transfer. The SSSF energy equation, Eq. 5.52, reduces to

$$h_i + \frac{\mathsf{V}_i^2}{2} = h_e + \frac{\mathsf{V}_e^2}{2}$$

If the fluid is a gas, the specific volume always increases in such a process, and, therefore, if the pipe is of constant diameter, the kinetic energy of the fluid increases. In many cases, however, this increase in kinetic energy is small (or perhaps the diameter of the exit pipe is larger than that of the inlet pipe) and we can say with a high degree of accuracy that in this process the final and initial enthalpies are equal.

It is for such a process that the Joule-Thomson coefficient, $\mu_J$, is significant. A positive Joule-Thomson coefficient means that the temperature drops during throttling, and when the Joule-Thomson coefficient is negative the temperature rises during throttling.

### Example 5.15

Steam at 800 kPa, 300°C is throttled to 200 kPa. Changes in kinetic energy are negligible for this process. Determine the final temperature of the steam, and the average Joule-Thomson coefficient.
For this process,

$$h_i = h_e = 3056.5 \text{ kJ/kg}$$

$$P_e = 200 \text{ kPa}$$

These two properties determine the final state. From the superheat table for steam

$$T_e = 292.4°C$$

$$\mu_{J(av)} = \left(\frac{\Delta T}{\Delta P}\right)_h = \frac{-7.6 \text{ K}}{-600} = 0.0127 \text{ K/kPa}$$

Frequently a throttling process involves a change in the phase of the fluid. A typical example is the flow through the expansion valve of a vapor compression refrigeration system. The following example deals with this problem.

### Example 5.16

Consider the throttling process across the expansion valve or through the capillary tube in a vapor compression refrigeration cycle. In this process the pressure of the refrigerant drops from the high pressure in the condenser to the low pressure in the evaporator, and during this process some of the liquid flashes into vapor. If we consider this process to be adiabatic, the quality of the refrigerant entering the evaporator can be calculated.

Consider the following process, in which ammonia is the refrigerant. The ammonia enters the expansion valve at a pressure of 1.50 MPa and a temperature of 32°C. Its pressure on leaving the expansion valve is 268 kPa. Calculate the quality of the ammonia leaving the expansion valve.

Considering a steady-state, steady-flow process for a control surface around the expansion valve or capillary tube, we conclude that $\dot{Q}_{\text{c.v.}} = 0$, $\dot{W}_{\text{c.v.}} = 0$, $\Delta KE = 0$, $\Delta PE = 0$, and therefore the statement of the first law reduces to

$$h_i = h_e$$

From the ammonia tables

$$h_i = 332.6 \text{ kJ/kg}$$

(The enthalpy of a slightly compressed liquid is essentially equal to the enthalpy of saturated liquid at the same temperature.)

$$h_e = h_i = 332.6 = 126.0 + x_e \,(1303.5)$$

$$x_e = 0.1585 = 15.85\%$$

## 5.13   The Uniform-State, Uniform-Flow Process

In Section 5.11 we considered the steady-state, steady-flow process and several examples of its application. Many processes of interest in thermodynamics involve unsteady flow, and do not fit into this category. A certain group of these—for example, filling closed tanks with a gas or liquid, or discharge from closed vessels—can be reasonably represented to a first approximation by another simplified model. We shall call this process the uniform-state, uniform-flow process, or for convenience, the USUF process. The basic assumptions are as follows:

1. The control volume remains constant relative to the coordinate frame.
2. The state of the mass within the control volume may change with time, but at any instant of time the state is uniform throughout the entire control volume (or over several identifiable regions that make up the entire control volume).

**3.** The state of the mass crossing each of the areas of flow on the control surface is constant with time although the mass flow rates may be time varying.

Let us examine the consequence of these assumptions and derive an expression for the first law that applies to this process. The assumption that the control volume remains stationary relative to the coordinate frame has already been discussed in Section 5.11. The remaining assumptions lead to the following simplifications for the continuity equation and the first law.

The overall process occurs during time $t$. At any instant of time during the process, the continuity equation is

$$\frac{dm_{\text{c.v.}}}{dt} + \Sigma \dot{m}_e - \Sigma \dot{m}_i = 0$$

where the summation is over all areas on the control surface through which flow occurs. Integrating over time $t$ gives the change of mass in the control volume during the overall process.

$$\int_0^t \left(\frac{dm_{\text{c.v.}}}{dt}\right) dt = (m_2 - m_1)_{\text{c.v.}}$$

The total mass leaving the control volume during time $t$ is

$$\int_0^t (\Sigma \dot{m}_e) dt = \Sigma m_e$$

and the total mass entering the control volume during time $t$ is

$$\int_0^t (\Sigma \dot{m}_i) dt = \Sigma m_i$$

Therefore, for this period of time $t$ we can write the continuity equation for the USUF process as

$$(m_2 - m_1)_{\text{c.v.}} + \Sigma m_e - \Sigma m_i = 0 \tag{5.55}$$

In writing the first law for the USUF process we consider Eq. 5.44, which applies at any instant of time during the process.

$$\dot{Q}_{\text{c.v.}} + \Sigma \dot{m}_i \left(h_i + \frac{\mathsf{V}_i^2}{2} + gZ_i\right) = \frac{dE_{\text{c.v.}}}{dt} + \Sigma \dot{m}_e \left(h_e + \frac{\mathsf{V}_e^2}{2} + gZ_e\right) + \dot{W}_{\text{c.v.}}$$

Since at any instant of time the state within the control volume is uniform, the first law for the USUF process becomes

$$\dot{Q}_{\text{c.v.}} + \Sigma \dot{m}_i \left( h_i + \frac{V_i^2}{2} + gZ_i \right) = \Sigma \dot{m}_e \left( h_e + \frac{V_e^2}{2} + gZ_e \right)$$

$$+ \frac{d}{dt} \left[ (m) \left( u + \frac{V^2}{2} + gZ \right) \right]_{\text{c.v.}} + \dot{W}_{\text{c.v.}}$$

Let us now integrate this equation over time $t$, during which time we have

$$\int_0^t \dot{Q}_{\text{c.v.}} dt = Q_{\text{c.v.}}$$

$$\int_0^t \left[ \Sigma \dot{m}_i \left( h_i + \frac{V_i^2}{2} + gZ_i \right) \right] dt = \Sigma m_i \left( h_i + \frac{V_i^2}{2} + gZ_i \right)$$

$$\int_0^t \left[ \Sigma \dot{m}_e \left( h_e + \frac{V_e^2}{2} + gZ_e \right) \right] dt = \Sigma m_e \left( h_e + \frac{V_e^2}{2} + gZ_e \right)$$

$$\int_0^t \dot{W}_{\text{c.v.}} dt = W_{\text{c.v.}}$$

$$\int_0^t \frac{d}{dt} \left[ (m) \left( u + \frac{V^2}{2} + gZ \right) \right]_{\text{c.v.}} dt$$

$$= \left[ m_2 \left( u_2 + \frac{V_2^2}{2} + gZ_2 \right) - m_1 \left( u_1 + \frac{V_1^2}{2} + gZ_1 \right) \right]_{\text{c.v.}}$$

Therefore, for this period of time $t$ we can write the first law for the uniform-state, uniform-flow process as

$$Q_{\text{c.v.}} + \Sigma m_i \left( h_i + \frac{V_i^2}{2} + gZ_i \right) = \Sigma m_e \left( h_e + \frac{V_e^2}{2} + gZ_e \right)$$

$$+ \left[ m_2 \left( u_2 + \frac{V_2^2}{2} + gZ_2 \right) - m_1 \left( u_1 + \frac{V_1^2}{2} + gZ_1 \right) \right]_{\text{c.v.}} + W_{\text{c.v.}} \qquad (5.56)$$

As an example of the type of problem for which these assumptions are valid and Eq. 5.56 is appropriate, let us consider the classic problem of flow into an evacuated vessel. This is the subject of Example 5.17.

**Example 5.17**

Steam at a pressure of 1.4 MPa, 300°C, is flowing in a pipe. Fig. 5.19. Connected to this pipe through a valve is an evacuated tank. The valve is

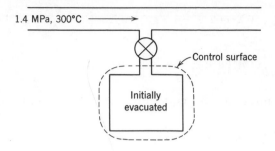

**Fig. 5.19**    Flow into an evacuated vessel – control volume analysis.

opened and the tank fills with steam until the pressure is 1.4 MPa, and then the valve is closed. The process takes place adiabatically and kinetic energies and potential energies are negligible. Determine the final temperature of the steam.

Consider first the control volume shown in Fig. 5.19. The assumptions of uniform state in the control volume and across the pipe are both reasonable, and therefore the first law can be written as stated in Eq. 5.56.

$$Q_{\text{c.v.}} + \sum m_i\left(h_i + \frac{V_i^2}{2} + gZ_i\right) = \sum m_e\left(h_e + \frac{V_e^2}{2} + gZ_e\right)$$

$$+ \left[m_2\left(u_2 + \frac{V_2^2}{2} + gZ_2\right) - m_1\left(u_1 + \frac{V_1^2}{2} + gZ_1\right)\right]_{\text{c.v.}} + W_{\text{c.v.}}$$

We note that $Q_{\text{c.v.}} = 0$, $W_{\text{c.v.}} = 0$, $m_e = 0$, $(m_1)_{\text{c.v.}} = 0$. We further assume that changes in kinetic and potential energy are negligible. Therefore, the statement of the first law for this process reduces to

$$m_i h_i = m_2 u_2$$

From the continuity equation for this process, Eq. 5.55, we conclude that

$$m_2 = m_i$$

Therefore, combining the continuity equation with the first law we have

$$h_i = u_2$$

That is, the final internal energy of the steam in the tank is equal to the enthalpy of the steam entering the tank.

From the steam tables

$$h_i = u_2 = 3040.4 \text{ kJ/kg}$$

Since the final pressure is given as 1.4 MPa, we know two properties of the final state and therefore the final state is determined. The temperature corresponding to a pressure of 1.4 MPa and an internal energy of 3040.4 kJ/kg is found to be 452°C. Had this problem involved a substance for which internal energies are not listed in the thermodynamic tables, it would have been necessary to calculate a few values for $u$ before an interpolation could be made for the final temperature.

This problem can also be solved by considering the steam that enters the tank and the evacuated space as a system, as indicated in Fig. 5.20.

The process is adiabatic, but we must examine the boundaries for work. If we visualize a piston between the steam that is included in the system and the steam that flows behind, we readily recognize that the boundaries of the system move and that the steam in the pipe does

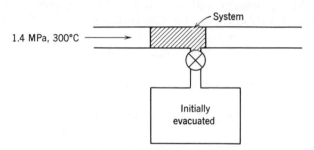

**Fig. 5.20**    Flow into an evacuated vessel – system analysis.

work on the steam that comprises the system. The amount of this work is

$$-W = P_1V_1 = mP_1v_1$$

Writing the first law for the system, Eq. 5.11, and noting that kinetic and potential energies can be neglected, we have

$$_1Q_2 = U_2 - U_1 + {_1W_2}$$
$$0 = U_2 - U_1 - P_1V_1$$
$$0 = mu_2 - mu_1 - mP_1v_1 = mu_2 - mh_1$$

Therefore,

$$u_2 = h_1$$

which is the same conclusion which was reached using a control volume analysis.

The two other examples that follow illustrate further the uniform-state, uniform-flow process.

**Example 5.18**

Let the tank of the previous example have a volume of 0.4 m³ and initially contain saturated vapor at 350 kPa. The valve is then opened and steam from the line at 1.4 MPa, 300°C, flows into the tank until the pressure is 1.4 MPa.

Calculate the mass of steam that flows into the tank.

Let us again consider a control surface around the tank as indicated in Fig. 5.19. In this case, however, the control volume initially contains the mass $m_1$ and the internal energy $m_1 u_1$.

Again we note that $Q_{c.v.} = 0$, $W_{c.v.} = 0$, $m_e = 0$, and we assume that changes in kinetic energy and potential energy are zero. The statement of the first law for this process, Eq. 5.56 reduces to

$$m_i h_i = m_2 u_2 - m_1 u_1$$

The continuity equation, Eq. 5.55 reduces to

$$m_2 - m_1 = m_i$$

Therefore, combining the continuity equation and the first law we have

$$(m_2 - m_1) h_i = m_2 u_2 - m_1 u_1$$

$$m_2 (h_i - u_2) = m_1 (h_i - u_1) \qquad (a)$$

There are two unknowns in this equation; namely, $m_2$ and $u_2$. However, we have one additional equation,

$$m_2 v_2 = V = 0.4 \ m^3 \qquad (b)$$

Substituting (b) into (a) and rearranging, we have

$$\frac{V}{v_2} (h_i - u_2) - m_1 (h_i - u_1) = 0 \qquad (c)$$

in which the only unknowns are $v_2$ and $u_2$, both functions of $T_2$ and $P_2$. Since $T_2$ is unknown, it means that there is only one value of $T_2$ for which Eq. (c) will be satisfied, and we must find it by trial and error. The correct solution is shown below:

$$v_1 = 0.5243 \text{ m}^3/\text{kg} \qquad m_1 = \frac{0.4}{0.5243} = 0.763 \text{ kg}$$

$$h_i = 3040.4 \text{ kJ/kg} \qquad u_1 = 2548.9 \text{ kJ/kg}$$

Assume

$$T_2 = 342°C$$

For this temperature and the known value of $P_2$,

$$v_2 = 0.1974 \text{ m}^3/\text{kg} \qquad u_2 = 2855.8 \text{ kJ/kg}$$

Substituting into (c),

$$\frac{0.4}{0.1974} (3040.4 - 2855.8) - 0.763(3040.4 - 2548.9) \approx 0$$

and we conclude that the assumed value for $T_2 = 342°C$ is correct. The final mass inside the tank is

$$m_2 = \frac{0.4}{0.1974} = 2.026 \text{ kg}$$

and the mass of steam that flows into the tank is

$$m_2 - m_1 = 2.026 - 0.763 = 1.263 \text{ kg}$$

**Example 5.19**

A tank of 2 m³ volume contains saturated ammonia at a temperature of 40°C. Initially the tank contains 50 per cent liquid and 50 per cent vapor by volume. Vapor is withdrawn from the top of the tank until the temperature is 10°C. Assuming that only vapor (i.e., no liquid) leaves and that the process is adiabatic, calculate the mass of ammonia that is withdrawn.

We consider a control volume around the tank and note that $Q_{c.v.} = 0$, $W_{c.v.} = 0$, $m_i = 0$, and we assume that changes in kinetic and potential energy are negligible. However, the enthalpy of saturated vapor varies with temperature, and therefore we cannot simply assume that the enthalpy of the vapor leaving the tank remains constant. However we note that at 40°C, $h_g = 1472.2$ kJ/kg and at 10°C, $h_g = 1453.3$ kJ/kg. Since the change in $h_g$ during this process is small, we may accurately assume that $h_e$ is the average of the two values given above. Therefore

$$(h_e)_{av} = 1462.8 \text{ kJ/kg}$$

With this assumption we can assume a uniform-state, uniform-flow process, and write:
First law (from Eq. 5.56):

$$m_e h_e + m_2 u_2 - m_1 u_1 = 0$$

Continuity equation (from Eq. 5.55):

$$(m_2 - m_1)_{c.v.} + m_e = 0$$

Combining these two equations we have

$$m_2(h_e - u_2) = m_1 h_e - m_1 u_1$$

The following values are from the ammonia tables:

$$v_{f1} = 0.001\ 726 \text{ m}^3\text{/kg} \qquad v_{g1} = 0.0833 \text{ m}^3\text{/kg}$$

$$v_{f2} = 0.001\ 601 \qquad\qquad v_{fg2} = 0.2040$$

$$u_{f1} = 371.7 - 1554.33 \times 0.001\ 726 = 369.0 \text{ kJ/kg}$$

$$u_{g1} = 1472.2 - 1554.33 \times 0.0833 = 1342.7$$

$$u_{f2} = 227.6 - 614.95 \times 0.001\ 601 = 226.6$$

$$u_{g2} = 1453.3 - 614.95 \times 0.2056 = 1326.9$$

$$u_{fg2} = 1326.9 - 226.6 = 1100.3$$

Calculating first the initial mass, $m_1$, in the tank: the mass of the liquid initially present, $m_{f1}$, is

$$m_{f1} = \frac{1.0}{0.001\ 726} = 579.4 \text{ kg}$$

Similarly, the initial mass of vapor, $m_{g1}$, is

$$m_{g1} = \frac{1.0}{0.0833} = 12.0 \text{ kg}$$

$$m_1 = m_{f1} + m_{g1} = 579.4 + 12.0 = 591.4 \text{ kg}$$

$$m_1 h_e = 591.4 \times 1462.8 = 865\ 100 \text{ kJ}$$

$$m_1 u_1 = (mu)_{f1} + (mu)_{g1} = 579.4 \times 369.0 + 12.0 \times 1342.7$$

$$= 229\ 910 \text{ kJ}$$

Substituting these into the first law,

$$m_2 (h_e - u_2) = m_1 h_e - m_1 u_1 = 865\ 100 - 229\ 910 = 635\ 190 \text{ kJ}$$

There are two unknowns, $m_2$ and $u_2$, in this equation. However,

$$m_2 = \frac{V}{v_2} = \frac{2.0}{0.001\ 601 + x_2\ (0.2040)}$$

and

$$u_2 = 226.6 + x_2\ (1100.3)$$

both functions only of $x_2$, the quality at the final state. Consequently,

$$\frac{2.0\,(1462.8 - 226.6 - 1100.3\;x_2)}{0.001\;601 + 0.204\;x_2} = 635\;190$$

Solving, $x_2 = 0.01104$

Therefore,

$$v_2 = 0.001\;601 + 0.01104 \times 0.2040 = 0.003\;854\;\text{m}^3/\text{kg}$$

$$m_2 = \frac{2}{0.003\;854} = 518.9\;\text{kg}$$

and the mass of ammonia withdrawn, $m_e$, is

$$m_e = m_1 - m_2 = 591.4 - 518.9 = 72.5\;\text{kg}$$

## PROBLEMS

5.1    Consider the following types of toy automobiles: $A$ is battery operated; $B$ has an electric motor that utilizes a-c current; $C$ has a spring that
can be wound with a key; and $D$ has a charged capsule of $CO_2$ gas and
operates like a rocket. Examine each of the automobiles as it operates
for heat, work, and changes in internal energy. In each case consider the
entire car as the system.

5.2    A thermoelectric refrigerator is used to maintain a constant temperature
in a cold space, as indicated in Fig. 5.21. The thermoelectric device operates

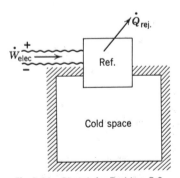

**Fig. 5.21**   Sketch for Problem 5.2.

on a current of 30 A at a voltage of 24 V, and rejects heat to the surroundings at the rate of 5 kW. If the temperature in the cold space remains constant, what is the rate of heat leak into the cold space?

5.3   The rate of heat transfer to the surroundings from a person at rest is about 400 kJ per hour. Suppose that the ventilation system fails in an auditorium containing 2000 people.

(a) How much does the internal energy of the air in the auditorium increase during the first 20 minutes after the ventilation system fails?

(b) Considering the auditorium and all the people as a system, and assuming no heat transfer to the surroundings, how much does the internal energy of the system change? How do you explain the fact that the temperature of the air increases?

5.4   A "bomb" type calorimeter is to be used to measure the energy released by a certain chemical reaction. The calorimeter is a closed vessel containing the chemicals and located in a large tank of water. When the chemicals react, heat is transferred from the bomb to the water, causing its temperature to rise. The power input to a stirrer used to circulate the water is 0.04 kW. In a 20-minute period the heat transfer from the bomb is 1200 kJ and the heat transfer from the water tank to the surrounding air is 60 kJ. Assuming no evaporation of water, determine the increase in the internal energy of the water.

5.5   A rigid vessel of 20 litres volume contains water at 90°C, 50% quality. The vessel is then cooled to −10°C. Calculate the heat transfer during the process.

5.6   A steam boiler has a total volume of 4 m³. The boiler initially contains 3 m³ of liquid water and 1 m³ of vapor in equilibrium at 0.1 MPa. The boiler is fired up and heat is transferred to the water and steam in the boiler. Somehow, the valves on the inlet and discharge of the boiler are both left closed. The relief valve lifts when the pressure reaches 5 MPa. How much heat was transferred to the water and steam in the boiler before the relief valve lifted?

5.7   A radiator of a steam heating system has a volume of 0.02 m³. When the radiator is filled with dry saturated steam at a pressure of 150 kPa all valves to the radiator are closed. How much heat will have been transferred to the room when the pressure of the steam is 75 kPa?

5.8   A rigid vessel having a volume of 0.5 m³ is filled with ammonia at 600 kPa, 70°C. Heat is transferred from the ammonia until it exists as saturated vapor. Calculate the heat transferred during this process.

5.9   A dewar vessel having a total volume of 100 litres contains liquid nitrogen at 77.35 K. The vessel is filled with 90% liquid and 10% vapor by volume. The dewar vessel is accidentally sealed off, so that as heat is transferred to the liquid nitrogen across the vacuum space, the pressure will increase. It is anticipated that the inner vessel will rupture when the pres-

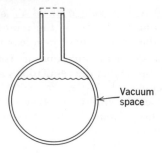

**Fig. 5.22**   Sketch for Problem 5.9.

sure reaches 400 kPa. How long will it take to reach this pressure if the heat leak into the dewar vessel is 5 J/s?

5.10   A sealed tube has a volume of 0.1 litres and contains a certain fraction of liquid and vapor $H_2O$ in equilibrium at 0.1 MPa. The fraction of liquid and vapor is such that when heated the steam will pass through the critical point. Calculate the heat transfer when the steam is heated from the initial state at 0.1 MPa to the critical state.

5.11   A tank is to be charged with ammonia. It is initially evacuated. A charging bottle filled with saturated liquid ammonia at 20°C is then connected to the tank, and the ammonia flows into the tank. The charging bottle remains connected and open to the tank, and the final temperature is 20°C. The charging bottle has a volume of 0.01 m³, and the tank has volume of 0.5 m³. What is the heat transfer to the ammonia?

5.12   Five kg of water at 15°C is contained in a vertical cylinder by a frictionless piston of a mass such that the pressure of the water is 700 kPa. Heat is

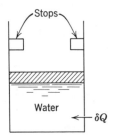

**Fig. 5.23**   Sketch for Problem 5.12.

transferred slowly to the water, causing the piston to rise until it reaches the stops, at which point the volume inside the cylinder is 0.5 m³. More heat is transferred to the water until it exists as saturated vapor.

(*a*)   Find the final pressure in the cylinder and the heat transfer and work done during the process.

(*b*)   Show this process on a *T-V* diagram.

5.13 Consider the insulated vessel indicated in Fig. 5.24. The vessel has an evacuated compartment separated by a membrane from a second compartment which contains 1 kg of water at 65°C, 700 kPa. The membrane then ruptures and the water fills the entire volume, with a resulting pressure of 15 kPa. Determine the final temperature of the water and the volume of the vessel.

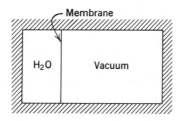

**Fig. 5.24**   Sketch for Problem 5.13.

5.14 Consider the piston-cylinder arrangement indicated in Fig. 5.25 in which a frictionless piston is free to travel between two sets of stops. When the piston is resting on the lower stops the enclosed volume is 0.4 m³. The enclosed volume is 0.6 m³ when the piston encounters the upper stops. Initially the cylinder is filled with water at 0.1 MPa and a quality of 20%. The water is then heated until it exists as a saturated vapor. If the mass of the piston is such that 0.3 MPa pressure is required to move it against atmospheric pressure, determine
   (a) The final pressure in the cylinder.
   (b) The heat transfer and work done during the process.

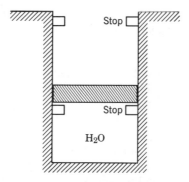

**Fig. 5.25**   Sketch for Problem 5.14.

5.15 An insulated cylinder fitted with a piston contains Freon-12 at 25°C with a quality of 90%. The volume at this state is 0.03 m³. The piston is then allowed to move, and the Freon expands until it exists as saturated vapor. During this process, the Freon does 4 kJ of work against the piston. Determine the final temperature if the process is adiabatic.

5.16 One kg of steam is confined inside a spherical elastic membrane or balloon which supports an internal pressure proportional to its diameter. The initial condition of the steam is saturated vapor at 110°C. Heat is transferred to the steam until the pressure reaches 200 kPa. Determine:

(a) The final temperature.

(b) The heat transfer.

5.17 Consider the system shown in Fig. 5.26. Tank $A$ has a volume of 0.1 m³ and contains saturated vapor Freon-12 at 25°C. When the valve is cracked open, Freon flows slowly into cylinder $B$. The piston mass is such that a pressure of 150 kPa in cylinder $B$ is required to raise the piston. The process ends when the pressure in tank $A$ has fallen to 150 kPa. During this process, heat is transferred with the surroundings such that the temperature of the Freon-12 always remains at 25°C. Calculate the heat transfer for this process.

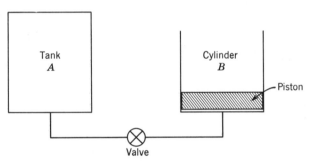

**Fig. 5.26** Sketch for Problem 5.17.

5.18 Two insulated tanks, $A$ and $B$, are connected by a valve. Tank $A$ has a volume of 0.6 m³ and contains steam at 200 kPa, 200°C. Tank $B$ has a volume of 0.3 m³ and contains steam at 500 kPa with a quality of 90%. The valve is then opened, and the two tanks come to a uniform state. If there is no heat transfer during the process, what is the final pressure?

5.19 An insulated cylinder containing water has a piston held by a pin, as shown in Fig. 5.27. The water is initially saturated vapor at 65°C, and the volume is 5 litres. The piston and weights have a total mass of 10 kg, the piston area is 0.003 m², and atmospheric pressure is 100 kPa. The pin is now released, allowing the piston to move. Determine the final state of the water, assuming the process to be adiabatic.

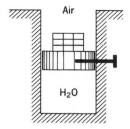

**Fig. 5.27**  Sketch for Problem 5.19.

5.20 A cylinder contains 0.1 kg of saturated vapor water at 105°C, as shown in Fig. 5.28. At this state the spring is touching the piston but exerts no force

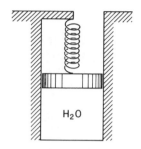

**Fig. 5.28**  Sketch for Problem 5.20

on it. Heat is then transferred to the $H_2O$ causing the piston to rise, during which process the spring resisting force is proportional to the distance moved, with a spring constant of 50 kN/m. The piston area is 0.05 m². 

(a) What is the temperature in the cylinder when the pressure reaches 300 kPa?

(b) Calculate the heat transfer for this process.

5.21 Consider the insulated vessel shown in Fig. 5.29, with compartment $A$ of

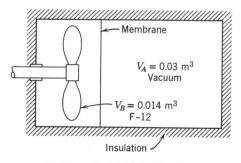

**Fig. 5.29**  Sketch for Problem 5.21.

volume of 0.03 m³ and separated by a membrane from compartment *B* of volume 0.014 m³, which contains 1.1 kg of Freon-12 at 25°C. The Freon-12 is stirred with a fan until the membrane ruptures. The membrane is designed to rupture at a pressure of 2 MPa.

(*a*) What is the temperature when the membrane ruptures?

(*b*) Calculate the work done by the fan.

(*c*) Find the pressure and temperature after the membrane ruptures and the Freon-12 reaches thermodynamic equilibrium.

5.22  A spherical aluminum vessel has an inside diameter of 0.5 m and a 10 mm thick wall. The vessel contains water at 25°C, with a quality of 1%. The vessel is then heated until the water inside is saturated vapor. Considering the vessel and water together as a system, calculate the heat transfer during this process. Use a density of 2700 kg/m³ and specific heat of 0.9 kJ/kg K for aluminum.

5.23  One kg of nitrogen gas is heated from 30°C to 1500°C. Calculate the change in enthalpy by the following methods:

(*a*) Constant specific heat, value from Table A.8.

(*b*) Constant specific heat, value at the average temperature from the equation, Table A.9.

(*c*) Variable specific heat, integrating the equation, Table A.9.

(*d*) Variable specific heat, values from the ideal gas tables, Table A.11.

5.24  An ideal gas is heated from 500 K to 1000 K. The enthalpy change per kg is to be calculated assuming constant specific heat, value from Table A.8. Calculate this value, and discuss the accuracy of the result, if the gas is

(*a*) Helium.

(*b*) Nitrogen.

(*c*) Carbon dioxide.

5.25  One kg of water at 30 kPa, quality of 50%, is heated to 1600°C, 200 kPa. Calculate the change in internal energy for the process.

5.26  Helium which is contained in a cylinder fitted with a piston expands slowly according to the relation $PV^{1.5} =$ constant. The initial volume of the helium is 0.1 m³, the initial pressure is 500 kPa, and the initial temperature is 300 K. After expansion the pressure is 150 kPa. Calculate the work done and heat transfer during the expansion.

5.27  Air contained in a cylinder fitted with a piston is compressed in a quasi-equilibrium process. During the compression process the relation between pressure and volume is $PV^{1.25} =$ constant. The mass of air in the cylinder is 0.1 kg. The initial pressure is 100 kPa, and the initial temperature is 20°C. The final volume is $\frac{1}{8}$ of the initial volume.

Determine the work and the heat transfer.

5.28  Heat is transferred at a given rate to a mixture of liquid and vapor in equilibrium in a closed container. Determine the rate of change of temperature as a function of the thermodynamic properties of the liquid and vapor and the mass of liquid and the mass of vapor.

5.29 Freon-12 vapor enters a compressor at 200 kPa, 20°C, and the mass rate of flow is 0.05 kg/s. What is the smallest diameter tubing that can be used if the velocity of refrigerant must not exceed 10 m/s?

5.30 Air is heated electrically in a constant-diameter tube in a steady-flow process. At the entrance, the air has a velocity of 5 m/s and is at 350 kPa, 25°C. The air exits at 320 kPa, 90°C. Calculate the velocity at the exit.

5.31 Nitrogen gas is heated in a steady-state, steady-flow process. The inlet conditions are 550 kPa, 35°C, and the exit conditions are 500 kPa, 1000°C. Kinetic and potential energy changes are negligible. Calculate the required heat transfer per kg of nitrogen.

5.32 The compressor of a large gas turbine receives air from the surroundings at 95 kPa, 20°C. At the compressor discharge, the pressure is 380 kPa, the temperature is 180°C, and the velocity is 120 m/s. The power input to the compressor is 3000 kW. Determine the mass flow rate of the air.

5.33 Steam enters the nozzle of a turbine with a low velocity at 3 MPa, 350°C, and leaves the nozzle at 1.6 MPa at a velocity of 550 m/s. The rate of flow of steam is 0.5 kg/s. Calculate the quality or temperature of the steam leaving the nozzle and the exit area of the nozzle.

5.34 A small, high-speed turbine operating on compressed air produces 0.1 kW. The inlet and exit conditions are 400 kPa, 25°C and 100 kPa, −50°C, respectively. Assuming the velocities to be low, find the required mass flow rate of air.

5.35 A number of years ago, a well-known architect designed a 1600 metre-tall building. Suppose that in such a building steam for the heating system enters a pipe at ground level as saturated vapor at 200 kPa. On the top floor of the building, the pressure in the pipe is 100 kPa, and the heat transfer from the steam as it flows up the pipe is 100 kJ/kg. What is the quality of the steam at the top of the pipe?

5.36 In an air liquefaction plant, air flows through an expansion engine at the rate of 0.075 kg/s. The pressure and temperature of the air entering the expansion engine are 1.5 MPa, −60°C, and on leaving the pressure is 170 kPa and the temperature is −110°C. The heat transfer to the air as it flows through the expansion engine is equal to 10% of the power output of the expansion engine. Determine the power output and the heat transfer from the expansion engine.

5.37 In a nuclear reactor steam generator 2 litres per second of water enters a 20 mm diameter tube at a pressure of 10 MPa and temperature of 30°C, and leaves the tube as a saturated vapor at 9 MPa. Find the heat transfer rate to the water.

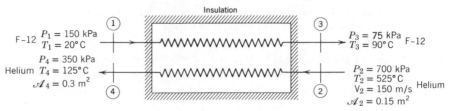

**Fig. 5.30** Sketch for Problem 5.38.

5.38 Consider the steady-state, steady-flow heat exchanger shown in Fig. 5.30, Freon-12 enters at point 1 and exits at point 3, while helium enters at point 2 and exits at point 4. The conditions are as shown in the figure. For this process, determine the following.

(*a*) The exit velocity of the helium.

(*b*) The mass flow rate of the Freon-12.

5.39 Liquid ammonia at a temperature of 20°C and a pressure of 1.2 MPa is mixed in a steady-state, steady-flow process with saturated ammonia vapor at a pressure of 1.2 MPa. The mass rates of flow of liquid and vapor are equal, and after mixing the pressure is 1.0 MPa and the quality is 85%. Determine the heat transfer per kg of mixture.

5.40 In certain situations, when only superheated steam is available, a need for saturated steam may arise for a specific purpose. This can be accomplished in a desuperheater, in which case water is sprayed into the superheated steam in such amounts that the steam leaving the superheater is dry and saturated. The following data apply to such a desuperheater, which operates as a steady-flow process. Superheated steam at the rate of 0.25 kg/s at 3 MPa, 350°C, enters the desuperheater. Water at 35°C, 3.2 MPa, also enters the desuperheater. The dry saturated vapor leaves at 2.5 MPa. Calculate the rate of flow of water.

5.41 The following data are for a simple steam power plant as shown in Fig. 5.31.

$P_1 = 6.2$ MPa

$P_2 = 6.1$ MPa, $T_2 = 45°C$

$P_3 = 5.9$ MPa, $T_3 = 175°C$

$P_4 = 5.7$ MPa, $T_4 = 500°C$

$P_5 = 5.5$ MPa, $T_5 = 490°C$

$P_6 = 10$ kPa, $x_6 = 0.92$, $V_6 = 200$ m/s

$P_7 = 9$ kPa, $T_7 = 40°C$

Rate of steam flow = 25 kg/s

Power to pump = 300 kW

Pipe diameters:

   Steam generator to turbine: 200 mm

   Condenser to steam generator: 75 mm

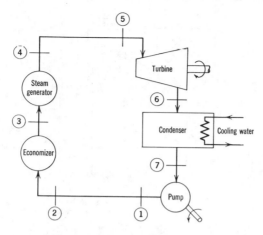

**Fig. 5.31**   Sketch for Problem 5.41.

*Calculate:*

(a) Power output of the turbine.

(b) Heat transfer rates in condenser, economizer, and steam generator.

(c) Diameter of pipe connecting the turbine to the condenser.

(d) Flow rate of cooling water through the condenser if the temperature of the cooling water increases from 15°C to 25°C in the condenser.

5.42  A somewhat simplified flow diagram for the nuclear power plant shown in Fig. 1.5 is given in Fig. 5.32. The mass flow rates and state of the steam at various points in the cycle are shown in the table below. This cycle involves a number of "heaters." In these units, heat is transferred from steam which leaves the turbine at some intermediate pressure to liquid water which is being pumped from the condenser to the steam drum. The rate of heat transfer to the $H_2O$ in the reactor is 157 MW.

(a) Assuming no heat transfer from the moisture separator between the high-pressure and low-pressure turbines, determine the enthalpy per kg of steam and the quality of the steam entering the low-pressure turbine.

(b) Determine the power output of the high-pressure turbine, assuming no heat transfer from the steam as it flows through the turbine.

(c) Determine the power output of the low-pressure turbine, assuming no heat transfer from the steam as it flows through the turbine.

(d) Determine the quality of the steam leaving the reactor.

(e) Determine the temperature of the water leaving the intermediate-pressure heater, assuming no heat transfer from the heater to the surroundings.

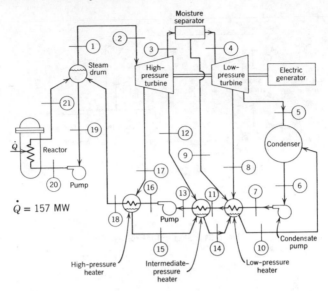

**Fig. 5.32**    Sketch for Problem 5.42.

| Point | $\dot{m}$ − kg/s | $P$ − kPa | $T$ − °C | $h$ − kJ/kg |
|-------|---------|---------|---------|---------|
| 1  | 75.6    | 7240 | sat vap |      |
| 2  | 75.6    | 6900 |         | 2765 |
| 3  | 62.874  | 345  |         | 2517 |
| 4  |         | 310  |         |      |
| 5  |         | 7    |         | 2279 |
| 6  | 75.6    | 7    | 33      |      |
| 7  |         | 415  |         | 140  |
| 8  | 2.772   | 35   |         | 2459 |
| 9  | 4.662   | 310  |         | 558  |
| 10 |         | 35   | 54      |      |
| 11 | 75.6    | 380  | 68      |      |
| 12 | 8.064   | 345  |         | 2517 |
| 13 | 75.6    | 330  |         |      |
| 14 |         |      |         | 349  |
| 15 | 4.662   | 965  | 139     | 584  |
| 16 | 75.6    | 7930 |         | 565  |
| 17 | 4.662   | 965  |         | 2593 |
| 18 | 75.6    | 7580 |         | 688  |
| 19 | 1386    | 7240 | 277     |      |
| 20 | 1386    | 7410 |         | 1221 |
| 21 | 1386    | 7310 |         |      |

(f) Determine the rate of heat transfer to the condenser cooling water.

(g) What is the ratio of the total power output of the two turbines to the heat transferred to the H₂O in the reactor?

5.43 A steam turbine is used to drive a nitrogen compressor, as shown in Fig. 5.33. The states of the steam and nitrogen are given in the diagram, and the mass flow rates are 0.125 kg/s and 0.02 kg/s through the turbine and compressor, respectively. Heat transfer from the turbine is small and can be neglected. The turbine delivers 12 kW to the compressor and the balance to an electric power generator.

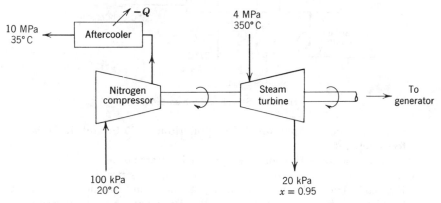

**Fig. 5.33** Sketch for Problem 5.43.

(a) Determine the power available from the turbine for driving the electric power generator.

(b) Determine the rate of heat transfer from the nitrogen as it flows through the compressor and aftercooler.

5.44 A schematic arrangement for a proposed procedure for producing fresh water from salt water which would operate in conjunction with a large steam power plant and utilize a flash evaporator is shown in Fig. 5.34. Cooling water at the rate of 300 kg/s enters the condenser, where its temperature is increased from 17°C to 30°C. It then enters a flash evaporator, where the pressure is reduced to that corresponding to a saturation temperature of 25°C. During this process some of the liquid flashes into vapor, and the remainder is pumped back to the sea. The vapor is then condensed at 23°C to form the desired fresh water. This takes place by utilizing the sea water which enters at 17°C and leaves at 21°C. Calculate:

(a) The amount of fresh water which is produced per hour. Assume that the mixture which is formed when the liquid at 30°C is throttled (as it

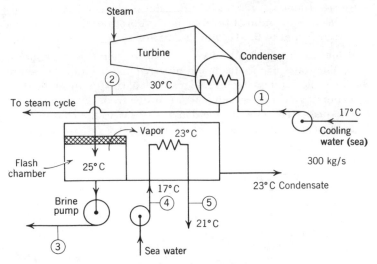

Fig. 5.34  Sketch for Problem 5.44.

enters the flash evaporator) is in equilibrium at 25°C, and that it is perfectly separated.

(b) The amount of cooling water which enters at 4.

5.45 A steam power plant for an automotive vehicle is proposed as shown in Fig. 5.35. The boiler has a volume of 25 litres and initially contains 90% liquid and 10% vapor by volume at 100 kPa. The burner is turned on, and when the pressure reaches 700 kPa the pressure regulating valve holds the pressure in the boiler constant at 700 kPa, and saturated vapor at this pressure passes to the turbine. Saturated vapor at a pressure of 100 kPa leaves the turbine and is discharged to the atmosphere. When the liquid in the boiler is depleted, the burner shuts off automatically. Find the total work done and the total heat transfer for one charge of the boiler.

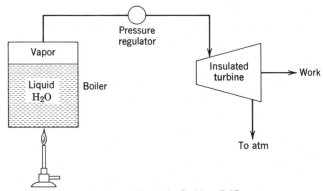

Fig. 5.35  Sketch for Problem 5.45.

5.46 Water at a pressure of 10 MPa, 140°C, is throttled to a pressure of 200 kPa in an adiabatic process. What is the quality after throttling?

5.47 Nitrogen at 300 K, 0.5 MPa, is throttled to 0.1 MPa. Calculate the exit temperature, assuming
    (a) Real gas behavior.
    (b) Ideal gas behavior.

5.48 Helium gas is throttled from a condition of 1.2 MPa, 25°C, to a pressure of 100 kPa. The diameter of the exit pipe is sufficiently larger than that of the inlet pipe so that the inlet and exit velocities are equal. Determine the exit temperature and the ratio of pipe diameters.

5.49 A throttling calorimeter is a device which is used to determine the quality of wet steam (i.e., steam which has a small amount of moisture present) flowing through a pipe. This device involves taking a small but continuous flow of steam from the pipe and throttling it in an adiabatic process to approximately atmospheric pressure. After this adiabatic throttling (constant enthalpy) process the pressure and temperature of the steam are measured, and thus the enthalpy of the wet steam in the line is known.

    A throttling calorimeter is used to measure the quality of steam in a pipe in which the absolute pressure is 1.2 MPa. A mercury manometer is used to measure the pressure in the calorimeter, and shows a pressure in the calorimeter of 8 kPa above atmospheric pressure. If a minimum of 5 degrees of superheat is required in the calorimeter, what is the minimum quality of steam that can be determined? The barometer reads 97 kPa.

5.50 The mixing process indicated in Fig. 5.36 is used to obtain saturated liquid nitrogen at 229 kPa. The high pressure nitrogen gas is throttled to the

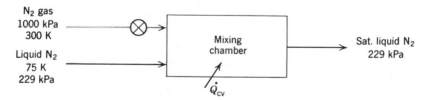

**Fig. 5.36** Sketch for Problem 5.50.

mixing chamber pressure and the mixed with the subcooled liquid which flows into the chamber at the rate of 0.07 kg/s. If the heat transfer to the chamber is 0.05 kW, what is the mass flow rate of nitrogen gas into the chamber?

5.51 A small steam turbine operating at part load produces 75 kW output with a flow rate of 0.17 kg/s. Steam at 1.4 MPa, 250°C, is throttled to 1.1 MPa before entering the turbine, and the exhaust pressure is 10 kPa. Find the quality (or temperature, if superheated) at the turbine outlet.

5.52 The following data are for the Freon-12 refrigeration cycle shown in Fig. 5.37:

$P_1 = 1240$ kPa, $T_1 = 115°C$
$P_2 = 1230$ kPa, $T_2 = 105°C$
$P_3 = 1200$ kPa, $T_3 = 35°C$
$P_4 = 200$ kPa,
$P_5 = 180$ kPa, $T_5 = -5°C$
$P_6 = 170$ kPa, $T_6 = 5°C$
Rate of flow of Freon $= 0.025$ kg/s
Power input to compressor $= 2$ kW

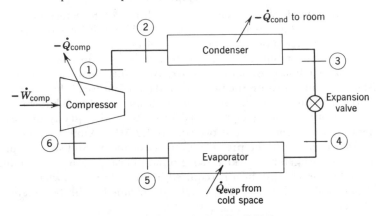

**Fig. 5.37**   Sketch for Problem 5.52.

Calculate:
(a) The heat transfer from the compressor.
(b) The heat transfer from the Freon in the condenser.
(c) The heat transfer to the Freon in the evaporator.

5.53 Consider the process indicated in Fig. 5.38 for producing liquid Freon-12. Freon-12 at 2.75 MPa, 100°C enters a heat exchanger and is cooled by the saturated vapor being withdrawn from the insulated liquid receiver. The high pressure gas is then throttled across a valve to the liquid receiver pressure. The liquid receiver contains liquid and vapor in equilibrium at −20°C and the vapor leaving the heat exchanger at 2 is at 80°C. Neglecting all pressure losses except across the expansion valve, determine:

(a) The fraction of high pressure gas which is liquified.
(b) The pressure and temperature (if superheated) or quality (if saturated) at the inlet and outlet of the valve.

5.54 Ammonia vapor flows through a pipe at a pressure of 1 MPa and a temperature of 70°C. Attached to the pipeline is an evacuated vessel having a volume of 25 litres. The valve in the line to this evacuated vessel is opened, and the pressure in the vessel comes to 1 MPa, at which time the valve is closed. If this process occurs adiabatically, how much ammonia flows into the vessel?

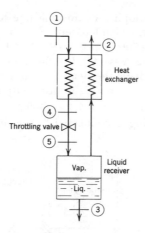

**Fig. 5.38**  Sketch for Problem 5.53.

5.55 Air flows in a pipeline at a pressure of 500 kPa and a temperature of 25°C. Connected to this tank is an evacuated vessel. When the valve on this tank is opened, air flows into the tank until the pressure is 500 kPa.

(a) If this process occurs adiabatically, what is the final temperature of the air?

(b) Determine a general expression that gives the relation between the temperature of a gas flowing into an evacuated vessel and the final temperature of the gas in the tank in terms of the thermodynamic properties of the gas.

5.56 Freon-12 is contained in a 25 litre tank at 25°C, 0.1 MPa. It is desired to fill the tank 60% full of liquid (by volume) at this temperature. The tank is connected to a line flowing F-12 at 0.8 MPa, 40°C, and the valve opened slightly.

(a) Calculate the final mass in the tank at 25°C.

(b) Determine the required heat transfer during the filling process if the temperature is to remain at 25°C.

5.57 A tank having a volume of 5 m³ contains saturated vapor steam at a pressure of 0.2 MPa. Attached to this tank is a line in which vapor at 0.6 MPa, 200°C, flows. Steam from this line enters the vessel until the pressure is 0.6 MPa. If there is no heat transfer from the tank and the heat capacity of the tank is neglected, calculate the mass of steam that enters the tank.

5.58 An evacuated 0.05 m³ tank is connected to a line flowing air at room temperature, 25°C and 7 MPa. The valve is then opened, allowing air to flow into the tank until the tank pressure reaches 5 MPa, at which time the valve is closed. This filling process occurs rapidly and is essentially adiabatic.

The tank is then allowed to sit for a long time with the valve closed, and eventually returns to room temperature. What is the final pressure inside the tank?

5.59 Steam is supplied through a connecting line to the power cylinder of a steam pump during the complete length of the stroke. At the initial position of the piston $P = 700$ kPa, $x = 0.95$, $V = 1.4$ litres, and at the end of the stroke $P = 700$ kPa, $x = 0.90$, $V = 30$ litres. During this process 28 kJ of heat is transferred from the steam. Steam in the connecting supply line is at 725 kPa, 170°C. Determine the work output for the process.

5.60 Steam is flowing in a line at 0.8 MPa, 250°C. An insulated vessel containing a piston and spring is connected to the line, as shown in Fig. 5.39.

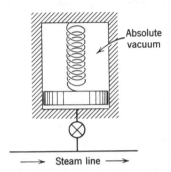

**Fig. 5.39** Sketch for Problem 5.60.

Initially, the spring force is zero, but when the valve is opened and steam enters, the resisting force is directly proportional to the distance moved. Find the temperature inside when the pressure reaches 0.8 MPa.

5.61 Consider the device shown in Fig. 5.40. Steam flows in a steam line at 0.8 MPa, 300°C. From this steam line, steam flows through a steam turbine. The steam exhausts into a large chamber having a volume of 50 m³. Initially, this chamber is evacuated. The turbine can operate until the pressure in the chamber is 0.8 MPa. At this point the steam temperature is 280°C. Assume the entire process to be adiabatic.

Determine the work done by the turbine during this process.

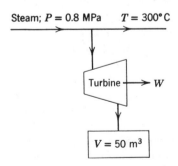

**Fig. 5.40** Sketch for Problem 5.61.

5.62 A 2 m³ insulated vessel contains saturated vapor steam at 3 MPa. A valve at the top of the vessel is then opened and steam is withdrawn. During the process, liquid collects at the bottom of the vessel, such that only saturated vapor leaves. Calculate the total mass withdrawn when the pressure inside reaches 0.8 MPa.

5.63 Freon-12 is contained in a 25 litre tank as indicated in Fig. 5.41. The tank initially contains 90% liquid, 10% vapor, by volume, at 25°C. The valve is then opened and Freon vapor flows out, leaving the nozzle at 100 kPa, −20°C, with a velocity of 180 m/s. During this process heat is transferred to the tank such that the Freon inside remains at 25°C. Determine:

   (a) The total mass exhausted when the tank contains saturated vapor.

   (b) The heat transferred to the tank during this process.

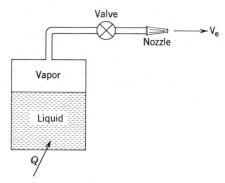

**Fig. 5.41**  Sketch for Problem 5.63.

5.64 A pressure vessel having a volume of 1 m³ contains saturated steam at 300°C. The vessel initially contains 50% vapor and 50% liquid by volume. Liquid is withdrawn slowly from the bottom of the tank, and heat is transferred to the tank in order to maintain constant temperature. Determine the heat transfer to the tank when half of the contents of the tank has been removed.

5.65 A 500 litre low-temperature storage tank initially contains nitrogen at 77.35 K, 80% liquid and 20% vapor by volume. Heat transfer to the tank from the surroundings is constant at the rate of 10 W, and causes the tank pressure to rise. The tank is fitted with a relief valve, such that once a pressure of 500 kPa is reached, saturated vapor will be discharged to the surroundings as necessary to maintain that pressure.

   (a) How much mass has been discharged from the tank by the time the tank contains 50% liquid, 50% vapor by volume at 500 kPa?

   (b) How long does it take to reach this state?

5.66 A popular demonstration involves making ice by pumping a vacuum over liquid water until the pressure is reduced to less than the triple-point pressure. Such an apparatus is shown schematically in Fig. 5.42. In this case

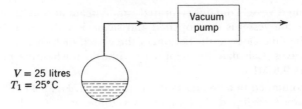

$V = 25$ litres
$T_1 = 25°C$

**Fig. 5.42** Sketch for Problem 5.66.

the tank has a total volume of 25 litres. Initially the tank contains 23 litres of saturated vapor ($H_2O$) and 2 litres of saturated liquid. The initial temperature is 25°C. Assume no heat transfer during this process.

(*a*) Determine what fraction of the intial mass will be pumped off when the liquid and vapor first reach 0°C.

(*b*) Determine what fraction of the initial mass can be solidified.

5.67 Nitrogen is to be stored at low temperature in an insulated, high-pressure spherical tank, 1 m in diameter. The rate of heat transfer to the tank from the room is 3 W, and may be assumed constant. One method proposed for venting the tank is shown in Fig. 5.43. The regulator maintains

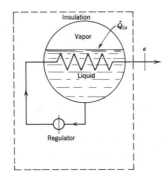

**Fig. 5.43** Sketch for Problem 5.67.

tank pressure at 500 kPa by throttling liquid to low pressure, after which it is passed through the tank and exhausts as a gas at 100 kPa, 90 K. At some particular time, the tank contains 50% liquid, 50% vapor by volume at 500 kPa. How much nitrogen will be lost during the next 5 days?

5.68 A spherical drop of liquid is suspended in an infinite atmosphere as shown in Fig. 5.44. Owing to differences in temperature it is exchanging heat with its surroundings. It also is losing mass by evaporation to the surroundings. The instantaneous rate of heat transfer $\dot{Q}$ is expressed by

$$\dot{Q} = K\mathscr{A}(T_0 - T)$$

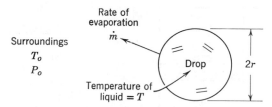

**Fig. 5.44**   Sketch for Problem 5.68.

where $K$ is a constant and $\mathscr{A}$ is the surface area. The instantaneous rate of mass transfer is given by the symbol $\dot{m}$. It may be assumed that at a given instant of time the sphere is uniform in temperature and always at a pressure equal to that of the surroundings, $P_0$. The vapor of the evaporating liquid in the region adjacent to the drop is saturated.

Derive an expression for the instantaneous time rate of change of the drop temperature, $\partial T/\partial t$ in terms of the significant physical quantities.

## SUPPLEMENTARY PROBLEMS

5.69  A spherical balloon initially 150 mm in diameter and containing Freon-12 at 0.1 MPa is connected to an uninsulated 0.03 m³ tank containing Freon-12 at 0.5 MPa. Both are at 20°C, the temperature of the surroundings.

The valve connecting the two is then opened very slightly and left open until the pressures become equal. During the process, heat is transferred with the surroundings such that the temperature of all the Freon remains at 20°C. It may also be assumed that the balloon diameter is proportional to the pressure inside the balloon at any point during the process. Calculate:
(a) The final pressure.
(b) Work done by the Freon during the process.
(c) Heat transfer to the Freon during the process.

5.70  Water at 1.5 MPa, 150°C, is throttled adiabatically through a valve to 0.2 MPa. The inlet velocity is 5 m/s and the inlet and exit cross-sectional areas are equal. Determine the state and the velocity of the water at the exit.

5.71  An insulated and evacuated vessel of 20 litres volume contains a capsule of water at 700 kPa, 150°C. The volume of the capsule is 1 litre. The capsule breaks and the contents fill the entire volume. What is the final pressure?

5.72  Consider the system shown in Fig. 5.45. Tank $B$ (insulated) has a volume of 0.3 m³ and is initially evacuated. Cylinder $A$ (uninsulated) has an initial volume of 0.15 m³ and contains air at 20°C, 3.5 MPa. The spring constant is 40 kN/m and the piston area is 0.03 m². It may be assumed that the spring force varies linearly with distance. The valve is now opened, and air flows into tank $B$ until $P_B = 1.5$ MPa, at which point the valve is closed. During this process heat is transferred to $A$ from the surroundings such

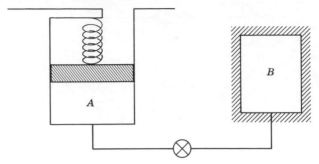

**Fig. 5.45**   Sketch for Problem 5.72.

that $T_A$ remains constant. Determine
  (a) The final temperature and mass in Tank $B$.
  (b) The final pressure and volume in cylinder $A$.
  (c) The heat transfer to $A$ during the process.
5.73 A frictionless, thermally conducting piston separates the air and water in the cylinder shown in Fig. 5.46. The initial volume of $A$ and $B$ are

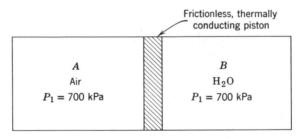

**Fig. 5.46**   Sketch for Problem 5.73.

equal, the volume of each being 0.5 m³. The initial pressure in both $A$ and $B$ is 700 kPa. The volume of the liquid in $B$ is 2% of the total volume of $B$. Heat is transferred to both $A$ and $B$ until all the liquid in $B$ evaporates.
  (a) Determine the total heat transfer during this process.
  (b) Determine the work done by the piston on the air and the heat transfer to the air.
5.74 Liquids are often transferred from a tank by pressurization with a gas. Consider the problem shown in Fig. 5.47. Tanks $A$ and $B$ both contain Freon-12. Tank $A$ has a volume of 0.09 m³, and $B$ a volume of 0.075 m³. The Freon in $A$ is initially at 25°C, saturated vapor, and that in $B$ is at −40°C with a quality of 0.01. The valve is opened slightly, allowing Freon to flow from $A$ to $B$. When the pressure in $B$ has built up to 150 kPa, liquid begins to flow out. The liquid transfer continues until the pressure in $A$ has dropped to 150 kPa. During the process, heat is transferred to the gas

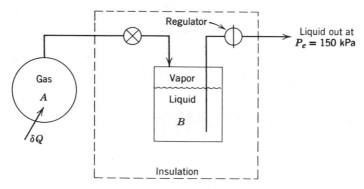

**Fig. 5.47**   Sketch for Problem 5.74.

inside tank $A$ such that its temperature always remains at 25°C, but tank $B$ is insulated. Determine

(a) The quality in tank $B$ at the end of the process.

(b) The mass of liquid transferred from tank $B$.

5.75 When a steam boiler is used in conjunction with a reactor in a nuclear power plant, the time required for a rapid blowdown is an important factor in safety considerations. In making this calculation it is necessary to know the pressure in the boiler as a function of the amount of liquid in the tank. To illustrate the procedure involved consider the following situation.

The boiler has a volume of 5 m³. Initially it is half filled with liquid and half with vapor. The initial pressure is 1.4 MPa. Assume that there is no heat transfer to the liquid or vapor during this process, and that the liquid and vapor are always in equilibrium.

Determine the pressure in the tank when the volume of the liquid is:

(a) $\frac{3}{8}$ of the volume of the boiler.

(b) $\frac{1}{4}$ of the volume of the boiler.

(c) $\frac{1}{8}$ of the volume of the boiler.

5.76 Write a computer program to solve the following problem. For one of the substances listed in Table A.9, it is desired to compare the enthalpy change between any two temperatures $T_1$ and $T_2$ as calculated by integrating the specific heat equation, by assuming constant specific heat at the average temperature, and by assuming constant specific heat at temperature $T_1$.

5.77 Write a computer program to solve the following problem. An insulated tank of volume $V$ contains a specified ideal gas (with constant specific heat) at $P_1$, $T_1$. A valve is opened, allowing the gas to flow out until the pressure inside drops to $P_2$. Determine $T_2$ and $m_2$ using a stepwise solution in increments of pressure between $P_1$ and $P_2$, where the number of increments is variable.

# 6

# The Second Law of Thermodynamics

The first law of thermodynamics states that during any cycle that a system undergoes, the cyclic integral of the heat is equal to the cyclic integral of the work. The first law, however, places no restrictions on the direction of flow of heat and work. A cycle in which a given amount of heat is transferred from the system and an equal amount of work is done on the system satisfies the first law just as well as a cycle in which the flows of heat and work are reversed. However, we know from our experience that the fact that a proposed cycle does not violate the first law does not insure that the cycle will actually occur. It is this kind of experimental evidence that has led to the formulation of the second law of thermodynamics. Thus a cycle will occur only if both the first and second laws of thermodynamics are satisfied.

In its broader significance the second law involves the fact that processes proceed in a certain direction but not in the opposite direction. A hot cup of coffee cools by virtue of heat transfer to the surroundings, but heat will not flow from the cooler surroundings to the hotter cup of coffee. Gasoline is used as a car drives up a hill, but on coasting down the hill, the fuel level in the gasoline tank cannot be restored to its original level. Such familiar observations as these, and a host of others, are evidence of the validity of the second law of thermodynamics.

We will first consider this second law for a system undergoing a cycle and in the next chapter will extend the principles to a system undergoing a change of state and to a control volume.

## 6.1   Heat Engines and Refrigerators

Consider the system and the surroundings previously cited in the development of the first law, as shown in Fig. 6.1. Let the gas constitute the system and, as in our discussion of the first law, let this system

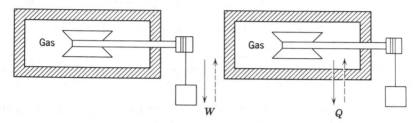

**Fig. 6.1**   A system that undergoes a cycle involving work and heat.

undergo a cycle in which work is first done on the system by the paddle wheel as the weight is lowered. Then let the cycle be completed by transferring heat to the surroundings.

We know from our experience, however, that we cannot reverse this cycle. That is, if we transfer heat to the gas, as shown by the dotted arrow, the temperature of the gas will increase, but the paddle wheel will not turn and raise the weight. With the given surroundings (the container, the paddle wheel, and the weight) this system can operate in a cycle in which the heat transfer and work are both negative, but it cannot operate in a cycle in which both the heat transfer and work are positive, even though this would not violate the first law.

Consider another cycle, which we know from our experience is impossible to actually accomplish. Let two systems, one at a high temperature and the other at a low temperature, undergo a process in which a quantity of heat is transferred from the high-temperature system to the low-temperature system. We know that this process can take place. We also know that the reverse process, in which heat is transferred from the low-temperature system to the high-temperature system, does not occur, and that it is impossible to complete the cycle by heat transfer only. This is illustrated in Fig. 6.2.

These two illustrations lead us to the consideration of the heat engine and refrigerator, which is also referred to as a heat pump. With the heat engine we can have a system that operates in a cycle and has a net positive work and a net positive heat transfer. With the heat pump we can have a system that operates in a cycle and has heat transferred to it from a low-temperature body and heat transferred from it to a high-temperature

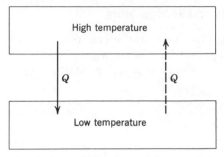

**Fig. 6.2**  An example showing the impossibility of completing a cycle by transferring heat from a low-temperature body to a high-temperature body.

body, though work is required to do this. Three simple heat engines and two simple refrigerators will be considered.

The first heat engine is shown in Fig. 6.3, and consists of a cylinder fitted with appropriate stops and a piston. Let the gas in the cylinder constitute the system. Initially the piston rests on the lower stops, with a weight on the platform. Let the system now undergo a process in which heat is transferred from some high-temperature body to the gas, causing it to expand and raise the piston to the upper stops. At this point the weight is removed. Now let the system be restored to its initial state by transferring heat from the gas to a low-temperature body, thus completing the cycle. Since the weight was raised during the cycle, it is evident that work was done by the gas during the cycle. From the first law we conclude that the net heat transfer was positive and equal to the work done during the cycle.

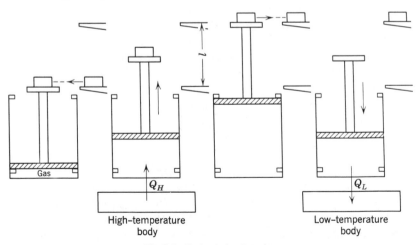

**Fig. 6.3**   A simple heat engine.

Such a device is called a heat engine, and the substance to which and from which heat is transferred is called the working substance or working fluid. A heat engine may be defined as a device that operates in a thermo-dynamic cycle and does a certain amount of net positive work as a result of heat transfer from a high-temperature body and to a low-temperature body. Often the term heat engine is used in a broader sense to include all devices that produce work, either through heat transfer or combustion, even though the device does not operate in a thermodynamic cycle. The internal-combustion engine and the gas turbine are examples of such devices, and calling these heat engines is an acceptable use of the term. In this chapter, however, we are concerned with the more restricted form of heat engine, as defined above, which operates on a thermodynamic cycle.

A simple steam power plant is an example of a heat engine in this restricted sense. Each component in this plant may be analyzed by a steady-state, steady-flow process, but considered as a whole it may be considered a heat engine (Fig. 6.4) in which water (steam) is the working fluid. An amount of heat, $Q_H$, is transferred from a high-temperature body, which may be the products of combustion in a furnace, a reactor, or a secondary fluid which in turn has been heated in a reactor. In Fig. 6.4 the turbine is shown schematically as driving the pump, indicating that what is significant is the net work that is delivered during the cycle. The quantity of heat $Q_L$ is rejected to a low-temperature body, which is usually the cooling water in a condenser. Thus, the simple steam power plant is a heat engine in the restricted sense, for it has a working fluid, to which and from which heat is transferred, and which does a certain amount of work as it undergoes a cycle.

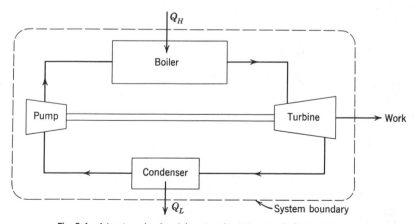

**Fig. 6.4**   A heat engine involving steady-state, steady-flow processes.

Another example of a heat engine is the thermoelectric power generation device that was discussed in Chapter 1 and shown schematically in Fig. 1.11. Heat is transferred from a high temperature body to the hot junction $(Q_H)$ and heat is transferred from the cold junction to the surroundings $(Q_L)$. Work is done in the form of electrical energy. Since there is no working fluid we usually do not think of this as a device that operates in a cycle. However, if one adopted a microscopic point of view one could think of a cycle as regards the flow of electrons. Furthermore, as in the case of the steam power plant, the state at each point in the thermoelectric power generator does not change with time under steady state conditions.

Thus, by means of a heat engine we are able to have a system operate in a cycle and have the net work and net heat transfer both positive, which we were not able to do with the system and surroundings of Fig. 6.1.

One should note that in using the symbols $Q_H$ and $Q_L$ we have departed from our sign connotation for heat, because for a heat engine $Q_L$ is negative when the working fluid is considered as the system. In this chapter it will be advantageous to use the symbol $Q_H$ to represent the heat transfer to or from the high-temperature body, and $Q_L$ the heat transfer to or from the low-temperature body. The direction of the heat transfer will be evident in each case from the context.

At this point it is appropriate to introduce the concept of thermal efficiency of a heat engine. In general we say that efficiency is the ratio of output (the energy sought) to input (the energy that costs), but these must be clearly defined. At the risk of oversimplification we may say that in a heat engine the energy sought is the work, and the energy that costs money is the heat from the high-temperature source (indirectly, the cost of the fuel). Thermal efficiency is defined as:

$$\eta_{\text{thermal}} = \frac{W(\text{energy sought})}{Q_H(\text{energy that costs})} = \frac{Q_H - Q_L}{Q_H} = 1 - \frac{Q_L}{Q_H} \qquad (6.1)$$

The second cycle we were not able to complete was the one that involved the impossibility of transferring heat directly from a low-temperature body to a high-temperature body. This can of course be done with a refrigerator or heat pump. A vapor-compression refrigerator cycle, which was introduced in Chapter 1 and shown in Fig. 1.8, is also shown in Fig. 6.5. The working fluid is the refrigerant, such as a Freon or ammonia, which goes through a thermodynamic cycle. Heat is transferred to the refrigerant in the evaporator, where its pressure and temperature are low. Work is done on the refrigerant in the compressor and heat is transferred from it in the condenser, where its pressure and

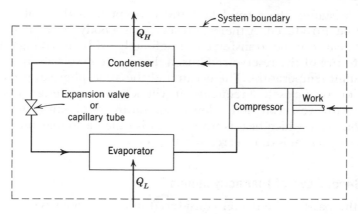

**Fig. 6.5**   A simple refrigeration cycle.

temperature are high. The pressure drop occurs as the refrigerant flows through the throttle valve or capillary tube.

Thus, in a refrigerator or heat pump we have a device that operates in a cycle, that requires work, and that accomplishes the objective of transferring heat from a low-temperature body to a high-temperature body.

The thermoelectric refrigerator, which was discussed in Chapter 1 and is shown schematically in Fig. 1.10, is another example of a device that meets our definition of a refrigerator. The work input to the thermo-electric refrigerator is in the form of electrical energy, and heat is transferred from the refrigerated space to the cold junction ($Q_L$) and from the hot junction to the surroundings ($Q_H$).

The "efficiency" of a refrigerator is expressed in terms of the coefficient of performance, which we designate with the symbol $\beta$. In the case of a refrigerator the objective (i.e., energy sought) is $Q_L$, the heat transferred from the refrigerated space, and the energy that costs is the work $W$. Thus the coefficient of performance, $\beta$,[1] is

$$\beta = \frac{Q_L(\text{energy sought})}{W(\text{energy that costs})} = \frac{Q_L}{Q_H - Q_L} = \frac{1}{Q_H/Q_L - 1} \qquad (6.2)$$

[1]It should be noted that a refrigeration or heat pump cycle can be used with either of two objectives in mind. It can be used as a refrigerator, in which case the primary objective is $Q_L$, the heat transferred to the refrigerant from the refrigerated space. It also can be used as a heating system (in which case it usually is referred to as a heat pump) the objective being $Q_H$, the heat transferred from the refrigerant to the high-temperature body, which is the space to be heated. $Q_L$ is transferred to the refrigerant from the ground, the

Before stating the second law, the concept of a thermal reservoir should be introduced. A thermal reservoir is a body to which and from which heat can be transferred indefinitely without change in the temperature of the reservoir. Thus, a thermal reservoir always remains at constant temperature. The ocean and the atmosphere approach this definition very closely. Frequently it will be useful to designate a high-temperature reservoir and a low-temperature reservoir. Sometimes a reservoir from which heat is transferred is called a source, and a reservoir to which heat is transferred is called a sink.

## 6.2   Second Law of Thermodynamics

On the basis of the matter considered in the previous section we are now ready to state the second law of thermodynamics. There are two classical statements of the second law, known as the Kelvin-Planck statement and the Clausius statement.

The Kelvin-Planck statement: It is impossible to construct a device that will operate in a cycle and produce no effect other than the raising of a weight and the exchange of heat with a single reservoir.

This statement ties in with our discussion of the heat engine, and, in effect, it states that it is impossible to construct a heat engine that operates in a cycle and receives a given amount of heat from a high-temperature body and does an equal amount of work. The only alternative is that some heat must be transferred from the working fluid at a lower temperature to a low-temperature body. Thus, work can be done by the transfer of heat only if there are two temperature levels involved, and heat is transferred from the high-temperature body to the heat engine and also from the heat engine to the low-temperature body. This implies that it is impossible to build a heat engine that has a thermal efficiency of 100 per cent.

The Clausius statement: It is impossible to construct a device that operates in a cycle and produces no effect other than the transfer of heat from a cooler body to a hotter body.

---

atmospheric air, or well water. The coefficient of performance in this case, $\beta'$ is

$$\beta' = \frac{Q_H(\text{energy sought})}{W(\text{energy that costs})} = \frac{Q_H}{Q_H - Q_L} = \frac{1}{1 - Q_L/Q_H}$$

It also follows that for a given cycle,

$$\beta' - \beta = 1$$

Unless otherwise specified, the term coefficient of performance will always refer to a refrigerator as defined by Eq. 6.2.

This statement is related to the refrigerator or heat pump, and in effect states that it is impossible to construct a refrigerator that operates without an input of work. This also implies that the coefficient of performance is always less than infinity.

In regard to these two statements, three observations should be made. The first is that both are negative statements. It is of course impossible to "prove" a negative statement. However, we can say that the second law of thermodynamics (like every other law of nature) rests on experimental evidence. Every relevant experiment that has been conducted has either directly or indirectly verified the second law, and no experiment has ever been conducted that contradicts the second law. The basis of the second law is therefore experimental evidence.

A second observation is that these two statements of the second law are equivalent. Two statements are equivalent if the truth of each statement implies the truth of the other, or if the violation of each statement implies the violation of the other. That a violation of the Clausius statement implies a violation of the Kelvin-Planck statement may be shown as follows. The device at the left in Fig. 6.6 is a refrigerator that requires no work, and thus violates the Clausius statement. Let an amount of heat $Q_L$ be transferred from the low-temperature reservoir to this refrigerator, and let the same amount of heat $Q_L$ be transferred to the high-temperature reservoir. Let an amount of heat $Q_H$, which is greater than $Q_L$, be transferred from the high-temperature reservoir to the heat engine, and let the engine reject the amount of heat $Q_L$ as it does an amount of work $W$ (which equals $Q_H - Q_L$). Since there is no net heat transfer to the low-temperature reservoir, the low-temperature reservoir, the heat engine, and the refrigerator can be considered together as a device that operates in a cycle and produces no effect other than the raising of a weight (work)

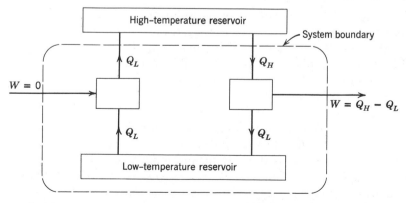

**Fig. 6.6**   Demonstration of the equivalence of the two statements of the second law.

and the exchange of heat with a single reservoir. Thus, a violation of the Clausius statement implies a violation of the Kelvin-Planck statement. The complete equivalence of these two statements is established when it is also shown that a violation of the Kelvin-Planck statement implies a violation of the Clausius statement. This is left as an exercise for the student.

The third observation is that frequently the second law of thermodynamics has been stated as the impossibility of constructing a perpetual motion machine of the second kind. A perpetual motion machine of the first kind would create work from nothing or create mass-energy, thus violating the first law. A perpetual motion machine of the second kind would violate the second law, and a perpetual-motion machine of the third kind would have no friction, and thus run indefinitely but would produce no work.

A heat engine that violated the second law could be made into a perpetual motion machine of the second kind as follows. Consider Fig. 6.7, which might be the power plant of a ship. An amount of heat $Q_L$ is transferred from the ocean to a high-temperature body by means of a heat pump. The work required is $W'$, and the heat transferred to the high-temperature body is $Q_H$; let the same amount of heat be transferred to a heat engine, which violates the Kelvin-Planck statement of the second law, and does an amount of work $W = Q_H$. Of this work an amount of work $Q_H - Q_L$ is required to drive the heat pump, leaving the net work ($W_{net} = Q_L$) available for driving the ship. Thus, we have a perpetual motion machine in the sense that work is done by utilizing freely available sources of energy such as the ocean or atmosphere.

## 6.3 The Reversible Process

The question that now logically arises is this. If it is impossible to have a heat engine of 100 per cent efficiency, what is the maximum efficiency one can have? The first step in the answer to this question is to define an ideal process, which is called a reversible process.

A reversible process for a system is defined as a process, which once having taken place, can be reversed and in so doing leave no change in either the system or surroundings.

Let us illustrate the significance of this definition for a gas contained in a cylinder that is fitted with a piston. Consider first Fig. 6.8, in which a gas (which we defined as the system) at high pressure is restrained by a piston that is secured by a pin. When the pin is removed, the piston is raised and forced abruptly against the stops. Some work is done by the system, since the piston has been raised a certain amount. Suppose we

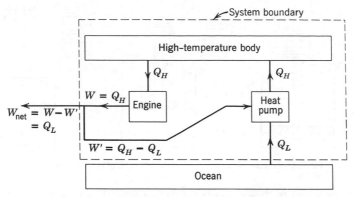

**Fig. 6.7**   A perpetual-motion machine of the second kind.

wish to restore the system to its initial state. One way of doing this would be to exert a force on the piston, thus compressing the gas until the pin could again be inserted in the piston. Since the pressure on the face of the piston is greater on the return stroke than on the initial stroke, the work done on the gas in this reverse process is greater than the work done by the gas in the initial process. An amount of heat must be transferred from the gas during the reverse stroke in order that the system have the same internal energy it had originally. Thus, the system is restored to its initial state, but the surroundings have changed by virtue of the fact that work was required to force the piston down and heat was transferred to the surroundings. Thus, the initial process is an irreversible one because it could not be reversed without leaving a change in the surroundings.

In Fig. 6.9 let the gas in the cylinder comprise the system and let the piston be loaded with a number of weights. Let the weights be slid off horizontally one at a time, allowing the gas to expand and do work in

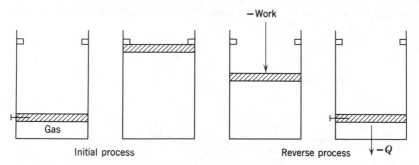

**Fig. 6.8**   An example of an irreversible process.

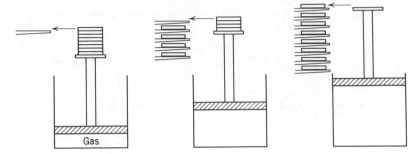

**Fig. 6.9**   An example of a process that approaches being reversible.

raising the weights that remain on the piston. As the size of the weights is made smaller and their number is increased, we approach a process that can be reversed, for at each level of the piston during the reverse process there will be a small weight that is exactly at the level of the platform and thus can be placed on the platform without requiring work. In the limit, therefore, as the weights become very small, the reverse process can be accomplished in such a manner that both the system and surroundings are in exactly the same state they were initially. Such a process is a reversible process.

## 6.4   Factors That Render Processes Irreversible

There are many factors that make processes irreversible, four of which are considered in this section.

### Friction

It is readily evident that friction makes a process irreversible, but a brief illustration may amplify the point. Let a block and an inclined plane comprise a system, Fig. 6.10, and let the block be pulled up the inclined plane by weights that are lowered. A certain amount of work is required to do this. Some of this work is required to overcome the friction between

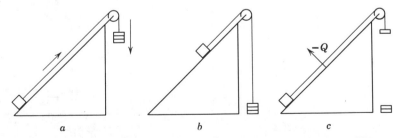

**Fig. 6.10**   Demonstration of the fact that friction makes processes irreversible.

the block and the plane, and some is required to increase the potential energy of the block. The block can be restored to its initial position by removing some of the weights, thus allowing the block to slide down the plane. Some heat transfer from the system to the surroundings will be required to restore the block to its initial temperature. Since the surroundings are not restored to their initial state at the conclusion of the reverse process, we conclude that friction has rendered the process irreversible. Another type of frictional effect is that associated with the flow of viscous fluids in pipes and passages and in the movement of bodies through viscous fluids.

## Unrestrained Expansion

The classic example of unrestrained expansion is shown in Fig. 6.11, in which a gas is separated from a vacuum by a membrane. Consider the process that occurs when the membrane breaks and the gas fills the entire vessel. It can be shown that this is an irreversible process by considering the process that would be necessary to restore the system to its original state. This would involve compressing the gas and transferring heat from the gas until its initial state was reached. Since the work and heat transfer involve a change in the surroundings, the surroundings are not restored to their initial state, indicating that the unrestrained expansion was an irreversible process. The process described in Fig. 6.8 is also an example of an unrestrained expansion.

In the reversible expansion of a gas there must be only an infinitesimal difference between the force exerted by the gas and the restraining force, so the rate at which the boundary moves will be infinitesimal. In accordance with our previous definition, this is a quasiequilibrium process. However, actual cases involve a finite difference in forces, which gives rise to a finite rate of movement of the boundary, and thus are irreversible in some degree.

## Heat Transfer Through a Finite Temperature Difference

Consider as a system a high-temperature body and a low-temperature body, and let heat be transferred from the high-temperature body to the

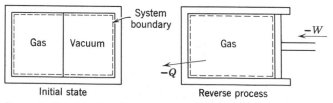

**Fig. 6.11** Demonstration of the fact that unrestrained expansion makes processes irreversible.

low-temperature body. The only way in which the system can be restored to its initial state is to provide refrigeration, which requires work from the surroundings, and some heat transfer to the surroundings will also be necessary. Because of the heat transfer and the work, the surroundings are not restored to their original state, indicating that the process was irreversible.

An interesting question now arises. Heat is defined as energy that is transferred due to a temperature difference. We have just shown that heat transfer through a temperature difference is an irreversible process. Therefore, how can we have a reversible heat-transfer process? A heat-transfer process approaches a reversible process as the temperature difference between the two bodies approaches zero. Therefore, we define a reversible heat-transfer process as one in which the heat is transferred through an infinitesimal temperature difference, which would require an infinite amount of time, or infinite area. Therefore, all actual heat-transfer processes are through a finite temperature difference and are therefore irreversible, and the greater the temperature difference the greater the irreversibility. We will find however, that the concept of reversible heat transfer is very useful in describing ideal processes.

### Mixing of Two Different Substances

This process is illustrated in Fig. 6.12 in which two different gases are separated by a membrane. Let the membrane break and a homogeneous mixture of oxygen and nitrogen fill the entire volume. This process will be considered in some detail in Chapter 11. We can say here that this may be considered as a special case of an unrestrained expansion, for each gas undergoes an unrestrained expansion as it fills the entire volume. A certain amount of work is necessary to separate these gases. Thus an air separation plant such as described in Chapter 1 requires an input of work in order that the separation may be accomplished.

### Other Factors

There are a number of other factors that make processes irreversible but they will not be considered in detail here. Hysteresis effects and the

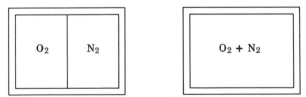

**Fig. 6.12**   Demonstration of the fact that the mixing of two different substances is an irreversible process.

$i^2R$ loss encountered in electrical circuits are both factors that make processes irreversible. A combustion process as it ordinarily takes place is also an irreversible process.

It is frequently advantageous to distinguish between internal and external irreversibility. Figure 6.13 shows two identical systems to which heat is transferred. Assuming each system to be a pure substance, the temperature remains constant during the heat-transfer process. In one the heat is transferred from a reservoir at a temperature $T + dT$, and in the other the reservoir is at a much higher temperature, $T + \Delta T$, than the system. The first is a reversible heat-transfer process and the second is an irreversible heat-transfer process. However, as far as the system itself is concerned, it passes through exactly the same states in both processes,

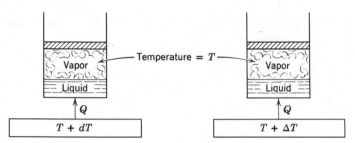

**Fig. 6.13**   Illustration of the difference between an internally and externally reversible process.

which we assume are reversible. Thus, we can say in the second case that the process is internally reversible but externally irreversible because the irreversibility occurs outside the system.

One should also note the general interrelation of reversibility, equilibrium, and time. In a reversible process, the deviation from equilibrium is infinitesimal, and therefore it occurs at an infinitesimal rate. Since it is desirable that actual processes proceed at a finite rate, the deviation from equilibrium must be finite, and therefore the actual process is irreversible in some degree. The greater the deviation from equilibrium, the greater the irreversibility, and the more rapidly the process will occur. It should also be noted that the quasiequilibrium process, which was described in Chapter 2, is a reversible process, and hereafter the term reversible process will be used.

## 6.5   The Carnot Cycle

Having defined the reversible process and considered some factors that make processes irreversible, let us again pose the question raised in Section 6.3, namely, if the efficiency of all heat engines is less than 100 per cent, what is the most efficient cycle we can have? Let us answer this question for a heat engine that receives heat from a high-temperature reservoir and rejects heat to a low-temperature reservoir. Since we are dealing with reservoirs we recognize that both the high temperature and the low temperature are constant and remain constant regardless of the amount of heat transferred.

Let us assume that this heat engine, which operates between the given high-temperature and low-temperature reservoirs, operates in a cycle in which every process is reversible. If every process is reversible, the cycle is also reversible, and if the cycle is reversed, the heat engine becomes a refrigerator. In the next section we will show that this is the most efficient cycle that can operate between two constant-temperature reservoirs. It is called the Carnot cycle, and is named after a French engineer, Nicolas Leonard Sadi Carnot (1796–1832), who stated the second law of thermodynamics in 1824.

We now turn our attention to a consideration of the Carnot cycle. Figure 6.14 shows a power plant that is similar in many respects to a simple steam power plant and which we assume operates on the Carnot cycle. Assume the working fluid to be a pure substance, such as steam. Heat is transferred from the high-temperature reservoir to the water (steam) in the boiler. For this to be a reversible heat transfer, the temperature of the water (steam) must be only infinitesimally lower than the temperature of the reservoir. This also implies, since the temperature of the reservoir remains constant, that the temperature of the water must remain constant. Therefore, the first process in the Carnot cycle is a reversible isothermal process in which heat is transferred from the high-temperature reservoir to the working fluid. A change of phase from liquid to vapor at constant pressure is of course an isothermal process for a pure substance.

The next process occurs in the turbine. It occurs without heat transfer and is therefore adiabatic. Since all processes in the Carnot cycle are reversible, this must be a reversible adiabatic process, during which the temperature of the working fluid decreases from the temperature of the high-temperatre reservoir to the temperature of the low-temperature reservoir.

In the next process heat is rejected from the working fluid to the low-temperature reservoir. This must be a reversible isothermal process in which the temperature of the working fluid is infinitesimally higher than that of the low-temperature reservoir. During this isothermal process some of the steam is condensed.

The final process, which completes the cycle, is a reversible adiabatic process in which the temperature of the working fluid increases from the low temperature to the high temperature. If this were to be done with water (steam) as the working fluid, it would involve taking a mixture of liquid and vapor from the condenser and compressing it. (This would be very inconvenient in practice and therefore in all power plants the working fluid is completely condensed in the condenser, and the pump handles only the liquid phase.)

Since the Carnot heat engine cycle is reversible, every process could be reversed, in which case it would become a refrigerator. The refrigerator is shown by the dotted lines and parentheses in Fig. 6.14. The temperature of the working fluid in the evaporator would be infinitesimally less than the temperature of the low-temperature reservoir, and in the condenser it is infinitesimally higher than that of the high-temperature reservoir.

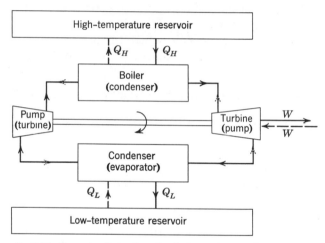

**Fig. 6.14** Example of a heat engine that operates on a Carnot cycle.

It should be emphasized that the Carnot cycle can be executed in many different ways. Many different working substances can be used, such as a gas, a thermoelectric device, or a paramagnetic substance in a magnetic field such as was described in Chapter 4. There are also various possible arrangements of machinery. For example, a Carnot cycle can be devised that takes place entirely within a cylinder, using a gas as a working substance, as shown in Fig. 6.15.

The important point to be made here is that the Carnot cycle, regardless of what the working substance may be, always has the same four basic processes. These are:

**1.** A reversible isothermal process in which heat is transferred to or from the high-temperature reservoir.

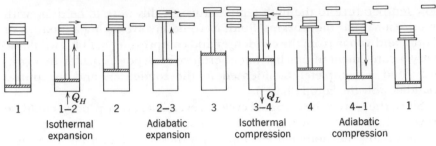

Isothermal    Adiabatic    Isothermal    Adiabatic
expansion    expansion    compression    compression

**Fig. 6.15** Example of a gaseous system operating on a Carnot cycle.

2. A reversible adiabatic process in which the temperature of the working fluid decreases from the high temperature to the low temperature.
3. A reversible isothermal process in which heat is transferred to or from the low-temperature reservoir.
4. A reversible adiabatic process in which the temperature of the working fluid increases from the low temperature to the high temperature.

## 6.6 Two Propositions Regarding the Efficiency of a Carnot Cycle

There are two important propositions regarding the efficiency of a Carnot cycle.

### First Proposition

It is impossible to construct an engine that operates between two given reservoirs and is more efficient than a reversible engine operating between the same two reservoirs.

The proof of this statement involves a "thought experiment." An initial assumption is made, and it is then shown that this assumption leads to impossible conclusions. The only possible conclusion is that the initial assumption was incorrect.

Let us assume that there is an irreversible engine operating between two given reservoirs that has a greater efficiency than a reversible engine operating between the same two reservoirs. Let the heat transfer to the irreversible engine be $Q_H$, the heat rejected be $Q_L'$, and the work be $W_{IE}$ (which equals $Q_H - Q_L'$) as shown in Fig. 6.16. Let the reversible engine operate as a refrigerator (since it is reversible this is possible) and let the heat transfer with the low-temperature reservoir be $Q_L$, the heat transfer with the high-temperature reservoir be $Q_H$, and the work required be $W_{RE}$ (which equals $Q_H - Q_L$).

Since the initial assumption was that the irreversible engine is more efficient, it follows (because $Q_H$ is the same for both engines) that $Q_L' <$

$Q_L$ and $W_{IE} > W_{RE}$. Now the irreversible engine can drive the reversible engine and still deliver the net work $W_{net}$ (which equals $W_{IE} - W_{RE} = Q_L - Q_L'$). However, if we consider the two engines and the high-temperature reservoir as a system, as indicated in Fig. 6.16, we have a system that operates in a cycle, exchanges heat with a single reservoir, and does

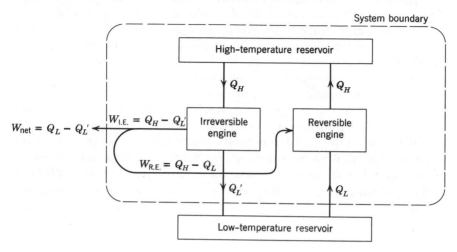

**Fig. 6.16**  Demonstration of the fact that the Carnot cycle is the most efficient cycle operating between two fixed temperature reservoirs.

a certain amount of work. However, this would constitute a violation of the second law and we conclude that our initial assumption (that the irreversible engine is more efficient than the reversible engine) is incorrect, and therefore we cannot have an irreversible engine that is more effcient than a reversible engine operating between the same two reservoirs.

### Second Proposition

All engines that operate on the Carnot cycle between two given constant-temperature reservoirs have the same efficiency. The proof of this proposition is similar to the proof outlined above, and involves the assumption that there is one Carnot cycle that is more efficient than another Carnot cycle operating between the same temperature reservoirs. Let the Carnot cycle with the higher efficiency replace the irreversible cycle of the previous argument, and the Carnot cycle with the lower efficiency operate as the refrigerator. The proof proceeds with the same line of reasoning as in the first proposition. The details are left as an exercise for the student.

## 6.7    The Thermodynamic Temperature Scale

In discussing the matter of temperature in Chapter 2 it was pointed out that the zeroth law of thermodynamics provides a basis for temperature measurement, but that a temperature scale must be defined in terms of a particular thermometer substance and device. A temperature scale that is independent of any particular substance, which might be called an absolute temperature scale, would be most desirable. In the last paragraph we noted that the efficiency of a Carnot cycle is independent of the working substance and depends only on the temperature. This fact provides the basis for such an absolute temperature scale, which we will call the thermodynamic temperature scale.

The concept of this temperature scale may be developed with the aid of Fig. 6.17, which shows three reservoirs and three engines that operate on the Carnot cycle. $T_1$ is the highest temperature, $T_3$ is the lowest temperature, and $T_2$ is an intermediate temperature, and the engines operate between the various reservoirs as indicated. $Q_1$ is the same for both $A$ and $C$, and since we are dealing with reversible cycles, $Q_3$ is the same for $B$ and $C$.

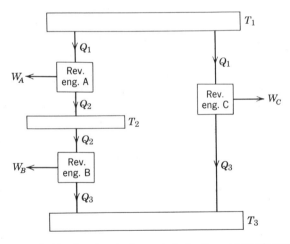

**Fig. 6.17**    Arrangement of heat engines to demonstrate the thermodynamic temperature scale.

Since the efficiency of a Carnot cycle is a function of only the temperature we can write

$$\eta_{\text{thermal}} = 1 - \frac{Q_L}{Q_H} = \psi(T_L, T_H) \qquad (6.3)$$

where $\psi$ designates a functional relation.

Let us apply this to the three Carnot cycles of Fig. 6.17.

$$\frac{Q_1}{Q_2} = \psi(T_1, T_2)$$

$$\frac{Q_2}{Q_3} = \psi(T_2, T_3)$$

$$\frac{Q_1}{Q_3} = \psi(T_1, T_3)$$

Since,

$$\frac{Q_1}{Q_3} = \frac{Q_1 Q_2}{Q_2 Q_3}$$

it follows that

$$\psi(T_1, T_3) = \psi(T_1, T_2) \times \psi(T_2, T_3) \tag{6.4}$$

Note that the left side is a function of $T_1$ and $T_3$ (and not $T_2$) and therefore the right side of this equation must also be a function of $T_1$ and $T_3$, (and not $T_2$). From this we can conclude that the form of the function $\psi$ is

$$\psi(T_1, T_2) = \frac{f(T_1)}{f(T_2)}$$

$$\psi(T_2, T_3) = \frac{f(T_2)}{f(T_3)}$$

for in this way $f(T_2)$ will cancel from the product of $\psi(T_1, T_2) \times \psi(T_2, T_3)$. Therefore, we conclude that

$$\frac{Q_1}{Q_3} = \psi(T_1, T_3) = \frac{f(T_1)}{f(T_3)} \tag{6.5}$$

In general terms,

$$\frac{Q_H}{Q_L} = \frac{f(T_H)}{f(T_L)} \tag{6.6}$$

Now there are several functional relations which will satisfy this equation. The one that has been selected, which was originally proposed by Lord Kelvin, for the thermodynamic scale of temperature, is the relation

$$\frac{Q_H}{Q_L} = \frac{T_H}{T_L} \tag{6.7}$$

With absolute temperatures so defined the efficiency of a Carnot cycle may be expressed in terms of the absolute temperatures.[2]

$$\eta_{\text{thermal}} = 1 - \frac{Q_L}{Q_H} = 1 - \frac{T_L}{T_H} \tag{6.8}$$

This means that if the thermal efficiency of a Carnot cycle operating between two given constant-temperature reservoirs is known, the ratio of the two absolute temperatures is also known.

It should be noted that Equation 6.7 gives us a ratio of absolute temperatures, but it does not give us information about the magnitude of the degree. Let us first consider a qualitative approach to this matter and then a more rigorous statement.

Suppose we had a heat engine operating on the Carnot cycle that received heat at the temperature of the steam point and rejected heat at the temperature of the ice point. (Since a Carnot cycle involves only reversible processes, it is impossible to construct such a heat engine and perform the proposed experiment. However, we can follow the reasoning as a "thought experiment" and gain a further understanding of the thermodynamic temperature scale.) If the efficiency of such an engine could be measured, it would be found to be 26.80 per cent. Therefore, from Eq. 6.8

[2]Lord Kelvin also proposed a logarithmic scale of the form

$$\frac{Q_H}{Q_L} = \frac{e^{"T_H"}}{e^{"T_L"}}$$

where "$T_H$" and "$T_L$" designate the absolute temperatures on this proposed logarithmic scale. This relation can also be written

$$\ln\frac{Q_H}{Q_L} = "T_H" - "T_L"$$

The form Kelvin actually proposed was

$$\log_{10}\frac{Q_H}{Q_L} = "T_H" - "T_L"$$

Thus, the relation between the scale in use and the proposed logarithmic scale is

$$"T" = \log_{10} T + L$$

where $L$ is a constant that determines the level of temperature that corresponds to zero on the logarithmic scale. On this logarithmic scale temperatures range from $-\infty$ to $+\infty$, whereas on the thermodynamic scale in use they vary from 0 to $+\infty$ for ordinary systems.

$$\eta_{th} = 1 - \frac{T_L}{T_H} = 1 - \frac{T_{\text{ice point}}}{T_{\text{steam point}}} = 0.2680$$

$$\frac{T_{\text{ice point}}}{T_{\text{steam point}}} = 0.7320$$

This gives us one equation involving the two unknowns $T_H$ and $T_L$. The second equation comes from an arbitrary decision regarding the magnitude of the degree on the thermodynamic temperature scale. If we wish to have the magnitude of the degree on the absolute scale correspond to the magnitude of the degree on the Celsius scale, we can write

$$T_{\text{steam point}} - T_{\text{ice point}} = 100$$

Solving these two equations simultaneously we find

$$T_{\text{steam point}} = 373.15 \text{ K} \qquad T_{\text{ice point}} = 273.15 \text{ K}$$

It follows that

$$T(°C) + 273.15 = T(K)$$

As already noted, the measurement of efficiencies of Carnot cycles is, however, not a practical way to approach the problem of temperature measurement on the thermodynamic scale of temperature. The actual approach used is based on the ideal gas thermometer and an assigned value for the triple point of water. At the Tenth Conference on Weights and Measures, which was held in 1954, the temperature of the triple point of water was assigned the value 273.16 K. (The triple point of water is approximately 0.01°C above the ice point. The ice point is defined as the temperature of a mixture of ice and water at a pressure of 1 atm (101.3 kPa) of air which is saturated with water vapor.)

Let us now briefly consider the ideal gas scale of temperature. This scale is based upon the fact that as the pressure of a gas approaches zero, its equation of state approaches the ideal gas equation of state, namely,

$$Pv = RT$$

Consider how an ideal gas might be used to measure temperature in a constant-volume gas thermometer, which is shown schematically in Fig. 6.18. Let the gas bulb be placed in the location where the temperature is to be measured, and let the mercury column be so adjusted that the level of mercury stands at the reference mark $A$. Thus the volume of the gas

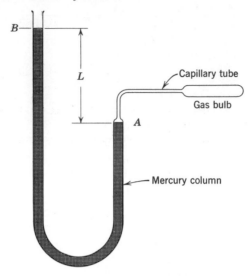

**Fig. 6.18**   Schematic diagram of a constant-volume gas thermometer.

remains constant. Assume that the gas in the capillary tube is at the same temperature as the gas in the bulb. Then the pressure of the gas, which is indicated by the height of the mercury column, is an indication of the temperature.

Let the pressure that is associated with the temperature of the triple point of water (273.16 K) also be measured and let us designate this pressure $P_{t.p.}$ Then, from the definition of an ideal gas, any other temperature $T$ could be determined from a pressure measurement $P$ by the relation

$$T = 273.16 \left(\frac{P}{P_{t.p.}}\right)$$

The temperature so measured is referred to as the ideal-gas temperature, and it can be shown that the temperature so measured is exactly equal to the thermodynamic temperature.

From a practical point of view we have the problem that no gas behaves exactly like an ideal gas. However, we do know that as the pressure approaches zero, the behavior of all gases approaches that of an ideal gas. Suppose then, that a series of measurements is made with varying amounts of gas in the gas bulb. This means that the pressure measured at the triple point, and also the pressure at any other temperature, will vary. If the indicated temperature $T_i$ (obtained by assuming that the gas is ideal) is plotted against the pressure of gas with the bulb at the triple point of water, a curve like the one shown in Fig. 6.19 is obtained. When this curve is extrapolated to zero pressure, the correct ideal-gas temperature is obtained. Different curves might result from different gases, but they would all indicate the same temperature at zero pressure.

We have outlined only the general features and principles for measuring temperature on the ideal gas scale of temperatures. Precision work in this field is difficult and laborious, and there are only a few laboratories

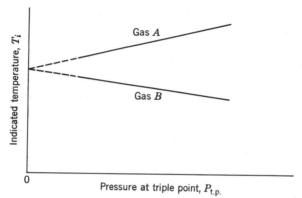

**Fig. 6.19**   Sketch showing how the ideal-gas temperature is determined.

in the world where this precision work is carried on. The International Scale of Temperature, which was presented in Chapter 2, closely approximates the thermodynamic temperature scale and is much easier to work with in actual temperature measurement.

The significance of absolute zero can be indicated by considering a Carnot cycle heat engine that receives a given amount of heat from a given high temperature reservoir. As the temperature at which heat is rejected from the cycle is lowered, the work increases and the amount of heat rejected decreases. In the limit, the heat rejected is zero, and the temperature of the reservoir corresponding to this limit is absolute zero.

Similarly, in the case of a Carnot cycle refrigerator, the amount of work required to produce a given amount of refrigeration increases as the temperature of the refrigerated space decreases. Absolute zero represents the limiting temperature that can be achieved, and the amount of work required to produce a finite amount of refrigeration approaches infinity as the temperature at which refrigeration is provided approaches zero.

## PROBLEMS

6.1 Calculate the thermal efficiency of the steam power plant described in Problem 5.41.

6.2 Calculate the coefficient of performance of the Freon-12 refrigeration cycle described in Problem 5.52.

6.3 Prove that a device that violates the Kelvin-Planck statement of the second law also violates the Clausius statement of the second law.

6.4 Discuss the factors that would make the cycle described in Problem 5.41 an irreversible cycle.

6.5 Calculate the thermal efficiency of a Carnot cycle heat engine operating between 500°C and 40°C, and compare the result with that of Problem 6.1.

6.6 Calculate the coefficient of performance of a Carnot cycle refrigerator operating between −5°C and 35°C, and compare the result with that of Problem 6.2.

6.7 The field of geothermal power, utilizing underground sources of hot water or steam, has received increased attention in recent years. Consider a supply of saturated liquid water at 150°C. What is the maximum possible thermal efficiency of a cyclic heat engine using this source of energy and operating in a 20°C environment?

6.8 A Carnot cycle refrigerator operates in a room in which the temperature is 25°C. It is required to transfer 100 kW from the cold space being held at −30°C. What power motor is required?

6.9 It is proposed to heat a house using a heat pump. The heat transfer from the house is 15 kW. The house is to be maintained at 22°C while the outside air is at a temperature of −10°C. What is the minimum power required to drive the heat pump?

6.10 It is proposed to construct a cyclic heat engine to operate in the ocean at a location where the water temperature is 20°C near the surface and 5°C at some greater depth. What is the maximum possible thermal efficiency of such an engine?

6.11 An inventor claims to have developed a refrigeration unit which maintains the refrigerated space at −10°C while operating in a room where the temperature is 25°C, and which has a coefficient of performance of 8.5. How do you evaluate his claim? How would you evaluate his claim of a coefficient of performance of 7.5?

6.12 Liquid sodium leaves a nuclear reactor of 800°C and is to be used as the energy source in a steam power plant. The condenser cooling water is recirculated using a cooling tower and leaves the tower at 15°C. Determine the limiting thermal efficiency of this power plant. Is it misleading to use the 800°C and 15°C temperatures to calculate this value?

6.13 A cyclic machine is used to transfer heat from a higher to a lower temperature reservoir, as shown in Fig. 6.20. Determine whether this machine is reversible, irreversible, or impossible.

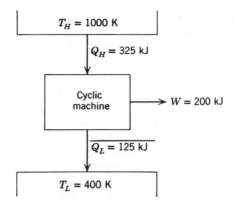

**Fig. 6.20**   Sketch for Problem 6.13.

6.14  Helium has the lowest normal boiling point of any of the elements, namely 4.2 K. At this temperature it has an enthalpy of evaporation of 83.3 J/mol.

A Carnot refrigeration cycle is to be used in the production of 1 mole of liquid helium at 4.2 K from saturated vapor at the same temperature. What is the work input to the refrigerator and the coefficient of performance of this refrigeration cycle? Assume an ambient temperature of 300 K.

6.15  A temperature of about 0.01 K can be achieved by the technique of magnetic cooling, which was described in Section 4.5. In this process a strong magnetic field is imposed on a paramagnetic salt which is maintained at 1 K by transferring heat to liquid helium which is boiling at very low pressure. The salt is then thermally isolated from the helium, the magnetic field is removed, and the temperature drops.

Assume that 1 mJ is to be removed from the paramagnetic salt at an average temperature of 0.1 K, and that the necessary refrigeration is produced by a Carnot refrigeration cycle. What is the work input to the refrigerator and the coefficient of performance of this refrigeration cycle? Assume an ambient temperature of 300 K.

6.16  The lowest temperature which has been achieved at the present time is about $1 \times 10^{-6}$ K. Achieving this temperature involved an additional stage to that described in Problem 6.15, namely nuclear cooling. This is similar to magnetic cooling, but involves the magnetic moment associated with the nucleus rather than that associated with certain ions in the paramagnetic salt.

Suppose that 10 $\mu$J were to be removed from a specimen at an average temperature of $10^{-5}$ K (10 $\mu$J is about the amount of energy associated with the dropping of a pin through a distance of 3 mm, and is about equal to the energy transferred as heat at this temperature in some experiments). If this amount of refrigeration at an average temperature of $10^{-5}$ K is produced by a Carnot refrigeration cycle, determine the work input and the coefficient of performance of the refrigeration cycle. Assume an ambient temperature of 300 K.

6.17  It is desired to produce refrigeration at −30°C. A reservoir is available at a temperature of 200°C and the ambient temperature is 30°C. Thus work can be done by a heat engine operating between the 200°C reservoir and the ambient, and this work can be used to drive the refrigerator. Determine the ratio of the heat transferred from the high temperature reservoir to the heat transferred from the refrigerated space, assuming all processes to be reversible.

6.18  A heat pump is to be used to heat a house in the winter and then reversed to cool the house in the summer. The interior temperature is to be maintained at 20°C. Heat transfer through the walls and roof is estimated to be 2400 kJ per hour per degree temperature difference between the inside and outside.

(a) If the outside temperature in the winter is 0°C, what is the minimum power required to drive the heat pump?

(b) If the power input is the same as that in part (a), what is the maximum outside temperature for which the inside can be maintained at 20°C?

6.19  Consider an engine in outer space which operates on the Carnot cycle. The only way in which heat can be transferred from the engine is by radiation. The rate at which heat is radiated is proportional to the fourth power of the absolute temperature and to the area of the radiating surface. Show that for a given power output and a given $T_H$, the area of the radiator will be a minimum when $T_L/T_H = \frac{3}{4}$.

6.20  It is proposed to build a one million kW electric power plant with steam as the working fluid. The condensers are to be cooled with river water (see Fig. 6.21). The maximum steam temperature will be 550°C and the pressure in the condensers will be 10 kPa. As an engineering consultant, you are asked to estimate the resulting temperature rise of the river far downstream of the plant. What is your estimate?

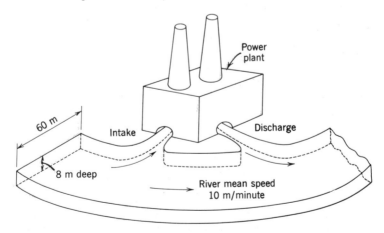

**Fig. 6.21**  Sketch for Problem 6.20.

# 7

# Entropy

Up to this point in our consideration of the second law of thermo-dynamics we have dealt only with thermodynamic cycles. Although this is a very important and useful approach, we are in many cases concerned with processes rather than cycles. Thus, we might be interested in the second-law analysis of processes we encounter daily, such as the combustion process in an automobile engine, the cooling of a cup of coffee, or the chemical processes that take place in our bodies. It would also be most desirable to be able to deal with the second law quantitatively as well as qualitatively.

In our consideration of the first law, we initially stated the law in terms of a cycle, but then defined a property, the internal energy, which enabled us to use the first law quantitatively for processes. Similarly we have stated the second law for a cycle, and we will now find that the second law leads to another property, entropy, which enables us to treat the second law quantitatively for processes. Energy and entropy are both abstract concepts that man has devised to aid in describing certain observations. As we noted in Chapter 2, thermodynamics can be described as the science of energy and entropy. The significance of this statement will now become increasingly evident.

## 7.1  Inequality of Clausius

The first step in our consideration of the property we call entropy is to establish the inequality of Clausius, which is

$$\oint \frac{\delta Q}{T} \le 0$$

The inequality of Clausius is a corollary or consequence of the second law of thermodynamics, and will be demonstrated to be valid for all possible cycles. This includes both reversible and irreversible heat engines and refrigerators. Since any reversible cycle can be represented by a series of Carnot cycles, in this analysis we need only consider a Carnot cycle that leads to the inequality of Clausius.

Consider first a reversible (Carnot) heat engine cycle, operating between reservoirs at temperatures $T_H$ and $T_L$, as shown in Fig. 7.1.

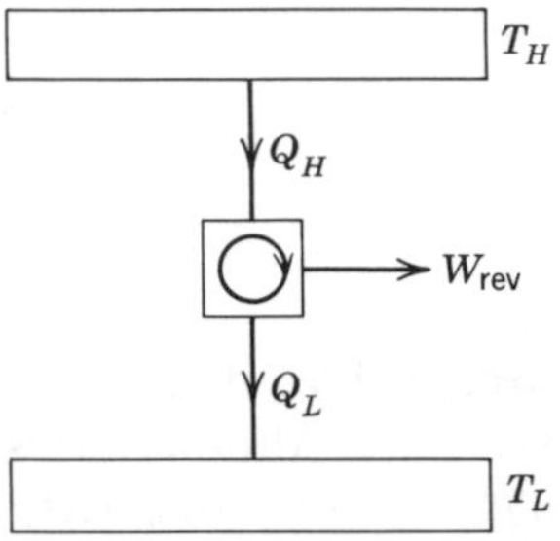

For this cycle, the cyclic integral of the heat transfer, $\oint \delta Q$, is greater than zero.

$$\oint \delta Q = Q_H - Q_L > 0$$

Since $T_H$ and $T_L$ are constant, it follows from the definition of the absolute temperature scale and the fact that this is a reversible cycle that

$$\oint \frac{\delta Q}{T} = \frac{Q_H}{T_H} - \frac{Q_L}{T_L} = 0$$

**Fig. 7.1**  Reversible heat engine cycle for demonstration of the inequality of Clausius.

If $\oint \delta Q$, the cyclic integral of $\delta Q$, is made to approach zero (by making $T_H$ approach $T_L$), while the cycle remains reversible, the cyclic integral of $\delta Q/T$ remains zero. Thus we conclude that for all reversible heat engine cycles

$$\oint \delta Q \geqslant 0$$

and

$$\oint \frac{\delta Q}{T} = 0$$

Now consider an irreversible cyclic heat engine operating between the same $T_H$ and $T_L$ as the reversible engine of Fig. 7.1, and receiving the same quantity of heat $Q_H$. Comparing the irreversible cycle with the reversible one, we conclude from the second law, that

$$W_{\text{irr}} < W_{\text{rev}}$$

Since $Q_H - Q_L = W$ for both the reversible and irreversible cycles, we conclude that

$$Q_H - Q_{L_{\text{irr}}} < Q_H - Q_{L_{\text{rev}}}$$

and therefore

$$Q_{L_{irr}} > Q_{L_{rev}}$$

Consequently, for the irreversible cyclic engine,

$$\oint \delta Q = Q_H - Q_{L_{irr}} > 0$$

$$\oint \frac{\delta Q}{T} = \frac{Q_H}{T_H} - \frac{Q_{L_{irr}}}{T_L} < 0$$

Suppose that we cause the engine to become more and more irreversible while keeping $Q_H$, $T_H$, and $T_L$ fixed. The cyclic integral of $\delta Q$ then approaches zero, while that for $\delta Q/T$ becomes a progressively larger negative value. In the limit, as the work output goes to zero,

$$\oint \delta Q = 0$$

$$\oint \frac{\delta Q}{T} < 0$$

Thus we conclude that for all irreversible heat engine cycles

$$\oint \delta Q \geqslant 0$$

$$\oint \frac{\delta Q}{T} < 0$$

To complete the demonstration of the inequality of Clausius we must perform similar analyses for both reversible and irreversible refrigeration cycles. For the reversible refrigeration cycle shown in Fig. 7.2,

$$\oint \delta Q = -Q_H + Q_L < 0$$

and

$$\oint \frac{\delta Q}{T} = -\frac{Q_H}{T_H} + \frac{Q_L}{T_L} = 0$$

As the cyclic integral of $\delta Q$ is made to approach zero reversibly ($T_H$ approaching $T_L$), the cyclic integral of $\delta Q/T$ remains at zero. In the limit,

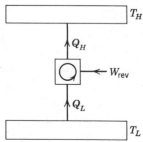

**Fig. 7.2** Reversible refrigeration cycle for demonstration of the inequality of Clausius.

$$\oint \delta Q = 0$$

$$\oint \frac{\delta Q}{T} = 0$$

Thus for all reversible refrigeration cycles

$$\oint \delta Q \leqslant 0$$

$$\oint \frac{\delta Q}{T} = 0$$

Finally let an irreversible cyclic refrigerator operate between temperatures $T_H$ and $T_L$ and receive the same amount of heat $Q_L$ as the reversible refrigerator of Fig. 7.2. From the second law, we conclude that the work input required will be greater for the irreversible refrigerator, or

$$W_{irr} > W_{rev}$$

Since $Q_H - Q_L = W$ for each cycle, it follows that

$$Q_{H_{irr}} - Q_L > Q_{H_{rev}} - Q_L$$

and therefore

$$Q_{H_{irr}} > Q_{H_{rev}}$$

That is, the heat rejected by the irreversible refrigerator to the high-temperature reservoir is greater than the heat rejected by the reversible refrigerator. Therefore, for the irreversible refrigerator,

$$\oint \delta Q = -Q_{H_{irr}} + Q_L < 0$$

$$\oint \frac{\delta Q}{T} = -\frac{Q_{H_{irr}}}{T_H} + \frac{Q_L}{T_L} < 0$$

By making this machine progressively more irreversible while keeping $Q_L$, $T_H$, and $T_L$ constant, the cyclic integrals of $\delta Q$ and $\delta Q/T$ both become larger in the negative direction. Consequently, a limiting case as the cyclic integral of $\delta Q$ approaches zero does not exist for the irreversible refrigerator.

Thus for all irreversible refrigeration cycles,

$$\oint \delta Q < 0$$

$$\oint \frac{\delta Q}{T} < 0$$

Summarizing, we note that, as regards the sign of $\oint \delta Q$, we have considered all possible reversible cycles (that is, $\oint \delta Q \lesseqgtr 0$), and for each of these reversible cycles

$$\oint \frac{\delta Q}{T} = 0$$

We have also considered all possible irreversible cycles for the sign of $\oint \delta Q$ (that is, $\oint \delta Q \lesseqgtr 0$), and for all these irreversible cycles

$$\oint \frac{\delta Q}{T} < 0$$

Thus for all cycles we can write

$$\oint \frac{\delta Q}{T} \leq 0 \tag{7.1}$$

where the equality holds for reversible cycles and the inequality for irreversible cycles. This relation, Eq. 7.1, is known as the inequality of Clausius.

The significance of the inequality of Clausius may be illustrated by considering the simple steam power plant cycle shown in Fig. 7.3. This

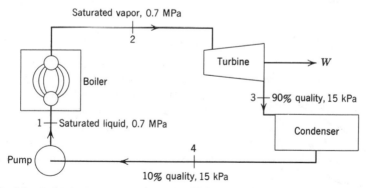

**Fig. 7.3.**   A simple steam power plant that demonstrates the inequality of Clausius.

cycle is slightly different from the usual cycle for steam power plants in that the pump handles a mixture of liquid and vapor in such proportions that saturated liquid leaves the pump and enters the boiler. Suppose that someone reports that the pressure and quality at various points in the cycle are as given in Fig. 7.3. Does this cycle satisfy the inequality of Clausius?

Heat is transferred in two places, the boiler and the condenser. Therefore

$$\oint \frac{\delta Q}{T} = \int \left(\frac{\delta Q}{T}\right)_{boiler} + \int \left(\frac{\delta Q}{T}\right)_{condenser}$$

Since the temperature remains constant in both the boiler and condenser this may be integrated as follows:

$$\oint \frac{\delta Q}{T} = \frac{1}{T_1}\int_1^2 \delta Q + \frac{1}{T_3}\int_3^4 \delta Q = \frac{_1Q_2}{T_1} + \frac{_3Q_4}{T_3}$$

Let us consider a 1 kg mass as the working fluid.

$$_1q_2 = h_2 - h_1 = 2066.3 \text{ kJ/kg } ; T_1 = 164.97°C$$

$$_3q_4 = h_4 - h_3 = 463.4 - 2361.8 = -1898.4 \text{ kJ/kg } ; T_3 = 53.97°C$$

Therefore

$$\oint \frac{\delta Q}{T} = \frac{2066.3}{164.97 + 273.15} - \frac{1898.4}{53.97 + 273.15} = -1.087 \text{ kJ/kg K}$$

Thus this cycle satisfies the inequality of Clausius, which is equivalent to saying that it does not violate the second law of thermodynamics.

## 7.2   Entropy—A Property of a System

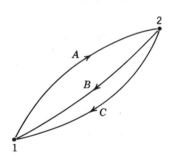

By the use of Eq. 7.1 and Fig. 7.4 it may be shown that the second law of thermodynamics leads to a property of a system which we call entropy. Let a system undergo a reversible process from state 1 to state 2 along path $A$, and let the cycle be completed along path $B$, which is also reversible.

Since this is a reversible cycle we can write

**Fig. 7.4**  Two reversible cycles demonstrating the fact that entropy is a property of a substance.

$$\oint \frac{\delta Q}{T} = 0 = \int_{1A}^{2A} \frac{\delta Q}{T} + \int_{2B}^{1B} \frac{\delta Q}{T}$$

Now consider another reversible cycle, which has the same initial process, but let the cycle be completed along path $C$. For this cycle we can write

$$\oint \frac{\delta Q}{T} = 0 = \int_{1A}^{2A} \frac{\delta Q}{T} + \int_{2C}^{1C} \frac{\delta Q}{T}$$

Subtracting the second equation from the first we have

$$\int_{2B}^{1B} \frac{\delta Q}{T} = \int_{2C}^{1C} \frac{\delta Q}{T}$$

Since the $\int \delta Q/T$ is the same for all reversible paths between states 2 and 1, we conclude that this quantity is independent of the path and is a function of the end states only, and is therefore a property. This property is called entropy, and is designated $S$. It follows that entropy may be defined as a property of a substance in accordance with the relation

$$dS \equiv \left( \frac{\delta Q}{T} \right)_{rev} \qquad (7.2)$$

Entropy is an extensive property, and the entropy per unit mass, is designated $s$. It is important to note that entropy is defined here in terms of a reversible process.

The change in the entropy of a system as it undergoes a change of state may be found by integrating Eq. 7.2. Thus

$$S_2 - S_1 = \int_1^2 \left( \frac{\delta Q}{T} \right)_{rev} \qquad (7.3)$$

In order to perform this integration, the relation between $T$ and $Q$ must be known, and illustrations will be given subsequently. The important point to note here is that since entropy is a property, the change in the entropy of a substance in going from one state to another is the same for all processes, both reversible and irreversible, between these two states. Equation 7.3 enables us to find the change in entropy only along a reversible path. However, once it has been evaluated, this is the magnitude of the entropy change for all processes between these two states.

Equation 7.3 enables us to determine changes of entropy, but tells us nothing about absolute values of entropy. However, from the third law of thermodynamics, which is discussed in Chapter 12, it follows that the entropy of all pure substances can be assigned the value of zero at the absolute zero of temperature. This gives rise to absolute values of entropy and is particularly important when chemical reactions are involved.

However, when no change of composition is involved, it is quite adequate to give values of entropy relative to some arbitrarily selected reference state. This is the procedure followed in most tables of thermodynamic properties, such as the steam tables and ammonia tables. Therefore, until absolute entropy is introduced in Chapter 12, values of entropy will always be given relative to some arbitrary reference state.

A word should be added here regarding the role of $T$ as an integrating factor. We noted in Chapter 4 that $Q$ is a path function, and therefore $\delta Q$ is an inexact differential. However, since $(\delta Q/T)_{rev}$ is a thermodynamic property, it is an exact differential. From a mathematical perspective we note that an inexact differential may be converted to an exact differential by the introduction of an integrating factor. Therefore, $1/T$ serves as the integrating factor in converting the inexact differential $\delta Q$ to the exact differential $\delta Q/T$ for a reversible process.

## 7.3.   The Entropy of a Pure Substance

Entropy is an extensive property of a system. Values of specific entropy (entropy per unit mass) are tabulated in tables of thermodynamic properties in the same manner as specific volume and specific enthalpy. The units of specific entropy in the steam tables, Freon-12 tables, and ammonia tables are kJ/kg K, and the values are given relative to an arbitrary reference state. In the steam tables the entropy of saturated liquid at $0.01\,°C$ is given the value of zero. For most refrigerants, such as Freon-12 and ammonia, the entropy of saturated liquid at $-40°C$ is assigned the value of zero.

In general, we shall use the term "entropy" to refer to both total entropy and entropy per unit mass, since the context or appropriate symbol will clearly indicate the precise meaning of the term.

In the saturation region the entropy may be calculated using the quality, the relations being similar to those for specific volume and enthalpy.

$$s = (1-x)s_f + xs_g$$
$$s = s_f + xs_{fg} \tag{7.4}$$
$$s = s_g - (1-x)s_{fg}$$

The entropy of a compressed liquid is tabulated in the same manner as the other properties. These properties are primarily a function of the temperature, and are not greatly different from those for saturated liquid at the same temperature. Table 4 of the steam tables, which is summarized in Table A.1.4 of the Appendix, gives the entropy of compressed liquid water in the same manner as for other properties, as has been discussed previously.

The thermodynamic properties of a substance are often shown on a temperature-entropy diagram, and an enthalpy-entropy diagram, which is also called a Mollier diagram, after Richard Mollier (1863–1935) of Germany. Figures 7.5 and 7.6 show the essential elements of

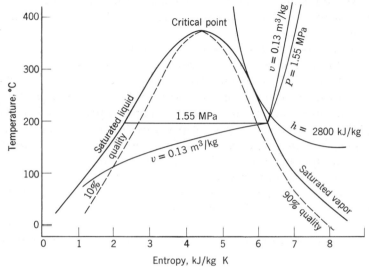

**Fig. 7.5**   Temperature-entropy diagram for steam.

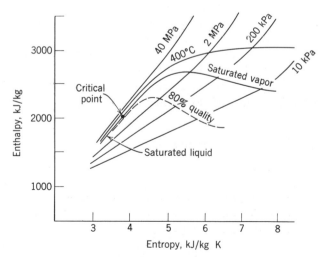

**Fig. 7.6**   Enthalpy-entropy diagram for steam.

temperature-entropy and enthalpy-entropy diagrams for steam. The general features of such diagrams are the same for all pure substances. A more complete temperature-entropy diagram for steam is shown in Fig. A.1.

These diagrams are valuable both as a means of presenting thermodynamic data and also because they enable one to visualize the changes of state that occur in various processes, and as our study progresses the student should acquire facility in visualizing thermodynamic processes on these diagrams. The temperature-entropy diagram is particularly useful for this purpose.

One more observation should be made here concerning the compressed-liquid lines on the temperature-entropy diagram for water. Reference to Table 4 of the steam tables (Appendix, Table A.1.4) indicates that the entropy of the compressed liquid is less than that of saturated liquid at the same temperature for all the temperatures listed except 0°C. It can be shown that each constant-pressure line crosses the saturated-liquid line at the point of maximum density, which is about 4°C for water (the exact temperature at which maximum density occurs varies with pressure). For temperatures less than the temperature at which the density is a maximum, the entropy of compressed liquid is greater than that of the saturated liquid. Thus, lines of constant pressure in the compressed liquid region appear (on a magnified scale) as shown in Fig. 7.7a. It is important to understand the general shape of these lines when showing the pumping process for liquids.

Having made this observation for water, it should be stated that for most substances the difference in the entropy between compressed liquid and saturated liquid at the same temperature is so small that usually a process in which liquid is heated at constant pressure is shown as coinciding with the saturated liquid line until the saturation temperature is reached (Fig. 7.7b). Thus, if water at 10 MPa is heated from 0°C to the saturation temperature, it would be shown by line *ABD*, which coincides with the saturated-liquid line.

## 7.4   Entropy Change in Reversible Processes

Having established the fact that entropy is a thermodynamic property of a system, its significance in various processes will now be considered. In this section we will limit ourselves to systems that undergo reversible processes, and consider the Carnot cycle, reversible heat-transfer processes, and reversible adiabatic processes.

Let the working fluid of a heat engine operating on the Carnot cycle comprise the system. The first process is the isothermal transfer of heat

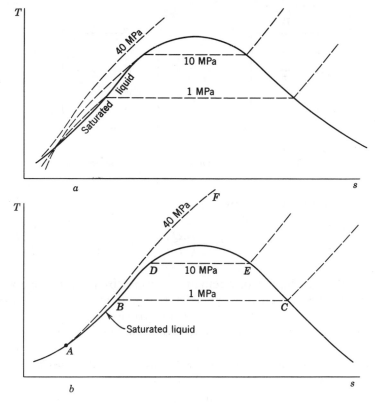

**Fig. 7.7**  Temperature-entropy diagram to show properties of a compressed liquid, water.

to the working fluid from the high-temperature reservoir. For this we can write

$$S_2 - S_1 = \int_1^2 \left(\frac{\delta Q}{T}\right)_{rev}$$

Since this is a reversible process in which the temperature of the working fluid remains constant, it can be integrated to give

$$S_2 - S_1 = \frac{1}{T_H} \int_1^2 \delta Q = \frac{{}_1 Q_2}{T_H}$$

This process is shown in Fig. 7.8$a$, and the area under line 1-2, area 1-2-$b$-$a$-1, represents the heat transferred to the working fluid during the process.

The second process of a Carnot cycle is a reversible adiabatic one.

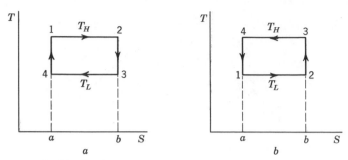

**Fig. 7.8** The Carnot cycle on the temperature-entropy diagram.

From the definition of entropy,

$$dS = \left(\frac{\delta Q}{T}\right)_{rev}$$

it is evident that the entropy remains constant in a reversible adiabatic process. A constant-entropy process is called an isentropic process. Line 2-3 represents this process, and this process is concluded at state 3 when the temperature of the working fluid reaches $T_L$.

The third process is the reversible isothermal process in which heat is transferred from the working fluid to the low-temperature reservoir. For this we can write

$$S_4 - S_3 = \int_3^4 \left(\frac{\delta Q}{T}\right)_{rev} = \frac{{}_3 Q_4}{T_L}$$

Since during this process the heat transfer is negative (as regards the working fluid) the entropy of the working fluid decreases during this process. Also, since the final process 4-1, which completes the cycle, is a reversible adiabatic process (and therefore isentropic), it is evident that the entropy decrease in process 3-4 must exactly equal the entropy increase in process 1-2. The area under line 3-4, area 3-4-$a$-$b$-3, represents the heat transferred from the working fluid to the low-temperature reservoir.

Since the net work of the cycle is equal to the net heat transfer, it is evident that area 1-2-3-4-1 represents the net work of the cycle. The efficiency of the cycle may also be expressed in terms of areas.

$$\eta_{th} = \frac{W_{net}}{Q_H} = \frac{\text{area 1-2-3-4-1}}{\text{area 1-2-}b\text{-}a\text{-1}}$$

Some statements made earlier about efficiencies may now be understood graphically. For example, increasing $T_H$ while $T_L$ remains constant increases the efficiency. Decreasing $T_L$ as $T_H$ remains constant increases the efficiency. It is also evident that the efficiency approaches 100 per cent as the absolute temperature at which heat is rejected approaches zero.

If the cycle is reversed, we have a refrigerator or heat pump, and the Carnot cycle for a refrigerator is shown in Fig. 7.8b. Notice in this case that the entropy of the working fluid increases at $T_L$, since heat is transferred to the working fluid at $T_L$. The entropy decreases at $T_H$ due to heat transfer from the working fluid.

Let us next consider reversible heat-transfer processes. Actually we are concerned here with processes that are internally reversible, i.e., processes that involve no irreversibilities within the boundary of the system. For such processes the heat transfer to or from a system can be shown as an area on a temperature-entropy diagram. For example, consider the change of state from saturated liquid to saturated vapor at constant pressure. This would correspond to the process 1-2 on the $T$-$s$ diagram of Fig. 7.9 (note that absolute temperature is required here), and

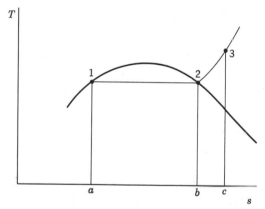

**Fig. 7.9** A temperature-entropy diagram to show areas which represent heat transfer for an internally reversible process.

the area 1-2-b-a-1 represents the heat transfer. Since this is a constant-pressure process, the heat transfer per unit mass is equal to $h_{fg}$. Thus,

$$s_2 - s_1 = s_{fg} = \frac{1}{m} \int_1^2 \left(\frac{\delta Q}{T}\right)_{\text{rev}} = \frac{1}{mT} \int_1^2 \delta Q = \frac{{}_1q_2}{T} = \frac{h_{fg}}{T}$$

This relation gives a clue as to how $s_{fg}$ is calculated for tabulation in tables of thermodynamic properties. For example, consider steam at 10 MPa.

From the steam tables we have

$$h_{fg} = 1317.1 \text{ kJ/kg}$$
$$T = 311.06 + 273.15 = 584.21 \text{ K}$$

Therefore,

$$s_{fg} = \frac{h_{fg}}{T} = \frac{1317.1}{584.21} = 2.2544 \text{ kJ/kg K}$$

This is the value listed for $s_{fg}$ in the steam tables.

If heat is transferred to the saturated vapor at constant pressure, the steam is superheated along line 2-3. For this process we can write

$$_2q_3 = \frac{1}{m} \int_2^3 \delta Q = \int_2^3 T \, ds$$

Since $T$ is not constant this cannot be integrated unless we know a relation between temperature and entropy. However, we do realize that the area under line 2-3, area 2-3-c-b-2, represents $\int_2^3 T \, ds$, and therefore represents the heat transferred during this reversible process.

The important conclusion to draw here is that for processes that are internally reversible, the area underneath the process line on a temperature-entropy diagram represents the quantity of heat transferred. This is not true for irreversible processes, as will be demonstrated later.

There are many situations in which essentially adiabatic processes take place. We have already noted that in such cases the ideal process, which is a reversible adiabatic process, is isentropic. We shall consider here an example of a reversible adiabatic process for a system and consider in the next chapter the reversible adiabatic process for a control volume. We shall also note in a later section of this chapter that by comparing an actual process with the ideal or isentropic process we have a basis for defining the efficiency of certain classes of machines.

**Example 7.1**

Consider a cylinder fitted with a piston which contains saturated Freon-12 vapor at $-10°C$. Let this vapor be compressed in a reversible adiabatic process until the pressure is 1.6 MPa. Determine the work per kilogram of Freon-12 for this process.

From the first law we conclude that

$$_1q_2 = u_2 - u_1 + _1w_2 = 0$$
$$_1w_2 = u_1 - u_2$$

State 1 is specified by the statement of the problem and therefore the initial entropy is known. We also conclude, from the second law that since the process is reversible and adiabatic, $s_1 = s_2$. Therefore we know entropy and pressure in the final state, which is sufficient to specify the final state, since we are dealing with a pure substance.

From the Freon-12 tables

$$u_1 = h_1 - P_1 v_1 = 183.058 - 219.1 \times 0.076\ 646 = 166.265\ \text{kJ/kg}$$

$$s_1 = s_2 = 0.7014\ \text{kJ/kg K}$$

$$P_2 = 1.6\ \text{MPa}$$

Therefore, from the superheat tables for Freon-12

$$T_2 = 72.2°C\ ;\ h_2 = 218.564\ ;\ v_2 = 0.011\ 382$$

$$u_2 = 218.564 - 1600 \times 0.011\ 382 = 200.352\ \text{kJ/kg}$$

$$_1 w_2 = u_1 - u_2 = 166.265 - 200.352 = -34.087\ \text{kJ/kg}$$

## 7.5   Two Important Thermodynamic Relations

At this point we will derive two important thermodynamic relations for a simple compressible substance. These relations are

$$T\ dS = dU + P\ dV$$
$$T\ dS = dH - V\ dP$$

The first of these relations can be derived by considering a simple compressible substance in the absence of motion or gravitational effects. The first law for a change of state under these conditions can be written

$$\delta Q = dU + \delta W$$

The equations we are deriving here deal first of all with those changes of state in which the state of the substance can be identified at all times. Thus we must consider a quasiequilibrium process, or, to use the term introduced in the last chapter, a reversible process. For a reversible process of a simple compressible substance we can write

$$\delta Q = T\ dS \quad \text{and} \quad \delta W = P\ dV$$

Substituting these relations into the first law equation we have

$$T\,dS = dU + P\,dV \tag{7.5}$$

which is the equation we set out to derive. Note that this equation was derived by assuming a reversible process, and thus this equation can be integrated for any reversible process, for during such a process the state of the substance can be identified at any point during the process. We also note that Eq. 7.5 deals only with properties. Suppose we have an irreversible process taking place between given initial and final states. The properties of a substance depend only on the state, and therefore the change in the properties during a given change of state are the same for an irreversible process as for a reversible process. Therefore Eq. 7.5 is often applied to an irreversible process between two given states, but the integration of Eq. 7.5 is performed along a reversible path between the same two states.

Since enthalpy is defined as

$$H = U + PV$$

it follows that

$$dH = dU + P\,dV + V\,dP$$

Substituting this relation into Eq. 7.5 we have

$$T\,dS = dH - V\,dP \tag{7.6}$$

which is the second equation that we set out to derive.

These equations can also be written for a unit mass,

$$\begin{aligned} T\,ds &= du + P\,dv \\ T\,ds &= dh - v\,dP \end{aligned} \tag{7.7}$$

or on a mole basis,

$$\begin{aligned} T\,d\bar{s} &= d\bar{u} + P\,d\bar{v} \\ T\,d\bar{s} &= d\bar{h} - \bar{v}\,dP \end{aligned} \tag{7.8}$$

These equations will be used extensively in certain subsequent sections of this book.

If we consider substances of fixed composition other than a simple compressible substance we can write other "$T\,dS$" equations than those given above for a simple compressible substance. In Chapter 4, Eq. 4.13, we noted that for a reversible process we can write the following expression for work.

$$\delta W = P \, dV - \mathscr{T} \, dL - \mathscr{S} \, d\mathscr{A} - \mu_0 \, \mathscr{H} \, d(V \, \mathscr{M}) - \mathscr{E} \, dZ + \cdots$$

It follows that a more general expression for the "$T \, dS$" equation would be

$$T \, dS = dU + P \, dV - \mathscr{T} \, dL - \mathscr{S} \, d\mathscr{A} - \mu_0 \, \mathscr{H} \, d(V \, \mathscr{M}) - \mathscr{E} \, dZ + \cdots \quad (7.9)$$

## 7.6   Entropy Change of a System During an Irreversible Process

Consider a system which undergoes the cycles shown in Fig. 7.10. The cycle made up of the reversible processes $A$ and $B$ is a reversible cycle. Therefore we can write

$$\oint \frac{\delta Q}{T} = \int_{1A}^{2A} \frac{\delta Q}{T} + \int_{2B}^{1B} \frac{\delta Q}{T} = 0$$

The cycle made up of the reversible process $A$ and the irreversible process $C$ is an irreversible cycle. Therefore, for this cycle the inequality of Clausius may be applied, giving the result

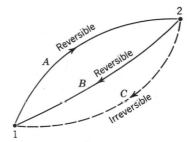

**Fig. 7.10**  Entropy change of a system during an irreversible process.

$$\oint \frac{\delta Q}{T} = \int_{1A}^{2A} \frac{\delta Q}{T} + \int_{2C}^{1C} \frac{\delta Q}{T} < 0$$

Subtracting the second equation from the first and rearranging, we have

$$\int_{2B}^{1B} \frac{\delta Q}{T} > \int_{2C}^{1C} \frac{\delta Q}{T}$$

Since path $B$ is reversible, and since entropy is a property,

$$\int_{2B}^{1B} \frac{\delta Q}{T} = \int_{2B}^{1B} dS = \int_{2C}^{1C} dS$$

Therefore,

$$\int_{2C}^{1C} dS > \int_{2C}^{1C} \frac{\delta Q}{T}$$

For the general case we can write

$$dS \geqslant \frac{\delta Q}{T}$$

$$S_2 - S_1 \geqslant \int_1^2 \frac{\delta Q}{T} \tag{7.10}$$

In these equations the equality holds for a reversible process and the inequality for an irreversible process.

This is one of the most important equations of thermodynamics and is used to develop a number of concepts and definitions. In essence this equation states the influence of irreversibility on the entropy of a system. Thus, if an amount of heat $\delta Q$ is transferred to a system at temperature $T$ in a reversible process the change of entropy is given by the relation

$$dS = \frac{\delta Q}{T}$$

However, if while the amount of heat $\delta Q$ is transferred to the system at temperature $T$ there are any irreversible effects occurring, the change of entropy will be greater than for the reversible process, and we would write

$$dS > \frac{\delta Q}{T}$$

Equation 7.10 holds when $\delta Q = 0$ or when $\delta Q < 0$ as well as when $\delta Q > 0$. If $\delta Q$ is negative, the entropy will tend to decrease as the result of the heat transfer. However, the influence of irreversibilities is still to increase the entropy of the system, and from the absolute numerical perspective we can still write for $\delta Q < 0$,

$$dS \geqslant \frac{\delta Q}{T}$$

## 7.7   Lost Work

The significance of the change of entropy for an irreversible process can be amplified by introducing the concept of lost work, to which we give the symbol $LW$. This concept can be described with the aid of Fig. 7.11, in which a gas is separated from a vacuum by a membrane. Let the membrane have a small rupture so that the gas fills the entire volume,

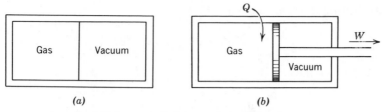

**Fig. 7.11** A process that demonstrates lost work.

and let the necessary heat transfer take place so that the final temperature is the same as the initial temperature. (How much heat transfer would be necessary if this were an ideal gas?). Since the work is zero for this process we conclude from the first law that

$$\delta Q = dU$$

We would like to compare this irreversible process with a reversible process between the same two states. This could be achieved by having the gas expand against a frictionless piston while heat is transferred to the gas in the amount necessary to maintain constant temperature. For this process we conclude from the first law that

$$\delta Q = dU + \delta W$$

and, since this is a reversible process we can write

$$\delta Q = T\,dS \quad \text{and} \quad \delta W = P\,dV$$

The major difference between these two processes is that in the irreversible process the work is zero, whereas in the reversible process the greatest possible work is done. Thus, we might speak of the lost work for an irreversible process. Every irreversible process has associated with it a certain amount of lost work. In Fig. 7.11 we have shown the two extreme cases, a reversible process and one in which the work is zero. Between these two extremes are processes that have various degrees of irreversibility, i.e., the lost work may vary from zero to some maximum value. For a simple compressible substance we can therefore write

$$P\,dV = \delta W + \delta LW \qquad (7.11)$$

where $\delta LW$ is the lost work and $\delta W$ the actual work.

Substituting the relation (Eq. 7.5)

$$T \, dS = dU + P \, dV$$

into Eq. 7.11 we have

$$T \, dS = dU + \delta W + \delta LW$$

But, from the first law

$$\delta Q = dU + \delta W$$

Therefore,

$$T \, dS = \delta Q + \delta LW$$

and

$$dS = \frac{\delta Q}{T} + \frac{\delta LW}{T} \tag{7.12}$$

Thus, we have an expression for the change of entropy for an irreversible process as an equality, whereas in the last section we had the inequality

$$dS \geqslant \frac{\delta Q}{T} \tag{7.13}$$

In a reversible process the lost work is zero and therefore both Eqs. 7.12 and 7.13 reduce to

$$dS = \left(\frac{\delta Q}{T}\right)_{\text{rev}}$$

for a reversible process.

Some important conclusions can now be drawn from Eqs. 7.12 and 7.13. First of all, there are two ways in which the entropy of a system can be increased, namely, by transferring heat to it and by having it undergo an irreversible process. Since the lost work cannot be less than zero, there is only one way in which the entropy of a system can be decreased, and that is to transfer heat from the system.

Secondly, the change in entropy of a system can be separated into the change due to heat transfer and the change due to internal irreversibilities. Frequently the increase in entropy due to irreversibilities is called the irreversible production of entropy.

Finally, as we have already noted, for an adiabatic process $\delta Q = 0$, and in this case the increase in entropy is always associated with the irreversibilities.

One other point, which involves the representation of irreversible processes on $P$-$V$ and $T$-$S$ diagrams, should be made. The work for an irreversible process is not equal to $\int P\,dV$ and the heat transfer is not equal to $\int T\,dS$. Therefore, the area underneath the path does not represent work and heat on the $P$-$V$ and $T$-$S$ diagrams, respectively. In fact, in many cases we are not certain of the exact state through which a system passes when it undergoes an irreversible process. For this reason it is advantageous to show irreversible processes as dotted lines and reversible processes as solid lines. Thus, the area underneath the dotted line will never represent work or heat. For example, the processes of Fig. 7.11 would be shown on $P$-$V$ and $T$-$S$ diagrams as in Fig. 7.12.

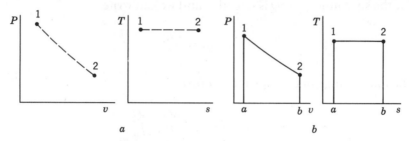

**Fig. 7.12**   Reversible and irreversible processes on pressure-volume and temperature-entropy diagrams.

Figure 7.12$a$ shows an irreversible process, and since the heat transfer and work for this process are zero, the area underneath the dashed line has no significance. Figure 7.12$b$ shows the reversible process, and area 1–2–$b$–$a$–1 represents the work on the $P$-$V$ diagram and the heat transfer on the $T$-$S$ diagram.

## 7.8   Principle of the Increase of Entropy

In this section we consider the total change in the entropy of a system and its surroundings when the system undergoes a change of state. This consideration leads to the principle of the increase of entropy.

Consider the process shown in Fig. 7.13 in which a quantity of heat $\delta Q$ is transferred from the surroundings at temperature $T_0$ to the system at temperature $T$, and let the work done by the system during this

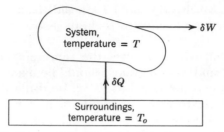

**Fig. 7.13** Entropy change for the system plus surroundings.

process be $\delta W$. For this process we can apply Eq. 7.10 to the system and write

$$dS_{\text{system}} \geqslant \frac{\delta Q}{T}$$

For the surroundings $\delta Q$ is negative and we can write

$$dS_{\text{surr}} = \frac{-\delta Q}{T_0}$$

The total change of entropy is therefore

$$dS_{\text{system}} + dS_{\text{surr}} \geqslant \frac{\delta Q}{T} - \frac{\delta Q}{T_0}$$

$$\geqslant \delta Q\left(\frac{1}{T} - \frac{1}{T_0}\right) \tag{7.14}$$

Since $T_0 > T$, the quantity $\left(\dfrac{1}{T} - \dfrac{1}{T_0}\right)$ is positive and we conclude that

$$dS_{\text{system}} + dS_{\text{surr}} \geqslant 0$$

If $T > T_0$, the heat transfer is from the system to the surroundings, and both $\delta Q$ and the quantity $(1/T) - (1/T_0)$ are negative, thus yielding the same result.

Thus we conclude that for all possible processes that a system in a given surroundings can undergo,

$$dS_{\text{system}} + dS_{\text{surr}} \geqslant 0 \tag{7.15}$$

where the equality holds for reversible processes and the inequality for irreversible processes. This is a very important equation, not only for thermodynamics, but also for philosophical thought, and is referred to as the principle of the increase of entropy. The great significance is that the only processes that can take place are those in which the net change in entropy of the system plus its surroundings increases (or in the limit, remains constant). The reverse process, in which both the system and surroundings are returned to their original state, can never be made to occur. In other words, Eq. 7.15 dictates the single direction in which any process can proceed. Thus, the principle of the increase of entropy can be considered as a quantitative general statement of the second law from the macroscopic point of view, and applies to the combustion of fuel in our automobile engines, the cooling of our coffee, and the processes that take place in our body.

Sometimes this principle of the increase of entropy is stated in terms of an isolated system, one in which there is no interaction between the system and its surroundings. In this case, there is no change in entropy of the surroundings, and one then concludes that

$$dS_{\text{isolated system}} \geq 0 \qquad (7.16)$$

That is, in an isolated system, the only processes that can occur are those that have an increase in entropy associated with them.

**Example 7.2**

Suppose 1 kg of saturated water vapor at 100°C is condensed to saturated liquid at 100°C in a constant-pressure process by heat transfer to the surrounding air, which is at 25°C. What is the net increase in entropy of the system plus surroundings?

For the system, from the steam tables

$$\Delta S_{\text{system}} = -ms_{fg} = -1 \times 6.0480 = -6.0480 \text{ kJ/K}$$

Considering the surroundings

$$Q_{\text{to surroundings}} = mh_{fg} = 1 \times 2257.0 = 2257 \text{ kJ}$$

$$\Delta S_{\text{surr}} = \frac{Q}{T_0} = \frac{2257}{298.15} = 7.5700 \text{ kJ/K}$$

$$\Delta S_{\text{system}} + \Delta S_{\text{surr}} = -6.0480 + 7.5700 = 1.5220 \text{ kJ/K}$$

This increase in entropy is in accordance with the principle of the increase of entropy, and tells, as does our experience, that this process can take place.

It is interesting to note how this heat transfer from the water to the surroundings might have taken place reversibly. Suppose that an engine operating on the Carnot cycle received heat from the water and rejected heat to the surroundings, as shown in Fig. 7.14. The decrease in the entropy of the water is equal to the increase in the entropy of the surroundings.

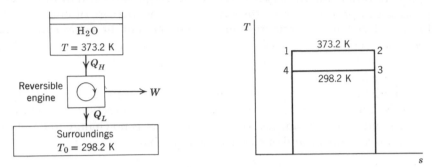

**Fig. 7.14**   Reversible heat transfer with the surroundings.

$$\Delta S_{\text{system}} = -6.0480 \text{ kJ/K}$$

$$\Delta S_{\text{surr}} = 6.0480 \text{ kJ/K}$$

$$Q_{\text{to surroundings}} = T_0 \Delta S = 298.15 \, (6.0480) = 1803.2 \text{ kJ}$$

$$W = Q_H - Q_L = 2257 - 1803.2 = 453.8 \text{ kJ}$$

Since this is a reversible cycle, the engine could be reversed and operated as a heat pump. For this cycle the work input to the heat pump would be 453.8kJ.

## 7.9   Entropy Change of an Ideal Gas

Two very useful equations for computing the entropy change of an ideal gas can be developed from Eq. 7.7 by substituting Eqs. 5.17 and 5.21, as follows:

$$T \, ds = du + P \, dv$$

For an ideal gas

$$du = C_{vo} \, dT \quad \text{and} \quad \frac{P}{T} = \frac{R}{v}$$

Therefore,

$$ds = C_{vo}\frac{dT}{T} + \frac{R\,dv}{v}$$

$$s_2 - s_1 = \int_1^2 C_{vo}\frac{dT}{T} + R\ln\frac{v_2}{v_1} \tag{7.18}$$

Similarly

$$Tds = dh - vdP$$

For an ideal gas

$$dh = C_{po}\,dT \quad \text{and} \quad \frac{v}{T} = \frac{R}{P}$$

Therefore,

$$ds = C_{po}\frac{dT}{T} - R\frac{dP}{P} \tag{7.19}$$

$$s_2 - s_1 = \int_1^2 C_{po}\frac{dT}{T} - R\ln\frac{P_2}{P_1} \tag{7.20}$$

In order to integrate Eqs. 7.18 and 7.20, the relation between specific heat and temperature must be known. It follows from Eqs. 7.18 and 7.20 that when $C_{po}$ and $C_{vo}$ are assumed constant that the change in entropy is given by the relations

$$s_2 - s_1 = C_{po}\ln\left(\frac{T_2}{T_1}\right) - R\ln\left(\frac{P_2}{P_1}\right) \tag{7.21}$$

and

$$s_2 - s_1 = C_{vo}\ln\left(\frac{T_2}{T_1}\right) + R\ln\left(\frac{v_2}{v_1}\right) \tag{7.22}$$

For those cases in which it is not reasonable to assume constant specific heat, Eqs. 7.18 and 7.20 can be integrated by using the empirical specific heat equations of Table A.9, as illustrated by the following example.

### Example 7.3

Consider Example 5.6, in which oxygen is heated from 300 K to 1500 K. Assume that during this process the pressure dropped from 200 kPa to 150 kPa. Calculate the change in entropy per kg.

Using Eq. 7.20 and the appropriate equation from Table A.9,

$$\bar{s}_2 - \bar{s}_1 = \int_{T_1}^{T_2} \bar{C}_{po} \frac{dT}{T} - \bar{R} \ln \frac{P_2}{P_1}$$

$$= \left( 37.432 \ln \theta + \frac{0.020102}{1.5} \theta^{1.5} \right.$$

$$\left. + \frac{178.57}{1.5} \theta^{-1.5} - \frac{236.88}{2} \theta^{-2} \right) \Bigg|_{\theta_1 = 3}^{\theta_2 = 15} - 8.31434 \ln \frac{150}{200}$$

$$= 52.726 + 2.392 = 55.118 \text{ kJ/kmol K}$$

$$s_2 - s_1 = \frac{\bar{s}_1 - \bar{s}_1}{M} = \frac{55.118}{32} = 1.7224 \text{ kJ/kg K}$$

Unless one is using a digital computer, it is most inconvenient to integrate these specific heat equations for each entropy calculation. To avoid this difficulty, an approach has been developed that involves integrating Eq. 7.19 along the one-tenth megapascal (one bar) pressure line, from a reference temperature $T_0$ (at which $s$ is taken as zero) to any other temperature $T$. The entropy value so calculated is called the standard-state entropy, designated $s^0$,

$$s_T^0 = \int_{T_0}^{T} \frac{C_{po}}{T} dT \tag{7.23}$$

which is a function of temperature only, and can be so tabulated. It follows that the entropy change between any two states 1 and 2 is then given by the relation

$$s_2 - s_1 = (s_{T_2}^0 - s_{T_1}^0) - R \ln \left( \frac{P_2}{P_1} \right) \tag{7.24}$$

Values of $s^0$ for air are given in Table A.10. Similarly, values of standard-state entropy for other ideal gases are given on a mole basis in Table A.11. Thus, an ideal gas entropy change that includes the variation of specific heat with temperature can be readily evaluated from these data and Eq. 7.24.

### Example 7.4

Calculate the change in entropy per kg as air is heated from 300 K to 600 K while pressure drops from 400 kPa to 300 kPa, assuming

1. constant specific heat
2. variable specific heat

1. From Table A.8 for air at 300 K,

$$C_{po} = 1.0035 \text{ kJ/kg K}$$

Therefore, using Eq. 7.21,

$$s_2 - s_1 = 1.0035 \ln\left(\frac{600}{300}\right) - 0.287 \ln\left(\frac{300}{400}\right) = 0.7781 \text{ kJ/kg K}$$

2. From Table A.10,

$$s^0_{T_1} = 2.5153 \text{ kJ/kg K} \qquad s^0_{T_2} = 3.2223 \text{ kJ/kg K}$$

Using Eq. 7.24,

$$s_2 - s_1 = 3.2223 - 2.5153 - 0.287 \ln\left(\frac{300}{400}\right) = 0.7896 \text{ kJ/kg K}$$

The *Air Tables* can be used for reversible adiabatic processes by employing the relative pressure $P_r$ and relative specific volume $v_r$. The definition of these terms and the derivation follow.

For the reversible adiabatic process

$$T\,ds = dh - v\,dP = 0$$

Therefore

$$dh = C_{po}\,dT = v\,dP = RT\frac{dP}{P}$$

$$\frac{dP}{P} = \frac{C_{po}}{R}\frac{dT}{T}$$

Let this equation be integrated between a reference state having a temperature $T_0$ and a pressure $P_0$, and a given arbitrary state having a temperature $T$ and a pressure $P$. Then

$$\ln\frac{P}{P_0} = \frac{1}{R}\int_{T_0}^{T} C_{po}\frac{dT}{T}$$

The right side of this equation is a function of temperature only. The relative pressure $P_r$ is defined as

$$\ln P_r \equiv \ln \frac{P}{P_0} = \frac{1}{R} \int_{T_0}^{T} C_{po} \frac{dT}{T} = \frac{s_T^0}{R} \qquad (7.25)$$

Thus, a value of $P_r$ can be tabulated as a function of temperature.

If we consider two states, 1 and 2, along a constant-entropy line it follows from Eq. 7.25 that

$$\frac{P_1}{P_2} = \left(\frac{P_{r1}}{P_{r2}}\right)_{s=\text{constant}} \qquad (7.26)$$

This equation states that the ratio of the relative pressures for two states having the same entropy is equal to the ratio of the absolute pressures.

The development of the relative specific volume is similar, and the ratio of the relative specific volumes $v_r$, in an isentropic process is equal to the ratio of the specific volumes, That is,

$$\frac{v_1}{v_2} = \left(\frac{v_{r1}}{v_{r2}}\right)_{s=\text{constant}} \qquad (7.27)$$

The isentropic functions $P_r$ and $v_r$ are tabulated for air, Table A.10, but are not listed in the other gas tables in the Appendix, A.11. In analyzing an isentropic process for these gases, it is necessary to use Eq. 7.24 with the left side of the equation equal to zero and the standard state entropies from Table A.11.

**Example 7.5**

One kg of air is contained in a cylinder fitted with a piston, at a pressure of 400 kPa and a temperature of 600 K. The air is now expanded to 150 kPa in a reversible, adiabatic process. Calculate the work done by the air.

Choosing the 1 kg of air as the system, we write

First law:

$$0 = u_2 - u_1 + w$$

Second law:

$$s_2 = s_1$$

Property relation: Air Tables
From Table A.10,

$$u_1 = 434.80 \text{ kJ/kg}; \quad P_{r1} = 16.278$$

From Eq. 7.26,

$$P_{r2} = P_{r1} \times \frac{P_2}{P_1} = 16.278 \times \frac{150}{400} = 6.104$$

From Table A.10,

$$T_2 = 457 \text{ K}; \qquad u_2 = 327.79$$

Therefore

$$w = 434.80 - 327.79 = 107.01 \text{ kJ/kg}$$

At this point it is advantageous to introduce the specific heat ratio $k$, which is defined as the ratio of the constant-pressure to the constant-volume specific heat at zero pressure.

$$k = \frac{C_{po}}{C_{vo}} \tag{7.28}$$

Since the difference between $C_{po}$ and $C_{vo}$ is a constant (Eq. 5.24) and since $C_{po}$ and $C_{vo}$ are functions of temperature, it follows that $k$ is also a function of temperature. However, when we consider the specific heat to be constant, $k$ is also constant.

From the definition of $k$ and Eq. 5.24 it follows that

$$C_{vo} = \frac{R}{k-1}; \quad C_{po} = \frac{kR}{k-1} \tag{7.29}$$

Some very useful and simple relations for the reversible adiabatic process can be developed when the specific heats are assumed to be constant.

For the reversible adiabatic process, $ds = 0$. Therefore,

$$T \, ds = du + P \, dv = C_{vo} \, dT + P \, dv = 0$$

From the equation of state for an ideal gas,

$$dT = \frac{1}{R}(P \, dv + v \, dP)$$

Therefore,

$$\frac{C_{vo}}{R}(P \, dv + v \, dP) + P \, dv = 0$$

Substituting Eq. 7.29 into the expression and rearranging,

$$\frac{1}{k-1}(P\,dv+v\,dP)+P\,dv=0$$

$$v\,dP+kP\,dv=0$$

$$\frac{dP}{P}+k\frac{dv}{v}=0$$

Since $k$ is constant when the specific heat is constant, this equation can be integrated under these conditions to give

$$Pv^k=\text{constant} \tag{7.30}$$

Equation 7.30 holds for all reversible adiabatic processes involving an ideal gas with constant specific heat. It is usually advantageous to express this constant in terms of the initial and final states.

$$Pv^k=P_1v_1{}^k=P_2v_2{}^k=\text{constant} \tag{7.31}$$

From this equation and the ideal gas equation of state the following expressions relating the initial and final states of an isentropic process can be derived.

$$\frac{P_2}{P_1}=\left(\frac{v_1}{v_2}\right)^k=\left(\frac{V_1}{V_2}\right)^k \tag{7.32}$$

$$\frac{T_2}{T_1}=\left(\frac{P_2}{P_1}\right)^{(k-1)/k}=\left(\frac{v_1}{v_2}\right)^{k-1} \tag{7.33}$$

With the assumption of constant specific heat some convenient equations can be derived for the work done by an ideal gas during an adiabatic process. Consider first a system consisting of an ideal gas that undergoes a process in which work is done only at the moving boundary.

$$_1Q_2=m(u_2-u_1)+{}_1W_2=0$$

$$_1W_2=-m(u_2-u_1)=-mC_{vo}(T_2-T_1)$$

$$=\frac{mR}{1-k}(T_2-T_1)=\frac{P_2V_2-P_1V_1}{1-k} \tag{7.34}$$

It should be noted that Eq. 7.34 applies to adiabatic processes only. Since no assumption was made regarding reversibility, it applies to both reversible and irreversible processes. Frequently Eq. 7.34 is derived for reversible processes by starting with the relation

$$_1W_2 = \int_1^2 P \, dV$$

for the system.

## 7.10   The Reversible Polytropic Process for an Ideal Gas

When a gas undergoes a reversible process in which there is heat transfer, the process frequently takes place in such a manner that a plot of $\log P$ vs. $\log V$ is a straight line as shown in Fig. 7.15. For such a process $PV^n = $ constant.

This is called a polytropic process. An example is the expansion of the combustion gases in the cylinder of a water-cooled reciprocating engine. If the pressure and volume during a polytropic process are measured during the expansion stroke, as might be done with an engine indicator, and the logarithms of the pressure and volume are plotted, the result would be similar to Fig. 7.15. From this figure it follows that

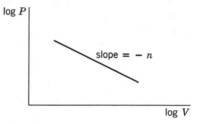

**Fig. 7.15**   Example of a polytropic process.

$$\frac{d \ln P}{d \ln V} = -n$$

$$d \ln P + nd \ln V = 0$$

If $n$ is a constant (which implies a straight line on the $\log P$ vs. $\log V$ plot), this can be integrated to give the following relation:

$$PV^n = \text{constant} = P_1 V_1{}^n = P_2 V_2{}^n \tag{7.35}$$

From this it is evident that the following relations can be written for a polytropic process.

$$\frac{P_2}{P_1} = \left(\frac{V_1}{V_2}\right)^n$$

$$\frac{T_2}{T_1} = \left(\frac{P_2}{P_1}\right)^{(n-1)/n} = \left(\frac{V_1}{V_2}\right)^{n-1} \tag{7.36}$$

For a system consisting of an ideal gas, the work done at the moving boundary during a reversible polytropic process can be derived from the relations

$$_1W_2 = \int_1^2 P \, dV \quad \text{and} \quad PV^n = \text{constant.}$$

$$_1W_2 = \int_1^2 P \, dV = \text{constant} \int_1^2 \frac{dV}{V^n}$$

$$= \frac{P_2 V_2 - P_1 V_1}{1-n} = \frac{mR(T_2 - T_1)}{1-n} \tag{7.37}$$

for any value of $n$ except $n = 1$.

The polytropic processes for various values of $n$ are shown in Fig. 7.16 on $P$-$v$ and $T$-$s$ diagrams. The values of $n$ for some familiar processes are given below.

| | |
|---|---|
| Isobaric process | $n = 0$ |
| Isothermal process | $n = 1$ |
| Isentropic process | $n = k$ |
| Isochoric process | $n = \infty$ |

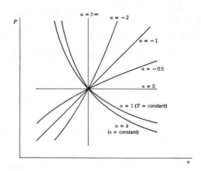

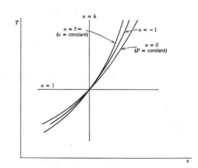

**Fig. 7.16**  Polytropic processes on $P$-$v$ and $T$-$s$ diagrams.

**Example 7.6**

Nitrogen is compressed in a reversible process in a cylinder from 100 kPa, 20°C, to 500 kPa. During the compression process the relation between pressure and volume is $PV^{1.3} = $ constant. Calculate the work and heat transfer per kilogram, and show this process on $P$-$v$ and $T$-$s$ diagrams.

$$P_1 = 100 \text{ kPa}; \qquad P_2 = 500 \text{ kPa}$$
$$T_1 = 293.2 \text{ K}$$

$T_2$ can be calculated from Eq. 7.36,

$$\frac{T_2}{T_1} = \left(\frac{P_2}{P_1}\right)^{(n-1)/n} = \left(\frac{500}{100}\right)^{(1.3-1)/1.3} = 1.4498$$

$$T_2 = 293.2 \times 1.4498 = 425 \text{ K}$$

For this process the work can be found using Eq. 7.37,

$$_1w_2 = \frac{R(T_2 - T_1)}{1 - n} = \frac{0.2968 \, (425 - 293.2)}{(1 - 1.3)} = -130.4 \text{ kJ/kg}$$

The heat transfer can be calculated using the first law. Let us assume $C_{vo}$ to be constant over this range in temperature.

$$_1q_2 = u_2 - u_1 + {}_1w_2 = C_{vo}(T_2 - T_1) + {}_1w_2$$

$$= 0.7448 \, (425 - 293.2) - 130.4 = -32.2 \text{ kJ/kg}$$

This process is shown on the $P$-$v$ and $T$-$s$ diagrams of Fig. 7.17.

The reversible isothermal process for an ideal gas is of particular interest. In this case

$$PV = \text{constant} = P_1V_1 = P_2V_2 \qquad (7.38)$$

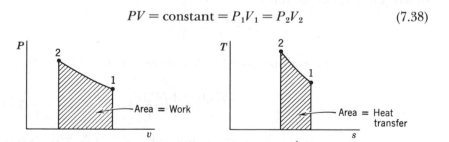

**Fig. 7.17**   Diagram for Example 7.6.

The work done at the boundary of a simple compressible system during a reversible isothermal process can be found by integrating the equation

$$_1W_2 = \int_1^2 P \, dV$$

The integration is as follows:

$$_1W_2 = \int_1^2 P \, dV = \text{constant} \int_1^2 \frac{dV}{V} = P_1 V_1 \ln \frac{V_2}{V_1} = P_1 V_1 \ln \frac{P_1}{P_2} \quad (7.39)$$

or

$$_1W_2 = mRT \ln \frac{V_2}{V_1} = mRT \ln \frac{P_1}{P_2} \quad (7.40)$$

Since there is no change in internal energy or enthalpy in an isothermal process, the heat transfer is equal to the work (neglecting changes in kinetic and potential energy). Therefore, we could have derived Eq. 7.39 by calculating the heat transfer.

For example, using Eq. 7.5

$$\int_1^2 T \, ds = {}_1q_2 = \int_1^2 du + \int_1^2 P \, dv$$

But $du = 0$ and $Pv = \text{constant} = P_1 v_1 = P_2 v_2$

$$_1q_2 = \int_1^2 P \, dv = P_1 v_1 \ln \frac{v_2}{v_1}$$

which yields the same result as Eq. 7.39.

### 7.11   The Second Law of Thermodynamics for a Control Volume

The second law of thermodynamics can be applied to a control volume by a similar procedure to that used in writing the first law for a control volume in Section 5.10. The second law for a system analysis has been stated (Eq. 7.12) in the form

$$dS = \frac{\delta Q}{T} + \frac{\delta LW}{T}$$

For a change in entropy, $S_2 - S_1$, that occurs during a time interval $\delta t$ we can write

$$\frac{S_2 - S_1}{\delta t} = \frac{1}{\delta t}\left(\frac{\delta Q}{T}\right) + \frac{1}{\delta t}\left(\frac{\delta LW}{T}\right) \quad (7.41)$$

Consider the system and control volume shown in Fig. 7.18. During time $\delta t$ the mass $\delta m_i$ enters the control volume across the discrete area

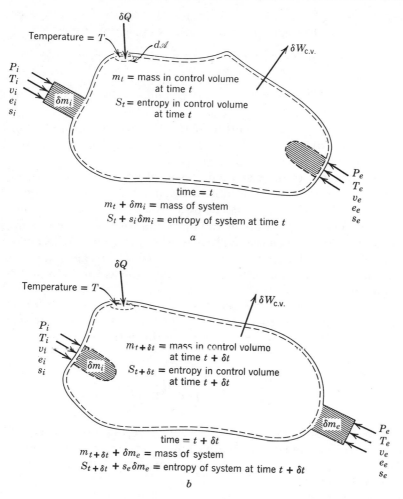

**Fig. 7.18**  Schematic diagram for a second law analysis of a control volume.

$\mathscr{A}_i$, the mass $\delta m_e$ leaves across the discrete area $\mathscr{A}_e$, an amount of heat $\delta Q$ is transferred to the system across an element of area where the surface temperature is $T$, and the work $\delta W$ is done by the system. As in the analysis of Section 5.10, we assume that the increment of mass $\delta m_i$ has uniform properties, and similarly that $\delta m_e$ has uniform properties.

Let

$$S_t = \text{the entropy in the control volume at time } t$$

$$S_{t+\delta t} = \text{the entropy in the control volume at time } t + \delta t$$

Then
$$S_1 = S_t + s_i \delta m_i = \text{entropy of the system at time } t$$

$$S_2 = S_{t+\delta t} + s_e \delta m_e = \text{entropy of the system at time } t + \delta t$$

Therefore

$$S_2 - S_1 = (S_{t+\delta t} - S_t) + (s_e \delta m_e - s_i \delta m_i) \tag{7.42}$$

The term $(s_e \delta m_e - s_i \delta m_i)$ represents the net flow of entropy out of the control volume during $\delta t$ as the result of the masses $\delta m_e$ and $\delta m_i$ crossing the control surface.

The significance of the remaining terms of Eq. 7.41 must be carefully considered when applied to a control volume. When we write the terms $(\delta Q/T)$ and $(\delta LW/T)$ for a system we normally consider a quantity of mass at uniform temperature at any instant of time. As noted before, in the case of a control volume analysis we deviate from a strictly classical concept of thermodynamics, and are prepared to consider a variation in temperature throughout the control volume, the assumption of "local" equilibrium.

As a consequence, for the heat transfer term we must consider each element of area on the surface across which heat flows and the surface temperature of this local area. Therefore this term in Eq. 7.41 must be written as the summation of all local $(\delta Q/T)$ terms over the control surface, or

$$\frac{1}{\delta t}\left(\frac{\delta Q}{T}\right) = \frac{1}{\delta t} \int_{\mathscr{A}} \left(\frac{\delta Q/\mathscr{A}}{T}\right) d\mathscr{A} \tag{7.43}$$

Similarly, for the lost work term, which represents irreversibilities occurring inside the control volume, the result must be written as the summation of all local $(\delta LW/T)$ terms throughout the interior of the control volume, or

$$\frac{1}{\delta t}\left(\frac{\delta LW}{T}\right) = \frac{1}{\delta t} \int_{V} \left(\frac{\delta LW/V}{T}\right) dV \tag{7.44}$$

Substituting Eqs. 7.42, 7.43, and 7.44 into Eq. 7.41, we have

$$\left(\frac{S_{t+\delta t} - S_t}{\delta t}\right) + \frac{s_e \delta m_e}{\delta t} - \frac{s_i \delta m_i}{\delta t} = \frac{1}{\delta t} \int_{\mathscr{A}} \left(\frac{\delta Q/\mathscr{A}}{T}\right) d\mathscr{A} + \frac{1}{\delta t} \int_{V} \left(\frac{\delta LW/V}{T}\right) dV \tag{7.45}$$

In order to reduce this expression to a rate equation, we establish the limit for each term as $\delta t$ is made to approach zero, at which time the system and control volume coincide.

$$\lim_{\delta t \to 0} \left( \frac{S_{t+\delta t} - S_t}{\delta t} \right) = \frac{dS_{\text{c.v.}}}{dt}$$

$$\lim_{\delta t \to 0} \left( \frac{s_e \delta m_e}{\delta t} \right) = \dot{m}_e s_e$$

$$\lim_{\delta t \to 0} \left( \frac{s_i \delta m_i}{\delta t} \right) = \dot{m}_i s_i$$

$$\lim_{\delta t \to 0} \left[ \frac{1}{\delta t} \int_{\mathcal{A}} \left( \frac{\delta Q / \mathcal{A}}{T} \right) d\mathcal{A} \right] = \int_{\mathcal{A}} \left( \frac{\dot{Q}_{\text{c.v.}} / \mathcal{A}}{T} \right) d\mathcal{A}$$

$$\lim_{\delta t \to 0} \left[ \frac{1}{\delta t} \int_{V} \left( \frac{\delta LW / V}{T} \right) dV \right] = \int_{V} \left( \frac{L \dot{W}_{\text{c.v.}} / V}{T} \right) dV$$

The assumptions of uniform properties for the incremental mass $\delta m_i$ entering the control volume across area $\mathcal{A}_i$ and also for $\delta m_e$ leaving across area $\mathcal{A}_e$ reduce in the limit to the restriction of uniform properties across area $\mathcal{A}_i$ and also across $\mathcal{A}_e$ at an instant of time. The heat transfer term in the limit becomes the integral over the control surface of the local rate of heat transfer divided by the local surface temperature. Similarly, the internal irreversibility term becomes the integral throughout the control volume of the local lost work rate divided by the local interior temperature.

In utilizing these limiting values to express the rate form of the entropy equation for a control volume, we again include summation signs on the flow terms, to account for the possibility of additional flow streams entering or leaving the control volume. The resulting expression is

$$\frac{dS_{\text{c.v.}}}{dt} + \sum \dot{m}_e s_e - \sum \dot{m}_i s_i = \int_{\mathcal{A}} \left( \frac{\dot{Q}_{\text{c.v.}} / \mathcal{A}}{T} \right) d\mathcal{A} + \int_{V} \left( \frac{L \dot{W}_{\text{c.v.}} / V}{T} \right) dV \quad (7.46)$$

which, for our purposes, is the general expression of the second law entropy equation. In words, this expression states that the rate of change of entropy inside the control volume plus the net rate of entropy flow out is equal to the sum of two terms, the integrated heat transfer surface term and the positive internal irreversibility production term. Since this latter term is necessarily positive (or zero) and is difficult if not impossible to evaluate quantitatively, Eq. 7.46 is frequently written in the form

$$\frac{dS_{\text{c.v.}}}{dt} + \sum \dot{m}_e s_e - \sum \dot{m}_i s_i \geqslant \int_{\mathcal{A}} \left( \frac{\dot{Q}_{\text{c.v.}} / \mathcal{A}}{T} \right) d\mathcal{A} \quad (7.47)$$

where the equality applies to internally reversible processes and the inequality to internally irreversible processes.

Note that if there is no mass flow into or out of the control volume and if temperature is considered uniform at any instant of time, then Eq. 7.46 reduces to a rate form of the system entropy equation, Eq. 7.12. In this sense, Eq. 7.46 (or 7.47) can be considered a general expression, as was also pointed out for the control volume form of the first law.

Representation of Eq. 7.46 in terms of local fluid properties follows readily from the corresponding representation of the first law. As in Eq. 5.45 for energy, the total entropy $S_{c.v.}$ inside the control volume at any time may be expressed as the following volume integral:

$$S_{c.v.} = \int_V s\rho \, dV \tag{7.48}$$

As in Eq. 5.46 for energy, the entropy flow across each area of flow $\mathscr{A}$ is

$$\int_{\mathscr{A}} s\delta\dot{m} = \int_{\mathscr{A}} s\rho V_{rn} \, d\mathscr{A} \tag{7.49}$$

Considering each $V_{rn}$ as the outward-directed normal, Eqs. 7.48 and 7.49 may be substituted into Eq. 7.46, with the result being

$$\frac{d}{dt}\int_V s\rho \, dV + \int_{\mathscr{A}} s\rho V_{rn} \, d\mathscr{A} = \int_{\mathscr{A}} \left(\frac{\dot{Q}_{c.v.}/\mathscr{A}}{T}\right) d\mathscr{A} + \int_V \left(\frac{L\dot{W}_{c.v.}/V}{T}\right) dV \tag{7.50}$$

the second law entropy equation for a control volume stated in terms of local fluid properties. This form of the equation should prove beneficial in pointing out the assumptions implicit in Eq. 7.46 and the significance of the notation and terms in that expression.

## 7.12    The Steady-State, Steady-Flow Process and the Uniform-State, Uniform-Flow Process

We now consider the application of the second law control volume equation, Eq. 7.46 or 7.47, to the two control volume model processes developed in Chapter 5.

For the steady-state, steady-flow process, which has been defined in Section 5.11, we conclude that there is no change with time of the entropy per unit mass at any point within the control volume, and therefore the first term of Eq. 7.47 equals zero. That is

$$\frac{dS_{c.v.}}{dt} = \frac{d}{dt}\int_V s\rho \, dV = 0 \tag{7.51}$$

so that, for the SSSF process

$$\sum \dot{m}_e s_e - \sum \dot{m}_i s_i \geq \int_{\mathcal{A}} \left( \frac{\dot{Q}_{c.v.}/\mathcal{A}}{T} \right) d\mathcal{A} \tag{7.52}$$

in which the various mass flows, heat transfer rate, and states are all constant with time.

If in a steady-state, steady-flow process there is only one area over which mass enters the control volume at a uniform rate and only one area over which mass leaves the control volume at a uniform rate we can write

$$\dot{m}(s_e - s_i) \geq \int_{\mathcal{A}} \left( \frac{\dot{Q}_{c.v.}/\mathcal{A}}{T} \right) d\mathcal{A} \tag{7.53}$$

For an adiabatic process with these assumptions it follows that

$$s_e \geq s_i \tag{7.54}$$

where the equality holds for a reversible adiabatic process.

## Example 7.7

Steam enters a steam turbine at a pressure of 1 MPa, a temperature of 300°C, and a velocity of 50 m/s. The steam leaves the turbine at a pressure of 150 kPa and a velocity of 200 m/s. Determine the work per kilogram of steam flowing through the turbine, assuming the process to be reversible and adiabatic.

In solving a problem such as this it is usually advisable to draw a schematic diagram of the apparatus, and to show the process or, in the case of irreversible process, the various states involved on a $T$-$s$ diagram. The next step in the solution is to write the continuity equation, the first and second laws of thermodynamics, and the property relation as they apply to the particular problem.

The schematic diagram and a $T$-$s$ diagram are shown in Fig. 7.19. The various equations that apply to this process are:

Continuity eq.: $\qquad\qquad \dot{m}_e = \dot{m}_i = \dot{m}$

First law: $\qquad\qquad h_i + \dfrac{V_i^2}{2} = h_e + \dfrac{V_e^2}{2} + w$

Second law: $\qquad\qquad s_e = s_i$

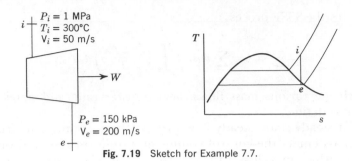

**Fig. 7.19** Sketch for Example 7.7.

Property relation: Steam tables
From the steam tables.

$$h_i = 3051.2 \text{ kJ/kg}; \qquad s_i = 7.1229 \text{ kJ/kg K}$$

The two properties known in the final state are pressure and entropy.

$$P_e = 0.15 \text{ MPa}; \qquad s_e = s_i = 7.1229 \text{ kJ/kg K}$$

Therefore, the quality and enthalpy of the steam leaving the turbine can be determined.

$$s_e = 7.1229 = s_g - (1 - x_e) \, s_{fg} = 7.2233 - (1 - x_e) \, 5.7897$$

$$(1 - x_e) = 0.01734$$

$$h_e = h_g - (1 - x_e) \, h_{fg} = 2693.6 - 0.01734 \, (2226.5)$$

$$= 2655.0 \text{ kJ/kg}$$

Therefore, the work per kg of steam for this isentropic process may be found using the equation for the first law as given above

$$w = 3051.2 + \frac{50 \times 50}{2 \times 1000} - 2655.0 - \frac{200 \times 200}{2 \times 1000} = 377.5 \text{ kJ/kg}$$

**Example 7.8**

Consider the reversible adiabatic flow of steam through a nozzle. Steam enters the nozzle at 1 MPa, 300°C, with a velocity of 30 m/s. The pressure of the steam at the nozzle exit is 0.3 MPa. Determine the exit velocity of the steam from the nozzle, assuming a reversible, adiabatic, steady-state, steady-flow process.

This process is shown schematically and on a *T-s* diagram in Fig. 7.20.

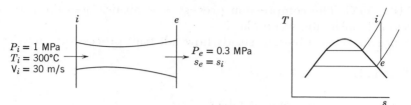

**Fig. 7.20**   Sketch for Example 7.8.

Since this is a steady-state, steady-flow process in which the work, the heat transfer, and the changes in potential energy are zero, we can write,

Continuity eq.:    $\dot{m}_e = \dot{m}_i = \dot{m}$

First law:    $$h_i + \frac{V_i^2}{2} = h_e + \frac{V_e^2}{2}$$

Second law:    $s_e = s_i$

Property relation: Steam tables

$$h_i = 3051.2 \text{ kJ/kg}; \qquad s_i = 7.1229 \text{ kJ/kg K}$$

The two properties that we know in the final state are

$$s_e = s_i = 7.1229 \text{ kJ/kg K} \qquad P_e = 0.3 \text{ MPa}$$

Therefore,

$$T_e = 159.6°C; \qquad h_e = 2781.2 \text{ kJ/kg}$$

Substituting into the equation for the first law as given above we have,

$$\frac{V_e^2}{2} = h_i - h_e + \frac{V_i^2}{2}$$

$$= 3051.2 - 2781.2 + \frac{30 \times 30}{2 \times 1000} = 270.5 \text{ kJ/kg}$$

$$V_e = 735.5 \text{ m/s}$$

**Example 7.9**

An inventor reports that he has a refrigeration compressor that receives saturated Freon-12 vapor at −20°C and delivers the vapor at

1 MPa, 50°C. The compression process is adiabatic. Does the process described violate the second law?

Since this is a steady-state, steady-flow, adiabatic process we can write:

Second law:      $s_e \geq s_i$

Property relation: Freon-12 tables
From the Freon-12 tables,

$$s_e = 0.7021 \text{ kJ/kg K}; \qquad s_i = 0.7082 \text{ kJ/kg K}$$

Therefore, $s_e < s_i$, whereas, for this process, the second law requires that $s_e \geq s_i$. The process described would involve a violation of the second law, and would not be possible.

For the uniform-state, uniform-flow process, which was described in Section 5.13 the second law for a control volume, Eq. 7.47, can be written in the following form.

$$\frac{d}{dt}[ms]_{\text{c.v.}} + \sum \dot{m}_e s_e - \sum \dot{m}_i s_i \geq \int_{\mathcal{A}} \left( \frac{\dot{Q}_{\text{c.v.}}/\mathcal{A}}{T} \right) d\mathcal{A} \qquad (7.55)$$

If this is integrated over the time interval $t$ we have

$$\int_0^t \frac{d}{dt}[ms]_{\text{c.v.}} \, dt = [m_2 s_2 - m_1 s_1]_{\text{c.v.}}$$

$$\int_0^t \left( \sum \dot{m}_e s_e \right) dt = \sum m_e s_e; \qquad \int_0^t \left( \sum \dot{m}_i s_i \right) dt = \sum m_i s_i \qquad (7.56)$$

Therefore, for this period of time $t$ we can write the second law for the uniform-state, uniform-flow process as

$$[m_2 s_2 - m_1 s_1]_{\text{c.v.}} + \sum m_e s_e - \sum m_i s_i \geq \int_0^t \left[ \int_{\mathcal{A}} \left( \frac{\dot{Q}_{\text{c.v.}}/\mathcal{A}}{T} \right) d\mathcal{A} \right] dt \qquad (7.57)$$

However, since in this process the temperature is uniform throughout the control volume at any instant of time, the integral on the right reduces to

$$\int_0^t \left[ \int_{\mathcal{A}} \left( \frac{\dot{Q}_{\text{c.v.}}/\mathcal{A}}{T} \right) d\mathcal{A} \right] dt = \int_0^t \left[ \frac{1}{T} \int_{\mathcal{A}} \left( \frac{\dot{Q}_{\text{c.v.}}}{\mathcal{A}} \right) d\mathcal{A} \right] dt = \int_0^t \left( \frac{\dot{Q}_{\text{c.v.}}}{T} \right) dt$$

and therefore the second law for the uniform-state, uniform-flow process can be written

$$[m_2 s_2 - m_1 s_1]_{\text{c.v.}} + \sum m_e s_e - \sum m_i s_i \geq \int_0^t \left(\frac{\dot{Q}_{\text{c.v.}}}{T}\right) dt \qquad (7.58)$$

By introducing the lost work we can write this as an equality. In doing so, we note that since the temperature is uniform throughout the control volume at any instant of time,

$$\int_0^t \left[ \int_V \left(\frac{L\dot{W}_{\text{c.v.}}/V}{T}\right) dV \right] dt = \int_0^t \left[ \frac{1}{T} \int_V \left(\frac{L\dot{W}_{\text{c.v.}}}{V}\right) dV \right] dt = \int_0^t \left(\frac{L\dot{W}_{\text{c.v.}}}{T}\right) dt$$

Therefore,

$$[m_2 s_2 - m_1 s_1]_{\text{c.v.}} + \sum m_e s_e - \sum m_i s_i = \int_0^t \left(\frac{\dot{Q}_{\text{c.v.}} + L\dot{W}_{\text{c.v.}}}{T}\right) dt \qquad (7.59)$$

## 7.13   The Reversible Steady-State, Steady-Flow Process

An expression can be derived for the work in a reversible steady-state, steady-flow process which is of great help in understanding the significant variables in such a process. We have noted that when a steady-state, steady-flow process involves a single flow of fluid into and out of the control volume, the first law can be written, Eq. 5.52,

$$q + h_i + \frac{V_i^2}{2} + gZ_i = h_e + \frac{V_e^2}{2} + gZ_e + w$$

and the second law, Eq. 7.53 is

$$\dot{m}(s_e - s_i) \geq \int \left(\frac{\dot{Q}_{\text{c.v.}}/\mathscr{A}}{T}\right) d\mathscr{A}$$

Let us now consider two types of flow, a reversible adiabatic process and a reversible isothermal process.

If the process is reversible and adiabatic, the second law equation above reduces to

$$s_e = s_i$$

and it follows from the property relation

$$T \, ds = dh - v \, dP$$

that

$$h_e - h_i = \int_i^e v\,dP \tag{7.60}$$

Substituting these relations into Eq. 5.52 and noting that $q = 0$ we have

$$w = (h_i - h_e) + \frac{(V_i^2 - V_e^2)}{2} + g(Z_i - Z_e)$$

$$= -\int_i^e v\,dP + \frac{(V_i^2 - V_e^2)}{2} + g(Z_i - Z_e) \tag{7.61}$$

If, instead, the process is reversible and isothermal, the second law reduces to

$$\dot{m}(s_e - s_i) = \frac{1}{T}\int_{\mathscr{A}} (\dot{Q}_{\text{c.v.}}/\mathscr{A})\,d\mathscr{A} = \frac{\dot{Q}_{\text{c.v.}}}{T}$$

or

$$T(s_e - s_i) = \frac{Q_{\text{c.v.}}}{\dot{m}} = q \tag{7.62}$$

and the property relation can be integrated to give

$$T(s_e - s_i) = (h_e - h_i) - \int_i^e v\,dP \tag{7.63}$$

Substituting Eqs. 7.62 and 7.63 into the first law, Eq. 5.52, then results in the same expression as for the reversible adiabatic process, Eq. 7.61. We further note that any other reversible process can be constructed, in the limit, from a series of alternate adiabatic and isothermal processes. Thus we may conclude that Eq. 7.61 is valid for any reversible, steady-state, steady-flow process, without the restriction that it be either adiabatic or isothermal.

This expression has a wide range of application. If we consider a reversible steady-state, steady-flow process in which the work is zero (such as flow through a nozzle) and the fluid is incompressible ($v = $ constant), Eq. 7.61 can be integrated to give

$$v(P_e - P_i) + \frac{(V_e^2 - V_i^2)}{2} + g(Z_e - Z_i) = 0 \tag{7.64}$$

This is known as the Bernoulli equation, (after Daniel Bernoulli) and is a very important equation in fluid mechanics.

Equation 7.61 is also frequently applied to the large class of flow processes involving work (such as turbines and compressors) in which changes in kinetic and potential energies of the working fluid are small. The model process for these machines is then a reversible, SSSF process with no change in kinetic or potential energy (and commonly, although not necessarily, adiabatic as well). For this process, Eq. 7.61 reduces to the form

$$w = -\int_i^e v\, dP \tag{7.65}$$

From this result, we conclude that the shaft work associated with this type of process is closely related to the specific volume of the fluid during the process. To further amplify this point, consider the simple steam power plant shown in Fig. 7.21. Suppose this is an ideal power plant, with no pressure drop in the piping, the boiler, or the condenser. Thus the pressure increase in the pump is equal to the pressure decrease in the turbine. Neglecting kinetic and potential energy changes, the work done in each of these processes is given by Eq. 7.65. Since the pump handles liquid, which has a very small specific volume as compared to the vapor

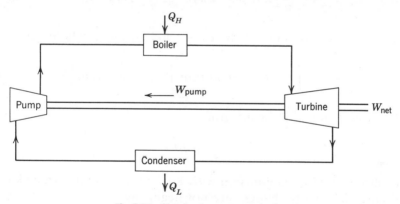

**Fig. 7.21**   Simple steam power plant.

that flows through the turbine, the power input to the pump is much less than the power output of the turbine, the difference being the net power output of the power plant.

This same line of reasoning can be qualitatively applied to actual devices that involve steady-state, steady-flow processes, even though the processes are not exactly reversible and adiabatic.

### Example 7.10

Calculate the work per kilogram to pump water isentropically from 100 kPa, 30°C to 5 MPa.

From the steam tables, $v_i = 0.001\ 004$ m³/kg. Assuming the specific volume to remain constant and using Eq. 7.65 we have,

$$-w = \int_1^2 v\, dP = v(P_2 - P_1) = 0.001\ 004\ (5000 - 100) = 4.92 \text{ kJ/kg}$$

As a final application of Eq. 7.61, we recall the reversible polytropic process for an ideal gas, discussed in Section 7.10 for a system process. For the SSSF process with no change in kinetic and potential energies, from the relations

$$w = -\int_i^e v\, dP \quad \text{and} \quad Pv^n = \text{constant} = C^n.$$

$$w = -\int_i^e v\, dP = -C \int_i^e \frac{dP}{P^{1/n}}$$

$$= -\frac{n}{n-1}(P_e v_e - P_i v_i) = -\frac{nR}{n-1}(T_e - T_i) \tag{7.66}$$

If the process is isothermal, then $n = 1$ and the integral becomes

$$w = -\int_i^e v\, dP = -\text{constant} \int_i^e \frac{dP}{P} = -P_i v_i \ln\frac{P_e}{P_i} \tag{7.67}$$

These evaluations of the integral

$$\int_i^e v\, dP$$

may also be used in conjunction with Eq. 7.61 in cases for which kinetic and potential energy changes are not negligibly small.

## 7.14   Principle of the Increase of Entropy for a Control Volume

The principle of the increase of entropy for a system analysis was discussed in Section 7.8. The same general conclusion is reached in the case of a control volume analysis. To demonstrate this, consider a control volume, Fig. 7.22, which exchanges both mass and heat with its surroundings. At the point in the surroundings where the heat transfer occurs the temperature is $T_0$. From Eq. 7.47 the second law for this process is

$$\frac{dS_{c.v.}}{dt} + \sum \dot{m}_e s_e - \sum \dot{m}_i s_i \geq \int_{\mathscr{A}} \left(\frac{\dot{Q}_{c.v.}/\mathscr{A}}{T}\right) d\mathscr{A}$$

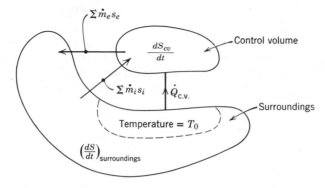

**Fig. 7.22**  Entropy change for a control volume plus surroundings.

We recall that the first term represents the rate of change of entropy within the control volume, and the next terms the net entropy flow out of the control volume as the result of the mass flow. Therefore, for the surroundings we can write

$$\frac{dS_{\text{surr}}}{dt} = \sum \dot{m}_e s_e - \sum \dot{m}_i s_i - \frac{\dot{Q}_{\text{c.v.}}}{T_0} \tag{7.68}$$

Adding Eqs. 7.47 and 7.68 we have

$$\frac{dS_{\text{c.v.}}}{dt} + \frac{dS_{\text{surr}}}{dt} \geqslant \int_{\mathcal{A}} \left( \frac{\dot{Q}_{\text{c.v.}}/\mathcal{A}}{T} \right) d\mathcal{A} - \frac{\dot{Q}_{\text{c.v.}}}{T_0} \tag{7.69}$$

Since $\dot{Q}_{\text{c.v.}} > 0$ when $T_0 > T$, and $\dot{Q}_{\text{c.v.}} < 0$ when $T_0 < T$, it follows that

$$\frac{dS_{\text{c.v.}}}{dt} + \frac{dS_{\text{surr}}}{dt} \geqslant 0 \tag{7.70}$$

which can be termed the general statement of the principle of the increase of entropy.

## 7.15  Efficiency

In Chapter 6 we noted that the second law of thermodynamics led to the concept of thermal efficiency for a heat engine cycle, namely

$$\eta_{\text{th}} = \frac{W_{\text{net}}}{Q_H}$$

where $W_{net}$ is the net work of the cycle and $Q_H$ is the heat transfer from the high-temperature body.

In this chapter we have extended our consideration of the second law to processes, and this leads us now to consider the efficiency of a process. For example, we might be interested in the efficiency of a turbine in a steam power plant, or the compressor in a gas turbine engine.

In general we can say that the efficiency of a machine in which a process takes place involves a comparison between the actual performance of the machine under given conditions and the performance that would have been achieved in an ideal process. It is in the definition of this ideal process that the second law becomes a major consideration. For example, a steam turbine is intended to be an adiabatic machine. The only heat transfer is the unavoidable heat transfer that takes place between the given turbine and the surroundings. Also, we note that for a given steam turbine operating in a steady-state, steady-flow manner, the state of the steam entering the turbine and the exhaust pressure are fixed. Therefore the ideal process would be a reversible adiabatic process, which is an isentropic process, between the inlet state and the turbine exhaust pressure. If we denote the actual work done per unit mass of steam flow through the turbine as $w_a$ and the work that would have been done in a reversible adiabatic process between the inlet state and the turbine exhaust pressure as $w_s$ the efficiency of the turbine is defined as

$$\eta_{turbine} = \frac{w_a}{w_s} \tag{7.71}$$

The same relation would hold for a gas turbine.

A few other examples may help to clarify this point. In a nozzle the objective is to have the maximum kinetic energy leaving the nozzle for the given inlet conditions and exhaust pressure. The nozzle is also an adiabatic device and therefore the ideal process is a reversible adiabatic or isentropic process. The efficiency of a nozzle is the ratio of the actual kinetic energy leaving the nozzle, $V_a^2/2$, to the kinetic energy for an isentropic process between the same inlet conditions and exhaust pressure, namely, $V_s^2/2$.

$$\eta_{nozzle} = \frac{V_a^2/2}{V_s^2/2} \tag{7.72}$$

In compressors for air or other gases, there are two ideal processes to which the actual performance can be compared. If no effort is made to cool the gas during compression (that is, when the process is adiabatic), the ideal process is a reversible adiabatic or isentropic process between the given inlet state and exhaust pressure. If we denote the work per unit mass of gas flow through the compressor for this isentropic process as $w_s$, and the actual work as $w_a$ (the actual work input will be greater than the work input for an isentropic process), the efficiency is defined by the relation

$$\eta_{\text{adiabatic compressor}} = \frac{w_s}{w_a} \tag{7.73}$$

If an attempt is made to cool the air during compression by use of a water jacket or fins, the ideal process is considered a reversible isothermal process. If $w_t$ is the work for the reversible isothermal process between the given inlet state and exhaust pressure, and $w_a$ the actual work, the efficiency is defined by the relation

$$\eta_{\text{cooled compressor}} = \frac{w_t}{w_a} \tag{7.74}$$

Thus we note that the efficiency of a device that involves a process (rather than a cycle) involves a comparison of the actual performance to that which would be achieved in a related, but well defined ideal process.

**Example 7.11**

A steam turbine receives steam at a pressure of 1 MPa, 300°C. The steam leaves the turbine at a pressure of 15 kPa. The work output of the turbine is measured and is found to be 600 kJ/kg of steam flowing through the turbine. Determine the efficiency of the turbine.

The efficiency of the turbine is given by Eq. 7.71.

$$\eta_{\text{turbine}} = \frac{w_a}{w_s}$$

Thus a determination of the turbine efficiency involves a calculation of the work which would be done in an isentropic process between the given inlet state and final pressure. For this isentropic process:

Continuity eq.:    $\dot{m}_1 = \dot{m}_2 = \dot{m}$

First law:    $h_1 = h_{2s} + w_s$

Second law:    $s_1 = s_{2s}$

Property relation: Steam tables

$$h_1 = 3051.2 \text{ kJ/kg}; \qquad s_1 = 7.1229 \text{ kJ/kg K}$$

$$s_{2s} = s_1 = 7.1229 = 8.0085 - (1 - x_{2s})\, 7.2536$$

$$(1 - x_{2s}) = 0.1221$$

$$h_{2s} = 2599.1 - 0.1221\,(2373.1) = 2309.4 \text{ kJ/kg}$$

$$w_s = h_1 - h_{2s} = 3051.2 - 2309.4 = 741.8 \text{ kJ/kg}$$

$$w_a = 600 \text{ kJ/kg}$$

$$\eta_{\text{turbine}} = \frac{w_a}{w_s} = \frac{600}{741.8} = 0.809 = 80.9\%$$

## 7.16  Some General Comments Regarding Entropy

It is quite possible at this point that a student may have a good grasp of the material that has been covered, and yet he may have only a vague understanding of the significance of entropy. In fact, the question "What is entropy?" is frequently raised by students with the implication that no one really knows! This section has been included in an attempt to give insight into the qualitative and philosophical aspects of the concept of entropy, and to illustrate the broad application of entropy to many different disciplines.

First of all, we recall that the concept of energy rises from the first law of thermodynamics and the concept of entropy from the second law of thermodynamics. Actually it is just as difficult to answer the question "What is energy?" as it is to answer the question "What is entropy?" However, since we regularly use the term energy and are able to relate this term to phenomena that we observe every day, the word energy has a definite meaning to us and thus serves as an effective vehicle for thought and communication. The word entropy could serve in the same capacity. If, when we observed a highly irreversible process (such as cooling coffee by placing an ice cube in it), we said, "That surely increases the entropy," we would soon be as familiar with the word *entropy* as we are with the word *energy*. In many cases when we speak about a higher efficiency we are actually speaking about accomplishing a given objective with a smaller total increase in entropy.

A second point to be made regarding entropy is that in statistical thermodynamics, the property entropy is defined in terms of probability. From this point of view the net increase in entropy that occurs during an irreversible process can be associated with a change of state from a less probable state to a more probable state. For example, to use a previous example, one is more likely to find gas on both sides of the ruptured membrane of Fig. 6.11 than to find a gas on one side and a vacuum on the other. Thus, when the membrane ruptures, the direction of the process is from a less probable state to a more probable state and associated with this process is an increase in entropy. Similarly, the more probable state is that a cup of coffee will be at the same temperature as its surroundings than at a higher (or lower) temperature. Therefore, as the coffee cools as the result of a transferring of heat to the surroundings, there is a change from a less probable to a more probable state, and associated with this is an increase in entropy.

The final point to be made is that the second law of thermodynamics and the principle of the increase of entropy have philosophical implications. Does the second law of thermodynamics apply to the universe as a whole? Are there processes unknown to us that occur somewhere in the universe, such as "continual creation," that have a decrease in entropy associated with them, and thus offset the continual increase in entropy that is associated with the natural processes that are known to us? If the second law is valid for the universe (we of course do not know if the universe can be considered as an isolated system) how did it get in the state of low entropy? On the other end of the scale, if all processes known to us have an increase in entropy associated with them, what is the future of the natural world as we know it?

Quite obviously it is impossible to give conclusive answers to these questions on the basis of the second law of thermodynamics alone. However, the authors see the second law of thermodynamics as man's description of the prior and continuing work of a creator, who also holds the answer to the future destiny of man and the universe.

## PROBLEMS

7.1  A Carnot cycle heat engine receives 500 kJ from a reservoir at 500°C, and rejects heat at 25°C.

(a) Show the cycle on a $T$-$s$ diagram, considering the working fluid as the system.

(b) Calculate the work and efficiency of the cycle.

(c) Calculate the change in entropy of the high-temperature and low-temperature reservoirs.

(d) Suppose the engine operates on the cycle indicated above, but the temperature of the high-temperature reservoir is increased to 1000°C. (Heat is then transferred from the reservoir at 1000°C to the working fluid at 500°C.) The low-temperature reservoir remains at 25°C. Determine the entropy change for each reservoir.

7.2   Consider a Carnot cycle heat engine using steam as the working fluid and having a thermal efficiency of 20%. Heat is transferred to the working fluid at 250°C and during this process the working fluid changes from saturated liquid to saturated vapor.

(a) Show this cycle on a *T-s* diagram that includes the saturated-liquid and saturated-vapor lines.

(b) Calculate the quality at the beginning and end of the heat-rejection process.

(c) Calculate the work per kg of steam.

7.3   A Carnot cycle refrigerator utilizes Freon-12 as the working fluid. Heat is rejected from the Freon at 35°C, and during this process the Freon changes from saturated vapor to saturated liquid. Heat transfer to the Freon is at −20°C.

(a) Show this cycle on a *T-s* diagram.

(b) What is the quality at the beginning and end of the isothermal process at −20°C?

(c) What is the coefficient of performance of this cycle?

7.4   A cylinder fitted with a piston contains steam initially at 1 MPa, 250°C, at which point the volume is 25 litres. The steam then expands in a reversible adiabatic process to a final pressure of 150 kPa. Determine the work done during the process.

7.5   Saturated vapor ammonia at 200 kPa is compressed in a cylinder by a piston to a pressure of 1.6 MPa in a reversible adiabatic process. Determine the work done per kg of ammonia.

7.6   A pressure vessel contains steam at 1.6 MPa, 400°C. A valve at the top of the pressure vessel is opened, allowing steam to escape. Assume that at any instant the steam that remains in the pressure vessel has undergone a reversible adiabatic process. Determine the fraction of steam that has escaped when the steam remaining in the pressure vessel is saturated vapor.

7.7   Freon-12 at 20°C, 500 kPa, is contained in a cylinder fitted with a piston. The volume at this condition is 10 litres. The Freon then undergoes a reversible, adiabatic expansion to −30°C. At the completion of this process, the Freon is allowed to warm back to 20°C at constant pressure. Show the two processes on a *T-s* diagram, and calculate the work and heat transfer for each process.

7.8   A mass of 40 kg is held in position in a length of 125 mm diameter tube by a pin, as shown in Fig. 7.23. Below this mass is a volume of 15 litres of steam at 5 MPa, 400°C. The pin is released and the mass is accelerated upwards, leaving the top of the tube with a certain velocity. If the steam undergoes a reversible adiabatic expansion, until a pressure of 600 kPa is reached, what is the exit velocity of the mass?

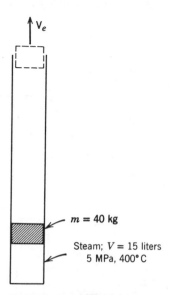

$v_e$

$m = 40$ kg

Steam; $V = 15$ liters
5 MPa, 400° C

**Fig. 7.23**  Sketch for Problem 7.8.

7.9  A cylinder fitted with a piston contains one kg of Freon-12 at 0.1 MPa, 90°C. The piston is slowly moved and the necessary heat transfer takes place so that the Freon-12 is compressed in a reversible isothermal process until the Freon-12 exists as saturated vapor.

(a) Show the process on a temperature-entropy diagram that is approximately to scale.

(b) Calculate the final pressure and specific volume of the Freon-12.

(c) Determine the work and the heat transfer for this process.

7.10  Consider Problem 4.12, involving steam expanding in a reversible isothermal process. Solve the problem without using the relation

$$_1W_2 = \int_1^2 P \, dV.$$

7.11  Tank $A$ initially contains 5 kg of steam at 800 kPa, 300°C, and is connected through a valve to a cylinder fitted with a frictionless piston, as shown in Fig. 7.24. A pressure of 200 kPa is required to balance the weight of the piston.

The connecting valve is opened until the pressure in $A$ equals 200 kPa. Assume the entire process to be adiabatic, and that the steam that finally remains in $A$ has undergone a reversible adiabatic process.

Determine the work done against the piston and the final temperature of the steam in cylinder $B$.

7.12  Repeat Problem 7.11, but let the piston initially rest on stops so that the initial volume of the cylinder $B$ is 0.3 m³, as shown in Fig. 7.25. Assume that this volume is initially evacuated.

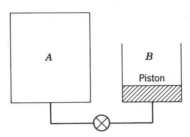

**Fig. 7.24**    Sketch for Problem 7.11.

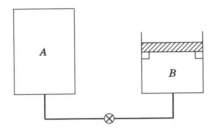

**Fig. 7.25**    Sketch for Problem 7.12.

7.13  A neophyte engineer has a problem to solve which involves the entropy of superheated steam at pressures below 10 kPa, but he finds that the steam table which he has does not give the properties of superheated steam below this pressure.

(a) How would you advise him to proceed?

(b) What assumption does this procedure involve?

(c) Set up, in accordance with the assumptions made, a line similar to the 10 kPa line in Table A.1.3 which gives $v$, $u$, $h$, and $s$ at 5 kPa at temperatures of 100°C, 400°C, 1000°C.

7.14  It is desired to quickly cool a given quantity of material to 10°C. The required heat transfer is 200 kJ. One possibility is to immerse it in a mixture of ice and water, in which case the heat transfer from the material results in the melting of ice. Another possibility is to cool the material by evaporating Freon-12 at −20°C, in which case the heat transfer results in changing Freon-12 from saturated liquid to saturated vapor. A third possibility is to do the same with liquid nitrogen at 101.3 kPa pressure.

(a) Calculate the change of entropy of the cooling medium in each of the three cases.

(b) What is the significance of these results?

7.15  Consider the process shown in Fig. 7.26. Tank $A$ is insulated, has a volume of 0.6 m³ and is initially filled with steam at 1.4 MPa, 300°C. Tank $B$ is uninsulated, has a volume of 0.3 m³ and is initially filled with steam at 0.2 MPa, 200°C. A valve interconnecting the two tanks is then opened and steam flows from $A$ to $B$ until the temperature in $A$ is 250°C, at which

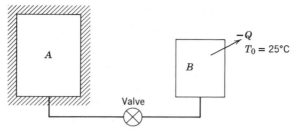

**Fig. 7.26** Sketch for Problem 7.15.

time the valve is closed. During this time, heat is transferred from $B$ to the surroundings at 25°C such that the temperature in $B$ remains at 200°C. It may be assumed that the steam remaining in $A$ has undergone a reversible adiabatic process. Determine

(a) The final pressure in each of the tanks.

(b) The final mass in $B$.

(c) The net entropy change for the process (system plus surroundings).

7.16 An insulated cylinder fitted with a piston has an initial volume of 0.15 m³ and contains steam at 400 kPa, 200°C. The steam is now expanded adiabatically, and during this process the work output is carefully measured and found to be 30 kJ. It is claimed that at the final state the steam is in the two-phase (liquid plus vapor) region. What is your evaluation of this claim?

7.17 Consider the scheme shown in Fig. 7.27 for raising a weight by heat transfer from the reservoir at 150°C to the Freon-12. The pressure on the Freon-12 due to the weight plus the atmosphere is 1.4 MPa. Initially the

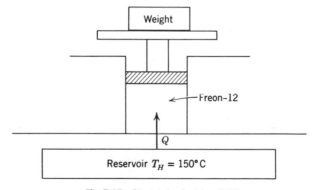

**Fig. 7.27** Sketch for Problem 7.17.

temperature of the Freon-12 is 70°C and its volume is 10 litres. Heat is transferred until the temperature of the Freon-12 is 150°C.

(a) Determine the heat transfer and work for this process.

(b) Determine the net change of entropy.

(c) Devise a scheme for accomplishing this with no net change of entropy.

7.18 A 10 kg bar of cast iron initially at 400°C is quenched in a 200 litre water tank initially at 25°C. Assuming no heat transfer to the surroundings, calculate the net entropy change for the process. Use a specific heat of 0.5 kJ/kg K for cast iron.

7.19 A cylinder contains 1 kg of ammonia at 100 kPa and 20°C, the temperature of the surroundings. The ammonia is then compressed isothermally to saturated vapor, with a work input of 340 kJ. During this process, any heat transfer takes place with the surroundings. Is this process reversible, irreversible, or impossible?

7.20 A Carnot engine has 1 kg of air as the working fluid. Heat is received at 800 K and rejected at 300 K. At the beginning of the heat addition process the pressure is 800 kPa, and during this process the volume triples. Calculate the net cycle work per kg of air.

7.21 One m³ of air at 4 MPa, 300°C expands in a reversible isothermal process in a cylinder to 0.1 MPa. Calculate
  (a) The heat transfer during the process.
  (b) The change of entropy of the air.

7.22 One kg of air contained in a cylinder at 500 kPa, 1000 K, expands in a reversible adiabatic process to 100 kPa. Calculate the final temperature and the work done during the process, using
  (a) The air tables, Table A.10.
  (b) Constant specific heat, value from Table A.8.

7.23 A large tank having a volume of 1 m³ is connected to a small tank having a volume of 0.3 m³. The large tank contains air initially at 700 kPa, 20°C and the small tank is initially evacuated. A valve in the connecting pipe between the two tanks is suddenly opened and is closed when the tanks come to pressure equilibrium. The air in the large tank may be assumed to have undergone a reversible process and the entire process is adiabatic. What is the final mass of air in the small tank? What is the final temperature of the air in the small tank?

7.24 It is desired to obtain a supply of cold helium gas by the following technique: helium contained in a cylinder at ambient conditions, 100 kPa, 25°C, is compressed in an isothermal process to 700 kPa, after which the gas is expanded back to 100 kPa in an adiabatic process. Both processes are reversible.
  (a) Show the process on T-s coordinates.
  (b) Calculate the final temperature and the net work per kg.
  (c) If a diatomic gas had been used instead of helium, would the final temperature be higher or lower?

7.25 An insulated tank having a volume of 1 m³ contains air at 800 kPa, 25°C. A valve on the tank is then opened and the pressure inside drops quickly to 150 kPa.
  (a) Assuming that the air remaining inside the tank has undergone a reversible adiabatic expansion, calculate the mass withdrawn.

(b) Calculate the mass withdrawn by a first law analysis, and compare the result with part a.

7.26  An insulated cylinder having an initial volume of 50 litres contains carbon dioxide at 100 kPa, 500 K. The gas is then compressed to 1.4 MPa in a reversible adiabatic process. Calculate the final temperature and the work, assuming

(a) Variable specific heat, Table A.11.

(b) Constant specific heat, value from Table A.8.

(c) Constant specific heat, value at an intermediate temperature from Table A.9.

7.27  Water at 20°C is pumped from a lake to an elevated storage tank. The average elevation of the water in the tank is 40 m above the surface of the lake, and the volume of the tank is 50 m³. Initially the tank contains air at 100 kPa, 20°C, and the tank is closed so that the air is compressed as the water enters the bottom of the tank. The pump is operated until the tank is three-quarters full. The temperature of the air and water remain constant at 20°C. Determine the work input to the pump.

7.28  Consider the following procedure for measuring the specific heat ratio $k$ of a gas. The gas is initially contained in a rigid vessel at pressure $P_1$ (slightly higher than ambient pressure $P_0$), and at the ambient temperature. A valve on the vessel is then opened and the gas pressure quickly falls to $P_0$, at which point the valve is closed. After a long time, the gas remaining inside the vessel returns to thermal equilibrium with the ambient, at which point the pressure inside is $P_2$. Develop an expression for $k$ in terms of $P_0$, $P_1$ and $P_2$. It may be assumed that there is no heat transfer to the gas during the time that gas is escaping from the vessel.

7.29  Argon gas at 100 kPa, 20°C is compressed reversibly in a cylinder to 400 kPa. Calculate the required work per kg if the process is

(a) Adiabatic.

(b) Isothermal.

(c) Polytropic with $n = 1.30$.

7.30  Air is contained in a cylinder fitted with a frictionless piston. Initially the cylinder contains 0.5 m³ of air at 150 kPa, 20°C. The air is then compressed reversibly according to the relationship $PV^n = $ constant until the final pressure is 600 kPa, at which point the temperature is 120°C. For this process determine

(a) The polytropic exponent $n$.

(b) The final volume of the air.

(c) The work done on the air and the heat transfer.

(d) The net change in entropy.

7.31  The power-stroke expansion of exhaust gases in an internal combustion engine can reasonably be represented as a polytropic expansion. Consider air contained in a cylinder volume of 0.1 litre at 7 MPa, 1650 K. The air then expands in a reversible polytropic process, with exponent $n = 1.30$, through a volume ratio of 8:1. Show the process on $P$-$v$ and $T$-$s$ coordinates, and calculate the work and heat transfer.

7.32 An ideal gas having a specific heat of

$$\overline{C}_{p_0} = 40 \text{ kJ/kmol K}$$

undergoes a reversible polytropic compression in which the polytropic exponent $n = 1.40$. Will the heat transfer for the process be positive, negative or zero?

7.33 The flow rate of Freon-12 in a refrigeration cycle is 0.02 kg/s. The compressor inlet conditions are 200 kPa, 0°C and the exit pressure is 1.2 MPa. Assuming the compression process to be reversible and adiabatic, what power motor is required to drive the compressor?

7.34 Nitrogen flows through a nozzle at the rate of 1 kg/s. The inlet conditions are 400 kPa, 200°C, the velocity at this point is 30 m/s, and the exit pressure is 100 kPa. Assuming the process to be reversible and adiabatic, find the exit velocity and the exit area of the nozzle.

7.35 A steam turbine receives steam at 800 kPa, 250°C. The steam expands in a reversible adiabatic process and leaves the turbine at 100 kPa. The power output of the turbine is 20 000 kW. What is the rate of steam flow to the turbine?

7.36 To aid in the reduction of pollutants from automotive exhaust, it has been proposed to utilize a small centrifugal compressor to take in ambient air at 100 kPa, 20°C, compress it by 40 kPa, and discharge the compressed air into the engine exhaust gases. It is estimated that the compressor must handle 20 litres/s of air at the inlet conditions. Determine the power required to drive the compressor.

7.37 Consider the turbine process in a gas turbine power plant, assuming the gas to have the same properties as air. The turbine inlet conditions are 1200 K, 500 kPa, and the exit pressure is 100 kPa. Calculate the turbine exit temperature and the work per kg, assuming
   (a) Variable specific heat, Table A.10.
   (b) Constant specific heat, value from Table A.8.

7.38 An ideal gas with constant specific heat enters a nozzle with a velocity $V_i$ and leaves with a velocity $V_e$ after undergoing a reversible adiabatic expansion. Show that the velocity leaving is given by the relation

$$V_e = \sqrt{V_i^2 + \frac{2kRT_i}{k-1}\left[1 - \left(\frac{P_e}{P_i}\right)^{(k-1)/k}\right]}$$

7.39 A diffuser is a device in which a fluid flowing at high velocity is decelerated in such a way that the pressure increases during the process. Steam at 0.2 MPa, 200°C enters the diffuser with a velocity of 700 m/s and leaves with a velocity of 70 m/s. If the process is reversible and occurs without heat transfer, what is the final pressure and temperature of the steam? A Mollier diagram may be helpful in solving this problem.

7.40 Air enters a turbine at 600 kPa, 1400 K, and expands in a reversible adiabatic process to 100 kPa. Calculate the exit temperature and the work done per kg of air, using

(a) The air tables, Table A.10.

(b) Constant specific heat, value at 300 K from Table A.8.

(c) Constant specific heat, value at an average temperature from Fig. 5.7. Discuss why the method of part (b) results in a poor value for the exit temperature and yet a relatively good value for the work.

7.41 A steam turbine can be operated at part-load conditions by throttling the steam to a lower pressure before it enters the turbine, as shown in Fig. 7.28. The line conditions are 1.4 MPa, 300°C, and the turbine exhaust pressure

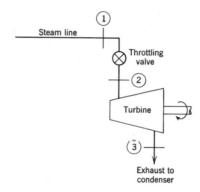

**Fig. 7.28** Sketch for Problem 7.41.

is fixed at 10 kPa. Assuming the expansion inside the turbine to be reversible and adiabatic, calculate

(a) The full-load work output of the turbine per kg of steam.

(b) The pressure to which the steam must be throttled to produce 75% of the full-load output. Show both processes on a *T-s* diagram.

7.42 Consider the two-stage steam turbine with interstage heating shown in Fig. 7.29. The data are as shown on the diagram. Assuming a constant pipe diameter in the interstage heater, and reversible adiabatic turbines, determine

(a) The power rating (kW) of the turbine.

(b) The rate of heat transfer in the interstage heater.

7.43 Consider the pump in a steam power plant cycle. Water leaves the condenser and enters the pump at 10 kPa, 35°C, and leaves the pump at 5 MPa. Assuming the process to be reversible and adiabatic, calculate

(a) The work required per kg.

(b) The water temperature leaving the pump.

7.44 Starting with the relation $w = -\int_{i}^{e} v \, dP$, show that the work done per kg

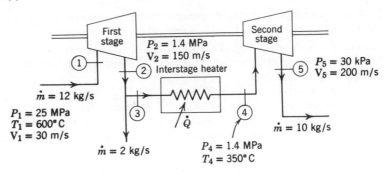

**Fig. 7.29**    Sketch for Problem 7.42.

of fluid flow in a steady-state, steady-flow, reversible adiabatic process, involving an ideal gas with constant specific heat, and with no changes in kinetic or potential energy, is given by the relation

$$w = \frac{kRT_i}{k-1}\left[1 - \left(\frac{P_e}{P_i}\right)^{(k-1)/k}\right]$$

**7.45** A centrifugal pump delivers liquid oxygen to a rocket engine at the rate of 40 kg/s. The oxygen enters the pump as saturated liquid at 90 K, and the discharge pressure is 3.5 MPa. Determine the power required to drive the pump if the process is reversible and adiabatic.

**7.46** A steam turbine powerplant operating at supercritical pressure is shown in Fig. 7.30. The turbine and pump processes are both reversible and

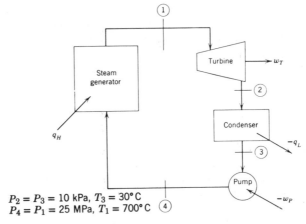

$P_2 = P_3 = 10$ kPa, $T_3 = 30°C$
$P_4 = P_1 = 25$ MPa, $T_1 = 700°C$ ④

**Fig. 7.30** Sketch for Problem 7.46.

adiabatic. Neglecting any changes in kinetic and potential energy, calculate

(a) The work output of the turbine, and the state at the turbine exit.

(b) The work input to the pump and the enthalpy of the liquid at the pump exit.

(c) The thermal efficiency of the cycle.

7.47 Air is compressed in a reversible steady-state, steady-flow process from 100 kPa, 25°C, to 800 kPa. Calculate the work of compression per kg, the change of entropy, and the heat transfer per kg of air compressed, assuming the following processes.

(a) Isothermal.

(b) Polytropic, $n = 1.25$.

(c) Adiabatic.

(d) Show all these processes on a P-$v$ and a T-$s$ diagram.

7.48 In a refrigeration plant liquid ammonia enters the expansion valve at 1.7 MPa, 32°C. The pressure leaving the expansion valve is 200 kPa, and changes in kinetic energy are negligible. What is the increase in entropy per kg? Show this process on a temperature-entropy diagram.

7.49 One type of feedwater heater for preheating the water before entering a boiler operates on the principle of mixing the water with steam. For the states as shown in Fig. 7.31, calculate the rate of increase in entropy per second, assuming the process to be steady-flow and adiabatic.

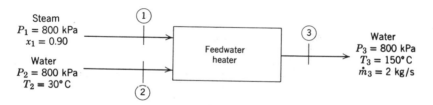

**Fig. 7.31** Sketch for Problem 7.49.

7.50 An uninsulated 0.5 m³ vessel contains steam at 1.4 MPa, 250°C. A valve on top of the vessel is opened and steam is expelled to the surroundings until the pressure reaches 400 kPa, at which time the valve is closed and the temperature inside is measured to be 175°C. Determine the net change in entropy for this process. The surroundings are at 250°C.

7.51 A salesman reports that he has a steam turbine available that delivers 2800 kW. The steam enters the turbine at 700 kPa, 250°C and leaves the turbine at a pressure of 15 kPa, and the required rate of steam flow is 3.8 kg/s.

(a) How do you evaluate his claim?

(b) Suppose he changed his claim and said the required steam flow was 4.3 kg/s.

7.52 Steam at 800 kPa, 300°C is flowing in a line. Connected to the line is a 2 m³ tank containing steam at 100 kPa, 200°C. The valve is opened, allowing steam to flow into the tank until the final pressure in the tank is 800 kPa. During this process, heat is transferred from the tank at such a rate that the temperature of its contents remains constant at 200°C. The surroundings are at a constant temperature of 25°C.

(a) Find the mass of steam that flows into the tank, and the heat transferred during the process.

(b) Find the change in entropy within the control volume (tank) during the process.

(c) Show that this process does not violate the second law of thermodynamics.

7.53 Steam enters a turbine at 0.8 MPa, 400°C, and exhausts at 0.2 MPa, 200°C. Any heat transfer is within the surroundings at 25°C, and changes in kinetic and potential energy are negligible. It is claimed that this turbine produces 100 kW with a mass flow of 0.3 kg/s. Does this process violate the second law of thermodynamics?

7.54 Freon-12 is contained in 1 m³ tank at the temperature of the surroundings, 20°C. The tank is three-quarters full of liquid by volume. The top of the withdrawal pipe is at the middle of the tank, as shown in Fig. 7.32. The

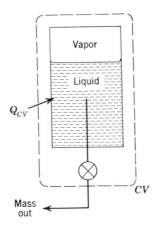

**Fig. 7.32** Sketch for Problem 7.54.

valve is now opened, and Freon-12 withdrawn slowly until the tank is one-quarter full of liquid. During this process the temperature inside remains constant, while the mass withdrawn is throttled across the valve to 0.2 MPa and then discharged. Calculate

(a) The heat transferred to the control volume.

(b) The entropy changes of the control volume and surroundings.

7.55 Steam enters a turbine at 600 kPa, 250°C and exhausts at a pressure of

10 kPa. The efficiency of the turbine is 75%. Determine:

(a) The work output per kg of steam.

(b) The state of the steam (quality if saturated, temperature if superheated) leaving the turbine.

7.56 Water enters a pump at 25°C, 100 kPa and exits at 3 MPa. If the efficiency (adiabatic) of the pump is 70%, determine the enthalpy of the water at the exit.

7.57 Steam enters an insulated nozzle at 800 kPa, 200°C, at low velocity and exits at 200 kPa. Calculate the exit velocity and temperature (quality if saturated) assuming a nozzle efficiency of 95%.

7.58 Air enters the compressor of a gas turbine at 95 kPa, 15°C at the rate of 25 m³/s, and leaves at 400 kPa. The process is adiabatic, and changes in kinetic and potential energy are negligible. Calculate the power required to drive this compressor and the exit temperature, assuming:

(a) That the process is reversible.

(b) That the compressor has an efficiency of 82%.

7.59 Repeat Problem 7.42, assuming that each stage of the two-stage turbine has an efficiency of 80%.

7.60 In one type of heat-powered refrigeration system, part of the working fluid is expanded through a turbine to drive the compressor of the refrigeration cycle. Consider the turbine-compressor unit of such a device, as shown in Fig. 7.33. The turbine produces just enough power to drive

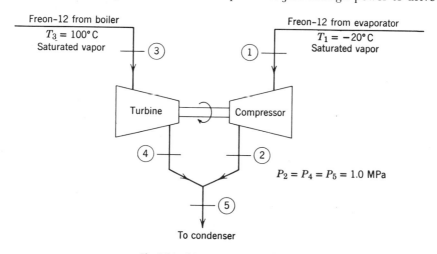

**Fig. 7.33**   Sketch for Problem 7.60.

the compressor, and both exit streams then mix together. Specifying any assumptions, find the ratio $\dot{m}_3/\dot{m}_1$ and $T_5$ ($x_5$ if in the two-phase region) if

(a) The turbine and compressor are reversible and adiabatic.

(b) The turbine and compressor each have an efficiency of 70%.

7.61 A jet ejector pump, shown schematically in Fig. 7.34, is a device in which a low-pressure (secondary) fluid is compressed by entrainment in a high-velocity (primary) fluid stream, the compression resulting from deceleration in a diffuser. For purposes of analysis, an ideal jet pump can be consid-

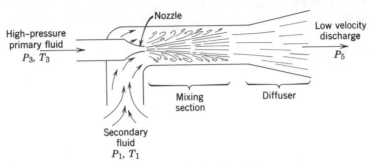

High-pressure primary fluid
$P_3, T_3$

Nozzle

Low velocity discharge
$P_5$

Mixing section

Diffuser

Secondary fluid
$P_1, T_1$

**Fig. 7.34**   Sketch for Problem 7.61.

ered as equivalent to the ideal (reversible adiabatic) turbine-compressor shown in Fig. 7.33, with states 1, 3, and 5 corresponding to those in Fig. 7.34.

Consider a steam jet pump in which state 1 is saturated vapor at 35 kPa, state 3 is 300 kPa, 150°C, and the discharge pressure $P_5$ is 100 kPa.

(a) Calculate the ideal mass ratio $(\dot{m}_1/\dot{m}_3)$,

(b) The efficiency of a jet pump is defined as

$$= \frac{(\dot{m}_1/\dot{m}_3) \text{ actual}}{(\dot{m}_1/\dot{m}_3) \text{ ideal}}$$

for the same inlet conditions and discharge pressure. For the pump considered above, determine the discharge temperature if the efficiency is 10%.

7.62 Repeat Problem 7.46, assuming that the turbine and pump each have an efficiency of 80%.

7.63 A small, high-speed air turbine having a 70% efficiency is required to produce 70 kJ/kg of work. The inlet temperature is 25°C and the turbine exhausts to the room. What is the required inlet pressure, and what is the exhaust air temperature?

7.64 The cycle shown in Fig. 7.35 has been proposed for an auxiliary power supply in a spacecraft. The working fluid is argon (ideal gas throughout the cycle). The compressor and turbine processes are both adiabatic and the heater and cooler processes are both constant-pressure. The following data are known:

$P_1 = P_4 = 35$ kPa
$P_2 = P_3 = 140$ kPa
$T_1 = 280$ K
$T_3 = 1100$ K

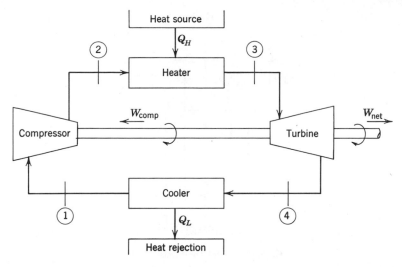

**Fig. 7.35** Sketch for Problem 7.64.

$n_{\text{turbine}} = 80\%$

$n_{\text{compressor}} = 80\%$

(a) Show this cycle on a T-s diagram.

(b) Calculate the net work output of the cycle and the thermal efficiency of the cycle.

## SUPPLEMENTARY PROBLEMS

7.65 The 50 litre tank shown in Fig. 7.36 is to be filled with Freon-12 for use as a constant-temperature bath in a research project. At the initial state 1 the tank is evacuated, and the desired conditions after filling are $T_2 = -40°C$, with liquid occupying three-quarters of the volume of the tank and vapor one-quarter of the volume. The Freon in the fill line, point 3, is at 0.7 MPa, 40°C. A supply of liquid nitrogen, to be used as the cooling medium during the filling process, enters at point 4 as saturated liquid at 100 kPa, passes through a coil and leaves at point 5 as a gas at 100 kPa, −40°C. It may be assumed that the pressures and temperatures at points 3, 4, and 5 remain constant throughout the process, and also that the tank and lines are well insulated. Calculate:

(a) The mass of Freon-12 in the tank at the final state.

(b) The enthalpy change of the nitrogen, per kg, between points 4 and 5.

(c) The total mass of nitrogen required for the process.

(d) The net entropy change for the process.

7.66 Consider the cylinder indicated in Fig. 7.37. The cylinder has a cross-sectional area of 0.013 m² and is fitted with a frictionless, nonheat-conducting

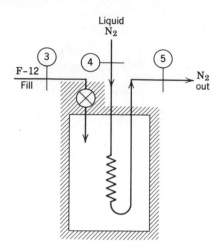

**Fig. 7.36**   Sketch for Problem 7.65.

piston weighing 20 kg. Compartment *A*, initially containing 0.1 kg of saturated vapor steam at 0.15 MPa, is connected through a valve to a supply line carrying steam at 0.8 MPa, 250°C. Compartment *B*, initially containing 0.01 kg of saturated vapor steam, contains a spring that is connected to the piston and has a spring constant of 22.5 kN/m. For the initial conditions stated the spring is touching but exerts no force on the piston. The valve connected to *A* is opened and steam enters until the pressure in *A* is 0.8 MPa, at which time the valve is closed. If the entire process is adiabatic, determine

(*a*) The final pressure in *B*.

(*b*) The mass entering *A*.

(*c*) The work done on the piston by the steam in *A*.

(*d*) The distance the piston travels.

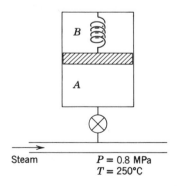

**Fig. 7.37**   Sketch for Problem 7.66.

7.67 Consider the system shown in Fig. 7.38. Tank $A$ has a volume of 0.3 m$^3$ and initially contains air at 700 kPa, 40°C. Cylinder $B$ is fitted with a frictionless piston resting on the bottom, at which position the spring is fully extended. The piston has a cross-sectional area of 0.065 m$^2$ and mass of 40 kg, and the spring constant $K_s$ is 17.5 kN/m. Atmospheric pressure is 100 kPa.

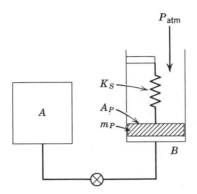

**Fig. 7.38**   Sketch for Problem 7.67.

The valve is opened and air flows into the cylinder until the pressures in $A$ and $B$ become equal, after which the valve is closed. During this process, the air finally remaining in $A$ may be considered to have undergone a reversible adiabatic process, and the entire process is adiabatic. The spring force is proportional to the displacement.

Determine the final pressure in the system and the temperature in cylinder $B$.

7.68 Consider the scheme shown in Fig. 7.39 for producing fresh water from saltwater. The conditions are as shown in the figure. Assume that the properties of saltwater are the same as for pure water, and also that the pump is reversible and adiabatic.

(a) Determine the ratio $\dot{m}_7/\dot{m}_1$, that is, the fraction of saltwater purified by the process.

(b) Determine the input quantities, $w_P$ and $q_H$.

(c) Make a second law analysis of the overall system.

7.69 An insulated cylinder is divided into two compartments $A$ and $B$ by a frictionless nonconducting piston as shown in Fig. 7.40. Compartment $A$ contains air at 100 kPa, 25°C and $B$ contains saturated water vapor at 100 kPa. Each side has an initial volume of 25 litres. Side $A$ is connected by a valve to a line in which air flows at 700 kPa, 25°C.

The valve is opened slightly and air from the line flows into $A$ until the pressure reaches 700 kPa. Calculate the final temperature in each of the compartments and the mass entering $A$.

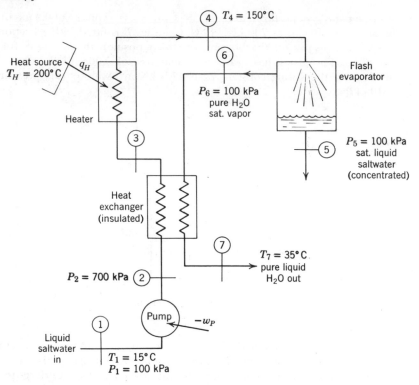

**Fig. 7.39** Sketch for Problem 7.68.

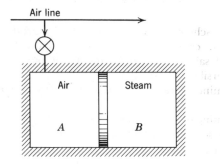

**Fig. 7.40** Sketch for Problem 7.69.

7.70 Two tanks, $A$ and $B$, contain steam in the amounts and at the condition indicated in Fig. 7.41. The tanks are connected to a common cylinder fitted with a piston of such a weight that a pressure of 1.4 MPa is required to support it. Intially the piston is at the bottom of the cylinder.

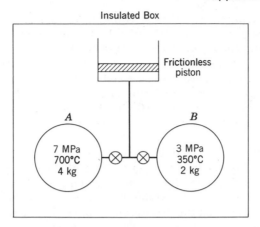

Insulated Box

**Fig. 7.41** Sketch for Problem 7.70.

Calculate the change of enthalpy, internal energy, and entropy of all the steam if the valves are opened and all the steam comes to a uniform pressure and temperature. Assume negligible volume in the pipes and valves and assume no heat transfer between the containing walls and the steam.

7.71 A turbo-supercharger is to be utilized to boost the inlet air pressure to an automobile engine. This device consists of an exhaust gas-driven turbine directly connected to an air compressor, as shown in Fig. 7.42. At a certain engine load, the conditions are as shown in the figure. It may be assumed that the exhaust gas properties are the same as those of air. If it is further assumed that the turbine and compressor are both reversible and adiabatic, calculate

(a) The turbine exit temperature and power output.

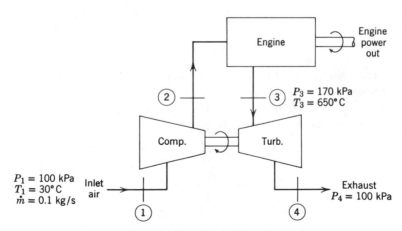

**Fig. 7.42** Sketch for Problem 7.71.

(b) The compressor exit pressure and temperature.

(c) Repeat parts (a) and (b), assuming that the turbine has an efficiency of 85% and the compressor an efficiency of 80%.

7.72 A 200 litre tank containing steam at 100 kPa pressure, 1% quality, is fitted with a relief valve. Heat is transferred to the tank from a large source at 250°C. When the pressure in the tank reaches 2 MPa, the relief valve opens, saturated vapor at 2 MPa is throttled across the valve and discharged at 100 kPa pressure. The process continues until the quality in the tank is 90%.

(a) Calculate the mass discharged from the tank.

(b) Determine the heat transfer to the tank during the process.

(c) Considering a control volume that contains the tank and valve, calculate the entropy change within the control volume and that of the surroundings. Show that the process does not violate the second law.

7.73 Consider the two tanks A and B as shown in Fig. 7.43, each tank having a volume of 25 litres. Tank A contains saturated vapor Freon-12 at 35°C, and B contains F-12 at −20°C, 0.1 MPa. Tank B, the line connecting the tanks, and the valve are heavily insulated. The valve is opened slightly and then closed when the two tanks have come to the same pressure. During the process, heat is exchanged between the surroundings and tank A such that the F-12 inside A at any time is at 35°C.

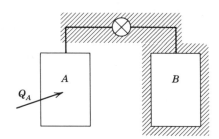

**Fig. 7.43** Sketch for Problem 7.73.

Calculate the final pressure, the final temperature in tank B, the heat transferred during the process, and the net entropy change.

7.74 A cylinder divided by a frictionless, thermally insulated piston contains air on one side and water on the other, as shown in Fig. 7.44. The cylinder

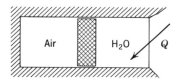

**Fig. 7.44** Sketch for Problem 7.74.

is well insulated, except for the end of the part containing the water. Each volume is initially 0.1 m³, at which point the air is at 40°C, and the water is at 90°C with a quality of 10%. Heat is now slowly transferred to the water, until at the final state it exists as saturated vapor. Calculate the final pressure and the amount of heat transferred.

7.75 A cylinder shown in Fig. 7.45 is so constructed that compartments $A$ and $B$ are separated by metal partition, and compartment $A$ is fitted with a frictionless piston. Assume that the temperatures in compartments $A$ and $B$

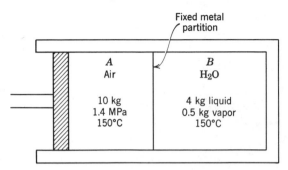

Fixed metal
partition

| $A$<br>Air | $B$<br>$H_2O$ |
|---|---|
| 10 kg<br>1.4 MPa<br>150°C | 4 kg liquid<br>0.5 kg vapor<br>150°C |

**Fig. 7.45** Sketch for Problem 7.75.

differ only by an infinitesimal. Compartment $A$ initially contains 10 kg of air at 1.4 MPa, 150°C. Compartment $B$ contains 4 kg of liquid $H_2O$ and 0.5 kg of vapor $H_2O$ at 150°C. The piston is moved to the right until the pressure in compartment $A$ reaches 2 MPa. Assume no heat transfer except across the metal partition. Determine the total change of entropy and internal energy.

7.76 An insulated closed cylinder fitted with a frictionless piston contains nitrogen at 100 kPa, 15°C, as shown in Fig. 7.46. The piston is thermally insulated and rests at the bottom of the cylinder, and the total volume of the cylinder is 50 litres. The valve is now opened to a 50 litre tank containing steam at 400 kPa, 200°C, and the valve is closed when the pressures

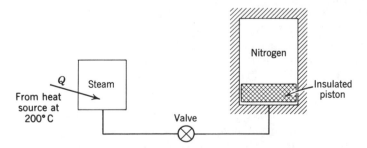

Nitrogen

$Q$    Steam

From heat
source at
200°C

Insulated
piston

Valve

**Fig. 7.46** Sketch for Problem 7.76.

equalize. During this process, heat is transferred to the steam inside the tank such that its temperature remains at 200°C. It may be assumed that the nitrogen is compressed reversibly in the cylinder. Determine

(a) The final pressure.

(b) The final temperatures of the nitrogen and the steam in the cylinder.

7.77 Steam is contained in tank $B$ and in cylinder $A$ at the conditions given in Fig. 7.47. Everything is thermally insulated. Now the valve is opened and the pin removed from the piston at the same time. When the pressure in $A$ and $B$ are equal, the valve is closed. Find the final state in $A$ and in $B$ in terms of the variables given.

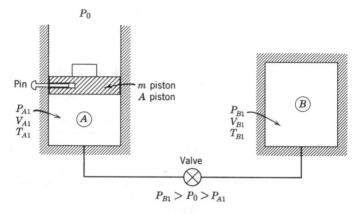

**Fig. 7.47** Sketch for Problem 7.77.

7.78 A tank of volume $V_T$ is filled with air at the initial condition $P_1$ and $T_1$. It is to be evacuated with a vacuum pump. The atmospheric pressure is $P_a$.

(a) Calculate the work of evacuation as a function of the pressure in the tank, and plot a curve of work of evacuation vs pressure.

(b) Considering the pump to handle a constant volume per unit time, determine power as a function of pressure in the tank and power as a function of time.

Assume both the tank and the pump to operate under isothermal conditions, and assume that the pumping process is reversible.

7.79 Side $A$ of the cylinder shown in Fig. 7.48 contains $H_2O$ and side $B$ contains argon, at the conditions given. The mass of the piston and weights is such that the piston is floating. Now heat is transferred from the source to the $H_2O$ in $A$ causing the piston to rise. (The piston can rise only until it hits the fixed stop, at which point the volume of $B$ would be 0.2 m³.) The process continues until the temperature in $A$ reaches that of the source, 600°C.

(a) Does the piston reach the fixed stop during the process?

(b) Calculate the final pressure in $A$, the heat transfer $Q_A$ and the work done by the $H_2O$ in $A$ during the process.

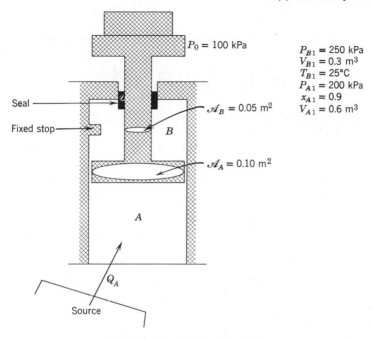

$P_0 = 100$ kPa

$P_{B1} = 250$ kPa
$V_{B1} = 0.3$ m$^3$
$T_{B1} = 25°C$
$P_{A1} = 200$ kPa
$x_{A1} = 0.9$
$V_{A1} = 0.6$ m$^3$

Seal

$\mathscr{A}_B = 0.05$ m$^2$

Fixed stop

$B$

$\mathscr{A}_A = 0.10$ m$^2$

$A$

$Q_A$

Source

**Fig. 7.48** Sketch for Problem 7.79.

7.80 Write a computer program to solve the following problem. One of the gases listed in Table A.9 undergoes a reversible adiabatic process in a cylinder from $P_1$, $T_1$, to $P_2$. It is desired to calculate the final temperature and the work for the process by integrating the specific heat equation, by assuming constant specific heat at temperature $T_1$, and finally by assuming constant specific heat at an average temperature (by iteration).

# 8

# Irreversibility and Availability

We now turn our attention to irreversibility and availability, two additional concepts that have found increasing use in recent years. These concepts are particularly applicable in the analysis of complex thermodynamic systems, for with the aid of a digital computer, irreversibility and availability are very powerful tools in design and optimization studies of such systems.

## 8.1   Reversible Work

In order to introduce the concepts of reversible work and irreversibility, let us consider Fig. 8.1, which shows a control volume undergoing a uniform-state, uniform flow process. There are irreversibilities present as the process takes place, as in every real process. The work crossing the control surface during the process is $W_{\text{c.v.}}$ and the heat transfer is $Q_{\text{c.v.}}$. All of the heat transfer is with the surroundings at temperature $T_0$.

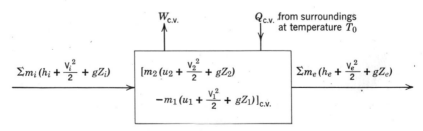

**Fig. 8.1**   A uniform-state, uniform-flow process.

We now ask the following questions: What if this working fluid had undergone exactly the same change in state, except in a completely reversible process? How much work would have been done in such a process? We realize, of course, that such a process does not really exist. Any process that actually occurs can do so only with some irreversibilities, of the types discussed in Chapter 6. Thus, we are imagining an ideal process, to which we can compare the real process shown in Fig. 8.1. This situation is similar to our discussion of cycles in Chapter 6. That led us to develop the ideal, completely reversible Carnot cycle, to which we can compare the performance of real cycles. The situation is also similar to that in Section 7.15, in which we compared the performance of real machines (turbines, compressors, nozzles) to idealized model processes. In that case, however, the ideal and real processes did not occur between the same end states, which was the question raised here.

We proceed then to Fig. 8.2, which shows a control volume identical to that of the real process, Fig. 8.1. All quantities and states of fluid entering and leaving the control volume are the same as in Fig. 8.1, and exactly the same change of state occurs within the control volume. In contrast to Fig. 8.1, however, in this case all processes are completely

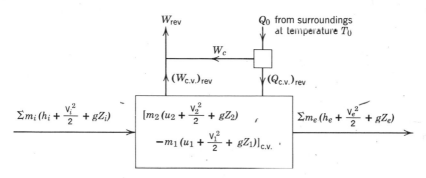

**Fig. 8.2**  A reversible process for the same change of state as in Fig. 8.1.

reversible. Thus, in Fig. 8.2 we have the (hypothetical) ideal process corresponding to exactly the same change of state that occurred in the real process of Fig. 8.1. The work and heat crossing the control surface for this ideal process will be different than in Fig. 8.1 (which involved irreversibilities) and these quantities are designated $(W_{c.v.})_{rev.}$ and $(Q_{c.v.})_{rev.}$. In order that the heat transfer between the control volume and the surroundings may occur reversibly when there is a difference in the temperature in the control volume and the surroundings, it is necessary that this heat transfer take place through a reversible heat engine. The work output of this reversible heat engine is designated $W_c$. The sum of the work crossing the control surface for the reversible case and the work output of the reversible heat engine is called the reversible work and is designated $W_{rev}$. That is,

$$W_{rev} = (W_{c.v.})_{rev} + W_c \qquad (8.1)$$

The difference between this reversible work $W_{rev}$, and that done in the first case, $W_{c.v.}$, when irreversible processes occur is called the irreversibility, which is designated $I$.

$$I = W_{rev} - W_{c.v.} \qquad (8.2)$$

Quite obviously, if real processes could take place in a completely reversible manner, the irreversibility would be zero.

While we could calculate the reversible work for each problem that we might consider, it is advantageous at this point to develop a general expression for the reversible work. Let us therefore analyze the reversible process shown in Fig. 8.2. We have assumed a uniform-state, uniform-flow process for this control volume. This process was described in Section 5.13, and for this process the first law can be written, Eq. 5.56, as

$$(Q_{c.v.})_{rev} + \sum m_i \left( h_i + \frac{\mathsf{V}_i^2}{2} + gZ_i \right)$$

$$= (W_{c.v.})_{rev} + \sum m_e \left( h_e + \frac{\mathsf{V}_e^2}{2} + gZ_e \right)$$

$$+ \left[ m_2 \left( u_2 + \frac{\mathsf{V}_2^2}{2} + gZ_2 \right) - m_1 \left( u_1 + \frac{\mathsf{V}_1^2}{2} + gZ_1 \right) \right]_{c.v.} \qquad (8.3)$$

The significance of each of these terms is shown in Fig. 8.2.

Since all processes are to be reversible, the heat transfer with the surroundings must also be reversible. Therefore, if the temperature within the control volume is different from the surroundings, this heat transfer must take place through a reversible heat engine. This is also shown schematically in Fig. 8.2, and the work done by the reversible engine is designated $W_c$.

For the reversible heat engine work, $W_c$, we can write:

First law:

$$W_c = Q_0 - (Q_{c.v.})_{rev}$$

Second law: Since

$$\frac{Q_0}{T_0} = \int_0^t \left(\frac{\dot{Q}_{\text{c.v.}}}{T}\right)_{\text{rev}} dt$$

it follows that

$$W_c = T_0 \int_0^t \left(\frac{\dot{Q}_{\text{c.v.}}}{T}\right)_{\text{rev}} dt - (Q_{\text{c.v.}})_{\text{rev}} \qquad (8.4)$$

The second law for the uniform-state, uniform-flow process, Eq. 7.58, for this reversible process is

$$m_2 s_2 - m_1 s_1 + \sum m_e s_e - \sum m_i s_i = \int_0^t \left(\frac{\dot{Q}_{\text{c.v.}}}{T}\right)_{\text{rev}} dt \qquad (8.5)$$

Substituting this into Eq. 8.4 we have

$$W_c = T_0 [m_2 s_2 - m_1 s_1 + \sum m_e s_e - \sum m_i s_i] - (Q_{\text{c.v.}})_{\text{rev}} \qquad (8.6)$$

We can now substitute the expression for $(W_{\text{c.v.}})_{\text{rev}}$ from Eq. 8.3 and for $W_c$ from Eq. 8.6 into Eq. 8.1.

$$W_{\text{rev}} = (Q_{\text{c.v.}})_{\text{rev}} + \sum m_i \left(h_i + \frac{\mathsf{V}_i^2}{2} + gZ_i\right)$$

$$- \sum m_e \left(h_e + \frac{\mathsf{V}_e^2}{2} + gZ_e\right)$$

$$- \left[m_2 \left(u_2 + \frac{\mathsf{V}_2^2}{2} + gZ_2\right) - m_1 \left(u_1 + \frac{\mathsf{V}_1^2}{2} + gZ_1\right)\right]_{\text{c.v.}}$$

$$+ T_0 [m_2 s_2 - m_1 s_1 + \sum m_e s_e - \sum m_i s_i] - (Q_{\text{c.v.}})_{\text{rev}}$$

On rearranging this equation, we have

$$W_{\text{rev}} = \sum m_i \left(h_i - T_0 s_i + \frac{\mathsf{V}_i^2}{2} + gZ_i\right) - \sum m_e \left(h_e - T_0 s_e + \frac{\mathsf{V}_e^2}{2} + gZ_e\right)$$

$$- \left[m_2 \left(u_2 - T_0 s_2 + \frac{\mathsf{V}_2^2}{2} + gZ_2\right) - m_1 \left(u_1 - T_0 s_1 + \frac{\mathsf{V}_1^2}{2} + gZ_1\right)\right]_{\text{c.v.}}$$

$$(8.7)$$

Note that $(Q_{\text{c.v.}})_{\text{rev}}$ cancels and does not appear in Eq. 8.7.

Thus we have an expression for the reversible work of a control volume that exchanges heat with the surroundings at temperature $T_0$. Note that the reversible work is a function of $T_0$, the temperature of the surroundings.

Let us now consider two special cases, namely, a system (fixed mass) and a steady-state, steady-flow process for a control volume.

For a system there is no flow across the control surface, and therefore, considering Eq. 8.7,

$$\sum m_i \left( h_i - T_0 s_i + \frac{V_i^2}{2} + g Z_i \right) = 0;$$

$$\sum m_e \left( h_e - T_0 s_e + \frac{V_e^2}{2} + g Z_e \right) = 0$$

Thus, for a system the reversible work is

$$_1\left( \frac{W_{rev}}{m} \right)_2 = {}_1 w_{rev\,2} = \left[ \left( u_1 - T_0 s_1 + \frac{V_1^2}{2} + g Z_1 \right) - \left( u_2 - T_0 s_2 + \frac{V_2^2}{2} + g Z_2 \right) \right]$$

(8.8)

For a steady-state, steady-flow process we note on consideration of Eq. 8.7 that

$$\left[ m_2 \left( u_2 - T_0 s_2 + \frac{V_2^2}{2} + g Z_2 \right) - m_1 \left( u_1 - T_0 s_1 + \frac{V_1^2}{2} + g Z_1 \right) \right]_{c.v.} = 0$$

Therefore, for a steady-state, steady-flow process, in rate form,

$$\dot{W}_{rev} = \sum \dot{m}_i \left( h_i - T_0 s_i + \frac{V_i^2}{2} + g Z_i \right) - \sum \dot{m}_e \left( h_e - T_0 s_e + \frac{V_e^2}{2} + g Z_e \right) \quad (8.9)$$

When there is a single flow of fluid into and out of the control volume in a steady-state, steady-flow process we can write

$$\frac{\dot{W}_{rev}}{\dot{m}} = w_{rev} = \left( h_i - T_0 s_i + \frac{V_i^2}{2} + g Z_i \right) - \left( h_e - T_0 s_e + \frac{V_e^2}{2} + g Z_e \right) \quad (8.10)$$

Each expression for reversible work involves the temperature of the surroundings, $T_0$. Therefore in solving problems this information must be available. Unless otherwise stated we will use 25°C for the temperature of the surroundings. This temperature has been selected because thermochemical data are frequently given relative to this base and it is also a reasonable temperature to assume for the surroundings.

One must keep in mind that one should use the actual temperature of the surroundings when this information is available.

## 8.2  Irreversibility

In the last section we considered the two control volumes shown in Figs. 8.1 and 8.2, and noted that the irreversibility is defined as

$$I = W_{rev} - W_{c.v.}$$

We also derived Eq. 8.7, which was a general expression for the reversible work for a control volume undergoing a uniform-state, uniform-flow process. We can also derive a general expression for irreversibility. We do so by noting that an expression for $W_{c.v.}$ can be found by applying the first law to the control volume that involves the irreversible process in Fig. 8.1. From Eq. 5.56,

$$W_{c.v.} = \sum m_i \left( h_i + \frac{V_i^2}{2} + gZ_i \right) - \sum m_e \left( h_e + \frac{V_e^2}{2} + gZ_e \right)$$

$$- \left[ m_2 \left( u_2 + \frac{V_2^2}{2} + gZ_2 \right) - m_1 \left( u_1 + \frac{V_1^2}{2} + gZ_1 \right) \right]_{c.v.} + Q_{c.v.} \qquad (8.11)$$

Substituting the expression for $W_{rev}$ from Eq. 8.7 and for $W_{c.v.}$ from Eq. 8.11 into Eq. 8.2 we have

$$I = \sum m_e T_0 s_e - \sum m_i T_0 s_i + m_2 T_0 s_2 - m_1 T_0 s_1 - Q_{c.v.} \qquad (8.12)$$

This is a general expression for the irreversibility in a uniform-state, uniform-flow process. Once again we consider two particular cases, namely, the system and the steady-state, steady-flow process. For the system,

$$\sum m_e T_0 s_e = 0; \qquad \sum m_i T_0 s_i = 0; \qquad m_2 = m_1 = m$$

and Eq. 8.12 reduces to

$$_1 I_2 = m T_0 (s_2 - s_1) - {}_1 Q_2 \qquad (8.13)$$

For the steady-state, steady-flow process,

$$m_2 T_0 s_2 = m_1 T_0 s_1$$

and Eq. 8.12 reduces to

$$I = \sum m_e T_0 s_e - \sum m_i T_0 s_i - Q_{c.v.} \qquad (8.14)$$

Returning to the general expression, Eq. 8.12, we can make an interesting comparison between the irreversibility $I$ and the principle of the increase of entropy discussed in Sections 7.8 and 7.14. Equation 8.12 in rate form becomes

$$\dot{I} = T_0 \left[ \frac{dS_{c.v.}}{dt} + \sum \dot{m}_e s_e - \sum \dot{m}_i s_i - \frac{\dot{Q}_{c.v.}}{T_0} \right]$$

But substituting Eq. 7.68 for the rate of entropy change of the surroundings, we have

$$\dot{I} = T_0 \left[ \frac{dS_{c.v.}}{dt} + \frac{dS_{surr}}{dt} \right] \qquad (8.15)$$

The quantity inside the brackets in Eq. 8.15 is the net rate of entropy change for the overall process, and as discussed in Chapter 7, is always greater than zero, approaching zero as the process becomes completely reversible. Thus, the irreversibility $I$ is equal to the product of the temperature of the surroundings and the net entropy change for the process, and is therefore directly related to the second law of thermodynamics. Equation 8.15 is a general result, and reduces to Eq. 8.13 or 8.14 in those special cases.

**Example 8.1**

**1.** Considering Fig. 8.3, tank $A$ has a volume of 4 m³ and initially contains Freon-12 at a pressure of 100 kPa and a temperature of 20°C. The compressor evacuates tank $A$ and charges tank $B$. Tank $B$ is initially evacuated and is of such a volume that the final pressure of the Freon-12 in tank $B$ is 500 kPa when the temperature reaches its final value of 20°C. The temperature of the surroundings is 20°C. Determine the minimum work input to the compressor.

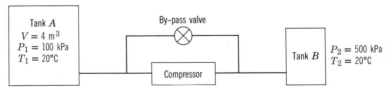

**Fig. 8.3**  Sketch for Example 8.1.

**2.** After tank $B$ is charged and tank $A$ is evacuated, a by-pass valve around the compressor is left open, and the two tanks come to a uniform pressure at a temperature of 20°C. Determine the irreversibility for this process.

The solution is as follows:

**1.** Consider a system consisting of the two tanks and the connecting piping. For this system there is no change in volume. Since changes in kinetic and potential energy are not significant, we have, from Eq. 8.8,

$$_1(w_{rev})_2 = (u_1 - u_2) - T_0(s_1 - s_2)$$

$$u_1 = h_1 - P_1 v_1 = 203.707 - 100 \times 0.197\ 277 = 183.979\ \text{kJ/kg}$$

$$s_1 = 0.8275\ \text{kJ/kg K}$$

$$u_2 = 196.935 - 500 \times 0.035\ 646 = 179.112\ \text{kJ/kg}$$

$$s_2 = 0.6999\ \text{kJ/kg K}$$

$$_1(w_{rev})_2 = (183.979 - 179.112) - 293.15\ (0.8275 - 0.6999)$$

$$= -32.54\ \text{kJ/kg}$$

$$m = \frac{V_1}{v_1} = \frac{4}{0.197\ 277} = 20.276\ \text{kg}$$

$$_1(W_{rev})_2 = 20.276\ (-32.54) = -659.8\ \text{kJ}$$

Note that the reversible work is a negative number. This means that the minimum work input to the compressor to accomplish this change of state is 659.8 kJ. If there are any irreversibilities this work input will increase, say to 800 kJ. Thus the reversible work can still be thought of as the maximum work, inasmuch as $-659.8$ is a larger number than $-800$.

**2.** When the Freon-12 flows from tank $B$ to tank $A$ through the by-pass valve, we recognize that this is an irreversible process. We could calculate this irreversibility either by calculating the reversible work from Eq. 8.8 and the irreversibility from Eq. 8.2 or we could calculate the irreversibility directly from Eq. 8.13. Let us do both calculations in order to gain further insight into these concepts. The state of the Freon-12 in tank $B$ is designated state 2 as above, and the final state is designated state 3.

We must first find the final pressure in the system. The two properties we know are specific volume and temperature. The volume of tank $B$ is found from the results of part **1**.

$$V_B = mv_{2B} = 20.276 \times 0.035\ 646 = 0.7228\ \text{m}^3$$

$$V_3 = V_A + V_B = 4 + 0.7228 = 4.7228\ \text{m}^3$$

$$v_3 = \frac{V_3}{m} = \frac{4.7228}{20.276} = 0.232\ 92\ \text{m}^3/\text{kg}$$

By interpolation from the Freon-12 table, $P_3 = 91.2$ kPa

$$h_3 = 203.842\ \text{kJ/kg}; \qquad s_3 = 0.8363\ \text{kJ/kg K}$$

$$u_3 = 203.842 - 91.2 \times 0.232\ 92 = 182.60\ \text{kJ/kg}$$

From Eq. 8.8

$$_2(w_{\text{rev}})_3 = u_2 - u_3 - T_0(s_2 - s_3)$$
$$= (179.112 - 182.60) - 293.15\ (0.6999 - 0.8363)$$
$$= 36.50\ \text{kJ/kg}$$
$$_2(W_{\text{rev}})_3 = 20.276\ (36.50) = 740.0\ \text{kJ}$$

The irreversibility can now be calculated from Eq. 8.2.

$$_2I_3 = {}_2(W_{\text{rev}})_3 - {}_2W_3 = 740.0 - 0 = 740.0\ \text{kJ}$$

The irreversibility can also be calculated from Eq. 8.13.

$$_2I_3 = mT_0(s_3 - s_2) - {}_2Q_3$$

In order to determine $_2Q_3$ we apply the equation for the first law to this process.

$$_2Q_3 = m(u_3 - u_2) + {}_2W_3$$
$$= 20.276\ (182.60 - 179.112) + 0 = 70.7\ \text{kJ}$$
$$_2I_3 = 20.276 \times 293.15\ (0.8363 - 0.6999) - 70.7 = 740.0\ \text{kJ}$$

**Example 8.2**

Referring to Example 7.11 determine the reversible work and the irreversibility for the actual change of state that takes place as the steam flows through the turbine (using a temperature of the surroundings of 25°C). In this solution it will be helpful to refer to the temperature entropy diagram of Fig. 8.4.

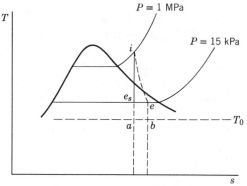

**Fig. 8.4** Sketch for Example 8.2.

Let us first determine the actual state of the steam as it leaves the turbine.

$$P_e = 15 \text{ kPa}$$

$$h_e = h_i - w$$

$$= 3051.2 - 600 = 2451.2 \text{ kJ/kg}$$

$$h_e = 2451.2 = 2599.1 - (1 - x_e) 2373.1$$

$$(1 - x_e) = 0.0623$$

$$s_e = 8.0085 - 0.0623 \, (7.2536) = 7.5564 \text{ kJ/kg K}$$

We calculate the reversible work using Eq. 8.10

$$\frac{\dot{W}_{\text{rev}}}{\dot{m}} = (h_i - T_0 s_i) - (h_e - T_0 s_e)$$

$$= (3051.2 - 2451.2) - 298.15 \, (7.1229 - 7.5564)$$

$$= 729.2 \text{ kJ/kg}$$

The irreversibility can be found using Equation 8.14

$$\frac{\dot{I}}{\dot{m}} = T_0(s_e - s_i) - \frac{\dot{Q}_{\text{c.v.}}}{\dot{m}}$$

$$= 298.15 \, (7.5564 - 7.1229) = 129.2 \text{ kJ/kg}$$

Note also that we could use our basic definition of the irreversibility.

$$\frac{\dot{I}}{\dot{m}} = \frac{\dot{W}_{\text{rev}}}{\dot{m}} - \frac{\dot{W}_{\text{c.v.}}}{\dot{m}}$$

$$= 729.2 - 600 = 129.2 \text{ kJ/kg}$$

It may be of help in understanding these concepts to conceive of a device by which we could achieve this amount of reversible work. Consider the device shown in Fig. 8.5 in which the steam expands in a

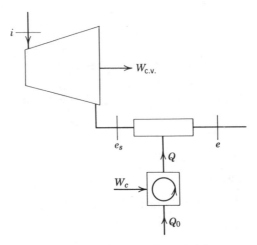

**Fig. 8.5** Sketch for Example 8.2.

reversible adiabatic process until the pressure reaches 15 kPa, followed by a heat transfer process in which heat is transferred to the steam in a constant pressure process. The heat is transferred from the surroundings, which requires a heat pump, with a work input, $W_c$.

We already found, from Example 7.11 that the work for the reversible adiabatic process is 741.8 kJ/kg. The work required for the heat pump is area $e_s a b e e_s$ on Fig. 8.4.

$$W_c = (T_e - T_0)(s_e - s_i)$$

$$= (327.12 - 298.15)(7.5564 - 7.1229) = 12.6 \text{ kJ/kg}$$

$$W_{\text{rev}} = 741.8 - 12.6 = 729.2 \text{ kJ/kg}$$

Thus, in deriving the expressions for reversible work and irreversibility we have only generalized the calculations, and one can always determine these quantities by applying the fundamentals directly to a given problem.

## 8.3   Availability

What is the maximum reversible work that can be done by a system in a given state? In Section 8.1 we developed an expression for the reversible work for a given change of state of a system. But the question that arises is, what final state will make this reversible work the maximum?

The answer to this question is that when a system is in equilibrium with the environment, no spontaneous change of state will occur, and the system will not be capable of doing any work. Therefore, if a system in a given state undergoes a completely reversible process until it reaches a state in which it is in equilibrium with the environment, the maximum reversible work will have been done by the system.

If a system is in equilibrium with the surroundings, it must certainly be in pressure and temperature equilibrium with the surroundings — that is, at pressure $P_0$ and temperature $T_0$. It must also be in chemical equilibrium with the surroundings, which implies that no further chemical reaction will take place. Equilibrium with the surroundings also requires that the system have zero velocity and minimum potential energy. Similar requirements could be set forth regarding magnetic, electrical, and surface effects if these are relevant to a given problem.

The same general remarks can be made in regard to a quantity of mass that undergoes a steady-state, steady-flow process. With a given state for the mass entering the control volume, the reversible work will be a maximum when this mass leaves the control volume in equilibrium with the surroundings. This means that as the mass leaves the control volume it must be at the pressure and temperature of the surroundings, in chemical equilibrium with the surroundings, and have minimum potential energy, and zero velocity. (The mass leaving the control volume must of necessity have some velocity but it can be made to approach zero.)

Let us first consider the availability associated with a steady-state, steady-flow process. In Eq. 8.10 we noted that when we consider a single flow,

$$w_{\text{rev}} = \left( h_i - T_0 s_i + \frac{V_i^2}{2} + g Z_i \right) - \left( h_e - T_0 s_e + \frac{V_e^2}{2} + g Z_e \right)$$

This reversible work will be a maximum when the mass leaving the control volume is in equilibrium with the surroundings. If we designate this state in which the fluid is in equilibrium with the surroundings with subscript 0, the reversible work will be a maximum when $h_e = h_0$, $s_e = s_0$, $V_e = 0$, and $Z_e = Z_0$. Designating this maximum reversible work per unit mass flow as the availability per unit mass flow, and assigning this the symbol $\psi$, we have

$$\psi = \left(h - T_0 s + \frac{V^2}{2} + gZ\right) - \left(h_0 - T_0 s_0 + gZ_0\right) \tag{8.16}$$

The initial state is designated without a subscript to indicate that this is the availability associated with a substance in any state as it enters a control volume in a steady-state, steady-flow process. It also follows that the reversible work per unit mass flow between any two states is equal to the decrease in availability between these two states.

$$w_{\text{rev}} = \psi_i - \psi_e \tag{8.17}$$

If we have more than one flow into and out of the control volume in a steady-state, steady-flow process we can write

$$\dot{W}_{\text{rev}} = \sum \dot{m}_i \psi_i - \sum \dot{m}_e \psi_e \tag{8.18}$$

The availability associated with a system is developed in a similar way, except for one factor. When the volume of a system increases, some work is done by the system against the surroundings, and this is not available for doing useful work.

To simplify this analysis let us also assume that the change in kinetic and potential energy of the system is negligible. In this case $w_{\text{rev}}$, as given by Eq. 8.8, is

$$_1(w_{\text{rev}})_2 = (u_1 - T_0 s_1) - (u_2 - T_0 s_2)$$

If the final state is in equilibrium with the surroundings $u_2 = u_0$ and $s_2 = s_0$. In this case the reversible work is a maximum. If we designate this maximum reversible work as $(w_{\text{rev}})_{\text{max}}$, and write it without subscript to indicate a general state we have

$$(w_{\text{rev}})_{\text{max}} = (u - T_0 s) - (u_0 - T_0 s_0) \tag{8.19}$$

The availability per unit mass for a system is equal to this maximum reversible work minus the work done against the surroundings. This work done against the surroundings, $W_{\text{surr}}$, is

$$W_{\text{surr}} = P_0(V_0 - V) = -mP_0(v - v_0) \tag{8.20}$$

The availability per unit mass for a system in the absence of kinetic and potential energy changes is designated $\phi$

$$
\begin{aligned}
\phi &= (w_{\text{rev}})_{\text{max}} - w_{\text{surr}} \\
&= (u - T_0 s) - (u_0 - T_0 s_0) + P_0 (v - v_0) \\
&= (u + P_0 v - T_0 s) - (u_0 + P_0 v_0 - T_0 s_0) \\
&= (u - u_0) + P_0 (v - v_0) - T_0 (s - s_0)
\end{aligned}
\tag{8.21}
$$

It follows that

$$
{}_1(w_{\text{rev}})_2 = \phi_1 - \phi_2 - P_0(v_1 - v_2) + \frac{V_1{}^2 - V_2{}^2}{2} + g(Z_1 - Z_2)
\tag{8.22}
$$

The use of availability and irreversibility in an actual thermodynamic problem is shown in Fig. 8.6. A theoretical analysis was made of a reciprocating internal-combustion automotive engine to see what happened to the availability of the air-fuel mixture that entered the engine and where the irreversibility occurred during the process. The abscissa on Fig. 8.6 is crank angle, the left side representing bottom dead center when the cylinder is assumed to be filled with an air-fuel mixture having the availability indicated on the ordinate. As the compression process takes place, the availability of this mixture increases as the result of the

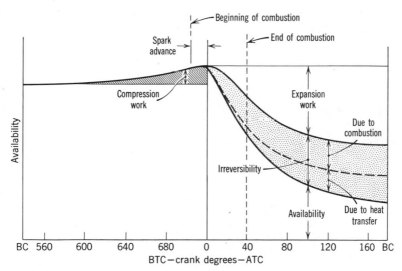

**Fig. 8.6** Availability vs. crank angle of the charge in a spark-ignited internal combustion engine. From D. J. Patterson and G. J. Van Wylen, "A Digital Computer Simulation for Spark Ignited Engine Cycles," *SAE Progress in Technology Series*, 7, p. 88. Published by SAE Inc., New York, 1964.

work done in compressing the mixture. When the piston passes top dead center the expansion process takes place. The beginning and end of combustion are also indicated on the diagram. During the combustion and expansion process irreversibilities occur. Those that are associated with the combustion process itself and those associated with heat transfer to the cooling water or surroundings are both indicated. The work done during the expansion process and availability at the end of the expansion are indicated on the right ordinate. Note that this availability that remains in the cylinder at the end of the expansion stroke is exhausted to the atmosphere.

The less the irreversibility associated with a given change of state, the greater the work that will be done (or the less work that will be required). This is significant in at least two regards. The first is that availability is one of our natural resources. This availability is found in such forms as oil reserves, coal reserves, and uranium reserves. Suppose we wish to accomplish a given objective that requires a certain amount of work. If this work is produced reversibly while drawing on one of the availability reserves, the decrease in availability would be exactly equal to the reversible work. However, since there are irreversibilities involved in producing this required amount of work, the actual work will be less than the reversible work, and the decrease in availability will be greater (by the amount of the irreversibility) than if this work had been produced reversibly. Thus the more irreversibilities we have in all of our processes, the greater will be the decrease in our availability reserves.[1] The conservation and effective use of these availability reserves is an important responsibility for all of us.

The second reason that it is desirable to accomplish a given objective with the smallest irreversibility is an economic one. Work costs money, and in many cases a given objective can be accomplished at less cost when the irreversibility is less. It should be noted, however, that many factors enter into the total cost of accomplishing a given objective, and an optimization process that involves consideration of many factors is often necessary in arriving at the most economical design. For example, in a heat transfer process, the smaller the temperature difference across which the heat is transferred, the less the irreversibility. However, for a given rate of heat transfer, a smaller temperature difference will require a larger (and therefore more expensive) heat exchanger, and these various factors must all be considered in the development of the optimum and most economical design.

[1]In many popular talks reference is made to our energy reserves. From a thermodynamic point of view availability reserves would be a much more acceptable term. There is much energy in the atmosphere and the ocean, but relatively little availability.

In many engineering decisions other factors, such as the impact on the environment (e.g., air pollution and water pollution) and the impact on society must be considered in developing the optimum design.

## Example 8.3

A steam turbine, Fig. 8.7, receives 30 kg of steam per second at 3 MPa, 350°C. At the point in the turbine where the pressure is 0.5 MPa,

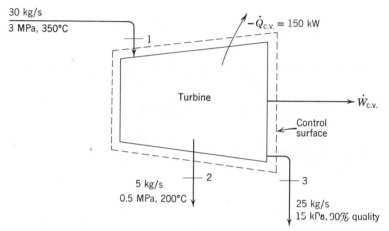

**Fig. 8.7**   Sketch for Example 8.3.

steam is bled off for use in processing equipment at the rate of 5 kg/s. The temperature of this bled steam is 200°C. The balance of the steam leaves the turbine at 15 kPa, 90 per cent quality. The heat transfer to the surroundings is 150 kW. Determine the availability per kilogram of the steam entering and at both points at which steam leaves the turbine and the reversible work per kilogram of steam for the given change of state.

Let us designate the states as shown in Fig. 8.7. The availability at any point for the steam entering or leaving the turbine is given by Eq. 8.16.

$$\psi = (h - h_0) - T_0(s - s_0) + \frac{V^2}{2} + g(Z - Z_0)$$

Since changes in kinetic and potential energy are not involved in this problem this equation reduces to

$$\psi = (h - h_0) - T_0(s - s_0)$$

At the pressure and temperature of the surroundings, namely, 0.1 MPa, 25°C, the water is a slightly compressed liquid, and the properties of the water are essentially equal to those for saturated liquid at 25°C.

$$h_0 = 104.9 \text{ kJ/kg}; \qquad s_0 = 0.3674 \text{ kJ/kg K}$$

$$\psi_1 = (3115.3 - 104.9) - 298.15 \, (6.7428 - 0.3674) = 1109.6 \text{ kJ/kg}$$

$$\psi_2 = (2855.4 - 104.9) - 298.15 \, (7.0592 - 0.3674) = 755.3 \text{ kJ/kg}$$

$$\psi_3 = (2361.8 - 104.9) - 298.15 \, (7.2831 - 0.3674) = 195.0 \text{ kJ/kg}$$

The reversible work can be found from Eq. 8.18.

$$\frac{W_{rev}}{m_1} = \psi_1 - \frac{m_2}{m_1}\psi_2 - \frac{m_3}{m_1}\psi_3$$

$$\frac{W_{rev}}{m_1} = 1109.6 - \frac{5}{30}\,(755.3) - \frac{25}{30}\,(195.0) = 821.2 \text{ kJ/kg}$$

**Example 8.4**

A lead storage battery of the type used in an automobile is able to deliver 5.2 MJ of electrical energy. This energy is available for starting the car.

Suppose we wish to use compressed air for doing an equivalent amount of work in starting the car. The compressed air is to be stored at 7 MPa, 25°C. What volume of tank would be required to have the compressed air have an availability of 5.2 MJ?

Let us first find the availability of air at 7 MPa, 25°C. From Eq. 8.21,

$$\phi = (u - u_0) - T_0(s - s_0) + P_0(v - v_0)$$

$$v = \frac{RT}{P} = \frac{0.287 \times 298.15}{7000} = 0.012\,22 \text{ m}^3/\text{kg}$$

$$v_0 = \frac{RT_0}{P_0} = \frac{0.287 \times 298.15}{100} = 0.8557 \text{ m}^3/\text{kg}$$

$$\phi = 0 - 298.15 \left( 0 - 0.287 \ \ln \frac{7000}{100} \right) + 100 \ (0.012 \ 22 - 0.8557)$$

$$= 0 + 363.5 - 84.3 = 279.2 \ \text{kJ/kg}$$

In order to have an availability of 5.2 MJ, the mass of air is

$$m = \frac{5200}{279.2} = 18.625 \ \text{kg}$$

$$V = mv = 18.625 \ (0.012 \ 22) = 0.2276 \ \text{m}^3$$

## 8.4   Consideration of Processes That Involve Heat Transfer with a Body Other than the Atmosphere

Up to this point we have considered process in which the only heat transfer is with the surroundings. Since many processes do involve heat transfer with bodies at a temperature above or below the temperature of the surroundings, a few remarks should be made regarding them.

The simplest way of handling such processes is to consider two (or more) sub-systems, one comprising the body from which heat is transferred, and the other the body to which heat is transferred. These two sub-systems comprise the system under consideration and the reversible work, availability and irreversibility for the system will be the sum of these quantities for the sub-systems. The same procedure could be used in the case of a control volume. The two examples that follow illustrate this approach. The first involves the irreversible transfer of heat and the second involves reversible heat transfer.

### Example 8.5

In a boiler, heat is transferred from the products of combustion to the steam. The temperature of the products of combustion decreases from 1100°C to 550°C while the pressure remains constant at 0.1 MPa. The average constant-pressure specific heat of the products of combustion is 1.09 kJ/kg K. The water enters at 0.8 MPa, 150°C and leaves at 0.8 MPa, 250°C. Determine the reversible work and the irreversibility for this process per kilogram of water evaporated.

We select the control volume shown in Fig. 8.8. Thus we are considering a steady-state, steady-flow process that involves the flow of products and flow of water across the control surface.

Assuming the products to be an ideal gas with constant specific heat, the change of entropy of the products for this constant-pressure process

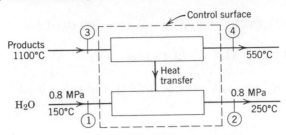

**Fig. 8.8**   Sketch for Example 8.5.

is given by the relation:

$$(s_e - s_i)_{\text{prod}} = C_{po} \ln \frac{T_e}{T_i}$$

For this control volume we can write the following governing equations:

Continuity eq.:

$$(\dot{m}_i)_{\text{H}_2\text{O}} = (\dot{m}_e)_{\text{H}_2\text{O}} \qquad\qquad (a)$$

$$(\dot{m}_i)_{\text{prod}} = (\dot{m}_e)_{\text{prod}} \qquad\qquad (b)$$

First law: (A steady-state, steady-flow process)

$$(\dot{m}_i h_i)_{\text{H}_2\text{O}} + (\dot{m}_i h_i)_{\text{prod}} = (\dot{m}_e h_e)_{\text{H}_2\text{O}} + (\dot{m}_e h_e)_{\text{prod}} \qquad\qquad (c)$$

Second law: (The process is adiabatic for the control volume shown)

$$(\dot{m}_e s_e)_{\text{H}_2\text{O}} + (\dot{m}_e s_e)_{\text{prod}} \geq (\dot{m}_i s_i)_{\text{H}_2\text{O}} + (\dot{m}_i s_i)_{\text{prod}}$$

From Eqs. $a$, $b$, and $c$, we can calculate ratio of the mass flow of products to the mass flow of water.

$$\dot{m}_{\text{prod}}(h_i - h_e)_{\text{prod}} = \dot{m}_{\text{H}_2\text{O}}(h_e - h_i)_{\text{H}_2\text{O}}$$

$$\frac{\dot{m}_{\text{prod}}}{\dot{m}_{\text{H}_2\text{O}}} = \frac{(h_e - h_i)_{\text{H}_2\text{O}}}{(h_i - h_e)_{\text{prod}}} = \frac{2950 - 632.2}{1.09\,(1100 - 550)} = 3.866$$

To find the reversible work we find $w_{\text{rev}}$ for the water and for the products and add these together.

We first calculate $w_{\text{rev}}$ for the change of state of the water.

$$(w_{\text{rev}})_{\text{H}_2\text{O}} = (h_1 - h_2) - T_0(s_1 - s_2)$$

$$= (632.2 - 2950) - 298.15\,(1.8418 - 7.0384)$$

$$= -768.4 \text{ kJ/kg}$$

Next we consider the products.

$$(w_{rev})_{prod} = \frac{\dot{m}_{prod}}{\dot{m}_{H_2O}} [(h_3 - h_4) - T_0(s_3 - s_4)]$$

(States 3 and 4 designate the initial and final states of the products.)

$$(w_{rev})_{prod} = 3.866 \left[ 1.09 (1100 - 550) - 298.15 \left( 1.09 \ln \frac{1373.15}{823.15} \right) \right]$$

$$= 1674.7 \text{ kJ/kg } H_2O$$

$$\frac{\dot{i}}{\dot{m}_{H_2O}} = (-768.4 + 1674.7) - 0 = 906.3 \text{ kJ/kg } H_2O$$

It is also of interest to determine the net change of entropy. The change in the entropy of the water is

$$(s_2 - s_1)_{H_2O} = 7.0384 - 1.8418 = 5.1966 \text{ kJ/kg } H_2O \text{ K}$$

The change in the entropy of the products is

$$\frac{\dot{m}_{prod}}{\dot{m}_{H_2O}} (s_4 - s_3)_{prod} = -3.866 \left( 1.09 \ln \frac{1373.15}{823.15} \right) = -2.1564 \text{ kJ/kg } H_2O \text{ K}$$

Thus, there is a net increase in entropy during the process. The irreversibility could also have been calculated from Eq. 8.14:

$$\dot{I} = \sum \dot{m}_e T_0 s_e - \sum \dot{m}_i T_0 s_i - \dot{Q}_{c.v.}$$

For the control volume selected, $\dot{Q}_{c.v.} = 0$, and therefore,

$$\frac{\dot{i}}{\dot{m}_{H_2O}} = T_0(s_2 - s_1)_{H_2O} + \frac{\dot{m}_{prod}}{\dot{m}_{H_2O}} T_0(s_4 - s_3)_{prod}$$

$$= 298.15 (5.1966) + 298.15 (-2.1564)$$

$$= 906.3 \text{ kJ/kg } H_2O$$

These two processes are shown on the $T$-$s$ diagram of Fig. 8.9. Line 3–4 represents the process for the 3.866 kg of products. Area 3–4–$c$–$d$–3 represents the heat transferred from the 3.866 kg of products of combustion, and area 3–4–$e$–$f$–3 represents the reversible work for the

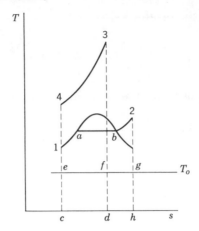

**Fig. 8.9** Temperature-entropy diagram for Example 8.5.

given change in state of these products. Area $1-a-b-2-h-c-1$ represents the heat transferred to the water, and this is equal to area $3-4-c-d-3$ which represents the heat transferred from the products of combustion. Area $1-a-b-2-g-e-1$ represents the reversible work for the given change in state of the water. The difference between area $3-4-e-f-3$ (the reversible work for the products) and area $1-a-b-2-g-e-1$ (the reversible work for the water) represents the net reversible work. It is readily shown that this net reversible work is equal to area $f-g-h-d-f$, or $T_0(\Delta s)_{net}$. Since the actual work is zero, this area also represents the irreversibility, which agrees with our calculation above.

It is essential to note that when the change of state involving heat transfer takes place reversibly, the net change in entropy is zero, and therefore the decrease in the entropy of the body from which heat is transferred must be equal to the increase in entropy of the body to which heat is transferred. This is best demonstrated by considering an example similar to Example 8.5.

**Example 8.6**

Repeat Example 8.5, but assume that the heat transfer could be made to be entirely reversible, i.e., that the heat transfer takes place through a reversible engine.

Schematically this would involve heat transfer from the products of combustion to reversible engines that reject heat to the water, as shown in Fig. 8.10.

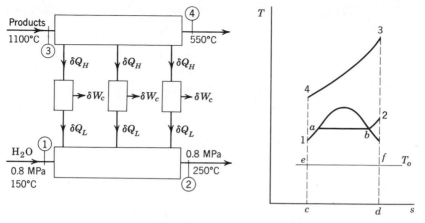

**Fig. 8.10**  Diagram for Example 8.6.

We again write the governing equations:

Continuity eq.:
$$(\dot{m}_i)_{H_2O} = (\dot{m}_e)_{H_2O}$$
$$(\dot{m}_i)_{prod} = (\dot{m}_e)_{prod}$$

First law:
$$(\dot{m}_i h_i)_{H_2O} + (\dot{m}_i h_i)_{prod} = (\dot{m}_e h_e)_{H_2O} + (\dot{m}_e h_e)_{prod} + \dot{W}_{c.v.}$$

Second law: Since this is a reversible adiabatic process,
$$(\dot{m}_e s_e)_{H_2O} + (\dot{m}_e s_e)_{prod} = (\dot{m}_i s_i)_{H_2O} + (\dot{m}_i s_i)_{prod}$$

From the continuity equation and second law we can determine flow of products per unit flow of water.

$$(s_2 - s_1)_{H_2O} + \frac{\dot{m}_{prod}}{\dot{m}_{H_2O}} (s_4 - s_3)_{prod} = 0$$

$$\frac{\dot{m}_{prod}}{\dot{m}_{H_2O}} (s_3 - s_4)_{prod} = (7.0384 - 1.8418)$$

$$\frac{\dot{m}_{prod}}{\dot{m}_{H_2O}} \left( 1.09 \ln \frac{1373.15}{823.15} \right) = 5.1966$$

$$\frac{\dot{m}_{prod}}{\dot{m}_{H_2O}} = 9.317$$

We now calculate the reversible work for the change of state of the water and the products:

$$(w_{rev})_{H_2O} = (h_1 - h_2) - T_0(s_1 - s_2)$$

$$(w_{rev})_{prod} = [(h_3 - h_4) - T_0(s_3 - s_4)] \frac{\dot{m}_{prod}}{\dot{m}_{H_2O}}$$

But, since

$$(s_2 - s_1)_{H_2O} = \frac{\dot{m}_{prod}}{\dot{m}_{H_2O}} (s_3 - s_4)_{prod}$$

it follows that

$$w_{rev} = (h_1 - h_2) + \frac{\dot{m}_{prod}}{\dot{m}_{H_2O}} (h_3 - h_4)$$

We note that this net reversible work is exactly equal to the work which would be determined from the first law. We would expect this, of course, since the process is completely reversible.

$$w_{rev} = (632.2 - 2950) + 9.317 \times 1.09 (1100 - 550)$$

$$= 3267.7 \text{ kJ/kg } H_2O$$

Since the net change in entropy is zero, the $T-s$ diagram is as shown in Fig. 8.10. Area $3-4-c-d-3$ represents the heat transferred from the products to the engines, and area $1-a-b-2-d-c-1$ represents the heat received by the water. Area $3-4-1-a-b-2-3$ represents the work done by the heat engines, which is equal to the reversible work for this process.

## 8.5    Processes Involving Chemical Reactions

Although chemical reaction will not be considered in detail until Chapter 12 some preliminary remarks regarding availability in such processes can be made here. In the first place, we observe that the reactants are often in pressure and temperature equilibrium with the surroundings before the reaction takes place, and the same is true of the products after the reaction. An automobile engine would be an example of such a process if we visualized the products being cooled to atmospheric temperature before being discharged from the engine.

Let us first consider for a system the implications of temperature equilibrium with the surroundings during a chemical reaction in a system.

The temperature of the system $T$ is equal to $T_0$, the temperature of the surroundings. Therefore, from Eq. 8.8 we can write (noting that in this case $T_0 = T$),

$$_1(w_{rev})_2 = \left(u_1 - T_1 s_1 + \frac{V_1^2}{2} + gZ_1\right) - \left(u_2 - T_2 s_2 + \frac{V_2^2}{2} + gZ_2\right)$$

The quantity $(U - TS)$ is a thermodynamic property of a substance and is called the Helmholtz function. It is an extensive property and we designate it by the symbol $A$. Thus

$$A = U - TS$$
$$a = u - Ts \tag{8.23}$$

Therefore, when a system undergoes a change of state while in temperature equilibrium with the surroundings, the reversible work is given by the relation

$$_1(w_{rev})_2 = \left(a_1 + \frac{V_1^2}{2} + gZ_1\right) - \left(a_2 + \frac{V_2^2}{2} + gZ_2\right)$$

In those cases where the kinetic and potential energy changes are not significant, this reduces to

$$_1(W_{rev})_2 = A_1 - A_2 = m(a_1 - a_2) \tag{8.24}$$

Let us now consider a system that undergoes a chemical reaction while in both pressure and temperature equilibrium with the surroundings. The availability function $\phi$ for a system has been defined as

$$\phi = (u + P_0 v - T_0 s) - (u_0 + P_0 v_0 - T_0 s_0)$$

If $P = P_0$ and $T = T_0$, then

$$\phi = (u + Pv - Ts) - (u_0 + P_0 v_0 - T_0 s_0)$$
$$= (h - Ts) - (h_0 - T_0 s_0) \tag{8.25}$$

The quantity $h - Ts$ is a thermodynamic property and is termed the Gibbs function, designated by the symbol $G$.

$$G = H - TS$$
$$g = h - Ts \tag{8.26}$$

Introducing the Gibbs function into Eq. 8.25 we have, when a system is in pressure and temperature equilibrium with the surroundings.

$$\phi = g - g_0$$

From Eq. 8.22 it follows that under these same conditions

$$_1(w_{rev})_2 = (g_1 - g_2) - P_0(v_1 - v_2) + \frac{V_1^2 - V_2^2}{2} + g\,(Z_1 - Z_2) \quad (8.27)$$

The Gibbs function is also of significance in a steady-state, steady-flow process that takes place in temperature equilibrium with the surroundings. Since in this case $T_i = T_e = T_0$, Eq. 8.9 reduces to

$$W_{rev} = m_i\left(h_i - T_i s_i + \frac{V_i^2}{2} + gZ_i\right) - m_e\left(h_e - T_e s_e + \frac{V_e^2}{2} + gZ_e\right)$$

Introducing the Gibbs function we have

$$W_{rev} = m_i\left(g_i + \frac{V_i^2}{2} + gZ_i\right) - m_e\left(g_e + \frac{V_e^2}{2} + gZ_e\right) \quad (8.28)$$

Thus we have introduced two new properties, the Helmholtz function, $A$, and the Gibbs function, $G$. Both these functions are very important in the thermodynamics of chemical reactions and will be used extensively in later chapters of this book.

## PROBLEMS

With no other information available assume that the surroundings are at 25°C, 0.1 MPa pressure.

8.1    Air enters the compressor of a gas turbine at 95 kPa, 15°C, with a velocity of 140 m/s. The air leaves the compressor at a pressure of 400 kPa, 200°C, and a velocity of 70 m/s. The process is adiabatic. Calculate the reversible work and irreversibility per kg of air for this process.

8.2    One means for controlling the power output of a steam turbine is a throttling governor, which throttles the supply steam to a lower pressure. Consider a case in which the supply steam is at 7 MPa at a temperature of 700°C, and the turbine load is such that the steam is throttled to 5 MPa in the governor. Determine the reversible work and irreversibility for this throttling process.

8.3    Freon-12 enters the expansion valve of a refrigerator at a pressure of 1 MPa and a temperature of 30°C. It leaves the expansion valve at a temperature of $-15°C$. Calculate the reversible work and irreversibility for this process. Do you think it would be worthwhile to attempt to replace the irreversibile throttling process with a reversible process in order to reduce the work required for operating the compressor?

8.4 Consider the process described in Problem 5.17. Determine the reversible work and the irreversibility for this process, assuming the surroundings to be at 25°C.

8.5 Consider the compressor of a gas turbine engine in which air enters at ambient conditions, 100 kPa, 5°C, and leaves at 400 kPa. The compressor has an efficiency of 84%. Determine the work of compression, the reversible work for the actual change of state, and the irreversibility of the process per kg of air.

8.6 Air enters an adiabatic compressor at ambient conditions. 100 kPa, 15°C, and leaves at 550 kPa. The mass flow rate is 0.01 kg/s and the efficiency of the compressor is 75%. After leaving the compressor, the air is cooled to 40°C in an aftercooler. Calculate:
   (a) The power required to drive the compressor.
   (b) The rate of irreversibility for the overall process (compressor plus cooler).

8.7 A commercial laundry requires a hot water supply of 2.5 kg/s at 85°C. This supply is to be produced in a steady-flow process by mixing the 10°C water supply with steam supplied from a boiler as saturated vapor at 0.1 MPa pressure. Determine the rate of irreversibility of the mixing process.

8.8 Consider the process described in Problem 5.53 for producing liquid Freon-12. Assume that the Freon-12 is compressed from 0.1 MPa, 25°C to the state at which it enters the heat exchanger, 2.75 MPa, 100°C, in a reversible, polytropic process.
   (a) Determine the compressor work required to liquefy 1 kg of Freon-12.
   (b) Determine the irreversibility per kg of Freon-12 liquefied.

8.9 What is the minimum work required to change 1 kg of oxygen from 0.1 MPa, 300 K to saturated liquid at 90 K? Sketch an apparatus which would be used in such a process and show the process on a temperature entropy diagram.

8.10 Steam flows in a pipe at 1.4 MPa, 300°C. An evacuated vessel having a volume of 0.4 m³ is attached to this line through a valve, which is initially closed. The valve is opened and steam flows into the vessel until the pressure reaches 1.4 MPa, at which point the valve is closed.
   Determine the irreversibility, assuming no heat transfer during the process.

8.11 A 200 litre tank containing saturated vapor Freon-12 at $-10°C$ is connected to a line flowing liquid Freon-12 at 25°C, 800 kPa. The valve is then opened and Freon flows into the tank. During the process, heat is transferred with the surroundings at 25°C such that when the valve is closed the tank contains 50% liquid, 50% vapor, by volume, at 25°C. Calculate the heat transfer and the irreversibility for this process.

8.12 In a certain chemical processing operation, it is necessary to compress 0.04 kg/s of ammonia from −26°C, 90% quality to 1200 kPa. It is estimated that the final temperature will be about 70°C and that a net heat loss to the surroundings of about 20 kJ/kg of ammonia will occur.

(*a*) Determine the compressor power requirement.

(*b*) Calculate the compressor efficiency, and the irreversibility of the process.

8.13 A solid aluminum sphere 0.1 m in diameter and initially at 200°C is allowed to cool to ambient temperature, 25°C. What is the irreversibility of this process? For aluminum, use a density of 2700 kg/m³ and specific heat of 0.9 kJ/kg K.

8.14 Figure 8.11 shows a closed feedwater heater used in a steam power plant. Determine the irreversibility per kg of feedwater leaving the heater at 2.7 MPa, 90°C.

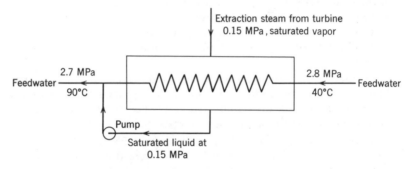

**Fig. 8.11**  Sketch for Problem 8.14.

8.15 Consider the system shown in Fig. 8.12 in which a refrigeration unit transfers heat from the water in cylinder *A* and rejects heat to the water in vessel *B* until *A* contains saturated liquid. Cylinder *A* is fitted with a frictionless piston of such a mass that 0.175 MPa is required to balance the force exerted by the piston and atmospheric pressure. Cylinder *A* initially contains 1 kg of vapor and 1 kg of liquid while vessel *B* with a volume of 2.4 m³ contains water at 0.2 MPa, 50% quality.

(*a*) If the refrigerator has a coefficient of performance of 0.5, determine the final temperature in *B*.

(*b*) If the refrigerator is reversible, determine the final temperature in *B*, and the coefficient of performance for the refrigerator.

8.16 An ideal gas with constant specific heat expands through an adiabatic turbine. Show that the irreversibility of the process can be expressed in terms of the following: specific heat, ambient temperature, turbine efficiency, and pressure ratio across the turbine.

8.17 A rigid 300 litre vessel contains air initially at 3 MPa and room temperature, 25°C. A valve is opened and the pressure quickly drops to 1.5 MPa before the valve is closed. It may be assumed that the air that remains

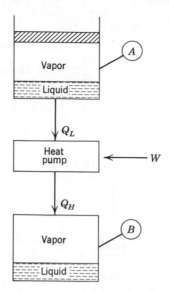

**Fig. 8.12** Sketch for Problem 8.15.

inside the vessel has undergone a reversible adiabatic expansion. Determine

(*a*) The initial availability of the air in the vessel.

(*b*) The availability of the air in the vessel immediately after the valve is closed.

(*c*) After a long time the air remaining inside the vessel returns to room temperature. Determine the availability of the air in the vessel at this state.

8.18 A pressure vessel has a volume of 1 m³ and contains air at 1.4 MPa, 175°C. The air is cooled to 25°C by heat transfer to the surroundings at 25°C. Calculate the availability in the initial and final states and the irreversibility of this process.

8.19 A nuclear reactor power plant utilizing water as the working fluid is shown schematically in Fig. 8.13. The high-temperature compressed liquid from the reactor is throttled into the flash evaporator. That fraction of the water which flashes into vapor flows to the turbine, while that which remains liquid flows to the mixing chamber. Determine

(*a*) The availability of the water leaving the reactor.

(*b*) The turbine output per kg of water leaving the reactor, assuming a reversible, adiabatic process in the turbine.

(*c*) The irreversibility in the flash evaporator per kg of water entering.

(*d*) The irreversibility in the mixing chamber per kg of water leaving.

8.20 Liquid nitrogen at 0.1013 MPa pressure is to be vaporized and delivered to a pipeline at 4 MPa, 275 K. Three possible schemes for doing this are shown in Fig. 8.14. The first scheme involves pumping the liquid and then vaporizing it. The second involves vaporizing the liquid and superheating it just enough so that it would leave a reversible adiabatic compressor at 4 MPa, 275 K. The third involves vaporizing it and superheating it to 275 K,

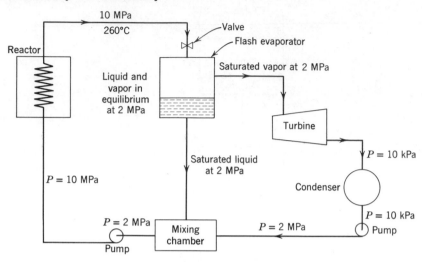

**Fig. 8.13**   Sketch for Problem 8.19.

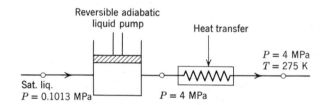

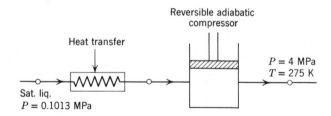

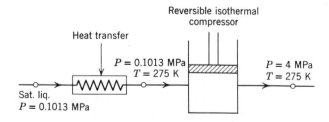

**Fig. 8.14**   Sketch for Problem 8.20.

followed by a reversible isothermal compressor. Assume an ambient temperature of 275 K.

(a) What is the availability of liquid nitrogen at 0.1013 MPa pressure?

(b) What is the minimum work necessary to deliver 1 kg of nitrogen at 4 MPa, 275 K?

Show a schematic arrangement of how you would do this with this minimum work input.

(c) Determine the work and the irreversibility per kilogram of nitrogen delivered for each of the three suggested ways of accomplishing this.

8.21 Helium at ambient conditions, 20°C, 0.1 MPa, is compressed in a reversible isothermal process to 0.5 MPa and then expanded back to 0.1 MPa in a reversible adiabatic process. Both are steady-state, steady-flow processes. What is the availability at the exit?

8.22 Two blocks of metal, each having a mass of 10 kg and a specific heat of 0.4 kJ/kg K, are at a temperature of 40°C. A reversible refrigerator receives heat from one block and rejects heat to the other. Calculate the work required to cause a temperature difference of 100°C between the two blocks.

8.23 Consider the process described in Problem 5.73. Calculate the irreversibility for this process assuming that the required heat transfer to the air and $H_2O$ is from a reservoir at 600°C.

8.24 At a certain location the temperature of the water supply is 15°C. Ice is to be made from this water supply by the process shown in Fig. 8.15. The final temperature of the ice is −10°C, and the final temperature of the water that is used as cooling water in the condenser is 30°C.

What is the minimum work required to produce 1000 kg of ice?

8.25 Consider two identical blocks of copper, A and B, each having a mass of 10 kg and specific heat of 0.4 kJ/kg K. The temperature of block A is 1200 K and that of block B is 300 K. The blocks are then brought to the same temperature.

(a) If the process is reversible, find the final temperature and the work done during the process.

(b) If the process is accomplished by simply bringing the blocks into thermal communication, find the final temperature and the irreversibility of the process.

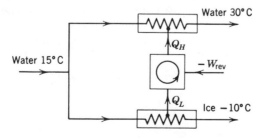

**Fig. 8.15**   Sketch for Problem 8.24.

8.26 An air preheater is used to cool the products of combustion from a furnace while heating the air to be used for combustion. The rate of flow of products is 12.5 kg/s, and the products are cooled from 300°C to 200°C, and for the products at this temperature $C_p = 1.09$ kJ/kg K. The rate of air flow is 11.5 kg/s, the initial air temperature is 40°C, and for the air $C_p = 1.004$ kJ/kg K.

   (a) What is the initial and final availability of the products?

   (b) What is the irreversibility for this process?

   (c) Suppose this heat transfer from the products took place reversibly through heat engines. What would be the final temperature of the air? What power would be developed by the heat engines?

8.27 The work output of a reversible cyclic heat engine is used to drive an air compressor, with the exiting air then being used as the heat-sink for the engine. Conditions are as shown in Fig. 8.16. Find the required reservoir temperature, $T_H$.

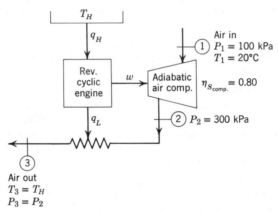

**Fig. 8.16**  Sketch for Problem 8.27.

8.28 One kmol of carbon dust is burned with 1 kmol of oxygen to form 1 kmol of carbon dioxide in a steady-flow process. The temperature of the carbon and oxygen before combustion is 25°C and the temperature of the carbon dioxide after combustion is also 25°C. The heat transferred during the process is −393 522 kJ, and the entropy of the carbon dioxide after combustion is 2.858 kJ/kmol K higher than the entropy of carbon and oxygen before combustion. Calculate the reversible work and the irreversibility for this process.

# 9

---

# Some Power and Refrigeration Cycles

Some power plants, such as the simple steam power plant, which we have considered several times, operate in a cycle. That is, the working fluid undergoes a series of processes and finally returns to the initial state. In other power plants, such as the internal combustion engine and gas turbine the working fluid does not go through a thermodynamic cycle, even though the engine itself may operate in a mechanical cycle. In this case the working fluid has a different composition or is in a different state at the conclusion of the process than at the beginning. Such equipment is sometimes said to operate on the open cycle (the word cycle is really a misnomer), whereas the steam power plant operates on a closed cycle. The same distinction between open and closed cycles can be made regarding refrigeration devices. For both the open- and closed-cycle type of apparatus, however, it is advantageous to analyze the performance of an idealized closed cycle similar to the actual cycle. Such a procedure is particularly advantageous in determining the influence of certain variables on performance. For example, the spark-ignition internal combustion engine is usually approximated by the Otto cycle. From an analysis of the Otto cycle one concludes that increasing the compression ratio increases the efficiency. This is also true for the actual engine, even though the Otto cycle efficiencies may deviate significantly from the actual efficiencies.

This chapter is concerned with these idealized cycles, both for power and refrigeration apparatus. The working fluids considered are both vapors and ideal gases. An attempt will be made to point out how the processes in actual apparatus deviate from the ideal. Consideration is also given to certain modifications of the basic cycles which are intended to improve performance. These involve the use of such devices as regenerators, multistage compressors and expanders, and intercoolers. The order in which the cycles will be considered is: (1) vapor power cycles, (2) vapor refrigeration cycles, (3) air-standard power cycles, and (4) air-standard refrigeration cycles.


## VAPOR POWER CYCLES

### 9.1   The Rankine Cycle

The ideal cycle for a simple steam power plant is the Rankine cycle, shown in Fig. 9.1. The processes that comprise the cycle are:

1–2: Reversible adiabatic pumping process in the pump.

2–3: Constant-pressure transfer of heat in the boiler.

3–4: Reversible adiabatic expansion in the turbine (or other prime mover such as a steam engine).

4–1: Constant-pressure transfer of heat in the condenser.

The Rankine cycle also includes the possibility of superheating the vapor, as cycle 1–2–3′–4′–1.

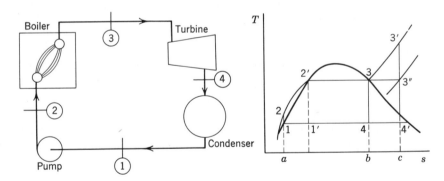

**Fig. 9.1**   Simple steam power plant which operates on the Rankine cycle.

If changes of kinetic and potential energy are neglected, heat transfer and work may be represented by various areas on the $T$-$s$ diagram. The heat transferred to the working fluid is represented by area $a$–2–2′–3–$b$–$a$, and the heat transferred from the working fluid by area $a$–1–4–$b$–$a$. From the first law we conclude that the area representing the work is the difference between these two areas, namely, area 1–2–2′–3–4–1. The thermal efficiency is defined by the relation

$$\eta_{\text{th}} = \frac{w_{\text{net}}}{q_H} = \frac{\text{area } 1\text{--}2\text{--}2'\text{--}3\text{--}4\text{--}1}{\text{area } a\text{--}2\text{--}2'\text{--}3\text{--}b\text{--}a} \tag{9.1}$$

In analyzing the Rankine cycle it is helpful to think of efficiency as depending on the average temperature at which heat is supplied and the average temperature at which heat is rejected. Any changes that increase the average temperature at which heat is supplied or decrease the average temperature at which heat is rejected will increase the Rankine cycle efficiency.

It should be stated that in analyzing the ideal cycles in this chapter the changes in kinetic and potential energies from one point in the cycle to another are neglected. In general this is a reasonable assumption for the actual cycles.

It is readily evident that the Rankine cycle has a lower efficiency than a Carnot cycle with the same maximum and minimum temperatures as a Rankine cycle, because the average temperature between 2 and 2′ is less than the temperature during evaporation. We might well ask, why choose the Rankine cycle as the ideal cycle? Why not select the Carnot cycle 1′–2′–3–4–1′? At least two reasons can be given. The first involves the pumping process. State 1′ is a mixture of liquid and vapor, and great difficulties are encountered in building a pump that will handle the mixture of liquid and vapor at 1′ and deliver saturated liquid at 2′. It is much easier to completely condense the vapor and handle only liquid in the pump, and the Rankine cycle is based on this fact. The second reason involves superheating the vapor. In the Rankine cycle the vapor is superheated at constant pressure, process 3–3′. In the Carnot cycle all the heat transfer is at constant temperature, and therefore the vapor is superheated in process 3–3″. Note, however, that during this process the pressure is dropping, which means that the heat must be transferred to the vapor as it undergoes an expansion process in which work is done. This also is very difficult to achieve in practice. Thus, the Rankine cycle is the ideal cycle that can be approximated in practice. In the sections that follow we will consider some variations on the Rankine cycle that enable one to more closely approach the Carnot-cycle efficiency.

Before discussing the influence of certain variables on the performance of the Rankine cycle, an example is given.

## Example 9.1

Determine the efficiency of a Rankine cycle utilizing steam as the working fluid in which the condenser pressure is 10 kPa. The boiler pressure is 2 MPa. The steam leaves the boiler as saturated vapor.

In solving Rankine-cycle problems we will let $w_p$ denote the work into the pump per kilogram of fluid flowing and $q_L$ the heat rejected from the working fluid per kilogram of fluid flowing.

In solving this problem we consider, in succession, a control surface around the pump, the boiler, the turbine, and the condenser. In each case the property relation used is the steam table.

Consider a control surface around the pump.

First law:     $w_p = h_2 - h_1$

Second law:     $s_2 = s_1$

Since

$$s_2 = s_1, \quad h_2 - h_1 = \int_1^2 v\,dP$$

Therefore, assuming the fluid to be incompressible,

$$w_p = v(P_2 - P_1) = 0.001\ 01\ (2000 - 10) = 2.0 \text{ kJ/kg}$$

$$h_2 = h_1 + w_p = 191.8 + 2.0 = 193.8$$

Next consider a control surface around the boiler.

First law:     $q_H = h_3 - h_2 = 2799.5 - 193.8 = 2605.7 \text{ kJ/kg}$

Consider a control surface around the turbine.

First law:     $w_t = h_3 - h_4$

Second law:     $s_3 = s_4$

We can determine the quality at state 4 as follows:

$$s_3 = s_4 = 6.3409 = 8.1502 - (1 - x_4)\ 7.5009, \qquad (1 - x_4) = 0.2412$$

$$h_4 = 2584.7 - 0.2412\ (2392.8) = 2007.5$$

$$w_t = 2799.5 - 2007.5 = 792.0 \text{ kJ/kg}$$

Finally, consider a control surface around the condenser.

First law:   $q_L = h_4 - h_1 = 2007.5 - 191.8 = 1815.7$ kJ/kg

We can now calculate the thermal efficiency

$$\eta_{th} = \frac{w_{net}}{q_H} = \frac{q_H - q_L}{q_H} = \frac{w_t - w_p}{q_H} = \frac{792.0 - 2.0}{2605.7} = 30.3\%$$

We could also write an expression for thermal efficiency in terms of properties at various points in the cycle.

$$\eta_{th} = \frac{(h_3 - h_2) - (h_4 - h_1)}{h_3 - h_2} = \frac{(h_3 - h_4) - (h_2 - h_1)}{h_3 - h_2}$$

$$= \frac{2605.7 - 1815.7}{2605.7} = \frac{792.0 - 2.0}{2605.7} = 30.3\%$$

## 9.2   Effect of Pressure and Temperature on the Rankine Cycle

Let us first consider the effect of exhaust pressure and temperature on the Rankine cycle. This effect is shown on the $T$-$s$ diagram of Fig. 9.2.

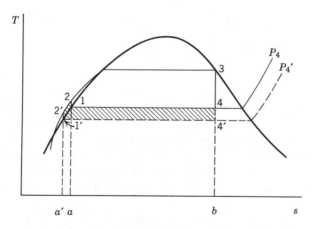

**Fig. 9.2**   Effect of exhaust pressure on Rankine-cycle efficiency.

Let the exhaust pressure drop from $P_4$ to $P_4'$, with the corresponding decrease in temperature at which heat is rejected. The net work is increased by area 1–4–4'–1'–2'–2–1 (shown by the crosshatching). The heat transferred to the steam is increased by area $a'$–2'–2–a–a'. Since these two areas are approximately equal, the net result is an increase in cycle efficiency. This is also evident from the fact that the average temperature at which heat is rejected is decreased. Note however, that lowering the back pressure causes an increase in the moisture content in the steam leaving the turbine. This is a significant factor because if the moisture in the low-pressure stages of the turbine exceeds about 10 per cent, not only is there a decrease in turbine efficiency, but also erosion of the turbine blades may be a very serious problem.

Next consider the effect of superheating the steam in the boiler, as shown in Fig. 9.3. It is readily evident that the work is increased by area

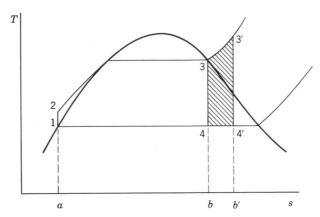

**Fig. 9.3**   Effect of superheating on Rankine-cycle efficiency.

3–3'–4'–4–3, and the heat transferred in the boiler is increased by area 3–3'–b'–b–3. Since this ratio of these two areas is greater than the ratio of net work to heat supplied for the rest of the cycle, it is evident that for given pressures, superheating the steam increases the Rankine-cycle efficiency. This would also follow from the fact that the average temperature at which heat is transferred to the steam is increased. Note also that when the steam is superheated the quality of the steam leaving the turbine increases.

Finally, the influence of the maximum pressure of the steam must be considered, and this is shown in Fig. 9.4. In this analysis the maximum temperature of the steam, as well as the exhaust pressure, is held constant. The heat rejected decreases by area b'–4'–4–b–b'. The net work increases by the amount of the single crosshatching and decreases by the amount of the double crosshatching. Therefore the net work tends to remain the same, but the heat rejected decreases, and therefore

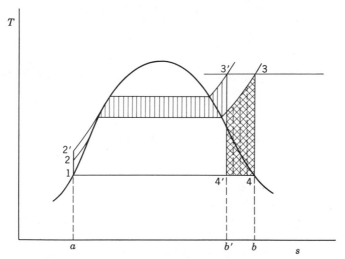

**Fig. 9.4**   Effect of boiler pressure on Rankine-cycle efficiency.

the Rankine-cycle efficiency increases with an increase in maximum pressure. Note that in this case also the average temperature at which heat is supplied increases with an increase in pressure. The quality of the steam leaving the turbine decreases as the maximum pressure increases.

To summarize this section we can say that the Rankine-cycle efficiency can be increased by lowering the exhaust pressure, increasing the pressure during heat addition, and by superheating the steam. The quality of the steam leaving the turbine is increased by superheating the steam, and decreased by lowering the exhaust pressure and by increasing the pressure during heat addition.

**Example 9.2**

In a Rankine cycle steam leaves the boiler and enters the turbine at 4 MPa, 400°C. The condenser pressure is 10 kPa. Determine the cycle efficiency.

To determine the cycle efficiency we must calculate the turbine work, the pump work, and the heat transfer to the steam in the boiler. We do this by considering a control surface around each of these components and assuming steady-state, steady-flow processes.

Consider a control surface around the pump.

First law:        $w_p = h_2 - h_1$

Second law:      $s_2 = s_1$

Since

$$s_2 = s_1, \quad h_2 - h_1 = \int_1^2 v \, dP = v(P_2 - P_1)$$

Therefore,

$$w_p = v(P_2 - P_1) = 0.001 \; 01 \; (4000 - 10) = 4.0 \text{ kJ/kg}$$
$$h_1 = 191.8$$
$$h_2 = 191.8 + 4.0 = 195.8$$

Consider a control surface around the turbine.

First law:    $w_t = h_3 - h_4$

Second law:    $s_4 = s_3$
$$h_3 = 3213.6 \quad ; \quad s_3 = 6.7690$$
$$s_4 = s_3 = 6.7690 = 8.1502 - (1 - x_4) \; 7.5009$$
$$(1 - x_4) = 0.1841$$
$$h_4 = 2584.7 - 0.1841 \; (2392.8) = 2144.1$$
$$w_t = h_3 - h_4 = 3213.6 - 2144.1 = 1069.5 \text{ kJ/kg}$$
$$w_\text{net} = w_t - w_p = 1069.5 - 4.0 = 1065.5 \text{ kJ/kg}$$

Consider a control surface around the boiler.

$$q_H = h_3 - h_2 = 3213.6 - 195.8 = 3017.8 \text{ kJ/kg}$$

$$\eta_\text{th} = \frac{w_\text{net}}{q_H} = \frac{1065.5}{3017.8} = 35.3\%$$

The net work could also be determined by calculating the heat rejected in the condenser, $q_L$, and noting, from the first law, that the net work for the cycle is equal to the net heat transfer. Considering a control surface around the condenser,

$$q_L = h_4 - h_1 = 2144.1 - 191.8 = 1952.3 \text{ kJ/kg}$$

Therefore,

$$w_\text{net} = q_H - q_L = 3017.8 - 1952.3 = 1065.5 \text{ kJ/kg}$$

## 9.3   The Reheat Cycle

In the last paragraph we noted that the efficiency of the Rankine cycle could be increased by increasing the pressure during the addition of heat. However, this also increases the moisture content of the steam in the low-pressure end of the turbine. The reheat cycle has been developed to take advantage of the increased efficiency with higher pressures, and yet avoid excessive moisture in the low-pressure stages of the turbine. This cycle is shown schematically and on a $T$-$s$ diagram in Fig. 9.5. The unique feature of this cycle is that the steam is expanded to some intermediate pressure in the turbine, and is then reheated in the boiler, after

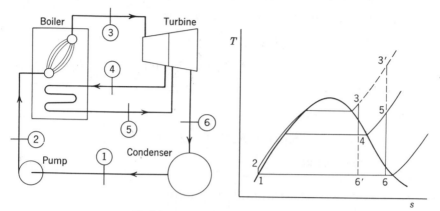

**Fig. 9.5**  The ideal reheat cycle.

which it expands in the turbine to the exhaust pressure. It is evident from the $T$-$s$ diagram that there is very little gain in efficiency from reheating the steam, because the average temperature at which heat is supplied is not greatly changed. The chief advantage is in decreasing the moisture content in the low-pressure stages of the turbine to a safe value. Note also, that if metals could be found that would enable one to superheat the steam to 3', the simple Rankine cycle would be more efficient than the reheat cycle, and there would be no need for the reheat cycle.

## Example 9.3

Consider a reheat cycle utilizing steam. Steam leaves the boiler and enters the turbine at 4 MPa, 400°C. After expansion in the turbine to 400 kPa, the steam is reheated to 400°C and then expanded in the low-pressure turbine to 10 kPa. Determine the cycle efficiency.

Designating the states in accordance with Fig. 9.5 we proceed as follows:

Consider a control surface around the turbine.

First law:     $w_t = (h_3 - h_4) + (h_5 - h_6)$

Second law:   $s_4 = s_3$

            $s_6 = s_5$

To find $w_t$ we proceed as follows:

$$h_3 = 3213.6 \quad ; \quad s_3 = 6.7690$$

$$s_4 = s_3 = 6.7690 = 6.8959 - (1 - x_4)\, 5.1193, \quad (1 - x_4) = 0.0248$$

$$h_4 = 2738.6 - 0.0248\,(2133.8) = 2685.7$$

$$h_5 = 3273.4 \quad ; \quad s_5 = 7.8985$$

$$s_6 = s_5 = 7.8985 = 8.1502 - (1 - x_6)\, 7.5009, \quad (1 - x_6) = 0.0336$$

$$h_6 = 2584.7 - 0.0336\,(2392.8) = 2504.4$$

$$w_t = (h_3 - h_4) + (h_5 - h_6)$$

$$= (3213.6 - 2685.7) + (3273.4 - 2504.4)$$

$$= 1296.9 \text{ kJ/kg}$$

Consider a control surface around the pump.

First law:      $w_p = h_2 - h_1$
Second law:     $s_2 = s_1$

Since

$$s_2 = s_1, \quad h_2 - h_1 = \int_1^2 v\, dP = v(P_2 - P_1)$$

Therefore,

$$w_p = v(P_2 - P_1) = 0.001\,01\,(4000 - 10) = 4.0 \text{ kJ/kg}$$

$$h_2 = 191.8 + 4.0 = 195.8$$

Consider a control surface around the boiler.

$$q_H = (h_3 - h_2) + (h_5 - h_4)$$

$$= (3213.6 - 195.8) + (3273.4 - 2685.7) = 3605.5 \text{ kJ/kg}$$

$$w_{\text{net}} = w_t - w_p = 1296.9 - 4.0 = 1292.9 \text{ kJ/kg}$$

$$\eta_{\text{th}} = \frac{w_{\text{net}}}{q_H} = \frac{1292.9}{3605.5} = 35.9\%$$

Note by comparison with Example 9.2 that the gain in efficiency from reheating is relatively small, but that the moisture content leaving the turbine is decreased as a result of reheating from 18.4 per cent to 3.4 per cent.

## 9.4   The Regenerative Cycle

Another important variation from the Rankine cycle is the regenerative cycle, which involves the use of feedwater heaters. The basic concepts of this cycle can be demonstrated by considering the Rankine cycle without superheat as shown in Fig. 9.6. During the process between states 2 and 2' the working fluid is heated while in the liquid phase, and

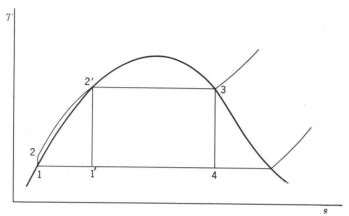

**Fig. 9.6**   Temperature-entropy diagram showing the relationship between Carnot cycle efficiency and Rankine-cycle efficiency.

the average temperature of the working fluid is much lower during this process than during the vaporization process 2'–3. This causes the average temperature at which heat is supplied in the Rankine cycle to be lower than in the Carnot cycle 1'–2'–3–4–1', and consequently the efficiency of the Rankine cycle is less than that of the corresponding Carnot cycle. In the regenerative cycle the working fluid enters the boiler at some state between 2 and 2', and consequently the average temperature at which heat is supplied is increased.

Consider first an idealized regenerative cycle, shown in Fig. 9.7. The unique feature of this cycle compared to the Rankine cycle is that after leaving the pump, the liquid circulates around the turbine casing, counterflow to the direction of vapor flow in the turbine. Thus, it is possible to transfer heat from the vapor as it flows through the turbine to the liquid flowing around the turbine. Let us assume for the moment that this is a reversible heat transfer; that is, at each point the temperature of the vapor is only infinitesimally higher than the temperature of the liquid. In this case line 4–5 on the $T$-$s$ diagram of Fig. 9.7, which represents the states of the vapor flowing through the turbine, is exactly

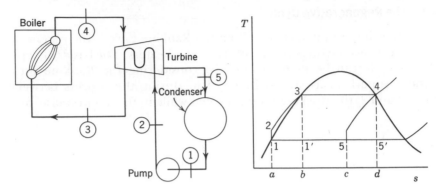

**Fig. 9.7**    The ideal regenerative cycle.

parallel to line 1–2–3, which represents the pumping process (1–2) and the states of the liquid flowing around the turbine. Consequently areas 2–3–b–a–2 and 5–4–d–c–5 are not only equal but congruous, and these areas respectively represent the heat transferred to the liquid and from the vapor. Note also that heat is transferred to the working fluid at constant temperature in process 3–4, and area 3–4–d–b–3 represents this heat transfer. Heat is transferred from the working fluid in process 5–1, and area 1–5–c–a–1 represents this heat transfer. Note that this area is exactly equal to area 1'–5'–d–b–1', which is the heat rejected in the related Carnot cycle 1'–3–4–5'–1'. Thus, this idealized regenerative cycle has an efficiency exactly equal to the efficiency of the Carnot cycle with the same heat-supply and heat-rejection temperatures.

Quite obviously this idealized regenerative cycle is not practical. First of all, it would not be possible to effect the necessary heat transfer from the vapor in the turbine to the liquid feedwater. Furthermore, the moisture content of the vapor leaving the turbine is considerably increased as a result of the heat transfer, and the disadvantage of this has been noted previously. The practical regenerative cycle involves the extraction of some of the vapor after it has partially expanded in the turbine and the use of feedwater heaters, as shown in Fig. 9.8.

Steam enters the turbine at state 5. After expansion to state 6, some of the steam is extracted and enters the feedwater heater. The steam that is not extracted is expanded in the turbine to state 7 and is then condensed in the condenser. This condensate is pumped into the feedwater heater where it mixes with the steam extracted from the turbine. The proportion of steam extracted is just sufficient to cause the liquid leaving the feedwater heater to be saturated at state 3. Note that the liquid has not been pumped to the boiler pressure, but only to the intermediate

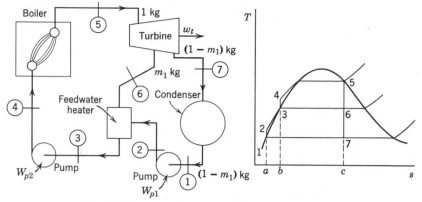

**Fig. 9.8** Regenerative cycle with open feedwater heater.

pressure corresponding to state 6. Another pump is required to pump the liquid leaving the feedwater heater to boiler pressure. The significant point is that the average temperature at which heat is supplied has been increased.

This cycle is somewhat difficult to show on a $T$-$s$ diagram because the mass of steam flowing through the various components is not the same. The $T$ $s$ diagram of Fig. 9.8 simply shows the state of the fluid at the various points.

Area 4–5–$c$–$b$–4 in Fig. 9.8 represents the heat transferred per kilogram of working fluid. Process 7–1 is the heat-rejection process, but since not all the steam passes through the condenser, area 1–7–$c$–$a$–1 represents the heat transfer per kilogram flowing through the condenser, which does not represent the heat transfer per kilogram of working fluid entering the turbine. Note also that between states 6 and 7 only part of the steam is flowing through the turbine. The example that follows illustrates the calculations involved in the regenerative cycle.

### Example 9.4

Consider a regenerative cycle utilizing steam as the working fluid. Steam leaves the boiler and enters the turbine at 4 MPa, 400°C. After expansion to 400 kPa some of the steam is extracted from the turbine for the purpose of heating the feedwater in an open feedwater heater. The pressure in the feedwater heater is 400 kPa and the water leaving it is saturated liquid at 400 kPa. The steam not extracted expands to 10 kPa. Determine the cycle efficiency.

The line diagram and $T$-$s$ diagram for this cycle are shown in Fig. 9.8.

From Examples 9.2 and 9.3 we have the following properties:

$$h_5 = 3213.6 \qquad h_6 = 2685.7$$

$$h_7 = 2144.1 \qquad h_1 = 191.8$$

Consider a control surface around the low pressure pump.

First law:    $w_{p1} = (h_2 - h_1)$
Second law:    $s_2 = s_1$
Therefore,

$$h_2 - h_1 = \int_1^2 v\, dP = v(P_2 - P_1)$$

$$w_{p1} = v(P_2 - P_1) = 0.001\ 01\ (400 - 10) = 0.4\ \text{kJ/kg}$$

$$h_2 = h_1 + w_p = 191.8 + 0.4 = 192.2$$

$$h_3 = 604.7$$

Consider a control surface around the turbine.

First law:    $w_t = (h_5 - h_6) + (1 - m_1)(h_6 - h_7)$
Second law:    $s_5 = s_6 = s_7$

From the second law, the values for $h_6$ and $h_7$ given above have been calculated in Examples 9.2 and 9.3.
Next consider a control surface around the feedwater heater.

First law:    $m_1(h_6) + (1 - m_1)h_2 = h_3$

$$m_1 = (2685.7) + (1 - m_1)\,(192.2) = 604.7$$

$$m_1 = 0.1654$$

We can now calculate the turbine work.

$$w_t = (h_5 - h_6) + (1 - m_1)(h_6 - h_7)$$

$$= (3213.6 - 2685.7) + (1 - 0.1654)\,(2685.7 - 2144.1)$$

$$= 979.9\ \text{kJ/kg}$$

Consider a control surface around the high pressure pump.

First law: $\qquad w_{p2} = (h_4 - h_3)$
Second law: $\qquad s_4 = s_3$

$$w_{p2} = v(P_4 - P_3) = 0.001\ 084\ (4000 - 400) = 3.9 \text{ kJ/kg}$$

$$w_{net} = w_t - (1 - m_1) w_{p1} - w_{p2}$$

$$= 979.9 - (1 - 0.1654)\ (0.4) - 3.9 = 975.7 \text{ kJ/kg}$$

Consider a control surface around the boiler.

$$h_4 = h_3 + w_{p2} = 604.7 + 3.9 = 608.6$$

$$q_H = h_5 - h_4 = 3213.6 - 608.6 = 2605.0 \text{ kJ/kg}$$

$$\eta_{th} = \frac{w_{net}}{q_H} = \frac{975.7}{2605.0} = 37.5\%$$

Note the increase in efficiency over the Rankine cycle of Example 9.2.

Up to this point the discussion and example have tacitly assumed that the extraction steam and feedwater are mixed in the feedwater heater. Another much-used type of feedwater heater, known as a closed heater, is one in which the steam and feedwater do not mix, but rather heat is transferred from the extracted steam as it condenses on the outside of tubes as the feedwater flows through the tubes. In a closed heater, a schematic sketch of which is shown in Fig. 9.9, the steam and feed-

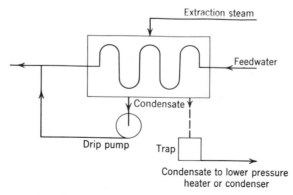

**Fig. 9.9** Schematic arrangement for a closed feedwater heater.

water may be at considerably different pressures. The condensate may be pumped into the feedwater line, or it may be removed through a trap (a device that permits liquid but no vapor to flow to a region of lower pressure) to a lower pressure heater or to the main condenser.

Open feedwater heaters have the advantage of being less expensive and having better heat-transfer characteristics compared to closed feedwater heaters. They have the disadvantage of requiring a pump to handle the feedwater between each heater.

In many power plants a number of stages of extraction are used, though only rarely more than five. The number is, of course, determined by economic considerations. It is evident that by using a very large number of extraction stages and feedwater heaters, the cycle efficiency would approach that of the idealized regenerative cycle of Fig. 9.7, where the feedwater enters the boiler as saturated liquid at the maximum pressure. However, in practice this could not be economically justified because the savings effected by the increase in efficiency would be more than offset by the cost of additional equipment (feedwater heaters, piping, etc.).

A typical arrangement of the main components in an actual power plant is shown in Fig. 9.10. Note that one open feedwater heater is a

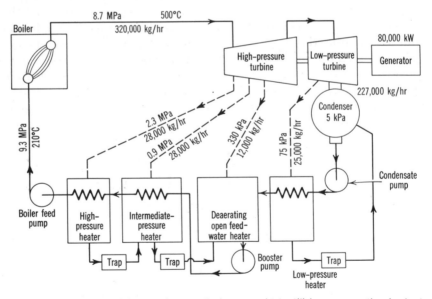

**Fig. 9.10** Arrangement of heaters in an actual power plant utilizing regenerative feedwater heaters.

deaerating feedwater heater, and this has the dual purpose of heating and removing the air from the feedwater. Unless the air is removed excessive corrosion occurs in the boiler. Note also that the condensate from the high-pressure heater drains (through a trap) to the intermediate heater, and the intermediate heater drains to the deaerating feedwater heater. The low-pressure heater drains to the condenser.

In many cases an actual power plant combines one reheat stage with a number of extraction stages. The principles already considered are readily applied to such a cycle.

## 9.5   Deviation of Actual Cycles from Ideal Cycles

Before leaving the matter of vapor power cycles, a few comments are in order regarding the ways in which an actual cycle deviates from an ideal cycle. (The losses associated with the combustion process are considered in a later chapter.) The most important of these are as follows.

*Piping Losses*

Pressure drop due to frictional effects and heat transfer to the surroundings are the most important piping losses. Consider for example the pipe connecting the turbine to the boiler. If only frictional effects occurred, the states $a$ and $b$ in Fig. 9.11 would represent the states of the

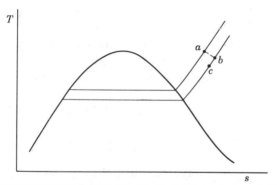

**Fig. 9.11**   Temperature-entropy diagram showing effect of losses between boiler and turbine.

steam leaving the boiler and entering the turbine respectively. Note that this causes an increase in entropy. Heat transferred to the surroundings at constant pressure can be represented by process $bc$. This effect causes a decrease in entropy. Both the pressure drop and heat transfer cause a decrease in the availability of the steam entering the turbine, and the irreversibility of this process can be calculated by the methods outlined in Chapter 8.

A similar loss is the pressure drop in the boiler. Because of this pressure drop, the water entering the boiler must be pumped to a much higher pressure than the desired steam pressure leaving the boiler, and this requires additional pump work.

*Turbine Losses*

The losses in the turbine are primarily those associated with the flow of the working fluid through the turbine. Heat transfer to the surroundings also represents a loss, but this is usually of secondary importance. The effects of these two losses are the same as those outlined for piping losses, and the process might be as represented in Fig. 9.12, where $4_s$ re-

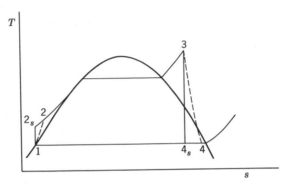

**Fig. 9.12** Temperature-entropy diagram showing effect of turbine and pump inefficiencies on cycle performance.

presents the state after an isentropic expansion and state 4 represents the actual state leaving the turbine. The governing procedures may also cause a loss in the turbine, particularly if a throttling process is used to govern the turbine.

The efficiency of the turbine has been defined (in Chapter 7) as

$$\eta_t = \frac{w_t}{h_3 - h_{4s}}$$

where the states are as designated in Fig. 9.12.

*Pump Losses*

The losses in the pump are similar to those of the turbine, and are primarily due to the irreversibilities associated with the fluid flow. Heat transfer is usually a minor loss.

The pump efficiency is defined as

$$\eta_p = \frac{h_{2s} - h_1}{w_p}$$

where the states are as shown in Fig. 9.12, and $w_p$ is the actual work input per kilogram of fluid.

*Condenser Losses*

The losses in the condenser are relatively small. One of these minor losses is the cooling below the saturation temperature of the liquid leaving the condenser. This represents a loss because additional heat transfer is necessary to bring the water to its saturation temperature.

The influence of these losses on the cycle is illustrated in the following example, which should be compared to Example 9.2.

**Example 9.5**

A steam power plant operates on a cycle with pressure and temperatures as designated in Fig. 9.13. The efficiency of the turbine is 86 per cent and the efficiency of the pump is 80 per cent. Determine the thermal efficiency of this cycle.

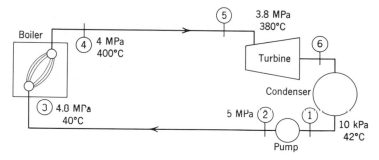

**Fig. 9.13**    Schematic diagram for Example 9.5.

From the steam tables

$$h_1 = 175.9 \qquad h_3 = 171.8$$

$$h_4 = 3213.6 \qquad s_4 = 6.7690$$

$$h_5 = 3169.1 \qquad s_5 = 6.7235$$

This cycle is shown on the $T$-$s$ diagram of Fig. 9.14. Consider a control surface around the turbine.

First law:    $w_t = h_6 - h_5$

Second law:    $s_{6s} = s_5$

$$\eta_t = \frac{w_t}{h_5 - h_{6s}} = \frac{h_5 - h_6}{h_5 - h_{6s}}$$

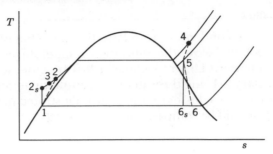

**Fig. 9.14**  Temperature-entropy diagram for Example 9.5.

$$s_{6s} = s_5 = 6.7235 = 8.1502 - (1 - x_{6s})\,7.5009, \qquad (1 - x_{6s}) = 0.1902$$

$$h_{6s} = 2584.7 - 0.1902\,(2392.8) = 2129.6$$

$$w_t = \eta_t(h_5 - h_{6s}) = 0.86\,(3169.1 - 2129.6) = 894.0 \text{ kJ/kg}$$

Consider a control surface around the pump.

First law:    $w_p = h_2 - h_1$
Second law:    $s_{2s} = s_1$

$$\eta_p = \frac{h_{2s} - h_1}{w_p} = \frac{h_{2s} - h_1}{h_2 - h_1}$$

Since

$$s_{2s} = s_1, \quad h_{2s} - h_1 = v(P_2 - P_1)$$

Therefore,

$$w_p = \frac{h_{2s} - h_1}{\eta_p} = \frac{v(P_2 - P_1)}{\eta_p} = \frac{0.001\,009\,(5000 - 10)}{0.80} = 6.3 \text{ kJ/kg}$$

$$w_{\text{net}} = w_t - w_p = 894.0 - 6.3 = 887.7 \text{ kJ/kg}$$

If we apply the first law to a control surface around the boiler, we have

$$q_H = h_4 - h_3 = 3213.6 - 171.8 = 3041.8 \text{ kJ/kg}$$

$$\eta_{\text{th}} = \frac{887.7}{3041.8} = 29.2\%$$

This compares to an efficiency of 35.3 per cent for the Rankine efficiency of the similar cycle of Example 9.2.

## VAPOR REFRIGERATION CYCLES

### 9.6    Vapor Compression Refrigeration Cycles

The ideal cycle for vapor compression refrigeration is shown in Fig. 9.15 as cycle 1–2–3–4–1. Saturated vapor at low pressure enters the

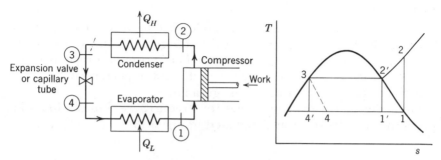

**Fig. 9.15**   The ideal vapor-compression refrigeration cycle.

compressor and undergoes a reversible adiabatic compression, 1–2. Heat is then rejected at constant pressure in process 2–3, and the working fluid leaves the condenser as saturated liquid. An adiabatic throttling process follows, process 3–4, and the working fluid is then evaporated at constant pressure, process 4–1, to complete the cycle.

The similarity between this cycle and the Rankine cycle is evident, for it is essentially the same cycle in reverse, except that an expansion valve replaces the pump. This throttling process is irreversible, whereas the pumping process of the Rankine cycle is reversible. The deviation of this ideal cycle from the Carnot cycle 1′–2′–3–4′–1′ is evident from the *T-s* diagram. The reason for the deviation is that it is much more expedient to have a compressor handle only vapor than a mixture of liquid and vapor as would be required in process 1′–2′ of the Carnot cycle. It is virtually impossible to compress (at a reasonable rate) a mixture such as that represented by state 1′ and maintain equilibrium between the liquid and vapor, because there must be a heat and mass transfer across the phase boundary. It is also much simpler to have the expansion process take place irreversibly through an expansion valve than to have an expansion device that receives saturated liquid and discharges a mixture of liquid and vapor, as would be required in process 3–4′. For these reasons the ideal cycle for vapor-compression refrigeration is as shown in Fig. 9.15 by cycle 1–2–3–4–1.

As we saw in Chapter 6, the performance of a refrigeration cycle is given in terms of the coefficient of performance, $\beta$, which is defined for a refrigeration cycle as

$$\beta = \frac{q_L}{w_c} \tag{9.2}$$

### Example 9.6

Consider an ideal refrigeration cycle that utilizes Freon-12 as the working fluid. The temperature of the refrigerant in the evaporator is $-20°C$ and in the condenser it is $40°C$. The refrigerant is circulated at the rate of 0.03 kg/s. Determine the coefficient of performance and the capacity of the plant in rate of refrigeration.

From the thermodynamic tables for Freon-12 we find the following properties for the states as designated in Fig. 9.15.

$$h_1 = 178.61 \qquad s_1 = 0.7082$$

$$P_2 = 0.9607 \text{ MPa}$$

Consider a control surface around the compressor.

First law:     $w_c = h_2 - h_1$
Second law:    $s_2 = s_1$

Therefore,

$$s_2 = s_1 = 0.7082$$

$$T_2 = 50.8°C \qquad \text{and} \qquad h_2 = 211.38$$

$$w_c = h_2 - h_1 = 211.38 - 178.61 = 32.77 \text{ kJ/kg}$$

Considering a control surface around the expansion valve, we conclude from the first law that

$$h_4 = h_3 = 74.53$$

Finally, consider a control surface around the evaporator.

First law:    $q_L = h_1 - h_4 = 178.61 - 74.53 = 104.08 \text{ kJ/kg}$

Therefore,

$$\beta = \frac{q_L}{w_c} = \frac{104.08}{32.77} = 3.18$$

$$\text{Capacity} = 104.08 \times 0.3 = 3.12 \text{ kW}$$

## 9.7  Working Fluids for Vapor-Compression Refrigeration Systems

A much larger number of different working fluids (refrigerants) are utilized in vapor-compression refrigeration systems than in vapor power cycles. Ammonia and sulfur dioxide were important in the early days of vapor-compression refrigeration. Today, however, the main refrigerants are the halogenated hydrocarbons, which are marketed under the trade names of Freon and Genatron. For example, dichlorodifluoromethane ($CCl_2F_2$) is known as Freon-12 and Genatron-12. Two important considerations in selecting a refrigerant are the temperature at which refrigeration is desired and the type of equipment to be used.

Since the refrigerant undergoes a change of phase during the heat transfer process, the pressure of the refrigerant will be the saturation pressure during the heat-supply and heat-rejection processes. Low pressures mean large specific volumes and correspondingly large equipment. High pressures mean smaller equipment but it must be designed to withstand higher pressure. In particular, the pressures should be well below the critical pressure. For extremely low temperature applications a binary fluid system may be used by cascading two separate systems.

The type of compressor used has a particular bearing on the refrigerant. Reciprocating compressors are best adapted to low specific volumes, which means higher pressures, whereas centrifugal compressors are most suitable for low pressures and high specific volumes.

It is also important that the refrigerants used in domestic appliances be nontoxic. Other important characteristics are tendency to cause corrosion, miscibility with compressor oil, dielectric strength, stability, and cost. Also, for given temperatures during evaporation and condensation, not all refrigerants have the same coefficient of performance for the ideal cycle. It is, of course, desirable to utilize the refrigerant with the highest coefficient of performance, other factors permitting.

### 9.8   Deviation of the Actual Vapor-Compression Refrigeration Cycle from the Ideal Cycle

The actual refrigeration cycle deviates from the ideal cycle primarily because of pressure drops associated with fluid flow and heat transfer to or from the surroundings. The actual cycle might approach the one shown in Fig. 9.16.

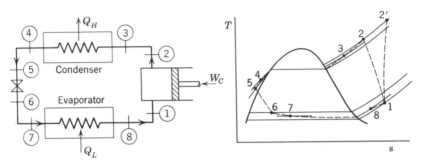

**Fig. 9.16**   The actual vapor-compression refrigeration cycle.

The vapor entering the compressor will probably be superheated. During the compression process there are irreversibilities and heat transfer either to or from the surroundings, depending on the temperature of the refrigerant and the surroundings. Therefore, the entropy might increase or decrease during this process, for the irreversibility and heat transfer to the refrigerant cause an increase in entropy, and heat transfer from the refrigerant causes a decrease in entropy. These possibilities are represented by the two dotted lines 1–2 and 1–2′. The pressure of the liquid leaving the condenser will be less than the pressure of the vapor entering, and the temperature of the refrigerant in the condenser will be somewhat above that of the surroundings to which heat is being transferred. Usually the temperature of the liquid leaving the condenser is lower than the saturation temperature, and it might drop somewhat more in the piping between the condenser and expansion valve. This represents a gain, however, because as a result of this heat transfer the refrigerant enters the evaporator with a lower enthalpy, thus permitting more heat transfer to the refrigerant in the evaporator.

There is some drop in pressure as the refrigerant flows through the evaporator. It may be slightly superheated as it leaves the evaporator, and due to heat transfer from the surroundings the temperature will increase in the piping between the evaporator and compressor. This heat transfer represents a loss, because it increases the work of the compressor as a result of the increased specific volume of the fluid entering it.

### Example 9.7

A refrigeration cycle utilizes Freon-12 as the working fluid. Following are the properties at various points of the cycle designated in Fig. 9.16.

$$P_1 = 125 \text{ kPa} \qquad T_1 = -10°C$$
$$P_2 = 1.2 \text{ MPa} \qquad T_2 = 100°C$$
$$P_3 = 1.19 \text{ MPa} \qquad T_3 = 80°C$$
$$P_4 = 1.16 \text{ MPa} \qquad T_4 = 45°C$$
$$P_5 = 1.15 \text{ MPa} \qquad T_5 = 40°C$$
$$P_6 = P_7 = 140 \text{ kPa} \qquad x_6 = x_7$$
$$P_8 = 130 \text{ kPa} \qquad T_8 = -20°C$$

The heat transfer from the Freon-12 during the compression process is 4 kJ/kg. Determine the coefficient of performance of this cycle. From the Freon-12 tables the following properties are found.

$$h_1 = 185.16 \qquad\qquad h_2 = 245.52$$
$$h_5 = h_6 = h_7 = 74.53 \qquad h_8 = 179.12$$

Consider a control surface around the compressor.

First law:  $q + h_1 = h_2 + w$
$$w_c = -w = h_2 - h_1 - q$$
$$= 245.52 - 185.16 - (-4) = 64.36 \text{ kJ/kg}$$

To find $q_L$, consider a control surface around the evaporator.

$$q_L = h_8 - h_7 = 179.12 - 74.53 = 104.59 \text{ kJ/kg}$$

Therefore,

$$\beta = \frac{q_L}{w_c} = \frac{104.59}{64.36} = 1.625$$

## 9.9   The Ammonia-Absorption Refrigeration Cycle

The ammonia-absorption refrigeration cycle differs from the vapor compression cycle in the manner in which compression is achieved. In the absorption cycle the low-pressure ammonia vapor is absorbed in water and the liquid solution is pumped to a high pressure by a liquid pump. Figure 9.17 shows a schematic arrangement of the essential elements of such a system.

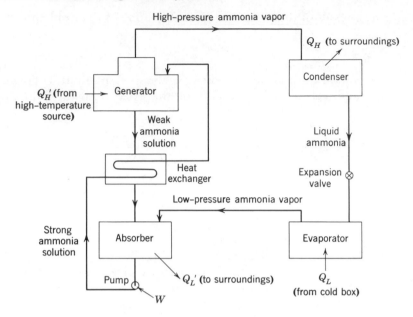

**Fig. 9.17**   The ammonia-absorption refrigeration cycle.

The low-pressure ammonia vapor leaving the evaporator enters the absorber where it is absorbed in the weak ammonia solution. This process takes place at a temperature slightly above that of the surroundings and heat must be transferred to the surroundings during this process. The strong ammonia solution is then pumped through a heat exchanger to the generator where a higher pressure and temperature are maintained. Under these conditions ammonia vapor is driven from the solution as a result of heat transfer from a high-temperature source. The ammonia vapor goes to the condenser where it is condensed, as in a vapor-compression system, and then to the expansion valve and evaporator. The weak ammonia solution is returned to the absorber through the heat exchanger.

The distinctive feature of the absorption system is that very little work input is required because the pumping process involves a liquid. This follows from the fact that for a reversible steady-flow process with negligible changes in kinetic and potential energy, the work is equal to $-\int v\, dP$, and the specific volume of the liquid is much less than the specific volume of the vapor. On the other hand, a relatively high-temperature source of heat must be available (100°C to 200°C). The equipment involved in an absorption system is somewhat greater than in a vapor-compression system and it can usually be economically justified only in those cases where a suitable source of heat is available that would otherwise be wasted.

This cycle brings out the important principle that since the work in a reversible steady-flow process with negligible changes in kinetic and potential energy is $- \int v \, dP$, a compression process should take place with the smallest possible specific volume.

## AIR-STANDARD POWER CYCLES

### 9.10   Air-Standard Cycles

Many work-producing devices (engines) utilize a working fluid that is always a gas. The spark-ignition automotive engine is a familiar example, and the same is true of the Diesel engine and conventional gas turbine. In all of these engines there is a change in the composition of the working fluid, because during combustion it changes from air and fuel to combustion products. For this reason these engines are called internal combustion engines. In contrast to this the steam power plant may be called an external-combustion engine, because heat is transferred from the products of combustion to the working fluid. External-combustion engines using a gaseous working fluid (usually air) have been built. To date they have had very limited application, but the use of the gas turbine cycle in conjunction with a nuclear reactor has been investigated extensively. Other external-combustion engines are currently receiving serious attention in an effort to combat the air pollution problem.

Because the working fluid does not go through a complete thermodynamic cycle in the engine (even though the engine operates in a mechanical cycle) the internal-combustion engine operates on the so-called open cycle. However, in order to analyze internal-combustion engines it is advantageous to devise closed cycles that closely approximate the open cycles. One such approach is the air-standard cycle, which is based on the following assumptions:

1. A fixed mass of air is the working fluid throughout the entire cycle, and the air is always an ideal gas. Thus there is no inlet process or exhaust process.
2. The combustion process is replaced by a heat-transfer process from an external source.
3. The cycle is completed by heat transfer to the surroundings (in contrast to the exhaust and intake process of an actual engine).
4. All processes are internally reversible.

**5.** The additional assumption is usually made that air has a constant specific heat.

The main value of the air-standard cycle is to enable us to examine qualitatively the influence of a number of variables on performance. The results obtained from the air-standard cycle, such as efficiency and mean effective pressure, will differ a great deal from those of the actual engine. The emphasis, therefore, in our consideration of the air-standard cycle will be primarily on the qualitative aspects.

The term "mean effective pressure," which is used in conjunction with reciprocating engines, is defined as the pressure which, if it acted on the piston during the entire power stroke, would do an amount of work equal to that actually done on the piston. The work for one cycle is found by multiplying this mean effective pressure by the area of the piston (minus the area of the rod on the crank end of a double-acting engine) and by the stroke.

## 9.11   The Air-Standard Carnot Cycle

The air-standard Carnot cycle is shown on the $P$-$v$ and $T$-$s$ diagrams of Fig. 9.18. Such a cycle could be achieved in either a reciprocating or steady-flow device, as shown in the same figure.

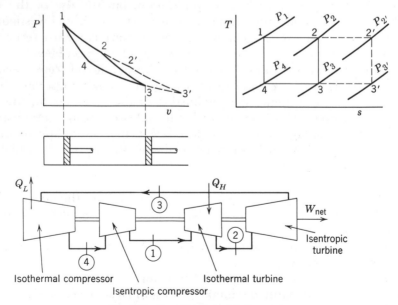

**Fig. 9.18**   The air-standard Carnot cycle.

In Chapter 6 it was noted that the efficiency of a Carnot cycle depends only on the temperatures at which heat is supplied and rejected, and is given by the relation

$$\eta_{th} = 1 - \frac{T_L}{T_H} = 1 - \frac{T_4}{T_1} = 1 - \frac{T_3}{T_2}$$

where the subscripts refer to Fig. 9.18. The efficiency may also be expressed by the pressure ratio or compression ratio during the isentropic processes. This follows from the fact that

$$\text{Isentropic pressure ratio} = r_{ps} = \frac{P_1}{P_4} = \frac{P_2}{P_3} = \left(\frac{T_3}{T_2}\right)^{k/(1-k)}$$

$$\text{Isentropic compression ratio} = r_{vs} = \frac{V_4}{V_1} = \frac{V_3}{V_2} = \left(\frac{T_3}{T_2}\right)^{1/(1-k)}$$

Therefore

$$\eta_{th} = 1 - r_{ps}^{(1-k)/k} = 1 - r_{vs}^{1-k} \qquad (9.3)$$

One other important variable in the air-standard Carnot cycle is the amount of heat transferred to the working fluid per cycle. It is evident from Fig. 9.18 that increasing the heat transfer per cycle at a given $T_H$ causes a larger change in volume during the cycle, and this causes a lower mean effective pressure in a reciprocating engine. In fact, for air-standard Carnot cycles having a minimum pressure between 0.1 and 1 MPa and a reasonable heat transfer per cycle, the mean effective pressure is so low that it would scarcely overcome friction forces.

Another practical difficulty of the Carnot cycle, which applies to both the reciprocating and steady-flow types of cycle, is the difficulty of transferring heat during the isothermal expansion and compression processes. It is virtually impossible to even approach this in an actual machine operating at a reasonable rate of speed. Thus, the air-standard Carnot cycle is not practical. Nonetheless it is of value as a standard for comparison with other cycles, and in a later paragraph we will consider the ideal gas turbine cycle with intercooling and reheating and note how its efficiency approaches that of the corresponding Carnot cycle.

### Example 9.8

In an air-standard Carnot cycle heat is transferred to the working fluid at 1200 K, and heat is rejected at 300 K. The heat transfer to the working fluid at 1200 K is 100 kJ/kg. The minimum pressure in the cycle is 0.1 MPa. Assuming constant specific heat of air, determine the cycle efficiency and the mean effective pressure.

Designating the states as in Fig. 9.18 we have

$$P_3 = 0.1 \text{ MPa} \qquad\qquad T_3 = T_4 = 300 \text{ K}$$

$$\frac{T_2}{T_3} = \left(\frac{P_2}{P_3}\right)^{(k-1)/k} = 4 \qquad\qquad T_1 = T_2 = 1200 \text{ K}$$

$$\frac{P_2}{P_3} = 128$$

$$P_2 = 0.1 \,(128) = 12.8 \text{ MPa}$$

Consider the process that occurs between states 1 and 2.

$$_1q_2 = RT \ln \frac{V_2}{V_1} = RT \ln \frac{P_1}{P_2} = 0.287 \times 1200 \ln \frac{P_1}{P_2} = 100$$

Therefore,

$$\frac{P_1}{P_2} = 1.337 \quad , \qquad\qquad P_1 = 17.11 \text{ MPa}$$

$$\frac{T_1}{T_4} = \left(\frac{P_1}{P_4}\right)^{(k-1)/k} = 4 \quad , \quad \frac{P_1}{P_4} = 128$$

$$P_4 = \frac{17.11}{128} = 0.134 \text{ MPa}$$

$$\eta_{\text{th}} = 1 - \frac{T_L}{T_H} = 1 - \frac{300}{1200} = 75.0\%$$

The product of mean effective pressure and the piston displacement is equal to the net work.

$$\text{mep}\,(v_3 - v_1) = w_{\text{net}} = \eta_{\text{th}} \times {_1q_2} = 0.75\,(100) = 75.0 \text{ kJ/kg}$$

$$v_3 = \frac{RT_3}{P_3} = \frac{0.287 \times 300}{100} = 0.861 \text{ m}^3/\text{kg}$$

$$v_1 = \frac{RT_1}{P_1} = \frac{0.287 \times 1200}{17\,110} = 0.0201 \text{ m}^3/\text{kg}$$

$$\text{mep} = \frac{75.0}{(0.861 - 0.0201)} = 89.2 \text{ kPa}$$

This is a much lower mean effective pressure than can be effectively utilized in a reciprocating engine.

## 9.12   The Air-Standard Otto Cycle

The air-standard Otto cycle is an ideal cycle that approximates a spark-ignition internal-combustion engine. This cycle is shown on the $P$-$v$ and $T$-$s$ diagrams of Fig. 9.19. Process 1–2 is an isentropic compression of the air as the piston moves from crank-end dead center to

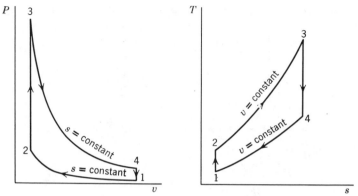

**Fig. 9.19**   The air-standard Otto cycle.

head-end dead center. Heat is then added at constant volume while the piston is momentarily at rest at head-end dead center. (This process corresponds to the ignition of the fuel-air mixture by the spark and the subsequent burning in the actual engine.) Process 3–4 is an isentropic expansion, and process 4–1 is the rejection of heat from the air while the piston is at crank-end dead center.

The thermal efficiency of this cycle is found as follows, assuming constant specific heat of air.

$$\eta_{\text{th}} = \frac{Q_H - Q_L}{Q_H} = 1 - \frac{Q_L}{Q_H} = 1 - \frac{mC_v(T_4 - T_1)}{mC_v(T_3 - T_2)}$$

$$= 1 - \frac{T_1\,(T_4/T_1 - 1)}{T_2\,(T_3/T_2 - 1)}$$

We note further that

$$\frac{T_2}{T_1} = \left(\frac{V_1}{V_2}\right)^{k-1} = \left(\frac{V_4}{V_3}\right)^{k-1} = \frac{T_3}{T_4}$$

Therefore,

$$\frac{T_3}{T_2} = \frac{T_4}{T_1}$$

and

$$\eta_{th} = 1 - \frac{T_1}{T_2} = 1 - (r_v)^{1-k} = 1 - \frac{1}{r_v^{k-1}} \qquad (9.4)$$

where $r_v$ = compression ratio = $\dfrac{V_1}{V_2} = \dfrac{V_4}{V_3}$

The important thing to note is that the efficiency of the air-standard Otto cycle is a function only of the compression ratio, and that the efficiency is increased by increasing the compression ratio. Figure 9.20

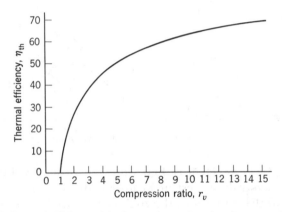

**Fig. 9.20**   Thermal efficiency of the Otto cycle as a function of compression ratio.

is a plot of the air-standard cycle thermal efficiency vs. compression ratio. It is also true of an actual spark-ignition engine that the efficiency can be increased by increasing the compression ratio. The trend toward higher compression ratios is prompted by the effort to obtain higher thermal efficiency. In the actual engine there is an increased tendency towards detonation of the fuel as compression ratio is increased. Detonation is characterized by an extremely rapid burning of the fuel and strong pressure waves present in the engine cylinder which give rise to the so-called spark knock. Therefore, the maximum compression ratio that can be used is fixed by the fact that detonation must be avoided. The advance in compression ratios over the years in the actual engine has been made possible by developing fuels with better antiknock characteristics, primarily through the addition of tetraethyl lead. More recently, however, non-leaded gasolines with fairly good antiknock characteristics have been developed in an effort to reduce atmospheric contamination.

Some of the most important ways in which the actual open-cycle spark-ignition engine deviates from the air-standard cycle are as follows:

1. The specific heats of the actual gases increase with an increase in temperature.
2. The combustion process replaces the heat-transfer process at high temperature, and combustion may be incomplete.
3. Each mechanical cycle of the engine involves an inlet and an exhaust process, and due to the pressure drop through the valves a certain amount of work is required to charge the cylinder with air and exhaust the products of combustion.
4. There will be considerable heat transfer between the gases in the cylinder and the cylinder walls.
5. There will be irreversibilities associated with pressure and temperature gradients.

**Example 9.9**

The compression ratio in an air-standard Otto cycle is 8. At the beginning of the compression stroke the pressure is 0.1 MPa and the temperature is 15°C. The heat transfer to the air per cycle is 1800 kJ/kg air. Determine:

1. The pressure and temperature at the end of each process of the cycle.
2. The thermal efficiency.
3. The mean effective pressure.

Designating the states as in Fig. 9.19 we have

$$P_1 = 0.1 \text{ MPa} \qquad\qquad T_1 = 288.2 \text{ K}$$

$$v_1 = \frac{0.287 \times 288.2}{100} = 0.827 \text{ m}^3/\text{kg}$$

$$\frac{T_2}{T_1} = \left(\frac{V_1}{V_2}\right)^{k-1} = 8^{0.4} = 2.3 \qquad T_2 = 662 \text{ K}$$

$$\frac{P_2}{P_1} = \left(\frac{V_1}{V_2}\right)^{k} = 8^{1.4} = 18.38 \quad , \quad P_2 = 1.838 \text{ MPa}$$

$$v_2 = \frac{0.827}{8} = 0.1034 \text{ m}^3/\text{kg}$$

$$_2q_3 = C_v(T_3 - T_2) = 1800 \text{ kJ/kg}$$

$$T_3 - T_2 = \frac{1800}{0.7165} = 2512 \quad, \quad T_3 = 3174 \text{ K}$$

$$\frac{T_3}{T_2} = \frac{P_3}{P_2} = \frac{3174}{662} = 4.795 \quad, \quad P_3 = 8.813 \text{ MPa}$$

$$\frac{T_3}{T_4} = \left(\frac{V_4}{V_3}\right)^{k-1} = 8^{0.4} = 2.3 \quad, \quad T_4 = 1380 \text{ K}$$

$$\frac{P_3}{P_4} = \left(\frac{V_4}{V_3}\right)^{k} = 8^{1.4} = 18.38 \quad, \quad P_4 = 0.4795 \text{ MPa}$$

$$\eta_{th} = 1 - \frac{1}{r_v^{k-1}} = 1 - \frac{1}{8^{0.4}} = 1 - \frac{1}{2.3} = 1 - 0.435 = 0.565 = 56.5\%$$

This can be checked by finding the heat rejected.

$$_4q_1 = C_v(T_1 - T_4) = 0.7165 \,(288.2 - 1380) = -782.3 \text{ kJ/kg}$$

$$\eta_{th} = 1 - \frac{782.3}{1800} = 1 - 0.435 = 0.565 = 56.5\%$$

$$w_{net} = 1800 - 782.3 = 1017.7 \text{ kJ/kg} = (v_1 - v_2)\, \text{mep}$$

$$\text{mep} = \frac{1017.7}{(0.827 - 0.1034)} = 1406 \text{ kPa}$$

Note how much higher this mean effective pressure is than for the Carnot cycle of Example 9.8. A low mean effective pressure means a large piston displacement for a given power output, and a large piston displacement means high frictional losses in an actual engine. In fact, the mean effective pressure of the Carnot cycle of Example 9.8 would scarcely overcome the friction in an actual engine.

### 9.13   The Air-Standard Diesel Cycle

The air-standard Diesel cycle is shown in Fig. 9.21. This is the ideal cycle for the Diesel engine, which is also called the compression-ignition engine.

In this cycle the heat is transferred to the working fluid at constant pressure. This process corresponds to the injection and burning of the

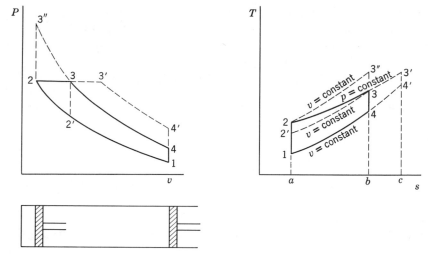

**Fig. 9.21**   The air-standard Diesel cycle.

fuel in the actual engine. Since the gas is expanding during the heat addition in the air-standard cycle, the heat transfer must be just sufficient to maintain constant pressure. When state 3 is reached the heat addition ceases and the gas undergoes an isentropic expansion, process 3–4, until the piston reaches crank-end dead center. As in the air-standard Otto cycle, a constant-volume rejection of heat at crank-end dead center replaces the exhaust and intake processes of the actual engine.

The efficiency of the Diesel cycle is given by the relation

$$\eta_{th} = 1 - \frac{Q_L}{Q_H} = 1 - \frac{C_v(T_4 - T_1)}{C_p(T_3 - T_2)} = 1 - \frac{T_1(T_4/T_1 - 1)}{kT_2(T_3/T_2 - 1)} \qquad (9.5)$$

It is important to note that the isentropic compression ratio is greater than the isentropic expansion ratio in the Diesel cycle. Also, for a given state before compression and a given compression ratio (i.e., given states 1 and 2) the cycle efficiency decreases as the maximum temperature increases. This is evident from the $T$-$s$ diagram, because the constant-pressure and constant-volume lines converge, and increasing the temperature from 3 to 3′ requires a large addition of heat (area 3–3′–$c$–$b$–3) and results in a relatively small increase in work (area 3–3′–4′–4–3).

There are a number of comparisons between the Otto cycle and the Diesel cycle, but here we will note only two. Consider Otto cycle 1–2–3″–4–1 and Diesel cycle 1–2–3–4–1, which have the same state at the beginning of the compression stroke and the same piston displacement and compression ratio. It is evident from the $T$-$s$ diagram that the Otto cycle has the higher efficiency. In practice, however, the Diesel engine can operate on a higher compression ratio than the spark-ignition engine. The reason is that in the spark-ignition engine an air-fuel mixture is compressed, and detonation (spark knock) becomes a serious problem if too high a compression ratio is used. This problem does not exist in the Diesel engine because only air is compressed during the compression stroke.

Therefore, we might compare an Otto cycle with a Diesel cycle and in each case select a compression ratio that might be achieved in practice. Such a comparison can be made by considering Otto cycle 1–2′–3–4–1 and Diesel cycle 1–2–3–4–1. The maximum pressure and temperature is the same for both cycles, which means that the Otto cycle has a lower compression ratio than the Diesel cycle. It is evident from the $T$-$s$ diagram that in this case the Diesel cycle has the higher efficiency. Thus, the conclusions drawn from a comparison of these two cycles must always be related to the basis on which the comparison has been made.

The actual compression-ignition open cycle differs from the air-standard Diesel cycle in much the same way that the spark-ignition open cycle differs from the air-standard Otto cycle.

### Example 9.10

An air-standard Diesel cycle has a compression ratio of 16, and the heat transferred to the working fluid per cycle is 1800 kJ/kg. At the beginning of the compression process the pressure is 0.1 MPa and the temperature is 15°C. Determine:

1. The pressure and temperature at each point in the cycle.
2. The thermal efficiency.
3. The mean effective pressure.

Designating the cycle as in Fig. 9.21 we have

$$P_1 = 0.1 \text{ MPa} \qquad T_1 = 288.2 \text{ K}$$

$$v_1 = \frac{0.287 \times 288.2}{100} = 0.827 \text{ m}^3/\text{kg}$$

$$v_2 = \frac{v_1}{16} = \frac{0.827}{16} = 0.0517 \text{ m}^3/\text{kg}$$

$$\frac{T_2}{T_1} = \left(\frac{V_1}{V_2}\right)^{k-1} = 16^{0.4} = 3.031 \quad , \quad T_2 = 873.5 \text{ K}$$

$$\frac{P_2}{P_1} = \left(\frac{V_1}{V_2}\right)^{k} = 16^{1.4} = 48.5 \quad , \quad P_2 = 4.85 \text{ MPa}$$

$$q_H = {}_2q_3 = C_p(T_3 - T_2) = 1800 \text{ kJ/kg}$$

$$T_3 - T_2 = \frac{1800}{1.0035} = 1794 \quad , \quad T_3 = 2667 \text{ K}$$

$$\frac{V_3}{V_2} = \frac{T_3}{T_2} = \frac{2667}{873.5} = 3.053 \quad , \quad v_3 = 0.1578 \text{ m}^3/\text{kg}$$

$$\frac{T_3}{T_4} = \left(\frac{V_4}{V_3}\right)^{k-1} = \left(\frac{0.827}{0.1578}\right)^{0.4} = 1.94 \quad , \quad T_4 = 1375 \text{ K}$$

$$q_L = {}_4q_1 = C_v(T_1 - T_4) = 0.7165\,(288.2 - 1375) = -778.7 \text{ kJ/kg}$$

$$w_{\text{net}} = 1800 - 778.7 = 1021.3 \text{ kJ/kg}$$

$$\eta_{\text{th}} = \frac{w_{\text{net}}}{q_H} = \frac{1021.3}{1800} = 56.7\%$$

$$\text{mep} = \frac{w_{\text{net}}}{v_1 - v_2} = \frac{1021.3}{(0.827 - 0.0517)} = 1317 \text{ kPa}$$

## 9.14  The Ericsson and Stirling Cycles

We shall briefly consider these two cycles, not so much because of their extensive use, but rather because they serve to demonstrate how a regenerator can often be incorporated in a cycle to give a significant increase in efficiency. This principle finds extensive application in turbines as well as in certain reciprocating devices. The Stirling cycle is shown on the P-v and T-s diagrams of Fig. 9.22. Heat is transferred to the working fluid during the constant-volume process 2–3 and during

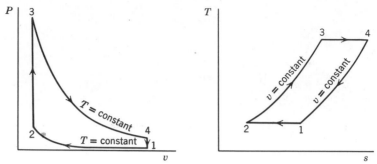

**Fig. 9.22**    The air-standard Stirling cycle.

the isothermal expansion process 3–4. Heat is rejected during the constant-volume process 4–1 and during the isothermal compression process 1–2. The significance of this cycle in conjunction with a regenerator is discussed in the next paragraph.

The Ericsson cycle is shown on the $P$-$v$ and $T$-$s$ diagrams of Fig. 9.23. This cycle differs from the Stirling cycle in that the constant-volume processes of the Stirling cycle are replaced by constant-pressure pro-

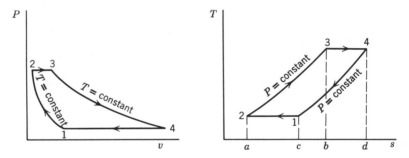

**Fig. 9.23**    The air-standard Ericsson cycle.

cesses. In both cycles there is an isothermal compression and expansion.

The importance of both cycles is the possibility of including a regenerator; by so doing the air-standard Stirling and Ericsson cycles may have an efficiency equal to that of a Carnot cycle operating between the same temperatures. This may be demonstrated by considering Fig. 9.24, in which the Ericsson cycle is accomplished in a device that is essentially a gas turbine. If we assume an ideal heat-transfer process in the regenerator, i.e., no pressure drop and an infinitesimal temperature difference

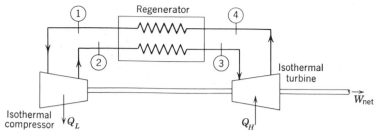

**Fig. 9.24** Schematic arrangement of an engine operating on the Ericsson cycle and utilizing a regenerator.

between the two streams, and reversible compression and expansion processes, then this device operates on the Ericsson cycle.

Note that the heat transfer to the gas between states 2 and 3, area $23ba2$, is exactly equal to the heat transfer from the gas between states 4 and 1, area $14dc1$. Thus $Q_H$ all takes place in the isothermal turbine between states 3 and 4, and $Q_L$ takes place in the isothermal compressor between states 1 and 2. Since all the heat is supplied and rejected isothermally, the efficiency of this cycle will equal the efficiency of a Carnot cycle operating between the same temperatures. A similar cycle could be developed which would approximate the Stirling cycle.

The difficulties in achieving such a cycle are primarily those associated with heat transfer. It is difficult to achieve an isothermal compression or expansion in a machine operating at a reasonable speed, and there will be pressure drops in the regenerator and a temperature difference between the two streams flowing through the regenerator. However, the gas turbine with intercooling and regenerators, which is described in Section 9.17, is a practical attempt to approach the Ericsson cycle. There have also been attempts to approach the Stirling cycle with the use of regenerators.

## 9.15  The Brayton Cycle

The air-standard Brayton cycle is the ideal cycle for the simple gas turbine. The simple open-cycle gas turbine utilizing an internal-combustion process and the simple closed-cycle gas turbine, which utilizes heat-transfer processes, are both shown schematically in Fig. 9.25. The air-standard Brayton cycle is shown on the $P$-$v$ and $T$-$s$ diagrams of Fig. 9.26.

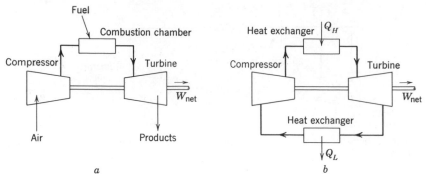

**Fig. 9.25**   A gas turbine operating on the Brayton cycle. (*a*) Open cycle. (*b*) Closed cycle.

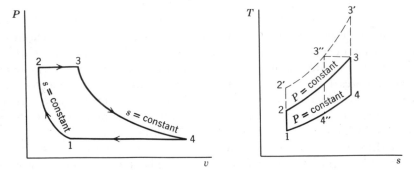

**Fig. 9.26**   The air-standard Brayton cycle.

The efficiency of the air-standard Brayton cycle is found as follows:

$$\eta_{th} = 1 - \frac{Q_L}{Q_H} = 1 - \frac{C_p(T_4 - T_1)}{C_p(T_3 - T_2)} = 1 - \frac{T_1(T_4/T_1 - 1)}{T_2(T_3/T_2 - 1)}$$

We note, however, that

$$\frac{P_3}{P_4} = \frac{P_2}{P_1}$$

$$\frac{P_2}{P_1} = \left(\frac{T_2}{T_1}\right)^{k/(k-1)} = \frac{P_3}{P_4} = \left(\frac{T_3}{T_4}\right)^{k/(k-1)}$$

$$\frac{T_3}{T_4} = \frac{T_2}{T_1} \quad \therefore \frac{T_3}{T_2} = \frac{T_4}{T_1} \quad \text{and} \quad \frac{T_3}{T_2} - 1 = \frac{T_4}{T_1} - 1$$

$$\eta_{th} = 1 - \frac{T_1}{T_2} = 1 - \frac{1}{(P_2/P_1)^{(k-1)/k}} \tag{9.6}$$

The efficiency of the air-standard Brayton cycle is therefore a function of isentropic pressure ratio; Figure 9.27 shows a plot of efficiency vs. pressure ratio. The fact that efficiency increases with pressure ratio is evident from the $T$-$s$ diagram of Fig. 9.26, because increasing the pressure ratio will change the cycle from 1–2–3–4–1 to 1–2'–3'–4–1. The

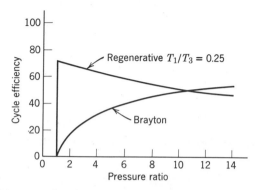

**Fig. 9.27**   Cycle efficiency as a function of pressure ratio for the Brayton and regenerative cycles.

latter cycle has a greater heat supply and the same heat rejected as the original cycle, and therefore it has a greater efficiency. Note further that the latter cycle has a higher maximum temperature ($T_3'$) than the original cycle ($T_3$). In an actual gas turbine the maximum temperature of the gas entering the turbine is fixed by metallurgical considerations. Therefore, if we fix the temperature $T_3$ and increase the pressure ratio, the resulting cycle is 1–2'–3''–4''–1. This cycle would have a higher efficiency than the original cycle, but the work per kilogram of working fluid is thereby changed.

With the advent of nuclear reactors the closed-cycle gas turbine has become more important. Heat is transferred, either directly or via a second fluid, from the fuel in the nuclear reactor to the working fluid in the gas turbine. Heat is rejected from the working fluid to the surroundings.

The actual gas-turbine engine differs from the ideal cycle primarily because of irreversibilities in the compressor and turbine, and because of pressure drop in the flow passages and combustion chamber (or in the heat exchanger of a closed-cycle turbine). Thus, the state points in a simple open-cycle gas turbine might be as shown in Fig. 9.28.

The efficiencies of the compressor and turbine are defined in relation to isentropic processes. Designating the states as in Fig. 9.28, the definitions of compressor and turbine efficiencies are as follows:

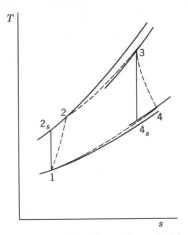

**Fig. 9.28**   Effect of inefficiencies on the gas-turbine cycle.

$$\eta_{\text{comp}} = \frac{h_{2s} - h_1}{h_2 - h_1} \tag{9.7}$$

$$\eta_{\text{turb}} = \frac{h_3 - h_4}{h_3 - h_{4s}} \tag{9.8}$$

One other important feature of the Brayton cycle is the large amount of compressor work (also called back work) compared to the turbine work. Thus, the compressor might require from 40 per cent to 80 per cent of the output of the turbine. This is particularly important when the actual cycle is considered, because the effect of the losses is to require a larger amount of compression work from a smaller amount of turbine work, and thus the over-all efficiency drops very rapidly with a decrease in the efficiencies of the compressor and turbine. In fact, if these efficiencies drop below about 60 per cent, all the work of the turbine will be required to drive the compressor, and the over-all efficiency will be zero. This is in sharp contrast to the Rankine cycle, where only 1 or 2 per cent of the turbine work is required to drive the pump. The reason for this is that for a reversible steady-state, steady-flow process with negligible change in kinetic and potential energy, the work is equal to $-\int v \, dP$. Because we are pumping a liquid in the Rankine cycle, the specific volume is very low compared to the specific volume of the gas in a gas turbine. This matter is illustrated in the following examples.

**Example 9.11**

In an air-standard Brayton cycle the air enters the compressor at 0.1 MPa, 15°C. The pressure leaving the compressor is 0.5 MPa and the maximum temperature in the cycle is 900°C. Determine:

**1.** The pressure and temperature at each point in the cycle.
**2.** The compressor work, turbine work, and cycle efficiency.

Designating the points as in Fig. 9.26 we have

$$P_1 = P_4 = 0.1 \text{ MPa} \qquad T_1 = 288.2 \text{ K}$$

$$P_2 = P_3 = 0.5 \text{ MPa} \qquad T_3 = 1173.2 \text{ K}$$

In the solution we consider successively a control surface around the compressor, the turbine, and the two heat exchangers. For the property relation we assume that air is an ideal gas with constant specific heat.

Control surface: Compressor

First law: $w_c = h_2 - h_1$. (Note that the compressor work $w_c$ is here defined as work input to the compressor.)

Second law: $\quad s_2 = s_1$

Therefore

$$\left(\frac{P_2}{P_1}\right)^{(k-1)/k} = 5^{0.286} = 1.5845 \quad , \quad T_2 = 456.6 \text{ K}$$

$$w_c = h_2 - h_1 = C_p(T_2 - T_1)$$

$$= 1.0035 \, (456.6 - 288.2) = 169.0 \text{ kJ/kg}$$

Control surface: Turbine

First law: $\qquad w_t = h_3 - h_4$

Second law: $\qquad s_3 = s_4$

Therefore,

$$\left(\frac{P_3}{P_4}\right)^{(k-1)/k} = 5^{0.286} = 1.5845 \quad , \quad T_4 = 740.4 \text{ K}$$

$$w_t = h_3 - h_4 = C_p(T_3 - T_4)$$

$$= 1.0035 \, (1173.2 - 740.4) = 434.3 \text{ kJ/kg}$$

$$w_{\text{net}} = w_t - w_c = 434.3 - 169.0 = 265.3 \text{ kJ/kg}$$

From a first law analysis of the two heat exchangers we have

$$q_H = h_3 - h_2 = C_p(T_3 - T_2) = 1.0035 \ (1173.2 - 456.6) = 719.1 \ \text{kJ/kg}$$

$$q_L = h_4 - h_1 = C_p(T_4 - T_1) = 1.0035 \ (740.4 - 288.2) = 453.8 \ \text{kJ/kg}$$

$$\eta_{th} = \frac{w_{net}}{q_H} = \frac{265.3}{719.1} = 36.9\%$$

This may be checked by using Eq. 9.6

$$\eta_{th} = 1 - \frac{1}{(P_2/P_1)^{(k-1)/k}} = 1 - \frac{1}{5^{0.286}} = 36.9\%$$

**Example 9.12**

Consider a gas turbine with air entering the compressor under the same conditions as in Example 9.11, and leaving at a pressure of 0.5 MPa. The maximum temperature is 900°C. Assume a compressor efficiency of 80 per cent, a turbine efficiency of 85 per cent, and a pressure drop between the compressor and turbine of 15 kPa. Determine the compressor work, turbine work, and cycle efficiency.

Designating the states as in Fig. 9.28 and proceeding as we did in Example 9.11 we have

Control surface: Compressor

First law:        $w_c = h_2 - h_1$

Second law:    $s_{2s} = s_1$

$$\eta_c = \frac{h_{2s} - h_1}{h_2 - h_1}$$

$$\left(\frac{P_2}{P_1}\right)^{(k-1)/k} = \frac{T_{2s}}{T_1} = 5^{0.286} = 1.5845 \quad , \quad T_{2s} = 456.6 \ \text{K}$$

$$\eta_c = \frac{h_{2s} - h_1}{h_2 - h_1} = \frac{T_{2s} - T_1}{T_2 - T_1} = \frac{456.6 - 288.2}{T_2 - T_1} = 0.80$$

$$T_2 - T_1 = \frac{456.6 - 288.2}{0.80} = 210.5 \quad , \quad T_2 = 498.7 \ \text{K}$$

$$w_c = h_2 - h_1 = C_p(T_2 - T_1)$$

$$= 1.0035 \ (498.7 - 288.2) = 211.2 \ \text{kJ/kg}$$

Control surface: Turbine

First law:     $w_t = h_3 - h_4$

Second law:  $s_3 = s_{4s}$

$$\eta_t = \frac{h_3 - h_4}{h_3 - h_{4s}}$$

$$P_3 = P_2 - \text{pressure drop} = 0.5 - 0.015 = 0.485 \text{ MPa}$$

$$\left(\frac{P_3}{P_4}\right)^{(k-1)/k} = \frac{T_3}{T_{4s}} = 4.85^{0.286} = 1.5708 \quad , \quad T_{4s} = 746.9 \text{ K}$$

$$\eta_t = \frac{h_3 - h_4}{h_3 - h_{4s}} = \frac{T_3 - T_4}{T_3 - T_{4s}} = 0.85$$

$$T_3 - T_4 = 0.85 \, (1173.2 - 746.9) = 362.4$$

$$T_4 = 810.8 \text{ K}$$

$$w_t = h_3 - h_4 = C_p(T_3 - T_4)$$

$$= 1.0035 \, (1173.2 - 810.8) = 363.7 \text{ kJ/kg}$$

$$w_{net} = w_t - w_c = 363.7 - 211.2 = 152.5 \text{ kJ/kg}$$

Control surface: Combustion Chamber

First law:     $q_H = h_3 - h_2 = C_p(T_3 - T_2)$

$$= 1.0035 \, (1173.2 - 498.7) = 676.9 \text{ kJ/kg}$$

$$\eta_{th} = \frac{w_{net}}{q_H} = \frac{152.5}{676.9} = 22.5\%$$

The following comparisons can be made between Examples 9.11 and 9.12.

|  | $w_c$ | $w_t$ | $w_{net}$ | $q_H$ | $\eta_{th}$ |
|---|---|---|---|---|---|
| Example 9.11 (Ideal) | 169.0 | 434.3 | 265.3 | 719.1 | 36.9 |
| Example 9.12 (Actual) | 211.2 | 363.7 | 152.5 | 676.9 | 22.5 |

As stated previously the result of the irreversibilities is to decrease the turbine work and increase the compressor work. Since the net work is the difference between these two it decreases very rapidly as compressor and turbine efficiencies decrease. The development of compressors and turbines of high efficiency is therefore an important aspect of the development of gas turbines.

Note also that in the ideal cycle (Example 9.11) about 39 per cent of the turbine work is required to drive the compressor and 61 per cent is delivered as net work. In the actual turbine (Example 9.12) 58 per cent of the turbine work is required to drive the compressor and 42 per cent is delivered as net work. Thus, if the net power of this unit is to be 10 000 kW, a 25 000 kW turbine and a 15 000 kW compressor are required. This demonstrates the statement that a gas turbine has a high back-work ratio.

## 9.16 The Simple Gas-Turbine Cycle with Regenerator

The efficiency of the gas-turbine cycle may be improved by introducing a regenerator. The simple open-cycle gas-turbine cycle with regenerator is shown in Fig. 9.29, and the corresponding ideal air-

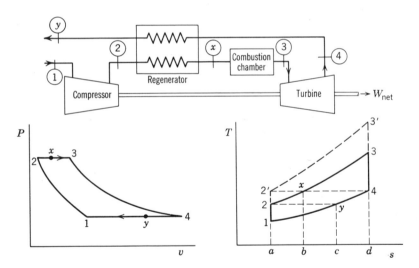

**Fig. 9.29** The ideal regenerative cycle.

standard cycle with regenerator is shown on the $P$-$v$ and $T$-$s$ diagrams. Note that in cycle 1–2–$x$–3–4–$y$–1, the temperature of the exhaust gas leaving the turbine in state 4 is higher than the temperature of the gas leaving the compressor. Therefore, heat can be transferred from the exhaust gases to the high-pressure gases leaving the compressor. If this is done in a counterflow heat exchanger, which is known as a regenerator, the temperature of the high-pressure gas leaving the regenerator, $T_x$, may, in the ideal case, have a temperature equal to $T_4$, the temperature of the gas leaving the turbine. In this case heat transfer from the external source is necessary only to increase the temperature from $T_x$ to $T_3$, and this heat transfer is represented by area $x$–$3$–$d$–$b$–$x$. Area $y$–$1$–$a$–$c$–$y$ represents the heat rejected.

The influence of pressure ratio on the simple gas-turbine cycle with regenerator is shown by considering cycle $1$–$2'$–$3'$–$4$–$1$. In this cycle the temperature of the exhaust gas leaving the turbine is just equal to the temperature of the gas leaving the compressor; therefore there is no possibility of utilizing a regenerator. This may be shown more exactly by determining the efficiency of the ideal gas-turbine cycle with regenerator.

The efficiency of this cycle with regeneration is found as follows, where the states are as given in Fig. 9.29.

$$\eta_{th} = \frac{w_{net}}{q_H} = \frac{w_t - w_c}{q_H}$$

$$q_H = C_p(T_3 - T_x)$$

$$w_t = C_p(T_3 - T_4)$$

But for ideal regenerator $T_4 = T_x$, and therefore $q_H = w_t$. Therefore,

$$\eta_{th} = 1 - \frac{w_c}{w_t} = 1 - \frac{C_p(T_2 - T_1)}{C_p(T_3 - T_4)}$$

$$= 1 - \frac{T_1(T_2/T_1 - 1)}{T_3(1 - T_4/T_3)} = 1 - \frac{T_1}{T_3}\frac{[(P_2/P_1)^{(k-1)/k} - 1]}{[1 - (P_1/P_2)^{(k-1)/k}]}$$

$$\eta_{th} = 1 - \frac{T_1}{T_3}\left(\frac{P_2}{P_1}\right)^{(k-1)/k}$$

Thus, we see that for the ideal cycle with regeneration the thermal efficiency depends not only upon the pressure ratio, but also upon the ratio of the minimum to maximum temperature. We also note, in contrast to the Brayton cycle, that the efficiency decreases with an increase in pressure ratio. The thermal efficiency vs. pressure ratio for this cycle is plotted in Fig. 9.27 for a value of

$$\frac{T_1}{T_3} = 0.25$$

The effectiveness or efficiency of a regenerator is given by the term regenerator efficiency. This may best be defined by reference to Fig. 9.30. State $x$ represents the high-pressure gas leaving the regenerator. In the ideal regenerator there would be only an infinitesimal temperature difference between the two streams, and the high-pressure gas would

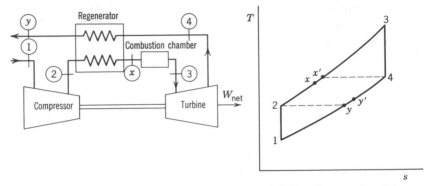

**Fig. 9.30**  Temperature-entropy diagram to illustrate the definition of regenerator efficiency.

leave the regenerator at temperature $T'_x$, and $T'_x = T_4$. In an actual regenerator, which must operate with a finite temperature difference, $T_x$, the actual temperature leaving the regenerator is therefore less than $T'_x$. The regenerator efficiency is defined by

$$\eta_{\text{reg}} = \frac{h_x - h_2}{h'_x - h_2}$$

If the specific heat is assumed to be constant, the regenerator efficiency is also given by the relation

$$\eta_{\text{reg}} = \frac{T_x - T_2}{T'_x - T_2}$$

It should also be pointed out that a higher efficiency can be achieved by using a regenerator with greater heat-transfer area. However, this also increases the pressure drop, which represents a loss, and both the pressure drop and the regenerator efficiency must be considered in determining which regenerator gives maximum thermal efficiency for the cycle. From an economic point of view, the cost of the regenerator must be weighed against the saving that can be effected by its use.

**Example 9.13**

If an ideal regenerator is incorporated into the cycle of Example 9.11, determine the thermal efficiency of the cycle.

Designating the states as in Fig. 9.30,

$$T_x = T_4 = 740.4 \text{ K}$$
$$q_H = h_3 - h_x = C_p(T_3 - T_x) = 1.0035 \ (1173.2 - 740.4) = 434.3 \text{ kJ/kg}$$
$$w_{net} = 265.3 \text{ kJ/kg (from Example 9.11)}$$
$$\eta_{th} = \frac{265.3}{434.3} = 61.1\%$$

## 9.17  The Ideal Gas-Turbine Cycle Using Multistage Compression with Intercooling, Multistage Expansion with Reheating, and Regenerator

In Section 9.14 it was pointed out that when an ideal regenerator was incorporated into the Ericsson cycle, an efficiency equal to the efficiency of the corresponding Carnot cycle could be attained. It was also pointed out that it is impossible to even closely approach in practice the reversible isothermal compression and expansion required in the Ericsson cycle.

The practical approach to this cycle is to use multistage compression with intercooling between stages, multistage expansion with reheat between stages, and a regenerator. Fig. 9.31 shows a cycle with two stages of compression and two stages of expansion. The air-standard cycle is shown on the corresponding $T$-$s$ diagram. It may be shown that for this cycle the maximum efficiency is obtained if equal pressure ratios are maintained across the two compressors and the two turbines. In this ideal cycle it is assumed that the temperature of the air leaving the intercooler, $T_3$, is equal to the temperature of the air entering the first stage of compression, $T_1$, and that the temperature after reheating, $T_8$, is equal to the temperature entering the first turbine, $T_6$. Further, in the ideal cycle it is assumed that the temperature of the high-pressure air leaving the regenerator, $T_5$, is equal to the temperature of the low-pressure air leaving the turbine, $T_9$.

If a large number of stages of compression and expansion are used, it is evident that the Ericsson cycle is approached. This is shown in Fig. 9.32. In practice the economical limit to the number of stages is usually two or three. The turbine and compressor losses and pressure drops that have already been discussed would be involved in any actual unit employing this cycle.

There are a variety of ways in which the turbines and compressors using this cycle can be utilized. Two possible arrangements for closed cycles are shown in Fig. 9.33. One advantage frequently sought in a given arrangement is ease of control of the unit under various loads. Detailed discussion of this point, however, is beyond the scope of this text.

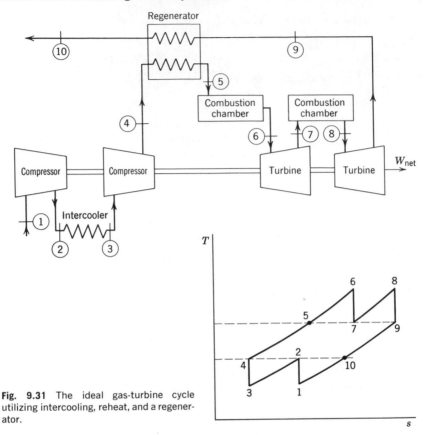

**Fig. 9.31** The ideal gas-turbine cycle utilizing intercooling, reheat, and a regenerator.

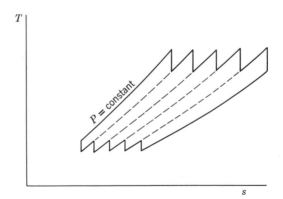

**Fig. 9.32** Temperature-entropy diagram which shows how the gas-turbine cycle with many stages approaches the Ericsson cycle.

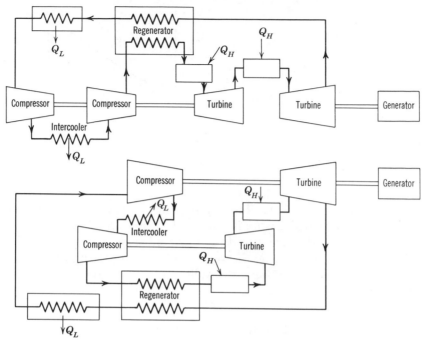

**Fig. 9.33** Some arrangements of components that may be utilized in stationary gas turbine power plants.

## 9.18   The Air-Standard Cycle for Jet Propulsion

The last air-standard power cycle we will consider is utilized in jet propulsion. In this cycle the work done by the turbine is just sufficient to drive the compressor. The gases are expanded in the turbine to a pressure such that the turbine work is just equal to the compressor work. The exhaust pressure of the turbine will then be above that of the surroundings, and the gas can be expanded in a nozzle to the pressure of the surroundings. Since the gases leave at a high velocity, the change in momentum the gases undergo results in a thrust upon the aircraft in which the engine is installed. The air-standard cycle for this is shown in Fig. 9.34.

Since all the principles involved in this cycle have already been covered, the example that follows will conclude our consideration of air-standard power cycles.

### Example 9.14

Consider an ideal cycle in which air enters the compressor at 0.1 MPa, 15°C. The pressure leaving the compressor is 0.5 MPa and the maxi-

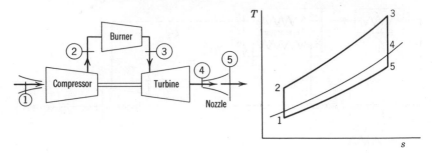

**Fig. 9.34** The ideal gas-turbine cycle for a jet engine.

mum temperature is 900°C. The air expands in the turbine to such a pressure that the turbine work is just equal to the compressor work. On leaving the turbine the air expands in a reversible adiabatic process in a nozzle to 0.1 MPa. Determine the velocity of the air leaving the nozzle.

From Example 9.11 we have the following, where the states are as designated in Fig. 9.34.

$$
\begin{aligned}
&P_1 = 0.1 \text{ MPa} \qquad T_1 = 288.2 \text{ K} \\
&P_2 = 0.5 \text{ MPa} \qquad T_2 = 456.6 \text{ K} \\
&w_c = 169.0 \text{ kJ/kg} \\
&P_3 = 0.5 \text{ MPa} \qquad T_3 = 1173.2 \text{ K} \\
&w_c = w_t = C_p(T_3 - T_4) = 169.0 \text{ kJ/kg}
\end{aligned}
$$

$$
T_3 - T_4 = \frac{169.0}{1.0035} = 168.4 \quad , \quad T_4 = 1004.8 \text{ K}
$$

$$
\frac{T_3}{T_4} = \left(\frac{P_3}{P_4}\right)^{(k-1)/k} = \frac{1173.2}{1004.8} = 1.1676
$$

$$
\frac{P_3}{P_4} = 1.720 \quad , \quad P_4 = 0.2907 \text{ MPa}
$$

Next consider a control surface around the nozzle. Let us assume that the velocity of the air entering the nozzle is low.

First law: $\qquad h_4 = h_5 + \dfrac{V_5^2}{2}$

Second law: $\qquad s_4 = s_5$

Therefore (from Example 9.11), $T_5 = 740.4$ K

$$V_5^2 = 2C_p (T_4 - T_5)$$
$$V_5^2 = 2 \times 1000 \times 1.0035 (1004.8 - 740.4)$$
$$V_5 = 728.5 \text{ m/s}$$

## AIR-STANDARD REFRIGERATION CYCLE

### 9.19   The Air-Standard Refrigeration Cycle

The final section of this chapter concerns the air-standard refrigeration cycle. Its main use in practice is in the liquefaction of air and other gases and in certain special situations that require refrigeration.

The simplest form of the air-standard refrigeration cycle, which is essentially the reverse of the Brayton cycle, is shown in Fig. 9.35. The

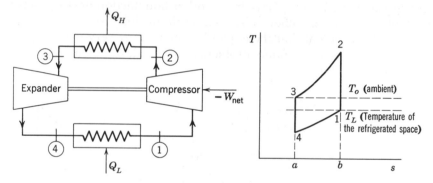

**Fig. 9.35**   The air-standard refrigeration cycle.

compressor and expander might be either reciprocating or rotary. After compression from 1 to 2, the air is cooled as a result of heat transfer to the surroundings at temperature $T_0$. The air is then expanded in process 3–4 to the pressure entering the compressor, and the temperature drops to $T_4$ in the expander. Heat may then be transferred to the air until temperature $T_L$ is reached. The work for this cycle is represented by area 1–2–3–4–1, and the refrigeration effect is represented by area 4–1–$b$–$a$–4. The coefficient of performance is the ratio of these two areas.

In practice this cycle has been utilized for the cooling of aircraft in an open cycle, a simplified form of which is shown in Fig. 9.36. Upon leaving the expander the cool air is blown directly into the cabin thus providing the cooling effect where needed.

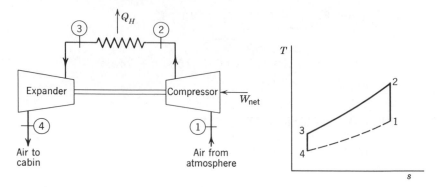

**Fig. 9.36** An air refrigeration cycle that might be utilized for aircraft cooling.

When counterflow heat exchangers are incorporated, very low temperatures can be obtained. This is essentially the cycle used in low-pressure air liquefaction plants and in other liquefaction devices such as the Collins helium liquefier. The ideal cycle in this case is as shown in Fig. 9.37. It is evident that the expander operates at very low temperature, which presents unique problems to the designer in regard to lubrication and materials.

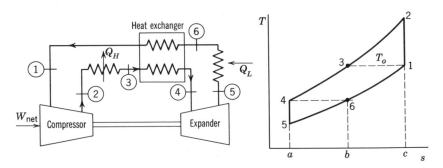

**Fig. 9.37** The air refrigeration cycle utilizing a heat exchanger.

### Example 9.15

Consider the simple air-standard refrigeration cycle of Fig. 9.35. Air enters the compressor at 0.1 MPa, $-20°C$, and leaves at 0.5 MPa. Air enters the expander at $15°C$. Determine

**1.** The coefficient of performance for this cycle.

**2.** The rate at which air must enter the compressor in order to provide one kW of refrigeration.

$$P_1 = P_4 = 0.1 \text{ MPa} \qquad T_1 = 253.2 \text{ K}$$
$$P_2 = P_3 = 0.5 \text{ MPa} \qquad T_3 = 288.2 \text{ K}$$

Control surface: Compressor

First law:      $w_c = h_2 - h_1$

Second law:    $s_1 = s_2$

Therefore,

$$\frac{T_2}{T_1} = \left(\frac{P_2}{P_1}\right)^{(k-1)/k} = 5^{0.286} = 1.5845 \quad , \quad T_2 = 401.2 \text{ K}$$

$$w_c = h_2 - h_1 = C_p(T_2 - T_1)$$
$$= 1.0035 \,(401.2 - 253.2) = 148.5 \text{ kJ/kg}$$

($w_c$ designates work into the compressor)

Control surface: Expander

First law:      $w_t = h_3 - h_4$

Second law:    $s_3 = s_4$

$$\frac{T_3}{T_4} = \left(\frac{P_3}{P_4}\right)^{(k-1)/k} = 5^{0.286} = 1.5845 \quad , \quad T_4 = 181.9 \text{ K}$$

$$w_t = h_3 - h_4 = 1.0035 \,(288.2 - 181.9) = 106.7 \text{ kJ/kg}$$

($w_t$ designates work done by expander)

From the first law as applied to a control surface around each of the heat exchangers we conclude that

$$q_H = h_2 - h_3 = C_p(T_2 - T_3) = 1.0035 \,(401.2 - 288.2) = 113.4 \text{ kJ/kg}$$
$$q_L = h_1 - h_4 = C_p(T_1 - T_4) = 1.0035 \,(253.2 - 181.9) = 71.6 \text{ kJ/kg}$$

$$w_{\text{net}} = w_c - w_t = 148.5 - 106.7 = 41.8 \text{ kJ/kg}$$

$$\beta = \frac{q_L}{w_{\text{net}}} = \frac{71.6}{41.8} = 1.713$$

In order to provide 1 kW of refrigeration capacity,

$$m = \frac{Q_L}{q_L} = \frac{1}{71.6} = 0.014 \text{ kg/s}$$

## PROBLEMS

9.1  Consider an ideal simple Rankine cycle using Freon-12 as the working fluid and a supply of geothermal hot water as the energy source. Saturated vapor leaves the Freon boiler at 85°C and the condenser temperature is 40°C.

(a) Calculate the thermal efficiency of this power cycle.

(b) If 1.5 kg/s of geothermal water is available at 95°C, what is the maximum power output of this cycle?

9.2  It is desired to determine the effect of turbine exhaust pressure on the performance of a steam Rankine cycle. Steam enters the turbine at 3.5 MPa, 350°C. Calculate the cycle thermal efficiency and the moisture content of the steam leaving the turbine for turbine exhaust pressures of 5 kPa, 10 kPa, 50 kPa, and 100 kPa. Plot thermal efficiency versus turbine exhaust pressure for the specified turbine inlet pressure and temperature.

9.3  It is desired to determine the effect of turbine inlet pressure on the performance of a steam Rankine cycle. Steam enters the turbine at 350°C and exhausts at 10 kPa. Calculate the cycle thermal efficiency and the moisture content of the steam leaving the turbine for turbine inlet pressures of 1 MPa, 3.5 MPa, 6 MPa, 10 MPa, and saturated vapor. Plot thermal efficiency versus turbine inlet pressure for the specified turbine inlet temperature and exhaust pressure.

9.4  It is desired to determine the effect of turbine inlet temperature on the performance of a steam Rankine cycle. Steam enters the turbine at 3.5 MPa and exhausts at 10 kPa. Calculate the cycle thermal efficiency and the moisture content leaving the turbine for turbine inlet temperatures of saturated vapor, 350°C, 500°C, and 800°C. Plot thermal efficiency versus turbine inlet temperature for the specified turbine inlet pressure and exhaust pressure.

9.5  Consider a reheat cycle utilizing steam as the working fluid. Steam enters the high-pressure turbine at 3.5 MPa, 350°C, and expands to 0.8 MPa. It is then reheated to 350°C and then expanded to 10 kPa in the low-pressure turbine. Calculate the cycle thermal efficiency and the moisture content leaving the low-pressure turbine.

9.6  Consider a regenerative cycle utilizing steam as the working fluid. Steam enters the turbine at 3.5 MPa, 350°C and exhausts to the condenser at 10 kPa. Steam is extracted at 0.8 MPa and also at 0.2 MPa for purposes of heating the feedwater in two open feedwater heaters. The feedwater leaves each heater at the temperature of the condensing steam. The appropriate pumps are used for the water leaving the condenser and the two feedwater heaters. Calculate the cycle thermal efficiency and the net work per kg of steam.

9.7 Repeat Problem 9.6 assuming closed instead of open feedwater heaters. A single pump is used to pump the water leaving the condenser to 3.5 MPa. Condensate from the high-pressure heater is drained through a trap to the low-pressure heater, and that from the low-pressure heater is drained through a trap to the condenser.

9.8 Consider a combined reheat and regenerative cycle utilizing steam as the working fluid. Steam enters the high-pressure turbine at 3.5 MPa, 350°C, and is extracted for purposes of feedwater heating at 0.8 MPa. The remainder of the steam is reheated at this pressure to 350°C and fed to the low-pressure turbine. Steam is extracted from the low-pressure turbine at 0.2 MPa for feedwater heating. The condenser pressure is 10 kPa and both feedwater heaters are open heaters. Calculate the cycle thermal efficiency and the net work per kg of steam.

9.9 It is desired to study the influence of the number of feedwater heaters on thermal efficiency for a cycle in which the steam leaves the steam generator at 17.5 MPa, 550°C, and which has a condenser pressure of 10 kPa. Assume that open feedwater heaters are used. Determine the thermal efficiency for each of the following cycles.
   (a) No feedwater heater.
   (b) One feedwater heater which operates at 1 MPa.
   (c) Two feedwater heaters, one operating at 3 MPa and the other at 0.2 MPa.

9.10 A nuclear reactor to produce power is so designed that the maximum temperature in the steam cycle is 450°C. The minimum temperature of the steam cycle is 40°C. It is planned to build a steam power plant which has one open feedwater heater.
   Select what you consider to be a reasonable cycle within these specifications, determine the ideal cycle efficiency, and explain why the particular pressures involved were selected.

9.11 Consider an ideal steam cycle that combines the reheat cycle and the regenerative cycle. The net power output of the turbine is 100 000 kW. Steam enters the high-pressure turbine at 8 MPa, 550°C. After expansion to 0.6 MPa, some of the steam goes to an open feedwater heater and the balance is reheated to 550°C, after which it expands to 10 kPa.
   (a) Draw a line diagram of the unit and show the state points on a *T-s* diagram.
   (b) What is the steam flow rate to the high-pressure turbine?
   (c) What power motor is required to drive each of the pumps?
   (d) If there is a 10°C rise in the temperature of the cooling water, what is the rate of flow of cooling water through the condenser?
   (e) The velocity of the steam flowing from the turbine to the condenser is limited to a maximum of 120 m/s. What is the diameter of this connecting pipe?

9.12 Steam leaves the boiler of a power plant at 3.5 MPa, 350°C, and when it enters the turbine it is at 3.4 MPa, 325°C. The turbine efficiency is 85% and the exhaust pressure is 10 kPa. Condensate leaves the condenser and enters the pump at 35°C, 10 kPa. The pump efficiency is 75% and the discharge pressure is 3.7 MPa. The feedwater enters the boiler at 3.6 MPa, 30°C. Calculate:

(a) The thermal efficiency of the cycle.

(b) The irreversibility of the process between the boiler exit and the turbine inlet, assuming an ambient temperature of 25°C.

9.13 Consider the following reheat-cycle power plant. Steam enters the high-pressure turbine at 3.5 MPa, 350°C, and expands to 0.5 MPa, after which it is reheated to 350°C. The steam is then expanded through the low-pressure turbines to 7.5 kPa. Liquid leaves the condenser at 30°C, is pumped to 3.5 MPa, and returned to the boiler. Each turbine is adiabatic, with an efficiency of 85%, and the pump efficiency is 80%. If the total power output of the turbines is 1000 kW, determine:

(a) Mass flow rate of steam.

(b) Pump power.

(c) Thermal efficiency of the unit.

9.14 Repeat Problem 9.9 assuming a turbine efficiency of 85%.

9.15 In a nuclear power plant heat is transferred in the reactor to liquid sodium. The liquid sodium is then pumped to a heat exchanger where heat is transferred to steam. The steam leaves this heat exchanger as saturated vapor at 5 MPa, and is then superheated in an external gas fired super-heater to 600°C. The steam then enters the turbine, which has one extraction point at 0.4 MPa, where steam flows to an open feedwater heater. The turbine efficiency is 75% and the condenser pressure is 7.5 kPa.

Determine the heat transfer in the reactor and in the superheater to produce a power output of 80 000 kW.

9.16 Consider the preliminary design for a supercritical steam power plant cycle. The maximum pressure will be 30 MPa, and the maximum tempera-ture will be 600°C. The cooling water temperature is such that 10 kPa pressure can be maintained in the condensers. Turbine efficiencies of at least 85% can be expected.

(a) Do you recommend any reheat for this cycle? If so, how many stages of reheat and at what pressures? Give reasons for your decisions.

(b) Do you recommend feedwater heaters? If so, how many and at what pressures would they operate? Would they be open or closed feedwater heaters?

(c) Estimate the thermal efficiency of the cycle which you recommend.

9.17 In both power and refrigeration cycles that operate over a wide tempera-ture range it is frequently advantageous to use more than one working fluid. In power plants the two fluids that have been used in combination are mercury and water, and this is referred to as a binary system. The advan-tage of this system is that mercury has a much lower vapor pressure than water, and it is possible to have an isothermal evaporation process in the mercury take place at a high temperature (much higher than the critical temperature of water) at a moderate pressure.

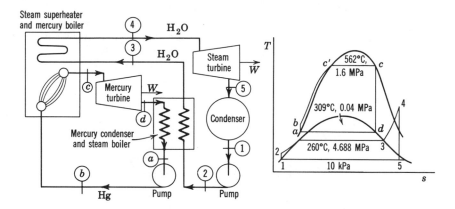

**Fig. 9.38**   Sketch for Problem 9.17.

Such a binary system is shown in Fig. 9.38, with the temperatures and pressures for an ideal cycle shown on the corresponding *T-s* diagram. For the steam, $T_4 = 500°C$.

(*a*) Determine the kilograms of mercury condensed in the mercury condenser per kilogram of water evaporated.

(*b*) Determine the thermal efficiency of this cycle.

9.18  In an ideal refrigeration cycle the temperature of the condensing vapor is 45°C and the temperature during evaporation is −15°C. Determine the coefficient of performance of this cycle for the working fluids Freon-12 and ammonia.

9.19  A study is to be made of the influence on power requirements of the difference between the temperature of the refrigerant in the condenser and the temperature of the surroundings. For this purpose consider the ideal cycle of Fig. 9.15 with Freon-12 as the refrigerant. The temperature of the surroundings is 40°C and the temperature of the refrigerant in the evaporator is −15°C. Plot a curve of power input per kW of refrigeration for temperature differences of 0 to 50°C between the refrigerant in the condenser and the surroundings.

9.20  A study is to be made of the influence on power requirements of the temperature difference between the cold box and the refrigerant in the evaporator. For this purpose consider the ideal cycle of Fig. 9.15 with Freon-12 as the refrigerant. The temperature of the cold box is −15°C and the temperature of the refrigerant in the condenser is 70°C. Plot a curve of power input per kW of refrigeration for temperature differences of 0 to 30°C between the cold box and the refrigerant in the evaporator.

9.21 In a conventional refrigeration cycle which uses Freon-12 as the refrigerant, the temperature of the evaporating fluid is $-20°C$. It leaves the evaporator as saturated vapor at $-20°C$ and enters the compressor. The pressure in the condenser is 1.2 MPa. The liquid leaves the condenser and enters the expansion valve at a temperature of 40°C.

It is proposed to modify this cycle as shown in Fig. 9.39. In this case an additional heat exchanger is supplied in which the cold vapor leaving the evaporator cools the liquid which enters the expansion valve.

Compare the coefficient of performance of these two cycles.

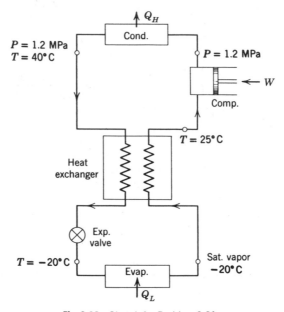

**Fig. 9.39** Sketch for Problem 9.21.

9.22 Consider an ideal dual-loop heat-powered refrigeration cycle utilizing Freon-12, as is shown schematically in Fig. 9.40. Saturated vapor at 105°C leaves the boiler and expands in the turbine to the condenser pressure. Saturated vapor at $-15°C$ leaves the evaporator and is compressed to the condenser pressure. The ratio of flows through the two loops is such that the turbine produces just enough power to drive the compressor. The two exiting streams mix together and enter the condenser. Saturated liquid at 45°C leaving the condenser is then separated into two streams in the necessary proportions. Calculate:

(a) The ratio of mass flow rate through the power loop to that through the refrigeration loop.

(b) The performance of the cycle, in terms of the ratio $\dot{Q}_L/\dot{Q}_H$.

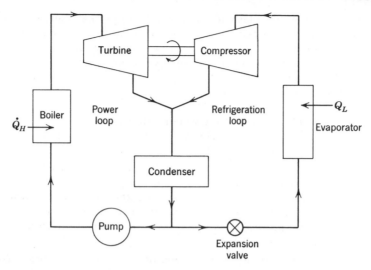

**Fig. 9.40**   Sketch for Problem 9.22.

9.23 In an actual refrigeration cycle using Freon-12 as a working fluid the rate of flow of refrigerant is 0.04 kg/s. The refrigerant enters the compressor at 0.15 MPa, −10°C and leaves at 1.2 MPa, 75°C. The power input to the compressor is 1.9 kW. The refrigerant enters the expansion valve at 1.15 MPa, 40°C and leaves the evaporator at 0.175 MPa, −15°C. Determine:

(a) The irreversibility during the compression process.

(b) The refrigeration capacity.

(c) The coefficient of performance for this cycle.

9.24 An ammonia-absorption system has an evaporator temperature of −10°C and a condenser temperature of 48°C. The generator temperature is 150°C. In this cycle, 0.42 kJ is transferred to the ammonia in the evaporator for each kJ transferred to the ammonia solution in the generator from the high-temperature source.

We wish to compare the performance of this cycle with the performance of a similar vapor-compression cycle. To do this, assume that a reservoir is available at 150°C, and that heat is transferred from this reservoir to a reversible engine which rejects heat to the surroundings at 25°C. This work is then used to drive an ideal vapor-compression system with ammonia as the refrigerant. Compare the amount of refrigeration that can be achieved per kJ from the high-temperature source in this case with the 0.42 kJ that can be achieved in the absorption system.

9.25 A stoichiometric mixture of fuel and air has an enthalpy of combustion of approximately −2800 kJ/kg of mixture. In order to approximate an actual spark-ignition engine using such a mixture, consider an air-standard

Otto cycle that has a heat addition of 2800 kJ per kg of air, a compression ratio of 8, and a pressure and temperature at the beginning of the compression process of 100 kPa, 15°C. Determine the following:

The maximum pressure and temperature for this cycle.

The thermal efficiency.

The mean effective pressure.

Assume:

(a) Constant specific heat as given in Table A.8.

(b) Variable specific heat. The Air Tables A.10 are recommended for this calculation (also $\bar{C}_{p_0}$ from Fig. 5.7 at high temperature).

9.26 It has been found that the power stroke in an internal combustion engine can be approximated reasonably as a polytropic process with a value of the polytropic exponent $n$ somewhat smaller than the specific heat ratio $k$. Repeat Problem 9.25, assuming that the expansion process is reversible and polytropic (instead of the isentropic expansion of the Otto Cycle) with a value of $n$ equal to 1.25.

9.27 In an Otto air-standard cycle, all the heat transfer $q_H$ occurs at constant volume. It would be more realistic to assume that part of the $q_H$ occurs after the piston has started to move down in the expansion stroke. Therefore, consider a cycle identical to the Otto cycle except that the first half of the total $q_H$ occurs at constant volume and the second half occurs at constant pressure. The total $q_H$ is 2400 kJ/kg. The pressure and temperature at the beginning of the compression process are 100 kPa, 20°C, and the compression ratio is 9 to 1.

(a) Show this cycle on $P$-$v$ and on $T$-$s$ coordinates.

(b) Calculate the maximum pressure and temperature and the thermal efficiency of the cycle.

9.28 An air-standard Diesel cycle has a compression ratio of 17. The pressure at the beginning of the compression stroke is 100 kPa, and the temperature is 15°C. The maximum temperature is 2500 K. Determine the thermal efficiency and the mean effective pressure for this cycle.

9.29 Consider an air-standard Ericsson cycle with an ideal regenerator incorporated. The temperature at which heat is supplied is 1200 K. Heat is rejected at 300 K. The pressure at the beginning of the isothermal compression process is 100 kPa. Determine the compressor and turbine work per kg of air and the thermal efficiency of the cycle. $q_H = 900$ kJ/kg.

9.30 Consider an air-standard Stirling cycle with an ideal regenerator. The maximum temperature in the cycle is 1100°C, the minimum temperature and pressure are 25°C and 100 kPa, and the compression ratio is 10. Analyze each of the four processes in this cycle for work and heat transfer, and determine the overall performance.

9.31 Consider a three-process air-standard power cycle in which process 1-2 is an isothermal compression, 2-3 is a constant-pressure heat addition, and 3-1 is an isentropic expansion. Given that $P_1 = 100$ kPa, $T_1 = 20°C$, and $P_2 = 600$ kPa, determine

(a) The work and heat transfer for each process and the thermal efficiency of the cycle.

(b) Show the cycle on T-s and P-v coordinates.

9.32 Consider a three-process air-standard power cycle in which process 1-2 is an isentropic compression, 2-3 is a constant pressure heat addition, and 3-1 is a constant-volume heat rejection. Given that $P_1 = 100$ kPa, $T_1 = 330$ K, and $P_2 = 800$ kPa, determine

(a) The work and heat transfer for each process and the thermal efficiency of the cycle.

(b) Show the cycle on T-s and P-v coordinates.

9.33 The pressure ratio across the compressor of an air-standard Brayton cycle is 4 to 1. The pressure of the air entering the compressor is 100 kPa and the temperature is 15°C. The maximum temperature in the cycle is 850°C. The rate of air flow is 10 kg/s. Determine the following, assuming constant specific heat:

(a) The compressor work, turbine work, and thermal efficiency of the cycle.

(b) If this cycle were utilized for a reciprocating machine, what would be the mean effective pressure? Would you recommend this cycle for a reciprocating machine?

9.34 Repeat Problem 9.33, assuming variable specific heat for the air.

9.35 An ideal regenerator is incorporated into the air-standard Brayton cycle of Problem 9.33. Determine the thermal efficiency of the cycle with this modification.

9.36 A stationary gas-turbine power plant operates on the Brayton cycle and delivers 20 000 kW to an electric generator. The maximum temperature is 1200 K and the minimum temperature is 290 K. The minimum pressure is 95 kPa and the maximum pressure is 380 kPa.

(a) What is the power output of the turbine?

(b) What fraction of the output of the turbine is used to drive the compressor?

(c) What is the mass rate of air flow to the compressor? What is the volume rate of air flow to the compressor?

9.37 Repeat Problem 9.36 assuming that a regenerator of 75% efficiency is added to the cycle, and the efficiency of the compressor is 80% and the efficiency of the turbine is 85%.

9.38 A gas turbine is to be used for pumping natural gas through a cross-country pipeline. The required power input to the natural gas compressor is 1000 kW. It has been decided to use a gas turbine to provide the necessary power input. A simple open cycle will be used with a regenerator of 50% efficiency. Since a supply of fuel is readily available, and since some of these units may be located in relatively isolated places, low maintenance costs are more important than a high efficiency.

Select a cycle which you would recommend, making appropriate assumptions for compressor and turbine efficiencies, and determine (for the gas turbine unit) the power output of the turbine, the power input to the compressor, and the efficiency.

9.39  Consider a gas-turbine cycle with two stages of compression and two stages of expansion. The pressure ratio across each turbine and each compressor is 2. The temperature entering each compressor is 15°C and the temperature entering each turbine is 900°C. An ideal regenerator is incorporated into the cycle. Determine the compressor work, turbine work, and thermal efficiency. $P_1 = 100$ kPa.

9.40  Repeat Problem 9.39 assuming that each compressor has an efficiency of 80%, each turbine has an efficiency of 85%, and the regenerator has an efficiency of 60%.

9.41  A gas turbine cycle for use as an automotive engine is shown in Fig. 9.41. In the first turbine, the gas expands to just a low enough pressure $P_5$ for that turbine to drive the compressor. The gas is then expanded through a second turbine connected to the drive wheels. The data for this engine are as shown in the figure. Consider the working fluid to be air throughout the entire cycle, and assume that all processes are ideal. Determine the following:

(a) Pressure $P_5$.
(b) Net work per kg and mass flow rate.
(c) Temperature $T_3$ and cycle thermal efficiency.
(d) The T-s diagram for the cycle.

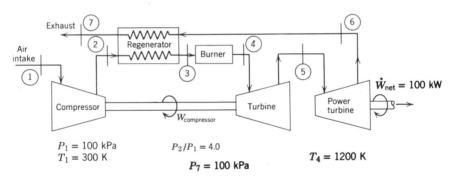

$P_1 = 100$ kPa
$T_1 = 300$ K

$P_2/P_1 = 4.0$

$P_7 = 100$ kPa

$T_4 = 1200$ K

$\dot{W}_{net} = 100$ kW

**Fig. 9.41**   Sketch for Problem 9.41.

9.42  Repeat Problem 9.41 assuming that the compressor has an efficiency of 82%, that both turbines have efficiencies of 87%, and that the regenerator efficiency is 70%. Also include the following pressure drops:

$$\Delta P_{23} = 0.02 P_2$$
$$\Delta P_{34} = 0.02 P_3$$
$$\Delta P_{67} = 0.02 P_6$$

9.43  A gas turbine engine operates on the Brayton Cycle. Air enters the compressor at 290 K, 100 kPa, the pressure ratio across the compressor is 5:1 and the maximum air temperature is 1200 K.

(*a*) If the engine is used in an electric power generation station, how much work (per kg of air) is available to drive the generator?

What is the thermal efficiency of the engine?

(*b*) If the engine is used to power a jet aircraft, at what pressure would the air leave the turbine? If the air leaving the turbine is expanded to 100 kPa in an an adiabatic nozzle, at what velocity will the air leave the nozzle?

9.44  Consider an air-standard cycle for a gas turbine-jet propulsion unit shown in Fig. 9.34. The pressure and temperature entering the compressor are 100 kPa, 290 K respectively. The pressure ratio across the compressor is 6 to 1 and the temperature at the turbine inlet is 1400 K. On leaving the turbine the air enters the nozzle and expands to 100 kPa. Determine the pressure at the nozzle inlet and the velocity of the air leaving the nozzle.

9.45  Repeat Problem 9.44, assuming that the efficiency of the compressor and turbine are both 85%, and that the nozzle efficiency is 95%.

9.46  A jet aircraft is flying at an altitude of 4900 m, where the ambient pressure is approximately 55 kPa and the ambient temperature is −18°C. The velocity of the aircraft is 220 m/s, and the pressure ratio across the compressor is 5.

   Devise an air-standard cycle that approximates this cycle, and determine the velocity (relative to the aircraft) of the air leaving the engine, assuming that it has been expanded to the ambient pressure. Assume a maximum gas temperature of 1200 K.

9.47  A heat exchanger is incorporated into an air-standard refrigeration cycle as shown in Fig. 9.42. Assume that both the compression and expansion are reversible adiabatic processes. Determine the coefficient of performance for this cycle.

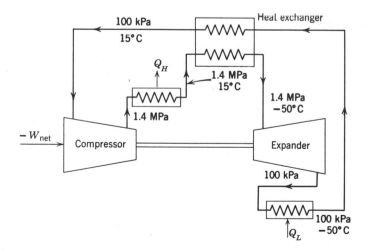

**Fig. 9.42**  Schematic arrangement of Problem 9.47.

9.48   Repeat Problem 9.47 assuming an isentropic efficiency for the compressor and expander of 75%.

9.49   Repeat Problems 9.47 and 48 assuming that helium is used as the working fluid instead of air. Discuss the results.

9.50   Write a computer program to solve the following problem. It is desired to determine the effects of varying parameters on the performance of an air-standard Brayton cycle. Consider a compressor inlet condition of 100 kPa, 20°C, and assume constant specific heat. The cycle thermal efficiency and net work output per kg are to be determined for all combinations of the following variables:

   (a) Compressor pressure ratio of 4, 6, 8, and 10.
   (b) Maximum cycle temperature of 700°C, 900°C, 1100°C and 1300°C.
   (c) Compressor and turbine efficiencies each 100%, 90%, 80%, and 70%.

9.51   Write a computer program to solve the following problem. Consider an ideal Otto cycle operating with nitrogen gas as the working fluid. The condition at the beginning of the compression process is 100 kPa, 20°C. Assume variable specific heat according to the equation in Table A.9. Determine the net work output per kg and the cycle thermal efficiency for various combinations of compression ratio and maximum cycle temperature, and compare the results with those found assuming constant specific heat.

9.52   Write a computer program to solve the following problem. It is desired to investigate the effect of the number of stages on the performance of an ideal gas turbine cycle having intercooling, reheat, and a regenerator, as shown in Fig. 9.31 (for two stages). Assume that the inlet condition, point 1, is 100 kPa, 20°C, and that the total pressure ratio, the maximum cycle temperature, and the number of stages (equal number of compressor and turbine stages) are all variable. Each stage is to have an equal pressure ratio. Determine the net work output per kg and the cycle thermal efficiency for various combinations of these variables.

# 10

# Thermodynamic Relations

We have already defined and used several thermodynamic properties. Among these are pressure, specific volume, density, temperature, mass, internal energy, enthalpy, entropy, constant-pressure and constant-volume specific heats, and Joule-Thomson coefficient. Two other properties, the Helmholtz function and the Gibbs function have been introduced and will be used more extensively in the following chapters. We have also had occasion to use tables of thermodynamic properties for a number of different substances.

One important question is now raised, namely, which of the thermodynamic properties can be experimentally measured? We can answer this question by reconsidering the measurements we can make in the laboratory. Some of the properties such as internal energy and entropy, cannot be measured directly, and must be calculated from other experimental data. If we carefully consider all these thermodynamic properties, we conclude that there are only four that can be directly measured; pressure, temperature, volume, and mass. Indirectly, through calorimetric measurements, we are able to determine experimental values for the specific heats $C_v$ and $C_p$.

This leads to a second question, namely, how can values of the thermodynamic properties that cannot be measured be determined from experimental data on those properties which can be measured? In answering this question we will develop certain general thermodynamic relations. In view of the fact that there are millions of such equations that can be written, our study will be limited to certain basic considerations, with particular reference to the determination of thermodynamic properties from experimental data. We shall also consider such related matters as generalized charts and equations of state.

## 10.1   Two Important Relations

This chapter involves partial derivatives, and two important relations are reviewed here. Consider a variable $z$ which is a continuous function of $x$ and $y$.

$$z = f(x, y)$$

$$dz = \left(\frac{\partial z}{\partial x}\right)_y dx + \left(\frac{\partial z}{\partial y}\right)_x dy$$

It is convenient to write this in the form:

$$dz = M\,dx + N\,dy \tag{10.1}$$

$M = \left(\dfrac{\partial z}{\partial x}\right)_y$ = partial derivative of $z$ with respect to $x$ (the variable $y$ being held constant)

$N = \left(\dfrac{\partial z}{\partial y}\right)_x$ = partial derivative of $z$ with respect to $y$ (the variable $x$ being held constant)

The physical significance of partial derivatives as they relate to the properties of a pure substance can be explained by referring to Fig. 10.1, which shows a $P$-$v$-$T$ surface of the superheated vapor region of a pure substance. It shows a constant-temperature, constant-pressure, and a constant-specific volume plane which intersect at point $b$ on the surface. Thus, the partial derivative $(\partial P/\partial v)_T$ is the slope of curve $abc$ at point $b$. Line $de$ represents the tangent to curve $abc$ at point $b$. A similar interpretation can be made of the partial derivatives $(\partial P/\partial T)_v$ and $(\partial v/\partial T)_P$.

It should also be noted that if we wish to evaluate the partial derivative along a constant-temperature line the rules for ordinary derivatives can be applied. Thus, we can write for a constant-temperature process:

$$\left(\frac{\partial P}{\partial v}\right)_T = \frac{dP_T}{dv_T}$$

and the integration can be performed as usual. This point will be demonstrated later in a number of examples.

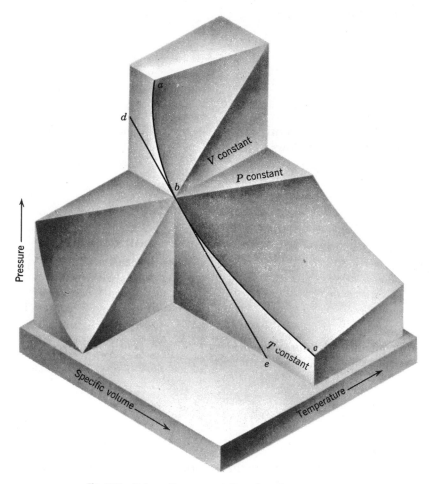

**Fig. 10.1** Schematic representation of partial derivatives.

Let us return to the consideration of the relation

$$dz = M\,dx + N\,dy$$

If $x$, $y$, and $z$ are all point functions (that is, quantities that depend only on the state and are independent of the path), the differentials are exact differentials. If this is the case, the following important relation holds:

$$\left(\frac{\partial M}{\partial y}\right)_x = \left(\frac{\partial N}{\partial x}\right)_y$$

The proof of this is as follows:

$$\left(\frac{\partial M}{\partial y}\right)_x = \frac{\partial^2 z}{\partial x\,\partial y}$$

$$\left(\frac{\partial N}{\partial x}\right)_y = \frac{\partial^2 z}{\partial y\,\partial x}$$

Since the order of differentiation makes no difference when point functions are involved, it follows that

$$\frac{\partial^2 z}{\partial x\,\partial y} = \frac{\partial^2 z}{\partial y\,\partial x}$$

$$\left(\frac{\partial M}{\partial y}\right)_x = \left(\frac{\partial N}{\partial x}\right)_y$$

The second important mathematical relation is

$$\left(\frac{\partial x}{\partial y}\right)_z\left(\frac{\partial y}{\partial z}\right)_x\left(\frac{\partial z}{\partial x}\right)_y = -1 \tag{10.2}$$

The proof of this relation is as follows. Consider three variables $x$, $y$, and $z$. Suppose there exists a relation between the variables of the form

$$x = f(y, z)$$

then

$$dx = \left(\frac{\partial x}{\partial y}\right)_z dy + \left(\frac{\partial x}{\partial z}\right)_y dz \tag{10.3}$$

If this relationship between the three variables is written in the form

$$y = f(x, z)$$

it follows that

$$dy = \left(\frac{\partial y}{\partial x}\right)_z dx + \left(\frac{\partial y}{\partial z}\right)_x dz \tag{10.4}$$

Substituting Eq. 10.4 into Eq. 10.3 we have

$$dx = \left(\frac{\partial x}{\partial y}\right)_z\left[\left(\frac{\partial y}{\partial x}\right)_z dx + \left(\frac{\partial y}{\partial z}\right)_x dz\right] + \left(\frac{\partial x}{\partial z}\right)_y dz$$

$$= \left(\frac{\partial x}{\partial y}\right)_z\left(\frac{\partial y}{\partial x}\right)_z dx + \left[\left(\frac{\partial x}{\partial y}\right)_z\left(\frac{\partial y}{\partial z}\right)_x + \left(\frac{\partial x}{\partial z}\right)_y\right] dz$$

Since there are only two independent variables, we may select $x$ and $z$. Suppose that $dz = 0$ and $dx \neq 0$. It then follows that

$$\left(\frac{\partial x}{\partial y}\right)_z \left(\frac{\partial y}{\partial x}\right)_z = 1 \tag{10.5}$$

Similarly, suppose that $dx = 0$ and $dz \neq 0$. It then follows that

$$\left(\frac{\partial x}{\partial y}\right)_z \left(\frac{\partial y}{\partial z}\right)_x + \left(\frac{\partial x}{\partial z}\right)_y = 0$$

$$\left(\frac{\partial x}{\partial y}\right)_z \left(\frac{\partial y}{\partial z}\right)_x = -\left(\frac{\partial x}{\partial z}\right)_y$$

$$\left(\frac{\partial x}{\partial y}\right)_z \left(\frac{\partial y}{\partial z}\right)_x \left(\frac{\partial z}{\partial x}\right)_y = -1$$

This is Eq. 10.2 which we set out to derive.

## 10.2 The Maxwell Relations

Consider a simple compressible system of fixed chemical composition. The Maxwell relations, which can be written for such a system, are four equations relating the properties $P$, $v$, $T$, and $s$.

The Maxwell relations are most easily derived by considering four relations involving thermodynamic properties. Two of these relations have already been derived and are:

$$du = T\,ds - P\,dv \tag{10.6}$$

$$dh = T\,ds + v\,dP \tag{10.7}$$

The other two are derived from the definition of the Helmholtz function, $a$, and the Gibbs function, $g$.

$$a = u - Ts$$
$$da = du - T\,ds - s\,dT$$

Substituting Eq. 10.6 into this relation gives the third relation.

$$da = -P\,dv - s\,dT \tag{10.8}$$

Similarly,

$$g = h - Ts$$
$$dg = dh - T\,ds - s\,dT$$

Substituting Eq. 10.7 yields the fourth relation.

$$dg = v \, dP - s \, dT \qquad (10.9)$$

Since Eqs. 10.6, 10.7, 10.8, and 10.9 are relations involving properties, we conclude that these are exact differentials, and, therefore, are of the general form

$$dz = M \, dx + N \, dy$$

Since

$$\left(\frac{\partial M}{\partial y}\right)_x = \left(\frac{\partial N}{\partial x}\right)_y \qquad (10.10)$$

it follows from Eq. 10.6 that

$$\left(\frac{\partial T}{\partial v}\right)_s = -\left(\frac{\partial P}{\partial s}\right)_v \qquad (10.11)$$

Similarly, from Eqs. 10.7, 10.8, and 10.9 we can write

$$\left(\frac{\partial T}{\partial P}\right)_s = \left(\frac{\partial v}{\partial s}\right)_P \qquad (10.12)$$

$$\left(\frac{\partial P}{\partial T}\right)_v = \left(\frac{\partial s}{\partial v}\right)_T \qquad (10.13)$$

$$\left(\frac{\partial v}{\partial T}\right)_P = -\left(\frac{\partial s}{\partial P}\right)_T \qquad (10.14)$$

These four equations are known as the Maxwell relations for a simple compressible system, and the great utility of these equations will be demonstrated in later sections of this chapter. In particular, it should be noted that pressure, temperature, and specific volume can be measured by experimental methods, whereas entropy cannot be determined experimentally. By using the Maxwell relations changes in entropy can be determined from quantities that can be measured, i.e., pressure, temperature, and specific volume.

There are a number of other very useful relations that can be derived from Eqs. 10.6 through 10.9. For example, from Eq. 10.6 we can write the relations

$$\left(\frac{\partial u}{\partial s}\right)_v = T \qquad \left(\frac{\partial u}{\partial v}\right)_s = -P \qquad (10.15)$$

Similarly, from the other equations we have the following:

$$\left(\frac{\partial h}{\partial s}\right)_P = T \qquad \left(\frac{\partial h}{\partial P}\right)_s = v$$

$$\left(\frac{\partial a}{\partial v}\right)_T = -P \qquad \left(\frac{\partial a}{\partial T}\right)_v = -s \qquad (10.16)$$

$$\left(\frac{\partial g}{\partial P}\right)_T = v \qquad \left(\frac{\partial g}{\partial T}\right)_P = -s$$

As already noted, the Maxwell relations just presented are written for a simple compressible substance. It is readily evident, however, that similar Maxwell relations can be written for substances involving other effects, such as electrical and magnetic effects. For example, Eq. 7.9 can be written in the form

$$dU = T\,dS - P\,dV + \mathcal{T}\,dL + \mathcal{S}\,d\mathcal{A} + \mu_0\mathcal{H}\,d(V\mathcal{M}) + \mathcal{E}\,dZ + \cdots \qquad (10.17)$$

Thus, at constant volume for a substance involving only magnetic effects we can write

$$dU = T\,dS + \mu_0 V\,\mathcal{H}\,d\mathcal{M}$$

and it follows that for such a substance

$$\left(\frac{\partial T}{\partial \mathcal{M}}\right)_{S,V} = \mu_0 V\left(\frac{\partial \mathcal{H}}{\partial S}\right)_{\mathcal{M},V}$$

Other Maxwell relations similar to Eqs. 10.12 through 10.14 could be written for such a substance. The extension of this approach to other systems, as well as to the interrelation between the various effects that may occur in a given system, is readily evident. For example, suppose a system involved both magnetic and surface effects. For such a system we could consider a constant entropy and volume process and write

$$\left(\frac{\partial \mathcal{S}}{\partial \mathcal{M}}\right)_{S,\mathcal{A},V} = \mu_0 V\left(\frac{\partial \mathcal{H}}{\partial \mathcal{A}}\right)_{S,\mathcal{M},V}$$

This matter becomes much more complex when we consider applying the property relation to a system of variable composition. This subject will be taken up in Chapter 11.

**Example 10.1**

From an examination of the properties of compressed liquid water, as given in Table A.1.4 of the Appendix, we find that the entropy of compressed liquid is greater than the entropy of saturated liquid for a temperature of 0°C, and is less than that of saturated liquid for all the other temperatures listed. Explain why this follows from other thermodynamic data.

Suppose we increase the pressure of liquid water that is initially saturated, while keeping the temperature constant. The change of entropy for the water during this process can be found by integrating the following Maxwell relation, Eq. 10.14,

$$\left(\frac{\partial s}{\partial P}\right)_T = -\left(\frac{\partial v}{\partial T}\right)_P$$

Therefore, the sign of the entropy change depends on the sign of the term $(\partial v/\partial T)_P$. The physical significance of this term is that it involves the change in specific volume of water as the temperature changes while the pressure remains constant. Now, as water at moderate pressures and 0°C is heated in a constant-pressure process, the specific volume decreases until the point of maximum density is reached at approximately 4°C, after which it increases. This is shown on a $v$-$T$ diagram in Fig. 10.2. Thus, the quantity $(\partial v/\partial T)_P$ is the slope of the curve in Fig.

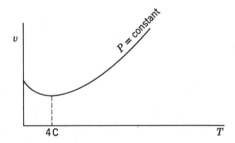

**Fig. 10.2**   Sketch for Example 10.1.

10.2. Since this slope is negative at 0°C, the quantity $(\partial s/\partial P)_T$ is positive at 0°C. At the point of maximum density the slope is zero, and therefore, the constant-pressure line shown in Fig. 7.7 crosses the saturated-liquid line at the point of maximum density.

## 10.3   Clapeyron Equation

The Clapeyron equation is an important relation involving the saturation pressure and temperature, the change of enthalpy associated with a change of phase, and the specific volumes of the two phases. In particular it is an example of how a change in a property that cannot be measured directly, the enthalpy in this case, can be determined from measurements of pressure, temperature, and specific volume. It can be derived in a number of ways. Here we proceed by considering one of the Maxwell relations, Eq. 10.13.

$$\left(\frac{\partial P}{\partial T}\right)_v = \left(\frac{\partial s}{\partial v}\right)_T$$

Consider, for example, the change of state from saturated liquid to saturated vapor of a pure substance. This is a constant-temperature process, and, therefore, Eq. 10.13 can be integrated between the saturated-liquid and saturated-vapor state. We also note that when saturated states are involved, pressure and temperature are independent of volume. Therefore,

$$\left(\frac{dP}{dT}\right)_{\text{sat}} = \frac{s_g - s_f}{v_g - v_f} = \frac{s_{fg}}{v_{fg}} = \frac{h_{fg}}{Tv_{fg}} \tag{10.18}$$

The significance of this equation is that $(dP/dT)_{\text{sat}}$ is the slope of the vapor-pressure curve. Thus, $h_{fg}$ at a given temperature can be determined from the slope of the vapor-pressure curve and the specific volume of saturated liquid and saturated vapor at the given temperature.

There are several different changes of phase that can occur at constant temperature and constant pressure. If we designate the two phases with superscripts $''$ and $'$, we can write the Clapeyron equation for the general case.

$$\left(\frac{dP}{dT}\right)_{\text{sat}} = \frac{s'' - s'}{v'' - v'}$$

We also note that $T(s'' - s') = h'' - h'$. Therefore

$$\left(\frac{dP}{dT}\right)_{\text{sat}} = \frac{h'' - h'}{T(v'' - v')} \tag{10.19}$$

If the phase designated $''$ is vapor, then at low pressure the equation is usually simplified by assuming that $v'' \gg v'$ and assuming also that $v'' = RT/P$. The relation then becomes

$$\left(\frac{dP}{dT}\right)_{\text{sat}} = \frac{h_g - h'}{T(RT/P)}$$

$$\left(\frac{dP}{P}\right)_{\text{sat}} = \frac{(h_g - h')}{R}\left(\frac{dT}{T^2}\right)_{\text{sat}} \tag{10.20}$$

**Example 10.2**

Determine the saturation pressure of water vapor at $-60°C$ using data available in the steam tables. Table 6 of the steam tables (Appendix Table A.1.5) does not give saturation pressures for temperatures less than $-40°C$. However, we do notice that $h_{ig}$ is relatively constant in this range, and, therefore, we proceed to use Eq. 10.20 and integrate between the limits $-40°C$ and $-60°C$.

$$\int_1^2 \frac{dP}{P} = \int_1^2 \frac{h_{ig}}{R}\frac{dT}{T^2} = \frac{h_{ig}}{R}\int_1^2 \frac{dT}{T^2}$$

$$\ln\frac{P_2}{P_1} = \frac{h_{ig}}{R}\left(\frac{T_2 - T_1}{T_1 T_2}\right)$$

Let

$$P_2 = 0.0129 \text{ kPa} \quad , \quad T_2 = 233.2 \text{ K}$$

$$P_1 = ? \quad , \quad T_1 = 213.2 \text{ K}$$

Then

$$\ln\frac{P_2}{P_1} = \frac{2838.9}{0.46152}\left(\frac{233.2 - 213.2}{233.2 \times 213.2}\right) = 2.4744$$

$$P_1 = 0.001\ 09 \text{ kPa}$$

## 10.4 Some Thermodynamic Relations Involving Enthalpy, Internal Energy, and Entropy

Let us first derive two equations, one involving $C_p$ and the other involving $C_v$.

We have defined $C_p$ as

$$C_p \equiv \left(\frac{\partial h}{\partial T}\right)_P$$

We have also noted that for a pure substance

$$T\ ds = dh - v\ dP$$

Therefore,

$$C_p = \left(\frac{\partial h}{\partial T}\right)_P = T\left(\frac{\partial s}{\partial T}\right)_P \tag{10.21}$$

Similarly, from the definition of $C_v$,

$$C_v \equiv \left(\frac{\partial u}{\partial T}\right)_v$$

and the relation

$$T \, ds = du + P \, dv,$$

it follows that

$$C_v = \left(\frac{\partial u}{\partial T}\right)_v = T\left(\frac{\partial s}{\partial T}\right)_v \qquad (10.22)$$

We will now derive a general relation for the change of enthalpy of a pure substance. We first note that for a pure substance

$$h = h(T, P)$$

Therefore,

$$dh = \left(\frac{\partial h}{\partial T}\right)_P dT + \left(\frac{\partial h}{\partial P}\right)_T dP$$

From the relation

$$T \, ds = dh - v \, dP$$

it follows that

$$\left(\frac{\partial h}{\partial P}\right)_T = v + T\left(\frac{\partial s}{\partial P}\right)_T$$

Substituting the Maxwell relation, Eq. 10.14, we have

$$\left(\frac{\partial h}{\partial P}\right)_T = v - T\left(\frac{\partial v}{\partial T}\right)_P \qquad (10.23)$$

On substituting this equation and Eq. 10.21 we have

$$dh = C_p \, dT + \left[v - T\left(\frac{\partial v}{\partial T}\right)_P\right] dP \qquad (10.24)$$

Along an isobar we have

$$dh_P = C_p \, dT_P$$

and along an isotherm,

$$dh_T = \left[v - T\left(\frac{\partial v}{\partial T}\right)_P\right] dP_T \qquad (10.25)$$

The significance of Eq. 10.24 is that this equation can be integrated to give the change in enthalpy associated with a change of state.

$$h_2 - h_1 = \int_1^2 C_p \, dT + \int_1^2 \left[ v - T \left( \frac{\partial v}{\partial T} \right)_P \right] dP \qquad (10.26)$$

The information needed to integrate the first term is a constant pressure specific heat along one (and only one) isobar. The integration of the second integral requires that an equation of state giving the relation between $P$, $v$, and $T$ be known. Furthermore, it is advantageous to have this equation of state explicit in $v$, for then the derivative $(\partial v / \partial T)_P$ is readily evaluated.

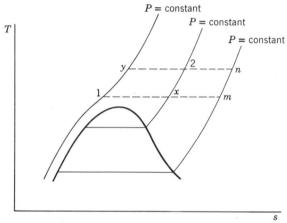

**Fig. 10.3**   Sketch showing various paths by which a given change of state can take place.

This can be further illustrated by reference to Fig. 10.3. Suppose we wished to know the change of enthalpy between states 1 and 2. We might do so along path 1–x–2, which consists of one isotherm (1–x) and one isobar (x–2). Thus we could integrate Eq. 10.26

$$h_2 - h_1 = \int_{T_1}^{T_2} C_p \, dT + \int_{P_1}^{P_2} \left[ v - T \left( \frac{\partial v}{\partial T} \right)_P \right] dP$$

Since $T_1 = T_x$ and $P_2 = P_x$ this can be written

$$h_2 - h_1 = \int_{T_x}^{T_2} C_p \, dT + \int_{P_1}^{P_x} \left[ v - T \left( \frac{\partial v}{\partial T} \right)_P \right] dP$$

The second term in this equation gives the change in enthalpy along the isotherm 1–x and the first term the change in enthalpy along the isobar x–2. When these are added together, the result is the net change in enthalpy between 1 and 2. It should be noted that in this case the constant-pressure specific heat must be known along the isobar passing through 2 and $x$. One could also find the change in enthalpy by following path 1–$y$–2, in which case the constant-pressure specific heat must be known along the 1–$y$ isobar. If the constant-pressure specific heat is known at another pressure, say the isobar passing through $m$–$n$, the change in enthalpy could be found by following path 1–$m$–$n$–2. This would involve calculating the change of enthalpy along two isotherms, namely, 1–$m$ and $n$–2.

Let us now derive a similar relation for the change of internal energy. All the steps in this derivation are given, but without detailed comment. Note that the starting point is to write $u = u(T, v)$ whereas in the case of enthalpy the starting point was $h = h(T, P)$.

$$u = f(T, v)$$

$$du = \left(\frac{\partial u}{\partial T}\right)_v dT + \left(\frac{\partial u}{\partial v}\right)_T dv$$

$$T\,ds = du + P\,dv$$

Therefore,

$$\left(\frac{\partial u}{\partial v}\right)_T = T\left(\frac{\partial s}{\partial v}\right)_T - P \tag{10.27}$$

Substituting the Maxwell relation, Eq. 10.13,

$$\left(\frac{\partial u}{\partial v}\right)_T = T\left(\frac{\partial P}{\partial T}\right)_v - P$$

Therefore,

$$du = C_v\,dT + \left[T\left(\frac{\partial P}{\partial T}\right)_v - P\right]dv \tag{10.28}$$

Along an isochore this reduces to

$$du_v = C_v\,dT_v,$$

and along an isotherm we have

$$du_T = \left[T\left(\frac{\partial P}{\partial T}\right)_v - P\right]dv_T \tag{10.29}$$

In a manner similar to that outlined above for changes in enthalpy, the change of internal energy for a given change of state for a pure substance can be determined from Eq. 10.28 if the constant-volume specific heat is known along one isochore and an equation of state explicit in $P$ (to obtain the derivative $(\partial P/\partial T)_v)$ is available in the region involved. A diagram similar to Fig. 10.3 could be drawn, with the isobars replaced with isochores, and the same general conclusions would be reached.

To summarize, we have derived Eqs. 10.24 and 10.28,

$$dh = C_p\, dT + \left[v - T\left(\frac{\partial v}{\partial T}\right)_P\right] dP$$

$$du = C_v\, dT + \left[T\left(\frac{\partial P}{\partial T}\right)_v - P\right] dv$$

The first of these involves the change of enthalpy, the constant-pressure specific heat, and is particularly suited to an equation of state explicit in $v$. The second involves the change of internal energy, the constant-volume specific heat, and is particularly suited to an equation of state explicit in $P$. If the first of these equations is used to determine the change of enthalpy, the internal energy is readily found by noting that

$$u_2 - u_1 = h_2 - h_1 - (P_2 v_2 - P_1 v_1)$$

If the second equation is used to find changes of internal energy, the change of enthalpy is readily found from this same relation. Which of these two equations is used to determine changes in internal energy and enthalpy will depend on the information available for specific heat and an equation of state (or other $P$-$v$-$T$ data).

Two parallel expressions can be found for the change of entropy.

$$s = s(T, P)$$

$$ds = \left(\frac{\partial s}{\partial T}\right)_P dT + \left(\frac{\partial s}{\partial P}\right)_T dP$$

Substituting Eqs. 10.21 and 10.14 we have

$$ds = C_p \frac{dT}{T} - \left(\frac{\partial v}{\partial T}\right)_P dP \tag{10.30}$$

$$s_2 - s_1 = \int_1^2 C_p \frac{dT}{T} - \int_1^2 \left(\frac{\partial v}{\partial T}\right)_P dP \tag{10.31}$$

Along an isobar we have

$$(s_2 - s_1)_P = \int_1^2 C_p \frac{dT_P}{T}$$

and along an isotherm,

$$(s_2 - s_1)_T = -\int_1^2 \left(\frac{\partial v}{\partial T}\right)_P dP_T$$

Note from Eq. 10.31 that if a constant-pressure specific heat is known along one isobar and an equation of state explicit in $v$ is available, the change of entropy can be evaluated. This is analogous to the expression for the change of enthalpy given in Eq. 10.24.

The second equation for the change of entropy can be found as follows.

$$s = s(T, v)$$

$$ds = \left(\frac{\partial s}{\partial T}\right)_v dT + \left(\frac{\partial s}{\partial v}\right)_T dv$$

Substituting Eqs. 10.22 and 10.13,

$$ds = C_v \frac{dT}{T} + \left(\frac{\partial P}{\partial T}\right)_v dv \qquad (10.32)$$

$$s_2 - s_1 = \int_1^2 C_v \frac{dT}{T} + \int_1^2 \left(\frac{\partial P}{\partial T}\right)_v dv \qquad (10.33)$$

This expression for change of entropy involves the change of entropy along an isochore where the constant-volume specific heat is known and along an isotherm where an equation of state explicit in $P$ is known, and thus it is analogous to the expression for change of internal energy given in Eq. 10.28.

### Example 10.3

Over a certain range of pressures and temperatures the equation of state of a certain substance is given with considerable accuracy by the relation

$$\frac{Pv}{RT} = 1 - C' \frac{P}{T^4}$$

or

$$v = \frac{RT}{P} - \frac{C}{T^3}$$

where $C$ and $C'$ are constants.

Derive an expression for the change of enthalpy and entropy of this substance in an isothermal process.

Since the equation of state is explicit in $v$, Eq. 10.25 would be particularly relevant to this problem. On integrating this equation we have,

$$(h_2 - h_1)_T = \int_1^2 \left[ v - T \left( \frac{\partial v}{\partial T} \right)_P \right] dP_T$$

From the equation of state,

$$\left( \frac{\partial v}{\partial T} \right)_P = \frac{R}{P} + \frac{3C}{T^4}$$

Therefore,

$$(h_2 - h_1)_T = \int_1^2 \left[ v - T \left( \frac{R}{P} + \frac{3C}{T^4} \right) \right] dP_T$$

$$= \int_1^2 \left[ \frac{RT}{P} - \frac{C}{T^3} - \frac{RT}{P} - \frac{3C}{T^3} \right] dP_T$$

$$(h_2 - h_1)_T = \int_1^2 -\frac{4C}{T^3} dP_T = -\frac{4C}{T^3} (P_2 - P_1)_T$$

For entropy we use Eq. 10.31, which is particularly relevant for an equation of state explicit in $v$.

$$(s_2 - s_1)_T = -\int_1^2 \left( \frac{\partial v}{\partial T} \right)_P dP_T = -\int_1^2 \left( \frac{R}{P} + \frac{3C}{T^4} \right) dP_T$$

$$(s_2 - s_1)_T = -R \ln \left( \frac{P_2}{P_1} \right)_T - \frac{3C}{T^4} (P_2 - P_1)_T$$

**Example 10.4**

The substance referred to in Example 10.3, having the equation of state

$$v = \frac{RT}{P} - \frac{C}{T^3}$$

has a constant-pressure specific heat along isobar $P_x$ (which lies within the region covered by the equation of state) given by the relation

$$C_p = A + BT$$

Derive a general expression for the change of enthalpy, internal energy, and entropy of this substance.

In solving this problem it may be helpful to refer to Fig. 10.4, where points 1 and 2 and the constant pressure line $P_x$ are arbitrarily selected.

The change in enthalpy between 1 and 2 is equal to the change in enthalpy along path 1–x–y–2.

$$h_2 - h_1 = (h_2 - h_y) + (h_y - h_x) + (h_x - h_1)$$

$$h_2 - h_1 = \int_{P_y = P_x}^{P_2} \left[ v - T\left(\frac{\partial v}{\partial T}\right)_P \right] dP_T + \int_{T_x = T_1}^{T_y = T_2} C_p \, dT_P$$

$$+ \int_{P_1}^{P_x = P_y} \left[ v - T\left(\frac{\partial v}{\partial T}\right)_P \right] dP_T$$

The first and last integrals have already been evaluated in Example 10.3.

$$h_2 - h_y = -\frac{4C}{T_2^{\,3}} (P_2 - P_x)$$

$$h_x - h_1 = -\frac{4C}{T_1^{\,3}} (P_x - P_1)$$

The second integral is readily evaluated since $C_p$ is known along $P_x$

$$h_y - h_x = \int_{T_x = T_1}^{T_y = T_2} C_p \, dT = \int_{T_x = T_1}^{T_y = T_2} (A + BT) \, dT$$

$$= A(T_2 - T_1) + \frac{B}{2} (T_2^{\,2} - T_1^{\,2})$$

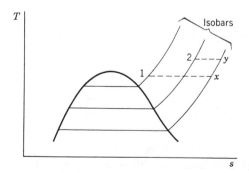

**Fig. 10.4** $T$-$s$ diagram for Example 10.4.

Therefore,

$$h_2 - h_1 = -\frac{4C}{T_2^3}(P_2 - P_x) + A(T_2 - T_1) + \frac{B}{2}(T_2^2 - T_1^2) - \frac{4C}{T_1^3}(P_x - P_1)$$

$$h_2 - h_1 = A(T_2 - T_1) + \frac{B}{2}(T_2^2 - T_1^2) - 4C\left[\frac{(P_2 - P_x)}{T_2^3} - \frac{(P_1 - P_x)}{T_1^3}\right]$$

$$u_2 - u_1 = h_2 - h_1 - (P_2 v_2 - P_1 v_1)$$

$$\therefore u_2 - u_1 = (A - R)(T_2 - T_1) + \frac{B}{2}(T_2^2 - T_1^2)$$

$$- 3C\left(\frac{P_2}{T_2^3} - \frac{P_1}{T_1^3}\right) + 4CP_x\left(\frac{1}{T_2^3} - \frac{1}{T_1^3}\right)$$

$$s_2 - s_1 = (s_2 - s_y) + (s_y - s_x) + (s_x - s_1)$$

From Eq. 10.30

$$s_2 - s_1 = -\int_{P_y = P_x}^{P_2}\left(\frac{\partial v}{\partial T}\right)_P dP_T + \int_{T_x = T_1}^{T_y = T_2} C_p\frac{dT_P}{T} - \int_{P_1}^{P_x}\left(\frac{\partial v}{\partial T}\right)_P dP_T$$

From the results of Example 10.3

$$s_2 - s_y = -R\ln\frac{P_2}{P_y} - \frac{3C}{T_2^4}(P_2 - P_y) = -R\ln\frac{P_2}{P_x} - \frac{3C}{T_2^4}(P_2 - P_x)$$

$$s_x - s_1 = -R\ln\frac{P_x}{P_1} - \frac{3C}{T_1^4}(P_x - P_1)$$

$$s_y - s_x = \int_{T_x = T_1}^{T_y = T_2} C_p\frac{dT}{T} = \int_{T_x = T_1}^{T_y = T_2}(A + BT)\frac{dT}{T} = A\ln\frac{T_2}{T_1} + B(T_2 - T_1)$$

Therefore,

$$s_2 - s_1 = -R\ln\frac{P_2}{P_x} - \frac{3C}{T_2^4}(P_2 - P_x) + A\ln\frac{T_2}{T_1} + B(T_2 - T_1)$$

$$- R\ln\frac{P_x}{P_1} - \frac{3C}{T_1^4}(P_x - P_1)$$

$$s_2 - s_1 = -R\ln\frac{P_2}{P_1} - 3C\left[\frac{(P_2 - P_x)}{T_2^4} - \frac{(P_1 - P_x)}{T_1^4}\right] + A\ln\frac{T_2}{T_1} + B(T_2 - T_1)$$

In this example we have selected relatively simple equations for the equation of state and the constant-pressure specific heat, and thus they would be accurate only over a limited range. When more complicated equations are used a digital computer can be utilized to excellent advantage in problems such as these.

## 10.5   Some Thermodynamic Relations Involving Specific Heat

Some important relations involving specific heats can also be developed. We have noted that the specific heat of an ideal gas is a function of the temperature only. For real gases the specific heat varies with pressure as well as temperature, and frequently we are interested in the variation of specific heat with pressure or volume, These relations can be derived as follows. Consider Eq. 10.30,

$$ds = \left(\frac{C_p}{T}\right) dT - \left(\frac{\partial v}{\partial T}\right)_P dP$$

Since this equation is of the general form $dz = M\, dx + N\, dy$, we can proceed as follows to find a relation that gives the variation of the constant-pressure specific heat with pressure at constant temperature.

$$\left(\frac{\partial (C_p/T)}{\partial P}\right)_T = -\left[\frac{\partial}{\partial T}\left(\frac{\partial v}{\partial T}\right)_P\right]_P$$

$$\left(\frac{\partial C_p}{\partial P}\right)_T = -T\left(\frac{\partial^2 v}{\partial T^2}\right)_P \tag{10.34}$$

The variation of the constant-volume specific heat with volume as the temperature remains constant can be found in a similar manner. Consider Eq. 10.32,

$$ds = \left(\frac{C_v}{T}\right) dT + \left(\frac{\partial P}{\partial T}\right)_v dv$$

Since this is of the form $dz = M\, dx + N\, dy$,

$$\left(\frac{\partial (C_v/T)}{\partial v}\right)_T = \left[\frac{\partial}{\partial T}\left(\frac{\partial P}{\partial T}\right)_v\right]_v$$

$$\left(\frac{\partial C_v}{\partial v}\right)_T = T\left(\frac{\partial^2 P}{\partial T^2}\right)_v \tag{10.35}$$

The important thing to note about Eqs. 10.34 and 10.35 is that the variation of the constant-volume and constant-pressure specific heats in a constant-temperature process can be found from the equation of state, although in each case it involves the second derivative of the equation, which is a severe test of the accuracy of any equation of state.

### Example 10.5

Determine the variation of $C_p$ with pressure at constant temperature for a substance such as the one in Example 10.3 over the range where the equation of state is given by the relation

$$v = \frac{RT}{P} - \frac{C}{T^3}$$

Using Eq. 10.34

$$\left(\frac{\partial C_p}{\partial P}\right)_T = -T\left(\frac{\partial^2 v}{\partial T^2}\right)_P$$

$$\left(\frac{\partial v}{\partial T}\right)_P = \frac{R}{P} + \frac{3C}{T^4}$$

$$\left(\frac{\partial^2 v}{\partial T^2}\right)_P = -\frac{12C}{T^5}$$

$$\left(\frac{\partial C_p}{\partial P}\right)_T = -T\left(-\frac{12C}{T^5}\right) = \frac{12C}{T^4}$$

A final interesting and useful relation involving the difference between $C_p$ and $C_v$ can be derived by equating Eqs. 10.30 and 10.32.

$$C_p\frac{dT}{T} - \left(\frac{\partial v}{\partial T}\right)_P dP = C_v\frac{dT}{T} + \left(\frac{\partial P}{\partial T}\right)_v dv$$

$$dT = \frac{T(\partial P/\partial T)_v}{C_p - C_v}\, dv + \frac{T(\partial v/\partial T)_P}{C_p - C_v}\, dP$$

But

$$T = f(v, P)$$

$$dT = \left(\frac{\partial T}{\partial v}\right)_P dv + \left(\frac{\partial T}{\partial P}\right)_v dP$$

Therefore,

$$\left(\frac{\partial T}{\partial v}\right)_P = \frac{T(\partial P/\partial T)_v}{C_p - C_v}$$

and

$$\left(\frac{\partial T}{\partial P}\right)_v = \frac{T(\partial v/\partial T)_P}{C_p - C_v}$$

When these equations are solved for $C_p - C_v$, they yield the same result.

$$C_p - C_v = T \left(\frac{\partial v}{\partial T}\right)_P \left(\frac{\partial P}{\partial T}\right)_v \qquad (10.36)$$

But from Eq. 10.2,

$$\left(\frac{\partial P}{\partial T}\right)_v = -\left(\frac{\partial v}{\partial T}\right)_P \left(\frac{\partial P}{\partial v}\right)_T$$

Therefore,

$$C_p - C_v = -T \left(\frac{\partial v}{\partial T}\right)_P^2 \left(\frac{\partial P}{\partial v}\right)_T \qquad (10.37)$$

From this equation we draw several conclusions:

1. For a liquid and solid $(\partial v/\partial T)_P$ is usually relatively small, and, there-fore, the difference between the constant-pressure and constant-volume specific heats of a liquid is small. For this reason many tables simply give the specific heat of a solid or a liquid without designating that it is at constant volume or pressure. Further, $C_p = C_v$ exactly when $(\partial v/\partial T)_P = 0$, as is true at the point of maximum density of water

2. $C_p \rightarrow C_v$ as $T \rightarrow 0$, and, therefore, we conclude that the constant-pressure and constant-volume specific heats are equal at absolute zero.

3. The difference between $C_p$ and $C_v$ is always positive because $(\partial v/\partial T)_P{}^2$ is always positive and $(\partial P/\partial v)_T$ is negative for all known substances.

## 10.6  Volume Expansivity and Isothermal and Adiabatic Compressibility

The student has most likely encountered the coefficient of linear ex-pansion in his studies of strength of materials. This coefficient indicates how the length of a solid body is influenced by a change in temperature while the pressure remains constant. In terms of the notation of partial derivatives the *coefficient of linear expansion*, $\delta_P$, is defined as follows:

$$\delta_P = \frac{1}{L} \left(\frac{\partial L}{\partial T}\right)_P \qquad (10.38)$$

A similar coefficient can be defined for changes in volume, and such a coefficient is applicable to liquids and gases as well as to solids. This coefficient of volume expansion $\alpha_P$, also called the volume expansivity, is an indication of the change in volume that results from a change in temperature while the pressure remains constant. The definition of *volume expansivity* is

$$\alpha_P \equiv \frac{1}{V}\left(\frac{\partial V}{\partial T}\right)_P = \frac{1}{v}\left(\frac{\partial v}{\partial T}\right)_P \qquad (10.39)$$

The isothermal compressibility $\beta_T$ is an indication of the change in volume that results from a change in pressure while the temperature remains constant. The definition of the *isothermal compressibility* is

$$\beta_T \equiv -\frac{1}{V}\left(\frac{\partial V}{\partial P}\right)_T = -\frac{1}{v}\left(\frac{\partial v}{\partial P}\right)_T \qquad (10.40)$$

The reciprocal of the isothermal compressibility is called the *isothermal bulk modulus* $B_T$.

$$B_T \equiv -v\left(\frac{\partial P}{\partial v}\right)_T \qquad (10.41)$$

The *adiabatic compressibility* $\beta_s$ is an indication of the change in volume that results from a change in pressure while the entropy remains constant, and is defined as follows:

$$\beta_s \equiv -\frac{1}{v}\left(\frac{\partial v}{\partial P}\right)_s \qquad (10.42)$$

The *adiabatic bulk modulus* $B_s$ is the reciprocal of the adiabatic compressibility.

$$B_s \equiv -v\left(\frac{\partial P}{\partial v}\right)_s \qquad (10.43)$$

Both the volume expansivity and isothermal compressibility are thermodynamic properties of a substance, and for a simple compressible substance are functions of two independent properties. Values of these properties are found in the standard handbooks of physical properties. The following example gives an indication of the use and significance of the volume expansivity and isothermal compressibility.

## Example 10.6

Show that $C_p - C_v$ can be expressed in terms of the volume expansivity $\alpha_P$, the specific volume $v$, the temperature $T$, and the isothermal compressibility $\beta_T$, by the relation

$$C_p - C_v = \frac{\alpha_P^2 v T}{\beta_T}$$

From Eq. 10.37

$$C_p - C_v = -T\left(\frac{\partial v}{\partial T}\right)_P^2 \left(\frac{\partial P}{\partial v}\right)_T$$

$$\alpha_P \equiv \frac{1}{v}\left(\frac{\partial v}{\partial T}\right)_P; \qquad \left(\frac{\partial v}{\partial T}\right)_P^2 = \alpha_P^2 v^2$$

$$\beta_T \equiv -\frac{1}{v}\left(\frac{\partial v}{\partial P}\right)_T; \qquad \left(\frac{\partial P}{\partial v}\right)_T = -\frac{1}{\beta_T v}$$

Therefore

$$C_p - C_v = -T(\alpha_P^2 v^2)\left(-\frac{1}{\beta_T v}\right) = \frac{\alpha_P^2 v T}{\beta_T}$$

**Example 10.7**

The pressure on a block of copper having a mass of 1 kg is increased in a reversible process from 0.1 MPa to 100 MPa while the temperature is held constant at 15°C. Determine the work done on the copper during this process, the change in entropy per kilogram of copper, the heat transfer, the change of internal energy per kilogram, and $C_p - C_v$ for this change of state.

Over the range of pressure and temperature involved in this problem the following data can be used:

Volume expansivity $= \alpha_p = 5.0 \times 10^{-5}$ K$^{-1}$
Isothermal compressibility $= \beta_T = 8.6 \times 10^{-12}$ m²/N
Specific volume $= 0.000\ 114$ m³/kg

The work done during the isothermal expansion is

$$w = \int P\, dv_T$$

The isothermal compressibility has been defined:

$$\beta_T - -\frac{1}{v}\left(\frac{\partial v}{\partial P}\right)_T$$

$$v\beta_T\, dP_T = -dv_T$$

Therefore, for this isothermal process

$$w = -\int_1^2 v\beta_T P\, dP_T$$

Since $v$ and $\beta_T$ remain essentially constant, this is readily integrated:

$$w = -\frac{v\beta_T}{2}(P_2^2 - P_1^2)$$

$$= -\frac{0.000\ 114 \times 8.6 \times 10^{-12}}{2}(100^2 - 0.1^2) \times 10^{12}$$

$$= -4.9 \text{ J/kg}$$

The change of entropy can be found by considering the Maxwell relation, Eq. 10.14, and the definition of volume expansivity.

$$\left(\frac{\partial s}{\partial P}\right)_T = -\left(\frac{\partial v}{\partial T}\right)_P = -\frac{v}{v}\left(\frac{\partial v}{\partial T}\right)_P = -v\alpha_P$$

$$ds_T = -v\alpha_P \, dP_T$$

This can be readily integrated as follows if we assume that $v$ and $\alpha_P$ remain constant:

$$(s_2 - s_1)_T = -v\alpha_P(P_2 - P_1)_T$$

$$= -0.000\ 114 \times 5.0 \times 10^{-5}\ (100 - 0.1) \times 10^6$$

$$= -0.5694 \text{ J/kg K}$$

The heat transfer for this reversible isothermal process is $T(s_2 - s_1)$.

$$q = T(s_2 - s_1) = -288.2 \times 0.5694 = -164.1 \text{ J/kg}$$

The change in internal energy follows directly from the first law.

$$(u_2 - u_1) = q - w = -164.1 - (-4.9) = -159.2 \text{ J/kg}$$

From Example 10.6,

$$C_p - C_v = \frac{\alpha_P{}^2 vT}{\beta_T}$$

Substituting the data for copper as given above,

$$C_p - C_v = (5.0 \times 10^{-5})^2 \times \frac{0.000\ 114 \times 288.2}{8.6 \times 10^{-12}}$$

$$= 9.55 \text{ J/kg K}$$

This is consistent with the earlier observation that $C_p - C_v$ is relatively small for solids.

## 10.7 Developing Tables of Thermodynamic Properties from Experimental Data

There are many ways in which tables of thermodynamic properties can be developed from experimental data. The purpose of this section is to convey some general principles and concepts by considering only the liquid and vapor phases.

Let us assume that the following data for a pure substance have been obtained in the laboratory.

a. Vapor-pressure data. That is, saturation pressures and temperatures have been measured over a wide range.

b. Pressure, specific volume, temperature data in the vapor region. These data are usually obtained by determining the mass of the substance in a closed vessel (which means a fixed specific volume) and then measuring the pressure as the temperature is varied. This is done for a large number of specific volumes.

c. Density of the saturated liquid and the critical pressure and temperature.

d. Zero-pressure specific heat for the vapor. This might be obtained either calorimetrically or from spectroscopic data.

From these data a complete set of thermodynamic tables for the saturated liquid, saturated vapor, and superheated vapor can be calculated. The first step is to determine an equation for the vapor-pressure curve that accurately fits the data. It may be necessary to use one equation for one portion of the vapor-pressure curve and a different equation for another portion of the curve.

One form of equation that has been used is

$$\ln P_{sat} = A + \frac{B}{T} + C \ln T + DT$$

Once an equation has been found that accurately represents the data, the saturation pressure for any given temperature can be found by solving this equation. Thus, the saturation pressures in Table 1 of the *Steam Tables* would be determined for the given temperatures. The second step is to determine an equation of state for the vapor region that accurately represents the *P-v-T* data. There are many possible forms of the equation of state which may be selected. The important considerations are that the equation of state accurately represents the data, and that it be of such a form that the differentiations required can be performed (i.e., in some cases it may be desirable to have an equation of state explicit in *v*, whereas on other occasions an equation of state that is explicit in *P* may be more desirable).

Once an equation of state has been determined, the specific volume of superheated vapor at given pressures and temperatures can be determined by solving the equation and tabulating the results as in the superheat tables for steam, ammonia, and Freon-12. The specific volume of saturated vapor at a given temperature may be found by finding the saturation pressure from the vapor-pressure curve and substituting this saturation pressure and temperature into the equation of state.

The procedure followed in determining enthalpy and entropy is best explained with the aid of Fig. 10.5. Let us assume that the enthalpy and

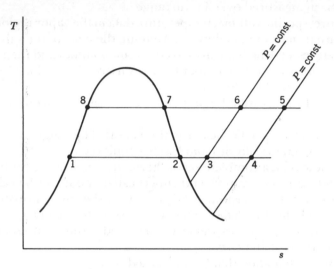

**Fig. 10.5**  Sketch showing procedure for developing a table of thermodynamic properties from experimental data.

entropy of saturated liquid in state 1 are zero. The enthalpy of saturated vapor in state 2 can be found from the Clapeyron equation.

$$\left(\frac{dP}{dT}\right)_{\text{sat}} = \frac{h_{fg}}{T(v_g - v_f)}$$

The left side of this equation is found by differentiating the vapor-pressure curve. The specific volume of the saturated vapor is found by the procedure outlined in the last paragraph, and it is assumed the specific volume of the saturated liquid has been measured. Thus, the enthalpy of evaporation, $h_{fg}$, can be found for this particular temperature, and the enthalpy at state 2 is equal to the enthalpy of evaporation (since the enthalpy in state 1 is assumed to be zero). The entropy at state 2 is readily found, since

$$s_{fg} = \frac{h_{fg}}{T}$$

From state 2 we now proceed along this isotherm into the superheated vapor region. The specific volume at 3 is found from the equation of state at this pressure, while the enthalpy and entropy are determined by integrating Eqs. 10.25 and 10.31,

$$h_3 - h_2 = \int_2^3 \left[ v - T \left( \frac{\partial v}{\partial T} \right)_P \right] dP_T$$

$$s_3 - s_2 = \int_2^3 - \left( \frac{\partial v}{\partial T} \right)_P dP_T$$

The properties at point 4 are found in exactly the same manner. Pressure $P_4$ is sufficiently low that the real superheated vapor behaves essentially as an ideal gas (perhaps 10 kPa). Thus, we use this constant-pressure line to make all temperature changes for our calculations, as for example to point 5. Since the specific heat $C_{po}$ is known as a function of temperature, the enthalpy and entropy at 5 are found by integrating the ideal gas relations

$$(h_5 - h_4)_P = \int_4^5 C_{po} \, dT_P$$

$$(s_5 - s_4)_P = \int_4^5 C_{po} \frac{dT_P}{T}$$

The properties at points 6 and 7 are found from those at 5 in the same manner as those at points 3 and 4 were found from 2 (the saturation pressure $P_7$ is calculated from the vapor pressure equation). Finally, the enthalpy and entropy for saturated liquid at point 8 are found from the properties at point 7 by applying the Clapeyron equation.

Thus, values for the pressure, temperature, specific volume, enthalpy, entropy, and internal energy of saturated liquid, saturated vapor, and superheated vapor can be tabulated for the entire region for which experimental data were obtained. It is evident that the accuracy of such a table depends both on the accuracy of the experimental data and the degree to which the equation for the vapor pressure and the equation of state represent the experimental data.

## 10.8  The Ideal Gas

The ideal gas has been discussed at various relevant points in prior chapters, particularly in Sections 3.4, 5.6, and 7.9. Since an understanding of the properties of an ideal gas is essential to a study of the behavior of real substances, equations of state, and tables of thermo-dynamic properties, a brief summary of the behavior of ideal gases is presented here.

The ideal gas was defined in Section 3.4 as a gas that follows the equation of state, Eq. 3.1,

$$P\bar{v} = \bar{R}T$$

From a microscopic viewpoint, such an equation results when there are no intermolecular forces, in other words when the molecules are widely separated from one another. Thus, the ideal gas can be a reasonable model of real gas behavior only at very low density.

In Section 5.6 it was stated that the internal energy of an ideal gas is a function only of temperature, as stated by Eq. 5.16. At this point it is appropriate to demonstrate that this is in fact the case. For an ideal gas, from Eq. 3.2,

$$P = \frac{RT}{v}$$

Therefore,

$$\left(\frac{\partial P}{\partial T}\right)_v = \frac{R}{v}$$

Substituting this relation into Eq. 10.29, we have

$$du_T = \left[T\left(\frac{\partial P}{\partial T}\right)_v - P\right]dv_T = \left[T\left(\frac{R}{v}\right) - P\right]dv_T = 0$$

and we find no variation of internal energy along an isotherm. In other words, for the ideal gas Eq. 10.28 reduces to Eq. 5.17,

$$du = C_{vo}\,dT$$

We can also examine various other general thermodynamic relations for the special case of the ideal gas. The procedure is in each case similar to that followed above for the internal energy, and the details are left as an exercise. To summarize the results:

Eqs. 10.34 and 10.35 demonstrate the validity of Eq. 5.23,

$$C_{vo} = f(T); \qquad C_{po} = f(T)$$

Eq. 10.24 reduces to Eq. 5.21,

$$dh = C_{po}\,dT$$

Eq. 10.37 reduces to Eq. 5.24,

$$C_{po} - C_{vo} = R$$

Eq. 10.32 reduces to Eq. 7.17,

$$ds = C_{vo} \frac{dT}{T} + \frac{R \, dv}{v}$$

and Eq. 10.30 reduces to Eq. 7.19,

$$ds = C_{po} \frac{dT}{T} - R \frac{dP}{P}$$

In discussing the accuracy of the ideal gas equation of state in Section 3.4, we introduced the compressibility factor

$$Z = \frac{P\bar{v}}{\bar{R}T} \tag{10.44}$$

as a useful and important parameter in expressing the nonideality of gases (or other phases). The compressibility factor is always unity (for all pressures and temperatures) for an ideal gas.

Another useful parameter in describing the behavior of a real gas relative to the ideal gas is the residual volume $\alpha$, which is defined as

$$\alpha = \frac{\bar{R}T}{P} - \bar{v} \tag{10.45}$$

We note that $\alpha$ is always zero for an ideal gas.

The Joule-Thomson coefficient $\mu_J$ was defined in Section 5.12 as

$$\mu_J = \left( \frac{\partial T}{\partial P} \right)_h \tag{10.46}$$

Since $h$ is a function only of $T$ for an ideal gas, it follows that the Joule-Thomson coefficient is always zero for an ideal gas.

## 10.9 The Behavior of Real Gases

In Section 3.4 we discussed briefly the nonideality in $P$-$v$-$T$ behavior of a gas, using nitrogen as an example, and in connection with that discussion constructed a skeleton compressibility diagram, Fig. 3.7. A more detailed compressibility diagram for $N_2$ is shown in Fig. 10.6. In examin-

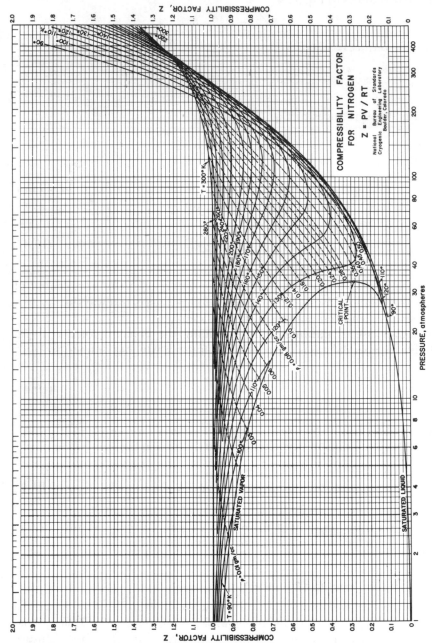

**Fig. 10.6**  Compressibility diagram for nitrogen.

ing this diagram, we note that $Z$ approaches unity for all isotherms as $P$ approaches zero, that the nonideality of the gas phase is especially severe in the vicinity of the critical point ($Z$ at the critical point is about 0.29), and that at high pressures $Z$ increases to values greater than unity for all isotherms.

If we examine compressibility diagrams for other pure substances, we find that the diagrams are all similar in the characteristics described above for nitrogen, at least in a qualitative sense. Quantitatively the diagrams are all different, since the critical temperatures and pressures of different substances vary over wide ranges, as evidenced from the values listed in Table A.7. Is there a way in which we can put all of these substances on a common basis? To do so, we "reduce" the properties with respect to the values at the critical point. The reduced properties are defined as

$$\text{Reduced pressure} = P_r = \frac{P}{P_c} \qquad P_c = \text{Critical pressure}$$

$$\text{Reduced temperature} = T_r = \frac{T}{T_c} \qquad T_c = \text{Critical temperature}$$

$$\text{Reduced specific volume} = v_r = \frac{v}{v_c} \qquad v_c = \text{Critical specific volume}$$

These equations state that the reduced property for a given state is the value of that property in this state divided by the value of that same property at the critical point.

If lines of constant $T_r$ are plotted on a $Z$ vs. $P_r$ diagram, a plot such as Fig. A.5 is obtained. The striking fact is that when such $Z$ vs. $P_r$ diagrams are prepared for a large number of different substances, these diagrams very nearly coincide. This leads to the development of a generalized compressibility chart. Figure A.5 is actually a generalized diagram, which means that it represents the average diagram for a number of different substances. It should be noted that when such a diagram is used for a particular substance, the results might be slightly in error. On the other hand, if $P$-$v$-$T$ information is required for a substance in a region where no experimental measurements have been made, the generalized compressibility diagram will in all probability give relatively accurate results. It is only necessary to know the critical pressure and the critical temperature in order to use the generalized chart. It should be emphasized that whenever one has accurate thermodynamic data for a given substance, these should be used rather than the generalized charts.

This behavior of pure substances that leads to the generalized compressibility chart is sometimes referred to as the rule of corresponding states, which may be expressed as

$$v_r = f(P_r, T_r)$$

That is, if the rule of corresponding states held exactly, there would be a single functional relation between $v_r$, $P_r$, and $T_r$ which would apply to all substances. The generalized chart is one way of expressing this relationship, and the fact that this is only an approximate relation indicates the approximate nature of the rule of corresponding states.

It also follows that the use of only two parameters, $P_c$ and $T_c$, in developing the generalized chart necessarily results in a common value of compressibility factor at the critical point for all substances. In Fig. A.5, $Z_c$ has been taken as 0.27, which is a good average value. Actually, $Z_c$ varies from about 0.23 to 0.33 for different substances, and therefore the generalized chart is usually not very accurate in the immediate vicinity of the critical point. More accurate three-parameter generalized compressibility charts that utilize $Z_c$ or another parameter as an independent variable have been developed, but will not be discussed further here. It should also be pointed out that the compressibility factor of liquids does not generalize well as a function of only $P_r$ and $T_r$. As a result, the liquid region is often not included on a generalized chart. Figure A.5 does show a saturated liquid line, and the compressed liquid region is shown as dashed lines. This region is considerably less accurate than the gas phase, and should not be used in preference to tables or data for any particular substance.

The following example illustrates the use of the generalized compressibility chart.

**Example 10.8**

1. Volume unknown. Calculate the specific volume of propane at a pressure of 7 MPa and a temperature of 150°C, and compare this with the specific volume given by the ideal-gas equation of state.

   For propane

$$T_c = 370 \text{ K} \qquad P_c = 4.26 \text{ MPa}$$

$$R = 0.18855 \text{ kJ/kg K}$$

$$T_r = \frac{423.2}{370} = 1.144 \quad , \quad P_r = \frac{7}{4.26} = 1.64$$

From the compressibility chart

$$Z = 0.54$$

$$v = \frac{ZRT}{P} = \frac{0.54 \times 0.18855 \times 423.2}{7000} = 0.006\ 16\ \text{m}^3/\text{kg}$$

The ideal-gas equation would give the value

$$v = \frac{0.18855 \times 423.2}{7000} = 0.0114\ \text{m}^3/\text{kg}$$

2. Pressure unknown. What pressure is required in order that propane have a specific volume of 0.006 16 m³/kg at a temperature of 150°C?

$$T_r = \frac{423.2}{370} = 1.144$$

$$P_r = \frac{P}{P_c} = \frac{ZRT}{vP_c} = Z \times \frac{0.18855 \times 423.2}{0.006\ 16 \times 4260} = 3.04\ Z$$

By a trial and error procedure, or by plotting a few points and drawing the curve representing this equation, $P_r$ is found to be 1.64 at the point where $T_r = 1.144$. Therefore,

$$P = P_r P_c = 1.64 \times 4260 = 7000\ \text{kPa} = 7\ \text{MPa}$$

3. Temperature unknown. What will be the temperature of propane when it has a specific volume of 0.006 16 m³/kg and a pressure of 7 MPa?

$$P_r = \frac{7}{4.26} = 1.64$$

$$T_r = \frac{T}{T_c} = \frac{Pv}{ZRT_c} = \frac{7000 \times 0.006\ 16}{Z \times 0.18855 \times 370} = \frac{0.6177}{Z}$$

When this line is plotted on the compressibility chart, the state point is given by the intersection of this line with the known reduced-pressure line ($P_r = 1.64$ in this case).

The reduced temperature is thus found to be

$$T_r = 1.144$$

$$T = T_r \times T_c = 1.144 \times 370 = 423.2 \text{ K}$$

In order to gain additional insight into the behavior of gases at low density, let us examine the low-pressure portion of the generalized compressibility chart in greater detail. This behavior is as shown in Fig. 10.7.

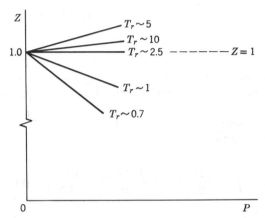

**Fig. 10.7**   Low-pressure region of compressibility chart.

The isotherms are essentially straight lines in this region, and their slope is of particular importance. Note that the slope increases as $T_r$ increases until a maximum value is reached at a $T_r$ of about 5, and then the slope decreases toward the $Z = 1$ line for higher temperatures. That single temperature, about 2.5 times the critical temperature, for which

$$\lim_{P \to 0} \left( \frac{\partial Z}{\partial P} \right)_T = 0 \qquad (10.47)$$

is defined as the Boyle temperature of the substance. This is the only temperature at which a gas really behaves exactly as an ideal gas at low, but finite pressures, since all other isotherms go to zero pressure on Fig. 10.7 with a nonzero slope.

To amplify this point, let us consider the residual volume $\alpha$, which was defined according to Eq. 10.45,

$$\alpha = \frac{\overline{R}T}{P} - \overline{v}$$

Multiplying this equation by $P$ we have

$$\alpha P = \bar{R}T - P\bar{v} \qquad (10.48)$$

Thus the quantity $\alpha P$ is the difference between $\bar{R}T$ and $P\bar{v}$. Now as $P \to 0$, $P\bar{v} \to \bar{R}T$. However, it does not necessarily follow that $\alpha \to 0$ as $P \to 0$. Instead, it is only required that $\alpha$ remain finite. The derivative in Eq. 10.47 can be written as

$$\lim_{P \to 0} \left( \frac{\partial Z}{\partial P} \right)_T = \lim_{P \to 0} \left( \frac{Z-1}{P-0} \right)$$

$$= \lim_{P \to 0} \frac{1}{\bar{R}T} \left( \bar{v} - \frac{\bar{R}T}{P} \right)$$

$$= -\frac{1}{\bar{R}T} \lim_{P \to 0} (\alpha) \qquad (10.49)$$

from which we find that $\alpha$ tends to zero as $P \to 0$ only at the Boyle temperature, since that is the only temperature for which the isothermal slope is zero on Fig. 10.7. It is perhaps a somewhat surprising result that in the limit as $P \to 0$, $P\bar{v} \to \bar{R}T$ but in general the quantity $(\bar{R}T/P - \bar{v})$ does not go to zero, but is instead a small difference between two large values. This does have an effect on certain other properties of the gas.

Another aspect of generalized behavior of gases is the behavior of isotherms in the vicinity of the critical point. If we plot experimental data on $P$-$v$ coordinates, it is found that the critical isotherm is unique in that it goes through a horizontal inflection point at the critical point as shown in Fig. 10.8.

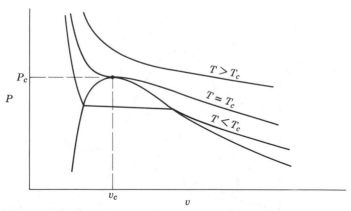

**Fig. 10.8**  Plot of isotherms in the region of the critical point on pressure-volume coordinates for a typical pure substance.

Mathematically, this means that the first two derivatives are zero at the critical point,

$$\left(\frac{\partial P}{\partial v}\right)_{T_C} = 0 \qquad \text{at C.P.} \tag{10.50}$$

$$\left(\frac{\partial^2 P}{\partial v^2}\right)_{T_C} = 0 \qquad \text{at C.P.} \tag{10.51}$$

a feature that is used to constrain many equations of state, as will be discussed in Section 10.10.

The Joule-Thomson coefficient, defined in connection with the throttling process, can be expressed in terms of $P$-$v$-$T$ behavior as

$$\mu_J = \left(\frac{\partial T}{\partial P}\right)_h = \frac{T\left(\frac{\partial v}{\partial T}\right)_P - v}{C_P} = \frac{RT^2}{PC_P}\left(\frac{\partial Z}{\partial T}\right)_P \tag{10.52}$$

From Eq. 10.52 and Fig. 10.7 it is evident that at low pressure $\mu_J$ is zero only at one temperature, $T_r$ about 5, and is positive at lower temperatures and negative at higher temperatures. By examining higher density $P$-$v$-$T$ data, we can determine the locus of points at which $\mu_J$ is zero, which is called the Joule-Thomson inversion curve. The result is as shown in Fig. 10.9. Inside the dome, $\mu_J$ is positive, which means that a substance cools upon being throttled to a lower pressure. Outside the dome, $\mu_J$ is negative, and the temperature increases during a throttling process.

There are also many other aspects of generalized behavior of gases,

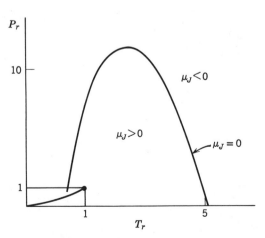

**Fig. 10.9**    Joule-Thomson inversion curve.

some of which are discussed in connection with equations of state in the
next section.

## 10.10   Equations of State

An accurate equation of state, which is an analytical representation of
$P$-$v$-$T$ behavior, is often desirable from a computational standpoint. Many
different equations of state have been developed. Most of these are
accurate only to some density less than the critical density, though a few
are reasonably accurate to approximately 2.5 times the critical density.
All equations of state fail badly when the density exceeds the maximum
density for which the equation was developed.

Three broad classifications of equations of state can be identified,
namely, generalized, empirical, and theoretical.

The best known of the generalized equations of state is also the oldest,
namely, the van der Waals equation, which was presented in 1873 as a
semi-theoretical improvement over the ideal gas equation. The van der
Waals equation of state is:

$$P = \frac{RT}{v-b} - \frac{a}{v^2} \tag{10.53}$$

The constant $b$ is intended to correct for the volume occupied by the
molecules, and the term $a/v^2$ is a correction that accounts for the intermolecular forces of attraction. As might be expected in the case of a
generalized equation, the constants $a$ and $b$ are evaluated from the
general behavior of gases. In particular these constants are evaluated by
noting that the critical isotherm passes through a point of inflection at
the critical point, and that the slope is zero at this point. Thus, for the
van der Waals equation of state we have

$$\left(\frac{\partial P}{\partial v}\right)_T = -\frac{RT}{(v-b)^2} + \frac{2a}{v^3} \tag{10.54}$$

$$\left(\frac{\partial^2 P}{\partial v^2}\right)_T = \frac{2RT}{(v-b)^3} - \frac{6a}{v^4} \tag{10.55}$$

Since both of these derivatives are equal to zero at the critical point we
can write

$$-\frac{RT_c}{(v_c-b)^2} + \frac{2a}{v_c{}^3} = 0$$

$$\frac{2RT_c}{(v_c-b)^3} - \frac{6a}{v_c{}^4} = 0 \tag{10.56}$$

$$P_c = \frac{RT_c}{(v_c-b)} - \frac{a}{v_c{}^2}$$

Solving these three equations we find

$$v_c = 3b$$

$$a = \frac{27}{64} \frac{R^2 T_c^2}{P_c} \tag{10.57}$$

$$b = \frac{RT_c}{8P_c}$$

The compressibility factor at the critical point for the van der Waals equation is

$$Z_c = \frac{P_c v_c}{RT_c} = \frac{3}{8}$$

which is considerably higher than the actual value for any substance.

Van der Waals' equation can be written in terms of the compressibility factor and the reduced pressure and reduced temperature as follows:

$$Z^3 - \left(\frac{P_r}{8T_r} + 1\right) Z^2 + \left(\frac{27P_r}{64T_r^2}\right) Z - \frac{27P_r^2}{512T_r^3} = 0 \tag{10.58}$$

It is significant to note that this is of the same form as the generalized compressibility chart, namely, $Z = f(P_r, T_r)$, though the functional relation is quite different from the generalized chart. This concept that different substances will have the same compressibility factor at the same reduced pressure and reduced temperature is another way of expressing the rule of corresponding states.

A simple equation of state that is considerably more accurate than the van der Waals equation is that proposed by Redlich and Kwong in 1949.

$$P = \frac{\bar{R}T}{\bar{v} - b} - \frac{a}{\bar{v}(\bar{v} + b)T^{1/2}} \tag{10.59}$$

with

$$a = 0.427\,48\frac{\bar{R}^2 T_c^{5/2}}{P_c} \tag{10.60}$$

$$b = 0.086\,64\frac{\bar{R}T_c}{P_c} \tag{10.61}$$

The numerical values in the constants have been determined by a procedure similar to that followed in the van der Waals equation. Because of its simplicity this equation could not be expected to be sufficiently

accurate to find use in the calculation of precision tables of thermodynamic properties. It has, however, been used frequently for mixture
calculations and phase equilibrium correlations with reasonably good
success. A number of modified versions of this equation have also been
utilized in recent years.

One of the best known empirical equations of state is the Beattie-
Bridgeman equation, proposed in 1928. This is a pressure-explicit
equation with five constants that are determined by graphical procedure
from experimental data for each substance. This equation was introduced
in Chapter 3 in the form

$$P = \frac{RT(1-\epsilon)}{v^2}(v+B) - \frac{A}{v^2} \tag{10.62}$$

where

$$A = A_0(1 - a/v)$$

$$B = B_0(1 - b/v)$$

$$\epsilon = c/vT^3$$

For use with thermodynamic relations, it is more convenient to rewrite
the equation in the form

$$P = \frac{RT}{v} + \frac{\beta}{v^2} + \frac{\gamma}{v^3} + \frac{\delta}{v^4} \tag{10.63}$$

where

$$\beta = B_0RT - A_0 - cR/T^2$$

$$\gamma = -B_0bRT + A_0a - B_0cR/T^2$$

$$\delta = B_0bcR/T^2$$

Values for the five constants have been given for a number of substances
in Table 3.3.

The Beattie-Bridgeman equation of state is quite accurate for densities
less than about 0.8 times the critical density. A more complex equation of
state, and one that is suitable for use at higher densities, is that of
Benedict, Webb and Rubin, proposed in 1940. This equation has 8
empirical constants and is essentially an extension of the Beattie-
Bridgeman equation through the addition of high density terms.
Strobridge in 1962 further extended this type of equation to 16 constants,
and many other empirical equations, some even more complex in form,
have been developed and used to represent the $P$-$v$-$T$ behavior of
various substances.

A different approach to this problem is from the theoretical point of
view. The theoretical equation of state, which is derived from kinetic

theory or statistical thermodynamics, is written here in the form of a power series in reciprocal volume:

$$Z = \frac{P\bar{v}}{RT} = 1 + \frac{B(T)}{\bar{v}} + \frac{C(T)}{\bar{v}^2} + \frac{D(T)}{\bar{v}^3} + \ldots \tag{10.64}$$

where $B(T)$, $C(T)$, $D(T)$ are temperature dependent and are called virial coefficients. $B(T)$ is termed the second virial coefficient and is due to binary interactions on the molecular level. The general temperature dependence of the second virial coefficient is as shown for nitrogen in Fig. 10.10.

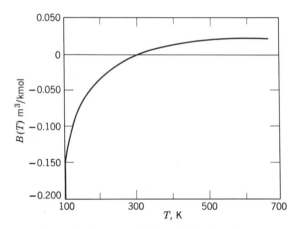

**Fig. 10.10** The second virial coefficient for nitrogen.

If we multiply Eq. 10.64 by $\bar{R}T/P$, the result can be rearranged to the form

$$\frac{\bar{R}T}{P} - \bar{v} = \alpha = -B(T)\frac{\bar{R}T}{P\bar{v}} - C(T)\frac{\bar{R}T}{P\bar{v}^2} - \ldots \tag{10.65}$$

In the limit, as $P \to 0$,

$$\lim_{P \to 0} \alpha = -B(T) \tag{10.66}$$

and we conclude from Eqs. 10.47 and 10.49 that the single temperature at which $B(T) = 0$, Fig. 10.10, is the Boyle temperature. The second virial coefficient can be viewed as the first order correction for nonideality of the gas, and consequently becomes of considerable importance and interest. In fact, the low-density behavior of the isotherms shown in Fig. 10.7 is directly attributable to the second virial coefficient. This is of further interest, since the virial coefficients can be expressed in terms of intermolecular forces from statistical mechanics, and evaluated upon selection of an empirical potential function model. The best known of these is the Lennard-Jones (6–12) potential. This function has two force constants, $\epsilon/k$ and $b_0$, values of which are listed for various substances in Table 10.1. The expression for $B(T)$ can be put into dimensionless form in terms of the parameter

$$T^* = \frac{T}{\epsilon/k} \tag{10.67}$$

and evaluated for the dimensionless second virial coefficient $B^*(T^*)$, defined from the relation

$$B^*(T^*) = \frac{B(T)}{b_0} \tag{10.68}$$

Values of the dimensionless parameter $B^*(T^*)$, as calculated using the Lennard-Jones (6–12) potential, are given in Table 10.2.

**Table 10.1**
Force Constants for the Lennard-Jones (6–12) Potential from Experimental Virial Coefficient Data

| Substance | $\epsilon/k$, K | $b_0$, m³/kmol |
|-----------|-----------------|----------------|
| Ne        | 35.8            | .0262          |
| Ar        | 119.0           | .0502          |
| Kr        | 173.0           | .0583          |
| Xe        | 225.3           | .0854          |
| $N_2$     | 95.05           | .0635          |
| $O_2$     | 117.5           | .0578          |
| CO        | 100.2           | .0675          |
| NO        | 131.0           | .0402          |
| $CO_2$    | 186.0           | .118           |
| $N_2O$    | 193.0           | .118           |
| $CH_4$    | 148.1           | .0698          |
| $CF_4$    | 152             | .131           |

From J. O. Hirschfelder, C. F. Curtiss, and R. B. Bird, *Molecular Theory of Gases and Liquids,* John Wiley and Sons, New York, Second Printing, Corrected, with Notes Added, March, 1964.

**Table 10.2**
The Reduced Second Virial Coefficient
$B*(T*)$ for the Lennard-Jones (6–12) Potential

| $T*$ | $B*(T*)$ | $T*$ | $B*(T*)$ |
|------|----------|------|----------|
| 0.3 | $-27.881$ | 2.6 | $-0.26613$ |
| 0.4 | $-13.799$ | 2.8 | $-0.18451$ |
| 0.5 | $-8.7202$ | 3.0 | $-0.11523$ |
| 0.6 | $-6.1980$ | 3.2 | $-0.05579$ |
| 0.7 | $-4.7100$ | 3.4 | $-0.00428$ |
| 0.8 | $-3.7342$ | 3.6 | $+0.04072$ |
| 0.9 | $-3.0471$ | 3.8 | $0.08033$ |
| 1.0 | $-2.5381$ | 4.0 | $0.11542$ |
| 1.1 | $-2.1464$ | 4.2 | $0.14668$ |
| 1.2 | $-1.8359$ | 4.4 | $0.17469$ |
| 1.3 | $-1.5841$ | 4.6 | $0.19990$ |
| 1.4 | $-1.3758$ | 4.8 | $0.22268$ |
| 1.5 | $-1.2009$ | 5.0 | $0.24334$ |
| 1.6 | $-1.0519$ | 6.0 | $0.32290$ |
| 1.7 | $-0.92362$ | 7.0 | $0.37609$ |
| 1.8 | $-0.81203$ | 8.0 | $0.41343$ |
| 1.9 | $-0.71415$ | 9.0 | $0.44060$ |
| 2.0 | $-0.62763$ | 10.0 | $0.46088$ |
| 2.2 | $-0.48171$ | 20.0 | $0.52537$ |
| 2.4 | $-0.36358$ | 30.0 | $0.52693$ |

**Example 10.9**

Use the virial equation of state to determine the error in specific volume if carbon dioxide is assumed to be an ideal gas at 300 K, 1 MPa pressure.

From Eq. 10.64 at low density,

$$Z = \frac{P\bar{v}}{\bar{R}T} = 1 + \frac{B(T)}{\bar{v}}$$

For carbon dioxide, from Table 10.1,

$$\epsilon/k = 186 \text{ K}, \qquad b_0 = 0.118 \text{ m}^3/\text{kmol}$$

Therefore,

$$T* = \frac{T}{\epsilon/k} = \frac{300}{186} = 1.612$$

From Table 10.2,

$$B^*(T^*) = -1.0365$$

Using Eq. 10.68,

$$B(T) = b_0 B^*(T^*) = 0.118 \, (-1.0365) = -0.122 \text{ m}^3/\text{kmol}$$

Substituting and solving for $\bar{v}$,

$$\frac{1000 \, \bar{v}}{8.31434 \times 300} = 1 - \frac{0.122}{\bar{v}}$$

$$\bar{v} = 2.366 \text{ m}^3/\text{kmol}$$

For an ideal gas,

$$\bar{v} = \frac{\bar{R}T}{P} = \frac{8.31434 \times 300}{1000} = 2.494 \text{ m}^3/\text{kmol}$$

which is in error by 5.4 per cent.

## 10.11 The Generalized Chart for Changes of Enthalpy at Constant Temperature

In Section 10.4, Eq. 10.25 was derived for the change of enthalpy at constant temperature.

$$(h_2 - h_1)_T = \int_1^2 \left[ v - T \left( \frac{\partial v}{\partial T} \right)_P \right] dP_T$$

The essential point about this equation for $(h_2 - h_1)_T$ is that the change of enthalpy at constant temperature is a function of the $P$-$v$-$T$ behavior of the gas. This means that if the equation of state is known for a substance, the change in enthalpy at constant temperature can be determined by the appropriate differentiation and integration. It also follows that since the generalized chart is a representation of the $P$-$v$-$T$ behavior of the gas, this chart can be used to prepare a generalized chart for the change of enthalpy in an isothermal process.

The procedure for the development of such a chart is as follows: It has been shown that

$$\left( \frac{\partial h}{\partial P} \right)_T = v - T \left( \frac{\partial v}{\partial T} \right)_P$$

But

$$v = \frac{ZRT}{P}$$

and

$$\left(\frac{\partial v}{\partial T}\right)_P = \frac{ZR}{P} + \frac{RT}{P}\left(\frac{\partial Z}{\partial T}\right)_P$$

Therefore

$$\left(\frac{\partial h}{\partial P}\right)_T = \frac{ZRT}{P} - \frac{ZRT}{P} - \frac{RT^2}{P}\left(\frac{\partial Z}{\partial T}\right)_P = -\frac{RT^2}{P}\left(\frac{\partial Z}{\partial T}\right)_P$$

$$dh_T = -\frac{RT^2}{P}\left(\frac{\partial Z}{\partial T}\right)_P dP_T$$

But $T = T_c T_r$ and $P = P_c P_r$. Therefore,

$$dT = T_c \, dT_r; \qquad dP = P_c \, dP_r$$

Substituting these values we have

$$dh_T = -\frac{RT_c^2 T_r^2}{P_c P_r}\left(\frac{\partial Z}{T_c \partial T_r}\right)_{P_r} P_c (dP_r)_{T_r}$$

$$= -RT_c T_r^2 \left(\frac{\partial Z}{\partial T_r}\right)_{P_r} (d\ln P_r)_{T_r}$$

Integrating at constant temperature we have

$$\frac{\Delta \bar{h}_T}{T_c} = -\bar{R} \int_1^2 T_r^2 \left(\frac{\partial Z}{\partial T_r}\right)_{P_r} d\ln P_r \tag{10.69}$$

Let us designate by an asterisk, $\bar{h}^*$, the enthalpy per mole at a given temperature and a very low pressure, so that $P_r \to 0$, and let us designate without superscript the enthalpy at any state at the same temperature and a given pressure $P_r$. Then the integration is as follows:

$$\frac{\bar{h}^* - \bar{h}}{T_c} = \bar{R} \int_{P_r=0}^{P_r} T_r^2 \left(\frac{\partial Z}{\partial T_r}\right)_{P_r} d\ln P_r \tag{10.70}$$

The right side of this equation can be obtained by graphical integration of the compressibility chart, and is a function of only the reduced pressure and temperature. The left side represents the change in enthalpy as the pressure is increased from zero to the given pressure while the temperature is held constant. Sometimes this quantity is called the enthalpy departure. Figure A.6 shows a plot of this quantity, $(\bar{h}^*-\bar{h})/T_c$, as a function of $P_r$ for various lines of constant $T_r$.

In many cases the specific heat of a substance is known at low pressure. This information, along with the generalized residual enthalpy chart, enables one to find the change of enthalpy for any given change of state.

### Example 10.10

Nitrogen is throttled from 20 MPa, $-70°C$, to 2 MPa in an adiabatic, steady-state, steady-flow process. Determine the final temperature of the nitrogen. This is a constant-enthalpy process, and in solving such a problem it is usually helpful to use a temperature-entropy diagram such as the one shown in Fig. 10.11.

$$P_1 = 20 \text{ MPa} \quad , \quad P_{r_1} = \frac{20}{3.39} = 5.9$$

$$T_1 = 203.2 \text{ K} \quad , \quad T_{r_1} = \frac{203.2}{126.2} = 1.61$$

$$P_2 = 2 \text{ MPa} \quad , \quad P_{r_2} = \frac{2}{3.39} = 0.59$$

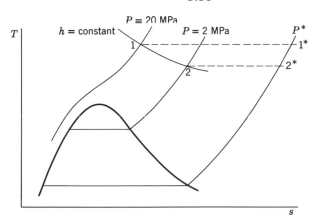

**Fig. 10.11**  Sketch for Example 10.10.

From the generalized chart for enthalpy change at constant temperature,

$$\frac{\bar{h}_1^*-\bar{h}_1}{T_c} = 14.2 \text{ kJ/kmol K}$$

$$h_1^*-h_1 = 14.2 \times \frac{126.2}{28} = 64.0 \text{ kJ/kg}$$

It is now necessary to assume a final temperature and check to see if the net change in enthalpy for the process is zero. Let us assume that $T_2 = 156$ K. Then the change in enthalpy between 1* and 2* can be found from the zero-pressure specific-heat data.

$$h_1^* - h_2^* = C_{po}(T_1^* - T_2^*) = 1.0416 \, (156 - 203.2) = -49.2 \text{ kJ/kg}$$

(The variation in $C_{po}$ with temperature can be taken into account when necessary.)

We now find the enthalpy change between 2* and 2.

$$T_{r_2} = \frac{156}{126.2} = 1.236 \quad , \quad P_{r_2} = 0.59$$

Therefore, from the enthalpy departure chart, Fig. A.6,

$$\frac{\bar{h}_2^* - \bar{h}_2}{T_c} = 3.3 \text{ kJ/kmol K}$$

$$h_2^* - h_2 = 3.3 \times \frac{126.2}{28} = 14.9 \text{ kJ/kg}$$

We now check to see if the net change in enthalpy for the process is zero.

$$h_1 - h_2 = 0 = - (h_1^* - h_1) + (h_1^* - h_2^*) + (h_2^* - h_2)$$
$$= -64 + 49.2 + 14.9 = 0.1 \simeq 0$$

This essentially checks, and we conclude that the final temperature is approximately 156 K.

## 10.12   Fugacity and the Generalized Fugacity Chart

At this point a new thermodynamic property, the fugacity, $f$, is introduced. The fugacity is particularly important when considering mixtures and equilibrium, which are discussed in two later chapters. However, in view of the fact that a generalized fugacity chart can be developed, the fugacity is introduced here.

The fugacity is essentially a pseudo-pressure. When the fugacity is substituted for pressure, one can, in effect, use the same equations for real gases that one normally uses for ideal gases. The concept of fugacity can be introduced as follows:

Consider the relation

$$dg = -s \, dT + v \, dP$$

At constant temperature

$$dg_T = v \, dP_T \tag{10.71}$$

For an ideal gas this last equation may be written

$$dg_T = \frac{RT}{P} dP_T = RT \, d(\ln P)_T \tag{10.72}$$

and for a real gas, with the equation of state $Pv = ZRT$, this equation becomes

$$dg_T = ZRT \frac{dP_T}{P} = ZRT \, d(\ln P)_T \tag{10.73}$$

Fugacity, $f$, is defined as

$$dg_T \equiv RT \, d(\ln f)_T \tag{10.74}$$

with the requirement that

$$\lim_{P \to 0} (f/P) = 1 \tag{10.75}$$

Therefore, as $P \to 0$, $f \to 0$.

Let us consider the change in the Gibbs function of a real gas during an isothermal process at temperature $T$ that involves a change in pressure from a very low pressure, $P^*$ (where ideal gas behavior can be assumed) to a higher pressure $P$. Let the Gibbs function at this low pressure be designated $g^*$. The value of the Gibbs function at this temperature $T$ and at the pressure $P$ can be found in terms of the fugacity at $P$ and $T$ and $g^*$. This is evident if one integrates Eq. 10.74 from the very low pressure $P^*$ to the pressure $P$.

$$\int_{g^*}^{g} dg_T = \int_{f^*=P^*}^{f} RT(d \ln f)_T$$
$$g = g^* + RT \ln (f/P^*) \tag{10.76}$$

A generalized fugacity coefficient chart based on the generalized compressibility chart, Fig. A.7, can be developed by the following procedure. From Eqs. 10.71 and 10.74.

$$dg_T = v \, dP_T = \frac{ZRT \, dP_T}{P} = RT \, d(\ln f)_T$$

Therefore,

$$Z \, d(\ln P)_T = d(\ln f)_T$$

But

$$\frac{dP}{P} = \frac{dP_r}{P_r}; \qquad d \ln P = d \ln P_r$$

Therefore,

$$Z(d \ln P_r)_T = d(\ln f)_T$$

$$Z(d \ln P_r)_T - d(\ln P_r)_T = d(\ln f)_T - d(\ln P_r)_T \qquad (10.77)$$

$$(Z - 1)(d \ln P_r)_T = d(\ln f/P)_T$$

Integrating at constant temperature from $P = 0$ to a finite pressure we have

$$\int_{f/P=1}^{f/P} d(\ln f/P)_T = \int_0^{P_r} (Z-1) \, d \ln P_{rT}$$

$$\ln(f/P) = \int_0^{P_r} (Z-1) \, d \ln P_{rT} \qquad (10.78)$$

At any temperature the right hand side of this equation can be integrated graphically using the generalized compressibility chart to find $Z$ at each $P_r$. The result is the generalized chart shown in Fig. A.7.

The reversible isothermal process is one type of problem in which the fugacity can be used to advantage.

### Example 10.11

Calculate the work of compression and the heat transfer per kilogram when ethane is compressed reversibly and isothermally from 0.1 MPa to 7 MPa at a temperature of 45°C in a steady-state, steady-flow process.

The first law for this process is,

$$q + h_1 + KE_1 + PE_1 = h_2 + KE_2 + PE_2 + w$$

Since the process is reversible and isothermal,

$$q = T(s_2 - s_1) = T_2 s_2 - T_1 s_1$$

Therefore,

$$-w = h_2 - h_1 - (T_2 s_2 - T_1 s_1) + \Delta KE + \Delta PE$$

$$-w = g_2 - g_1 + \Delta KE + \Delta PE$$

We assume that the changes in kinetic and potential energy are negligible.

$$-w = g_2 - g_1 = RT \ln(f_2/f_1)$$

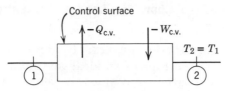

**Fig. 10.12**  Sketch for Example 10.11.

The fugacity in state 1 and state 2 can be found from the generalized fugacity chart in the Appendix, Fig. A.7.

For ethane,

$$P_c = 4.88 \text{ MPa} \quad , \quad T_c = 305.5 \text{ K}$$

Therefore,

$$P_{r_1} = \frac{0.1}{4.88} = 0.0205 \quad , \quad P_{r_2} = \frac{7.0}{4.88} = 1.434$$

$$T_{r_1} = T_{r_2} = \frac{318.2}{305.5} = 1.042$$

From the generalized enthalpy departure chart, Fig. A.6,

$$\frac{\bar{h}_2^* - \bar{h}_2}{T_c} = 26.0 \qquad \frac{\bar{h}_1^* - \bar{h}_1}{T_c} < 0.1$$

Since $h_1^* = h_2^* = h_1$

$$h_2^* - h_2 = h_1 - h_2 = 26.0 \times \frac{305.5}{30.07} = 264.2 \text{ kJ/kg}$$

From the generalized fugacity chart, Fig. A.7,

$$f_2/P_2 = 0.6 \qquad f_2 = 0.6 \times 7.0 = 4.2 \text{ MPa}$$

$$f_1/P_1 = 1.0 \qquad f_1 = 1.0 \times 0.1 = 0.1 \text{ MPa}$$

Therefore,

$$-w = g_2 - g_1 = RT \ln f_2/f_1$$

$$= 0.2765 \times 318.2 \ln \frac{4.2}{0.1} = 328.8 \text{ kJ/kg}$$

$$g_2 - g_1 = (h_2 - h_1) - T(s_2 - s_1) = (h_2 - h_1) - q$$

$$328.8 = -264.2 - q$$

$$q = -264.2 - 328.8 = -593.0 \text{ kJ/kg}$$

$$s_2 - s_1 = \frac{-593.0}{318.2} = -1.8636 \text{ kJ/kg K}$$

### 10.13  The Generalized Chart for Changes of Entropy at Constant Temperature

A generalized chart for entropy can be developed that gives the difference between the entropy of an ideal gas and an actual gas as the pressure is increased from zero pressure along an isotherm. The thermodynamic analysis for the development of such a chart is as follows.

From Eq. 10.30

$$ds_T = -\left(\frac{\partial v}{\partial T}\right)_P dP_T$$

Integrating at constant temperature, $T$, from $P = 0$ to $P$ we have

$$(s_P - s_0^*)_T = -\left[\int_{P=0}^{P} \left(\frac{\partial v}{\partial T}\right)_P dP_T\right] \tag{10.79}$$

The entropy at zero pressure can be designated with an * (which we use to refer to ideal gas behavior) since ideal gas behavior is approached as $P \to 0$. However, $s \to \infty$ as $P \to 0$, and therefore this equation is not particularly useful in this form.

If we repeat this integration for an ideal gas we have

$$(s_P^* - s_0^*)_T = -\left[\int_{P=0}^{P} \left(\frac{\partial v}{\partial T}\right)_P dP_T\right] = -\left[\int_{P=0}^{P} R\frac{dP_T}{P}\right] \tag{10.80}$$

Subtracting Eq. 10.79 from Eq. 10.80,

$$(s_P^* - s_P) = -\int_{P=0}^{P} \left[\frac{R}{P} - \left(\frac{\partial v}{\partial T}\right)_P\right] dP_T$$

Therefore, with the equation of state $Pv = ZRT$,

$$(s_P^* - s_P) = -\int_{P=0}^{P} \left[\frac{R}{P} - \frac{ZR}{P} - \frac{RT}{P}\left(\frac{\partial Z}{\partial T}\right)_P\right] dP_T$$

or,

$$(\bar{s}_P^* - \bar{s}_P) = \bar{R} \int_0^{P_r} (Z-1)\frac{dP_r}{P_r} + \bar{R}T_r \int_0^{P_r} \left(\frac{\partial Z}{\partial T_r}\right)_{P_r} \frac{dP_r}{P_r}$$

But, from Eq. 10.78

$$\bar{R} \int_0^{P_r} (Z-1)\frac{dP_r}{P_r} = \bar{R} \ln (f/P)$$

and from Eq. 10.70,

$$\bar{R}T_r \int_0^{P_r} \left(\frac{\partial Z}{\partial T_r}\right)_{P_r} \frac{dP_r}{P_r} = \frac{\bar{h}^* - \bar{h}}{T_c T_r}$$

Therefore,

$$(\bar{s}_P^* - \bar{s}_P) = \bar{R} \ln f/P + \frac{\bar{h}^* - \bar{h}}{T_c T_r} \qquad (10.81)$$

Thus, for the equation of state $Pv = ZRT$, the quantity $(s_P^* - s_P)$ can be obtained from the generalized enthalpy departure and the generalized fugacity coefficient charts. Figure A.8 shows such a chart.

### Example 10.12

Nitrogen at 8 MPa, 150 K is throttled to 0.5 MPa. After passing through a short length of pipe the temperature is measured and found to be 125 K. Determine the heat transfer and the change of entropy using the generalized charts and compare these results with those

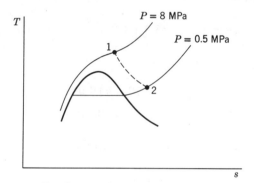

**Fig. 10.13**   Sketch for Example 10.12.

obtained by using the nitrogen tables.

$$P_{r_1} = \frac{8}{3.39} = 2.36 \quad , \quad T_{r_1} = \frac{150}{126.2} = 1.189$$

$$P_{r_2} = \frac{0.5}{3.39} = 0.147 \quad , \quad T_{r_2} = \frac{125}{126.2} = 0.99$$

From the first law

$$q = h_2 - h_1 = -(h_2^* - h_2) + (h_2^* - h_1^*) + (h_1^* - h_1)$$

From Fig. A.6 in the Appendix,

$$\frac{\bar{h}_1^* - \bar{h}_1}{T_c} = 20.0 \text{ kJ/kmol K}; \quad h_1^* - h_1 = 20.0 \times \frac{126.2}{28} = 90.1 \text{ kJ/kg}$$

$$\frac{\bar{h}_2^* - \bar{h}_2}{T_c} = 1.3 \text{ kJ/kmol K}; \quad h_2^* - h_2 = 1.3 \times \frac{126.2}{28} = 5.9 \text{ kJ/kg}$$

Assuming a constant specific heat for the ideal gas,

$$h_2^* - h_1^* = C_{p0}(T_2 - T_1) = 1.0416 \,(125 - 150) = -26.0 \text{ kJ/kg}$$
$$q = -5.9 - 26.0 + 90.1 = 58.2 \text{ kJ/kg}$$

From the nitrogen tables, Table A.5, we can find the change of enthalpy directly.

$$q = h_2 - h_1 = 123.82 - 61.90 = 61.92 \text{ kJ/kg}$$

To calculate the change of entropy using the generalized charts we proceed as follows:

$$\bar{s}_2 - \bar{s}_1 = -(\bar{s}_{P_2,T_2}^* - \bar{s}_2) + (\bar{s}_{P_2,T_2}^* - \bar{s}_{P_1,T_1}^*) + (\bar{s}_{P_1,T_1}^* - \bar{s}_1)$$

From Fig. A.8 in the Appendix,

$$\bar{s}_{P_1,T_1}^* - \bar{s}_1 = 12.6 \text{ kJ/kmol K}$$
$$\bar{s}_{P_2,T_2}^* - \bar{s}_2 = 1.3 \text{ kJ/kmol K}$$

Assuming a constant specific heat for the ideal gas,

$$\bar{s}_{P_2,T_2}^* - \bar{s}_{P_1,T_1}^* = \bar{C}_{p0} \ln \frac{T_2}{T_1} - \bar{R} \ln \frac{P_2}{P_1}$$
$$= 1.0416 \times 28.013 \ln \frac{125}{150} - 8.31434 \ln \frac{0.5}{8}$$
$$= 17.73 \text{ kJ/kmol K}$$

$$\bar{s}_2 - \bar{s}_1 = -1.3 + 17.73 + 12.6 = 29.03 \text{ kJ/kmol K}$$

$$s_2 - s_1 = \frac{29.03}{28.013} = 1.0363 \text{ kJ/kg K}$$

From Table A.5

$$s_2 - s_1 = 5.4343 - 4.3518 = 1.0825 \text{ kJ/kg K}$$

# PROBLEMS

10.1 Using the first Maxwell relation derived, Eq. 10.11, and the mathematical relation Eq. 10.2, derive the other three Maxwell relations, Eqs. 10.12–10.14.

10.2 Show that on a Mollier diagram ($h$-$s$ diagram) the slope of a constant-pressure line increases with temperature in the superheat region and that the constant-pressure line is straight in the two-phase region.

10.3 A transfer line for liquid nitrogen is so designed that it has a double walled pipe as indicated in Fig. 10.14. The annular space is so constructed

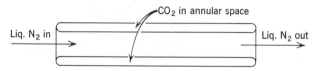

**Fig. 10.14**  Sketch for Problem 10.3.

that when the pipeline is at ambient temperature it contains carbon dioxide at 101.3 kPa pressure. When liquid nitrogen at 101.3 kPa pressure flows through the pipe the carbon dioxide will freeze on the inner walls. The pressure of the carbon dioxide in the annular space will essentially be equal to the vapor pressure of carbon dioxide at the saturation temperature of nitrogen at 101.3 kPa pressure. For $CO_2$ the tables of thermodynamic properties which are available do not go down to this temperature. However, at 38.1 kPa pressure the saturation temperature is −90°C and the enthalpy of sublimation is 574.5 kJ/kg. Estimate the pressure of the carbon dioxide in the annular space when liquid nitrogen is flowing through the tube.

10.4 Using triple point data for water from Table A.1.5 and assuming $h_{ig}$ constant, extrapolate the sublimation pressure from the triple point to −40°C. What is the percent error in this calculated value at −40°C?

10.5 At 0.1013 MPa pressure helium boils at 4.22 K, and the latent heat of vaporization is 83.3 J/mol. Suppose one wishes to determine how low a temperature could be obtained by producing a vacuum over liquid helium, thus causing it to boil at a lower pressure and temperature. Determine what pressure would be necessary to produce a temperature of 1 K and 0.1 K.

What do you think is the lowest temperature that could be obtained by pumping a vacuum over liquid helium?

10.6 One of the serious pollutants in combustion products appears in the form of very small solid particles. In order to analyze the problem it is necessary to know the sublimation pressure of this substance at various temperatures. Suppose that the only experimental data available are as follows: normal (0.101325 MPa) boiling point temperature and enthalpy of vaporization $h_{fg}$; normal melting point temperature and enthalpy of fusion

$h_{if}$. Develop a procedure for estimating sublimation pressure as a function of temperature based on this information.

10.7 The following data are available at the triple point of water:

| | |
|---|---|
| Pressure: | 0.6113 kPa |
| Temperature: | 0.01°C. |
| Enthalpy: | Liquid: 0.01 kJ/kg |
| | Solid: −333.4 kJ/kg |
| Specific Volume: | Liquid: 0.001 m³/kg |
| | Solid: 0.001 091 m³/kg |

A man weighing 80 kg is ice skating on "hollow ground" blades which have a total area in contact with the ice of 20 mm². The temperature of the ice is −2°C. Will the ice melt under the blades?

10.8 An experiment is to be conducted at −120°C using Freon-12 as the bath liquid in a closed tank (with some vapor at the top of the tank).
(a) What is the pressure in the tank?
(b) When the experiment is completed, the tank will be allowed to warm to room temperature, 25°C. If the pressure in the tank is never to exceed 700 kPa, what is the maximum per cent liquid (by volume) that can be used in the tank at −120°C?

10.9 Derive expressions for $(\partial u/\partial P)_T$ and $(\partial u/\partial T)_P$ that do not contain the properties $h$, $u$, or $s$.

10.10 Derive expressions for $(\partial T/\partial v)_u$ and $(\partial h/\partial s)_v$ that do not contain the properties $h$, $u$, or $s$.

10.11 Consider a homogeneous phase of a pure substance.
(a) Develop a general expression (in terms of specific heat and $P$-$V$-$T$ behavior only) for the slope of a constant entropy line on a $P$–$T$ diagram.
(b) Assuming a monatomic ideal gas ($C_{p0} = \frac{5}{2}R$) what is the relationship between the slope of a constant entropy line and that of a constant volume line on a $P$-$T$ diagram?

10.12 Show that if the volume expansivity $\alpha_p$ is constant, the following relation is valid:

$$\left(\frac{\partial C_p}{\partial P}\right)_T = -vT\alpha_p{}^2$$

10.13 From the data given for water in Table A.1.4 of the appendix determine the average value of the isothermal compressibility at 0°C, 100°C and 260°C. Does the isothermal compressibility vary significantly with pressure at a given temperature?

10.14 It can be shown that the velocity of sound $c$ in a fluid is given by the relation

$$c^2 = \left(\frac{\partial P}{\partial \rho}\right)_s$$

(a) Using this relation, show that the velocity of sound can be written as

$$c = \sqrt{\frac{1}{\beta_s \rho}}$$

where $\beta_s$ is the adiabatic compressibility.

(b)  At a temperature of 35°C the velocity of sound in water is 1530 m/s. Verify this by determining the adiabatic compressibility of water at 35°C by using Tables A.1.1 and A.1.4. (Note that the state of water after an isentropic process can be determined by using Tables A.1.1 and A.1.4.)

10.15 The following data are available for ethyl alcohol at 20°C.

$$\rho = 788 \text{ kg/m}^3$$
$$\beta_s = 943 \text{ } \mu m^2/N$$
$$\beta_T = 1117 \text{ } \mu m^2/N$$

(a)  What is the velocity of sound (see Problem 10.14) in ethyl alcohol?

(b)  The pressure on a cylinder having a volume of 10 litres containing ethyl alcohol at 20°C is increased from 0.1 MPa to 20 MPa while the temperature remains constant. Determine the work done on the ethyl alcohol during this process.

10.16 Suppose that the following data are available for a given pure substance:
A vapor-pressure curve.
An equation of state of the form $P - f(v, T)$.
Specific volume of saturated liquid.
Critical pressure and temperature.
Constant-volume specific heat of the vapor at one specific volume.
Outline the steps you would follow in order to develop a table of thermodynamic properties comparable to Tables 1, 2, and 3 of the steam tables.

10.17 Fifty litres of ethylene at 2 MPa, 25°C, is cooled at constant volume to saturated vapor, after which the $C_2H_4$ is condensed at constant pressure to saturated liquid. Determine the final pressure, temperature, and volume using the generalized compressibility chart.

10.18 A 0.1 m³ rigid tank contains propane at 9 MPa, 280°C. The propane is then allowed to cool to 60°C due to heat transfer with the surroundings. Determine the quality at the final state and the mass of liquid in the tank.

10.19 Saturated liquid carbon dioxide at the temperature of the surroundings, 15°C, is contained in a cylinder as shown in Fig. 10.15. The volume at this state is 0.3 litres. The piston has a mass of 5 kg, area of 0.015 m², and ambient pressure is 100 kPa. The pin is then released from the piston, allowing it to rise and hit the stops, at which point the volume is 4.3 litres. Eventually, the temperature of the $CO_2$ returns to that of the surroundings.

(a)  Calculate the pressure (if superheated) or quality (if saturated at the final state.

(b)  Calculate the work done during the process.

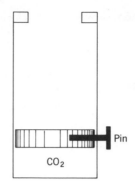

**Fig. 10.15** Sketch for Problem 10.19.

10.20 Show that the Joule-Thomson coefficient $\mu_J$ is given by Eq. 10.52,

$$\mu_J = \frac{1}{C_p}\left[T\left(\frac{\partial v}{\partial T}\right)_P - v\right] = \frac{RT^2}{PC_p}\left(\frac{\partial Z}{\partial T}\right)_P$$

10.21 Two uninsulated tanks of equal volume are connected by a valve. One tank contains a gas at moderate pressure $P_1$ and the other tank is evacuated. The valve is now opened, and remains open for a long time. Is the final pressure $P_2$ greater than, equal to, or less than $\frac{1}{2}P_1$?

10.22 A method of determining the compressibility $(Pv/RT)$ of gases by measuring only pressure and temperature is shown in Fig. 10.16. Tanks $A$ and $B$ of unknown volumes $V_A$ and $V_B$ are held in a uniform and constant temperature $T$. Initially, tank $A$ contains a gas at high pressure $P_0$, and tank $B$ is evacuated. Valve I is then opened, and remains open until the pressure has stabilized at a lower value $P_1$. Valve I is then closed and tank $B$ is reevacuated through valve II. This expansion process is then repeated by opening valve I, allowing the pressure to come to a new value $P_2$, then closing valve I and reevacuating tank $B$. A continued series of these expansion − reevacuation processes results in the set of experimental data:

$$P_0, P_1, P_2, \ldots, P_j \text{ at temperature } T.$$

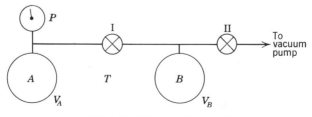

**Fig. 10.16** Sketch for Problem 10.22.

Develop a graphical-analytical procedure for determining the compressibility at each of these finite pressures (the final value $P_j$ is low but not ideal gas). It may be assumed that volumes $V_A$ and $V_B$ do not change with pressure.

10.23 Show that for van der Waals' equation the following relations apply.

(a) $(h_2 - h_1)_T = (P_2 v_2 - P_1 v_1) + a \left( \dfrac{1}{v_1} - \dfrac{1}{v_2} \right)$

(b) $(s_2 - s_1)_T = R \ln \left( \dfrac{v_2 - b}{v_1 - b} \right)$

(c) $\left( \dfrac{\partial C_v}{\partial v} \right)_T = 0$

(d) $C_p - C_v = \dfrac{R}{1 - 2a(v-b)^2/RTv^3}$

10.24 Use the Redlich-Kwong equation of state to develop expressions equivalent to those developed for van der Waals' equation in Problem 10.23.

10.25 Derive expressions for the changes in enthalpy and entropy at constant temperature for a substance that follows the Beattie-Bridgeman equation of state.

10.26 Oxygen is contained in a rigid 1-litre vessel at 0°C, 10 MPa pressure. The vessel is then cooled to $-100$°C. Use the virial equation of state to find the final pressure.

10.27 Determine the Boyle temperature and also the temperature at which

$$\lim_{p \to 0} \mu_J = 0$$

for
   (a) van der Waals equation
   (b) Redlich-Kwong equation

10.28 A generalized plot of the Joule-Thomson inversion curve is shown in Fig. 10.9. Use the result of Problem 10.20 to determine this curve as predicted by
   (a) van der Waals equation
   (b) Redlich-Kwong equation

10.29 Consider a straight line connecting the point $P = 0$, $Z = 1$, to the critical point on a $Z$ vs. $P$ diagram. This line will be tangent to one isotherm at low pressure.
   (a) What reduced temperature does the van der Waals equation of state predict this isotherm to be?
   (b) Repeat part (a) for the Redlich-Kwong equation.

10.30 Using the virial equation of state (2nd virial coefficient) calculate the difference in internal energy between 3 MPa pressure and ideal-gas for carbon dioxide at 20°C.

10.31 Methane at a pressure of 8 MPa, 40°C is throttled in a steady-flow process to 2 MPa. Determine the final temperature of the methane.

10.32 Ethane at 5.5 MPa, 150°C is cooled in a heat exchanger to 50°C. The volume rate of flow entering the heat exchanger is 0.1 m³/s, and the pressure drop through the heat exchanger is small. Determine the heat transfer.

10.33 A 50 litre vessel contains 25 kg of carbon dioxide at 15°C. A valve on the vessel is opened such that carbon dioxide escapes slowly. When the final pressure reaches half of the initial pressure the valve is closed and the temperature is measured to be 0°C. Determine the mass of carbon dioxide that has escaped from the vessel and the heat transfer for the process.

10.34 Saturated liquid nitrogen at 110 K is throttled to 600 kPa. Calculate the quality at the outlet, using
   (a) The generalized charts.
   (b) The nitrogen tables.

10.35 A number of tanks are to be filled with ethylene for use in another location. The procedure is as follows: a 0.1 m³ evacuated tank is connected to a line carrying $C_2H_4$ at 14.5 MPa and room temperature, 25°C. The valve is then opened, and ethylene flows in rapidly, such that the process is essentially adiabatic. The valve is to be closed when the pressure reaches a certain value $P_2$, and the tank is then disconnected. After a period of time, the temperature inside the tank will return to room temperature because of heat transfer with the surroundings. At this time, the pressure inside the tank must be 10 MPa. What is the pressure $P_2$ at which the valve should be closed while filling the tank?

10.36 Carbon dioxide is to be liquefied in a steady-flow insulated condenser. $CO_2$ enters the condenser at 75°C, 6 MPa at the rate of 0.5 litres/s, and leaves at 6 MPa. Cooling water enters at 5°C and leaves at 20°C. The flow rate of the cooling water is 0.24 kg/s.

   Determine the temperature of the $CO_2$ leaving the condenser and show the process for the $CO_2$ on a $T$-$s$ diagram.

10.37 Propane flows through a pipe of constant cross-sectional area. At the inlet section of this pipe the velocity is 25 m/s, the pressure is 2.75 MPa, and the temperature is 100°C. At the exit section the pressure is 0.7 MPa. There is no heat transfer to or from the gas as it flows in the pipe. Determine the temperature and velocity of the propane leaving the pipe.

10.38 High-pressure liquid methane is to be produced at the rate of 0.25 kg/s. The available supply of $CH_4$ gas at 25°C, 0.7 MPa is cooled and condensed in a heat exchanger at constant pressure, after which the liquid is pumped to 3.5 MPa.
   (a) Calculate the required heat transfer from the methane in flowing through the heat exchanger.
   (b) Assuming a pump efficiency of 80%, calculate the power required to drive the pump.

10.39 An uninsulated cylinder with a volume of 0.1 m³ contains ethylene at 7 MPa, 25°C, the temperature of the surroundings. The cylinder valve leaks very slightly so that after a considerable length of time the cylinder pressure drops to 3.5 MPa. Determine:

(a) The mass of ethylene which escaped from the cylinder.

(b) The heat transferred during the process.

10.40 Consider a 1 m³ tank in which methane is stored at low temperature. The pressure is 700 kPa and the tank contains 25% liquid, 75% vapor on a volume basis.

(a) The tank now slowly warms because of heat transfer from the ambient. What is the temperature inside when the pressure reaches 10 MPa?

(b) Calculate the heat transferred during this process.

10.41 Ethane is contained in an insulated cylinder, as shown in Fig. 10.17. The initial conditions are 0°C, quality of 90%, and volume of 0.3 litres. The piston area is 300 mm² and the total mass of piston and weights is 200 kg. Ambient pressure is 0.1 MPa. The pin holding the piston is now removed. Determine the final pressure, volume, and temperature.

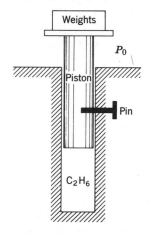

**Fig. 10.17**   Sketch for Problem 10.41.

10.42 Freon-12, which has a molecular weight of 121, expands in a cylinder from 2 MPa, 80°C to 200 kPa in an isothermal process. During this process the heat transfer to the Freon-12 is 45 kJ/kg. The process is not reversible.

Determine, for this process, the work per kg of Freon-12 using the following sources for thermodynamic data:

(a) The Freon-12 tables.

(b) The generalized charts.

10.43 Repeat Problem 10.42 using the van der Waals equation of state and also using the Redlich-Kwong equation of state. Which of the four answers is the best?

10.44 A 0.1 m³ evacuated tank is to be filled from a supply line containing ethane at 45°C, 14 MPa. Gas flows into the tank until the pressure reaches 12 MPa, at which time the valve is closed. The tank is well insulated so that heat transfer is negligible. Using the generalized charts, determine the final temperature in the tank and the mass of ethane entering.

10.45 An insulated tank having a volume of 50 litres contains carbon dioxide at 1.4 MPa, 40°C. It is pressurized from a line in which carbon dioxide is flowing at 8 MPa, 40°C. The valve is closed when the pressure in the tank reaches 8 MPa.

Using the generalized charts determine the amount of carbon dioxide which flows into the tank.

10.46 Propane is compressed in a reversible isothermal steady-flow process from 1 MPa pressure, 120°C to 5 MPa pressure. Determine the work of compression and heat transfer per kg of propane.

10.47 Propane is compressed in a reversible adiabatic steady-flow process from 1 MPa pressure, 120°C to 5 MPa pressure. Determine the work of compression per kg of propane.

10.48 Determine the minimum work of refrigeration required to perform the process described in Problem 10.17, assuming the ambient temperature to be 25°C.

10.49 Using the generalized charts and the nitrogen tables, estimate the per cent correction (from saturated liquid values) in $v, h, s$ for liquid nitrogen at 95 K, 8 MPa.

10.50 Ten kg of propane ($C_3H_8$) at 25°C, 80% quality, is contained in an uninsulated vessel. This vessel is connected by a valve to a second uninsulated, evacuated vessel. The valve is opened and the propane comes to equilibrium in the vessels at 25°C by transferring heat with the surroundings, at which time the pressure is 90% of the initial pressure. Determine the irreversibility for this process.

10.51 Consider the process described in Problem 10.18. Calculate:

(a) The heat transferred during the process.

(b) The irreversibility of the process, assuming that heat is rejected from the tank to the surroundings at 25°C.

10.52 Propane gas expands in a turbine from 2.8 MPa, 150°C to 350 kPa, 100°C. Assume the process to be adiabatic. Using the generalized charts determine:

(a) The work done per kg of propane entering the turbine.

(b) The increase of entropy per kg of propane as it flows through the turbine.

(c) The isentropic turbine efficiency.

10.53 Carbon dioxide is flowing at low velocity in a line at 115°C, 7 MPa. The velocity of the carbon dioxide is to be increased to 250 m/s by having it flow through an appropriate nozzle. Assuming the flow to be isentropic, determine the final pressure and temperature by use of the generalized charts.

10.54  An uninsulated 0.2 m³ tank contains $CO_2$ at 15°C with an initial quality of 10%. Saturated vapor is withdrawn and expanded through a turbine to 1 MPa in a reversible adiabatic process as shown in Fig. 10.18. If the process ends when the tank quality reaches 100%, calculate the total work output of the turbine during the process.

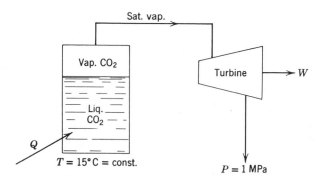

**Fig. 10.18**  Sketch for Problem 10.54.

10.55  It is desired to evaluate a recently developed compound for possible use as a working fluid in a portable, closed-cycle power plant, shown schematically in Fig. 10.19.

The only information available on this compound is the following:
$$T_c = 160°C$$
$$P_c = 2.0 \text{ MPa}$$
$$M = 100$$
$$k = 1.10 \text{ (low-pressure value)}$$

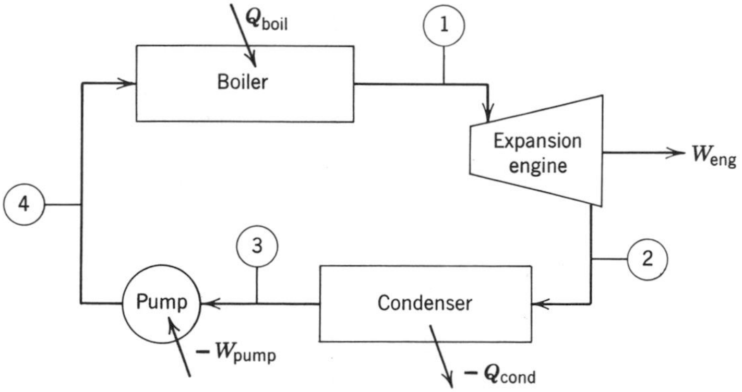

**Fig. 10.19** Sketch for Problem 10.55.

The condenser temperature is fixed at 50°C, and the maximum cycle temperature at 180°C. The isentropic efficiency of the expansion engine is estimated to be 80% and the minimum allowable quality of the exiting fluid (actual state) is 90%.

(a) Calculate the heat transfer from the condenser, assuming saturated liquid at the exit.

(b) Determine the maximum cycle pressure, based on the specifications listed above.

10.56 Write a computer program to solve the following problem. It is desired to obtain a plot of pressure versus volume at various temperatures (all on a generalized reduced basis) as predicted by the van der Waals equation of state. (Temperatures less than the critical should be included in the results.)

10.57 Write a computer program to solve the following problem. For one of the substances listed in Table 3.3, calculate enthalpy changes along several isotherms at various integral pressures, using the Beattie–Bridgeman equation of state.

# 11

---

# Mixtures and Solutions

Up to this point in our development of thermodynamics we have limited our consideration primarily to pure substances. A large number of thermodynamic problems involve mixtures of different pure substances. Sometimes these mixtures are referred to as solutions, particularly in the liquid and solid phases.

In this chapter we shall turn our attention to various thermodynamic considerations of mixtures and solutions. We begin with a consideration of a rather simple problem, mixtures of ideal gases. This leads to a consideration of a simplified but very useful model of certain mixtures, such as air and water vapor, which may involve a condensed (solid or liquid) phase of one of the components. This is followed by certain considerations of mixtures and solutions in general.

An understanding of the material in this chapter is a necessary foundation for the consideration of chemical reactions and chemical and phase equilibrium. These topics are covered in subsequent chapters.

## 11.1 General Considerations and Mixtures of Ideal Gases.

In any gaseous mixture, the mole fraction $y_i$ of component $i$ is defined as

$$y_i = \frac{n_i}{n} \tag{11.1}$$

where $n_i$ is the number of moles of component $i$, and $n$ the total number of moles in the mixture.

Similarly, in any gaseous mixture, the mass fraction $mf_i$ is defined as

$$mf_i = \frac{m_i}{m} \tag{11.2}$$

where $m_i$ is the mass of component $i$ and $m$ is the total mass of the mixture.

Consider a mixture of two gases (not necessarily ideal gases) such as shown in Fig. 11.1.

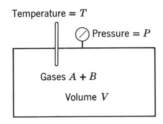

**Fig. 11.1**    A mixture of two gases.

What properties can we experimentally measure for such a mixture? Certainly we can measure the pressure, temperature, volume, and mass of the mixture. We can also experimentally measure the composition of the mixture, and thus determine the mole and mass fractions.

Suppose that this mixture undergoes a process or a chemical reaction and we wish to perform a thermodynamic analysis of this process or reaction. What type of thermodynamic data would we use in performing such an analysis? One possibility would be to have tables of thermodynamic properties of mixtures. However, the number of different mixtures that is possible, both as regards the substances involved and the relative amounts of each is such that one would need a library full of tables of thermodynamic properties to handle all possible situations. It would be much simpler if we could determine the thermodynamic properties of a mixture from the properties of the pure components. This is in essence the approach that is used in dealing with ideal gases and certain other simplified models of mixtures.

One exception to this procedure is the case where a particular mixture is encountered very frequently, the most familiar being air. Tables and charts of the thermodynamics properties of air are available. However, even in this case it is necessary to define the composition of the "air" for which the tables are given, since the composition of the atmosphere varies with altitude, with the number of pollutants, and with other variables at a given location. The composition of air on which air tables are usually based is as follows:

| Component | % on Mole Basis |
|---|---|
| Nitrogen | 78.09 |
| Oxygen | 20.95 |
| Argon | 0.93 |
| $CO_2$ & trace elements | 0.03 |

Consider again the mixture of Fig. 11.1. For the general case the properties of the mixture are defined in terms of the partial molal properties of the individual components. A partial molal property is defined as the value of a property, such as internal energy, for a given component as it exists in the mixture. With this definition, the internal energy of the mixture of Fig. 11.1 would be

$$U_{\text{mix}} = n_A \bar{U}_A + n_B \bar{U}_B \qquad (11.3)$$

where $\bar{U}$ designates the partial molal internal energy. Similar equations can be written for other properties, and this matter will be further developed in Section 11.7.

In this section we focus on mixtures of ideal gases. We assume that each component is uninfluenced by the presence of the other components, and that each component can be treated as an ideal gas. In an actual case of a gaseous mixture at high pressure this assumption would probably not be true because of the nature of the interaction between the molecules of the different components.

There are two models used in conjunction with the mixtures of gases, namely the Dalton model and the Amagat model.

### Dalton Model

In the case of the Dalton model, the properties of each component are considered as though each component existed separately at the volume and temperature of the mixture, as shown in Fig. 11.2.

Consider this model for the special case in which both the mixture and the separated components can be considered an ideal gas.

For the mixture:
$$PV = n\bar{R}T$$
$$n = n_A + n_B \qquad (11.4)$$

For the components:
$$P_A V = n_A \bar{R}T$$
$$P_B V = n_B \bar{R}T \qquad (11.5)$$

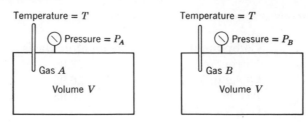

**Fig. 11.2**   The Dalton model.

On substituting we have

$$n = n_A + n_B$$

$$\frac{PV}{\bar{R}T} = \frac{P_A V}{\bar{R}T} + \frac{P_B V}{\bar{R}T}$$

$$(11.6)$$

or

$$P = P_A + P_B$$

where $P_A$ and $P_B$ are referred to as partial pressures.

Thus for a mixture of ideal gases, the pressure is the sum of the partial pressures of the individual components.

It should be stressed that the term partial pressure has relevance only for ideal gases. In effect this concept assumes that the molecules of each component are uninfluenced by the other components, and that the total pressure is the sum of partial pressures of the individual components. It should also be noted that partial pressure is not a partial molal property in the sense as defined by Eq. 11.3, since partial molal properties relate only to extensive properties.

### Amagat Model

In the Amagat model the properties of each component are considered as though each component existed separately at the pressure and temperature of the mixture, as shown in Fig. 11.3. The volumes of $A$ and $B$ under these conditions are $V_A$ and $V_B$ respectively.

In the general case the sum of the volumes when separated, namely, $V_A + V_B$, need not be equal to the volume of the mixture. However, let us consider the special case in which both the separated components and the mixture are considered to be ideal gases. In this case we can write:

For the mixture:           $$PV = n\bar{R}T$$

$$n = n_A + n_B$$           $$(11.7)$$

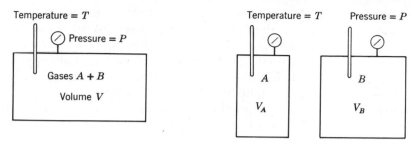

**Fig. 11.3**   The Amagat model.

For the components:    $$PV_A = n_A \bar{R} T$$
$$PV_B = n_B \bar{R} T \tag{11.8}$$

On substituting we have,

$$n = n_A + n_B$$

$$\frac{PV}{\bar{R}T} = \frac{PV_A}{\bar{R}T} + \frac{PV_B}{\bar{R}T}$$

Therefore

$$V_A + V_B = V$$

or

$$\frac{V_A}{V} + \frac{V_B}{V} = 1 \tag{11.9}$$

$V_A/V$ and $V_B/V$ are referred to as the volume fractions.

Thus in the case of ideal gases, Amagat's model leads to the conclusion that the sum of the volume fractions is unity, and that there would be no volume change if the components were mixed while holding the temperature and pressure constant.

From Eqs. 11.4, 11.5, 11.7 and 11.8 it is evident that

$$\frac{V_A}{V} = \frac{n_A}{n} = \frac{P_A}{P}$$

$$\frac{V_A}{V} = y_A = \frac{P_A}{P} \tag{11.10}$$

That is, for each component of a mixture of ideal gases, the volume fraction, the mole fraction, and the ratio of the partial pressure to the total pressure are equal.

In determining the internal energy, enthalpy, and entropy of a mixture of ideal gases, the Dalton model proves useful because the assumption is made that each constituent behaves as though it occupied the entire volume by itself. Thus the internal energy, enthalpy, and entropy can be evaluated as the sum of the respective properties of the constituent gases at the condition at which the component exists in the mixture. Since for ideal gases the internal energy and enthalpy are functions only of temperature, it follows that

$$U = n\bar{u} = n_A \bar{u}_A + n_B \bar{u}_B \qquad (11.11)$$

$$H = n\bar{h} = n_A \bar{h}_A + n_B \bar{h}_B \qquad (11.12)$$

where $\bar{u}_A$ and $\bar{h}_A$ are the internal energy and enthalpy per mole for pure $A$ and $\bar{u}_B$ and $\bar{h}_B$ are the same quantities for pure $B$, all at the temperature of the mixture.

The entropy of an ideal gas is a function of pressure as well as temperature. Since each component exists in the mixture at its partial pressure,

$$S = n\bar{s} = n_A \bar{s}_A + n_B \bar{s}_B \qquad (11.13)$$

where $\bar{s}_A$ is the entropy per mole for pure $A$ at $T, P_A$ (the partial pressure of $A$), and $\bar{s}_B$ that for pure $B$ at $T$ and $P_B$.

**Example 11.1**

A volumetric analysis of a gaseous mixture yields the following results:

| | |
|---|---|
| $CO_2$ | 12.0% |
| $O_2$ | 4.0 |
| $N_2$ | 82.0 |
| $CO$ | 2.0 |

Determine the analysis on a mass basis, and the molecular weight and the gas constant on a mass basis for the mixture. Assume ideal gas behavior.

The table shown below is a very convenient way to solve this problem.

If the analysis had been given on a mass basis, and the mole faction or volumetric analysis was desired, the procedure shown in Table 11.2 could be used.

## Table 11.1

| Constit-uent | Per Cent by Volume | Mole Fraction | Molecular Weight | Mass kg per kmol of Mixture | Analysis on Mass Basis, Per Cent |
|---|---|---|---|---|---|
| $CO_2$ | 12 | $0.12 \times 44.0 =$ | | 5.28 | $\dfrac{5.28}{30.08} = 17.55$ |
| $O_2$ | 4 | $0.04 \times 32.0 =$ | | 1.28 | $\dfrac{1.28}{30.08} = 4.26$ |
| $N_2$ | 82 | $0.82 \times 28.0 =$ | | 22.96 | $\dfrac{22.96}{30.08} = 76.33$ |
| $CO$ | 2 | $0.02 \times 28.0 =$ | | 0.56 | $\dfrac{0.56}{30.08} = 1.86$ |
| | | | | 30.08 | 100.00 |

Molecular weight of mixture $= 30.08$

$$R \text{ for mixture} = \frac{\bar{R}}{M} = \frac{8.31434}{30.08} = 0.2764 \text{ kJ/kg K}$$

## Table 11.2

| Constit-uent | Mass Fraction | Molecular Weight | kmol per kg of Mixture | Mole Fraction | Volumetric Analysis, Per Cent |
|---|---|---|---|---|---|
| $CO_2$ | $0.1755 \div 44.0 =$ | | 0.00399 | 0.120 | 12.0 |
| $O_2$ | $0.0426 \div 32.0 =$ | | 0.00133 | 0.040 | 4.0 |
| $N_2$ | $0.7633 \div 28.0 =$ | | 0.02726 | 0.820 | 82.0 |
| $CO$ | $0.0186 \div 28.0 =$ | | 0.00066 | 0.020 | 2.0 |
| | | | 0.03324 | 1.000 | 100.0 |

$$M = \frac{1}{\text{kmol/kg mixture}} = \frac{1}{0.03324} = 30.08$$

$$R = \frac{\bar{R}}{M} = \frac{8.31434}{30.08} = 0.2764 \text{ kJ/kg K}$$

### Example 11.2

Let $n_A$ moles of gas $A$ at a given pressure and temperature be mixed with $n_B$ moles of gas $B$ at the same pressure and temperature in an adiabatic constant-volume process, shown in Fig. 11.4. Determine the increase in entropy for this process.

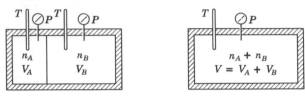

**Fig. 11.4**   Sketch for Example 11.2.

The final partial pressure of gas $A$ is $P_A$ and for gas $B$ it is $P_B$. Since there is no change in temperature, Eq. 7.21 reduces to

$$(S_2 - S_1)_A = -n_A \bar{R} \ln \frac{P_A}{P} = -n_A \bar{R} \ln y_A$$

$$(S_2 - S_1)_B = -n_B \bar{R} \ln \frac{P_B}{P} = -n_B \bar{R} \ln y_B$$

The total change in entropy is the sum of the entropy changes for gases $A$ and $B$.

$$S_2 - S_1 = -\bar{R}(n_A \ln y_A + n_B \ln y_B) \tag{11.14}$$

This equation can be written for the general case of mixing any number of components at the same pressure and temperature as

$$S_2 - S_1 = -\bar{R} \sum_k n_k \ln y_k \tag{11.15}$$

The interesting thing about this equation is that the increase in entropy depends only on the number of moles of component gases, and is independent of the composition of the gas. For example, when 1 mole of oxygen and 1 mole of nitrogen are mixed, the increase in entropy is the same as when 1 mole of hydrogen and 1 mole of nitrogen are mixed. But we also know that if 1 mole of nitrogen is "mixed" with another mole of nitrogen there is no increase in entropy. The question that arises is how dissimilar must the gases be in order to have an increase in entropy? The answer lies in our ability to distinguish between the two gases. The entropy increases whenever we can distinguish between the gases being mixed. When we cannot distinguish between the gases, there is no increase in entropy.

## 11.2  A Simplified Model of a Mixture Involving Gases and a Vapor

Let us now consider a simplification, which in many cases is a reasonable one, of the problem involving a mixture of ideal gases that is in contact with a solid or liquid phase of one of the components. The most familiar example is a mixture of air and water vapor in contact with liquid water or ice, such as the problems encountered in air conditioning or drying. We are all familiar with the condensation of water from the atmosphere when it is cooled on a summer day.

This problem, and a number of similar problems, can be analyzed quite simply and with considerable accuracy if the following assumptions are made:

1. The solid or liquid phase contains no dissolved gases.
2. The gaseous phase can be treated as a mixture of ideal gases.
3. When the mixture and the condensed phase are at a given pressure and temperature, the equilibrium between the condensed phase and its vapor is not influenced by the presence of the other component. This means that when equilibrium is achieved the partial pressure of the vapor will be equal to the saturation pressure corresponding to the temperature of the mixture.

Since this approach is used extensively and with considerable accuracy, let us give some attention to the terms that have been defined and the type of problems for which this approach is valid and relevant. In our discussion we will refer to this as a gas-vapor mixture.

The dew point of a gas-vapor mixture is the temperature at which the vapor condenses or solidifies when it is cooled at constant pressure. This is shown on the $T$-$s$ diagram for the vapor shown in Fig. 11.5. Suppose that the temperature of the gaseous mixture and the partial pressure of the vapor in the mixture are such that the vapor is initially superheated at state 1. If the mixture is cooled at constant pressure the partial pressure of the vapor remains constant until point 2 is reached, and then condensation begins. The temperature at state 2 is the dew-point temperature. Line 1–3 on the diagram indicates that if the mixture is cooled at constant volume the condensation begins at point 3, which is slightly lower than the dew-point temperature.

If the vapor is at the saturation pressure and temperature, the mixture is referred to as a saturated mixture, and for an air-water vapor mixture, the term "saturated air" is used.

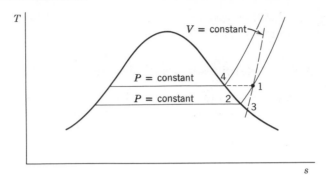

**Fig. 11.5** Temperature-entropy diagram to show definition of the dew point.

The relative humidity $\phi$ is defined as the ratio of the mole fraction of the vapor in the mixture to the mole fraction of vapor in a saturated mixture at the same temperature and total pressure. Since the vapor is considered an ideal gas, the definition reduces to the ratio of the partial pressure of the vapor as it exists in the mixture $P_v$, to the saturation pressure of the vapor at the same temperature $P_g$.

$$\phi = \frac{P_v}{P_g}$$

In terms of the numbers on the $T$-$s$ diagram of Fig. 11.5, the relative humidity $\phi$ would be

$$\phi = \frac{P_1}{P_4}$$

Since we are considering the vapor to be an ideal gas, the relative humidity can also be defined in terms of specific volume or density.

$$\phi = \frac{P_v}{P_g} = \frac{\rho_v}{\rho_g} = \frac{v_g}{v_v} \tag{11.16}$$

The humidity ratio $\omega$ of an air-water vapor mixture is defined as the ratio of the mass of water vapor $m_v$ to the mass of dry air $m_a$. The term "dry air" is used to emphasize that this refers only to air and not to the water vapor. The term "specific humidity" is used synonymously with humidity ratio.

$$\omega = \frac{m_v}{m_a} \tag{11.17}$$

This definition is identical for any other gas-vapor mixture, and the subscript $a$ refers to the gas, exclusive of the vapor. Since we are considering both the vapor and the mixture to be ideal gases, a very useful expression for humidity ratio in terms of partial pressures and molecular weights can be developed.

$$m_v = \frac{P_v V}{R_v T} = \frac{P_v V M_v}{\bar{R} T} \qquad m_a = \frac{P_a V}{R_a T} = \frac{P_a V M_a}{\bar{R} T}$$

Then

$$\omega = \frac{P_v V / R_v T}{P_a V / R_a T} = \frac{R_a P_v}{R_v P_a} = \frac{M_v P_v}{M_a P_a} \tag{11.18}$$

For an air-water vapor mixture this reduces to

$$\omega = 0.622 \frac{P_v}{P_a} \tag{11.19}$$

The degree of saturation is defined as the ratio of the actual humidity ratio to the humidity ratio of a saturated mixture at the same temperature and total pressure.

An expression for the relation between the relative humidity $\phi$ and the humidity ratio $\omega$ can be found by solving Eqs. 11.16 and 11.19 for $P_v$ and equating them. The resulting relation for an air-water vapor mixture is

$$\phi = \frac{\omega P_a}{0.622 P_g} \tag{11.20}$$

A few words should also be said about the nature of the process that occurs when a gas-vapor mixture is cooled at constant pressure. Suppose that the vapor is initially superheated at state 1 in Fig. 11.6. As the mix-

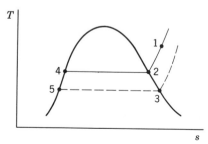

**Fig. 11.6**   Temperature-entropy diagram to show the cooling of a gas-vapor mixture at a constant pressure.

ture is cooled at constant pressure the partial pressure of the vapor remains constant until the dew point is reached at point 2, at which point the vapor in the mixture is saturated. The initial condensate is at state 4, and is in equilibrium with the vapor at state 2. As the temperature is lowered further, more of the vapor condenses, which lowers the partial pressure of the vapor in the mixture. The vapor that remains in the mixture is always saturated, and the liquid or solid is in equilibrium with it. For example, when the temperature is reduced to $T_3$, the vapor in the mixture is at state 3, and its partial pressure is the saturation pressure corresponding to $T_3$. The liquid in equilibrium with it is at state 5.

**Example 11.3**

Consider 100 m³ of an air-water vapor mixture at 0.1 MPa, 35°C, 70 per cent relative humidity. Calculate the humidity ratio, dew point, mass of air, and mass of vapor.
From Eq. 11.16 and the steam tables,

$$\phi = 0.70 = \frac{P_v}{P_g}$$

$$P_v = 0.70\ (5.628) = 3.94 \text{ kPa}$$

The dew point is the saturation temperature corresponding to this pressure, which is 28.6°C.
The partial pressure of the air is

$$P_a = P - P_v = 100 - 3.94 = 96.06 \text{ kPa}$$

The humidity ratio can be calculated from Eq. 11.19.

$$\omega = 0.622 \times \frac{P_v}{P_a} = 0.622 \times \frac{3.94}{96.06} = 0.0255$$

The mass of air is

$$m_a = \frac{P_a V}{R_a T} = \frac{96.06 \times 100}{0.287 \times 308.2} = 108.6 \text{ kg}$$

The mass of the vapor can be calculated by using the humidity ratio or by using the ideal gas equation of state.

$$m_v = \omega m_a = 0.0255\ (108.6) = 2.77 \text{ kg}$$

$$m_v = \frac{3.94 \times 100}{0.46152 \times 308.2} = 2.77 \text{ kg}$$

**Example 11.4**

Calculate the amount of water vapor condensed if the mixture of Example 11.3 is cooled to 5°C in a constant-pressure process.

At 5°C the mixture is saturated, since this is below the dew-point temperature. Therefore

$$P_{v2} = P_{g2} = 0.8721 \text{ kPa}$$

$$P_{a2} = 100 - 0.8721 = 99.128 \text{ kPa}$$

$$\omega_2 = 0.622 \times \frac{0.8721}{99.128} = 0.0055$$

The amount of water vapor condensed is equal to the difference between the initial and final mass of water vapor.

$$\text{mass of vapor condensed} = m_a \, (\omega_1 - \omega_2) = 108.6 \, (0.0255 - 0.0055)$$
$$= 2.172 \text{ kg}$$

## 11.3  The First Law Applied to Gas-Vapor Mixtures

In applying the first law of thermodynamics to gas-vapor mixtures it is helpful to realize that because of our assumption that ideal gases are involved, the various components can be treated separately when calculating changes of internal energy and enthalpy. Therefore, in dealing with air-water vapor mixtures the changes in enthalpy of the water vapor can be found from the steam tables and the ideal gas relations can be applied to the air. This is illustrated by the examples that follow.

**Example 11.5**

An air-conditioning unit is shown in Fig. 11.7, with pressure, temperature, and relative humidity data. Calculate the heat transfer per

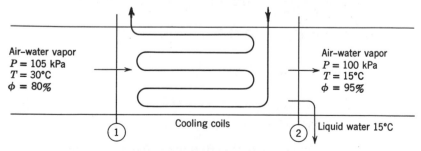

Air-water vapor
$P = 105$ kPa
$T = 30°C$
$\phi = 80\%$

Air-water vapor
$P = 100$ kPa
$T = 15°C$
$\phi = 95\%$

Cooling coils

Liquid water 15°C

①      ②

**Fig. 11.7**  Sketch for Example 11.5.

kg of dry air, assuming that changes in kinetic energy are negligible.

Let us consider a steady-state, steady-flow process for a control volume that excludes the cooling coils.

Since this is a steady-flow process, the continuity equation for the air and water (vapor and liquid) are

$$m_{a1} = m_{a2}$$

$$m_{v1} = m_{v2} + m_{l2}$$

The first law for this steady-state, steady-flow process is

$$Q_{c.v.} + \sum m_i h_i = \sum m_e h_e$$

$$Q_{c.v.} + m_a h_{a1} + m_{v1} h_{v1} = m_a h_{a2} + m_{v2} h_{v2} + m_{l2} h_{l2}$$

If we divide this equation by $m_a$, introduce the continuity equation for the $H_2O$, and note that $m_v = \omega m_a$, we can write the first law in the form

$$\frac{Q_{c.v.}}{m_a} + h_{a1} + \omega_1 h_{v1} = h_{a2} + \omega_2 h_{v2} + (\omega_1 - \omega_2) h_{l2}$$

The air can be considered an ideal gas with constant specific heat, and the values of enthalpy for the $H_2O$ can be taken from the steam tables. (Since the water vapor at these low pressures is being considered an ideal gas, the enthalpy of the water vapor is a function of the temperature only. Therefore, the enthalpy of slightly superheated water vapor is equal to the enthalpy of saturated vapor at the same temperature.)

$$P_{v1} = \phi P_{g1} = 0.80\ (4.246) = 3.397 \text{ kPa}$$

$$\omega_1 = \frac{R_a}{R_v} \frac{P_{v1}}{P_{a1}} = 0.622 \times \left(\frac{3.397}{105 - 3.4}\right) = 0.0208$$

$$P_{v2} = \phi_2 P_{g2} = 0.95\ (1.7051) = 1.620 \text{ kPa}$$

$$\omega_2 = \frac{R_a}{R_v} \times \frac{P_{v2}}{P_{a2}} = 0.622 \times \left(\frac{1.62}{100 - 1.62}\right) = 0.0102$$

$$Q_{c.v.}/m_a + h_{a1} + \omega_1 h_{v1} = h_{a2} + \omega_2 h_{v2} + (\omega_1 - \omega_2) h_{l2}$$

$$Q_{c.v.}/m_a = 1.0035\ (15 - 30) + 0.0102\ (2528.9)$$

$$- 0.0208\ (2556.3) + (0.0208 - 0.0102)\ (62.99)$$

$$= -41.76 \text{ kJ/kg dry air}$$

**Example 11.6**

A tank has a volume of 0.5 m³ and contains nitrogen and water vapor. The temperature of the mixture is 50°C and the total pressure is 2 MPa. The partial pressure of the water vapor is 5 kPa. Calculate the heat transfer when the contents of the tank are cooled to 10°C.

This is a constant-volume process, and since the work is zero the first law reduces to

$$Q = U_2 - U_1 = m_{N_2} C_{v(N_2)} (T_2 - T_1) + (m_2 u_2)_v + (m_2 u_2)_l - (m_1 u_1)_v$$

This equation assumes that some of the vapor condensed. This must be checked, however, as shown below.

The mass of nitrogen and water vapor can be calculated using the ideal-gas equation of state.

$$m_{N_2} = \frac{P_{N_2} V}{R_{N_2} T} = \frac{1995 \times 0.5}{0.2968 \times 323.2} = 10.39 \text{ kg}$$

$$m_{v1} = \frac{P_{v1} V}{R_v T} = \frac{5 \times 0.5}{0.46152 \times 323.2} = 0.016 \ 76 \text{ kg}$$

If condensation takes place the final state of the vapor will be saturated vapor at 10°C. In this case

$$m_{v2} = \frac{P_{v2} V}{R_v T} = \frac{1.2276 \times 0.5}{0.46152 \times 283.2} = 0.004 \ 70 \text{ kg}$$

Since this is less than the original mass of vapor there must have been condensation.

The mass of liquid that is condensed, $m_{l2}$, is

$$m_{l2} = m_{v1} - m_{v2} = 0.016 \ 76 - 0.004 \ 70 = 0.012 \ 06 \text{ kg}$$

The internal energy of the water vapor is equal to the internal energy of saturated water vapor at the same temperature. Therefore,

$$u_{v1} = 2443.5 \text{ kJ/kg}$$

$$u_{v2} = 2389.2 \text{ kJ/kg}$$

$$u_{l2} = 42.0 \text{ kJ/kg}$$

$$\begin{aligned} Q_{c.v.} &= 10.39 \times 0.7448 \ (10 - 50) + 0.0047 \ (2389.2) \\ &\quad + 0.01206 \ (42.0) - 0.01676 \ (2443.5) \\ &= -338.8 \text{ kJ} \end{aligned}$$

## 11.4 The Adiabatic Saturation Process

An important process involving an air-water vapor mixture is the adiabatic saturation process, in which an air-vapor mixture comes in contact with a body of water in a well-insulated duct (Fig. 11.8). If the initial relative humidity is less than 100 per cent some of the water will

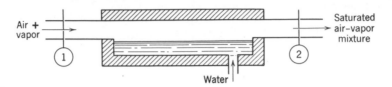

**Fig. 11.8** The adiabatic saturation process.

evaporate and the temperature of the air-vapor mixture will decrease. If the mixture leaving the duct is saturated and if the process is adiabatic, the temperature of the mixture on leaving is known as the adiabatic saturation temperature. In order for this to take place as a steady-flow process, make-up water at the adiabatic saturation temperature is added at the same rate at which water is evaporated. The pressure is assumed to be constant.

Considering the adiabatic saturation process to be a steady-state, steady-flow process, and neglecting changes in kinetic and potential energy, the first law reduces to

$$h_{a1} + \omega_1 h_{v1} + (\omega_2 - \omega_1) h_{l2} = h_{a2} + \omega_2 h_{v2}$$

$$\omega_1 (h_{v1} - h_{l2}) = C_{pa}(T_2 - T_1) + \omega_2 (h_{v2} - h_{l2})$$

$$\omega_1 (h_{v1} - h_{l2}) = C_{pa}(T_2 - T_1) + \omega_2 h_{fg2} \qquad (11.21)$$

The most significant point to be made about the adiabatic saturation process is that the adiabatic saturation temperature, the temperature of the mixture when it leaves the duct, is a function of the pressure, temperature, and relative humidity of the entering air-vapor mixture and of the exit pressure. Thus, the relative humidity and the humidity ratio of the entering air-vapor mixture can be determined from the measurements of the pressure and temperature of the air-vapor mixture entering and leaving the adiabatic saturator. Since these measurements are relatively easy to make, this is one means of determining the humidity of an air-vapor mixture.

**Example 11.7**

The pressure of the mixture entering and leaving the adiabatic saturator is 0.1 MPa, the entering temperature is 30°C, and the temperature leaving is 20°C, which is the adiabatic saturation temperature. Calculate the humidity ratio and relative humidity of the air-water vapor mixture entering.

Since the water vapor leaving is saturated, $P_{v2} = P_{g2}$ and $\omega_2$ can be calculated.

$$\omega_2 = 0.622 \times \left(\frac{2.339}{100 - 2.34}\right) = 0.0149$$

$\omega_1$ can be calculated using Eq. 11.21.

$$\omega_1 = \frac{C_{pa}(T_2 - T_1) + \omega_2 h_{fg2}}{(h_{v1} - h_{l2})}$$

$$\omega_1 = \frac{1.0035\ (20 - 30) + 0.0149 \times 2454.1}{2556.3 - 83.96} = 0.0107$$

$$\omega_1 = 0.0107 = 0.622 \times \left(\frac{P_{v1}}{100 - P_{v1}}\right)$$

$$P_{v1} = 1.691\ \text{kPa}$$

$$\phi_1 = \frac{P_{v1}}{P_{g1}} = \frac{1.691}{4.246} = 0.398$$

## 11.5   Wet-Bulb and Dry-Bulb Temperatures

The humidity of an air-water vapor mixture is usually found from dry-bulb and wet-bulb data. These data are obtained by use of a psychrometer, which involves the flow of air past wet-bulb and dry-bulb thermometers. The bulb of the wet-bulb thermometer is covered with a cotton wick that is saturated with water. The dry-bulb thermometer is used simply to measure the temperature of the air. The flow of air may be maintained by a fan, as in the continuous-flow psychrometer shown in Fig. 11.9, or by moving the thermometer through the air, as in the sling psychrometer, which consists of wet-bulb and dry-bulb thermometers mounted so that they can be whirled.

The processes that take place at the wet-bulb thermometer are somewhat involved. First of all, if the air-water vapor mixture is not saturated, some of the water in the wick evaporates and diffuses into the surrounding air. A drop in the temperature of the water in the wick will

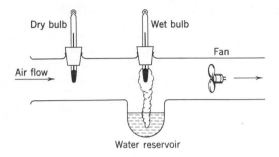

**Fig. 11.9**   Steady-flow apparatus for measuring wet- and dry-bulb temperatures.

be associated with this evaporation. However, as soon as the temperature of the water drops, heat is transferred to the water from both the air and the thermometer. Finally, a steady state, determined by heat and mass transfer rates, will be reached. In general, air velocities exceeding 3 m/s are required in order that the convective heat transfer be large in comparison to the radiant heat transfer.

The psychrometric chart is the most convenient method of determining relative humidity and humidity ratio from wet-bulb and dry-bulb data, although equations that have been developed can also be used. A psychrometric chart is included in the Appendix, Fig. A.4.

The difference between the wet-bulb temperature and adiabatic saturation temperature should be carefully noted. The wet-bulb temperature is influenced by heat and mass transfer rates, whereas the adiabatic saturation temperature simply involves equilibrium between the entering air-vapor mixture and water at the adiabatic saturation temperature. However, it happens that the wet-bulb temperature and the adiabatic saturation temperature are approximately equal for air-water vapor mixtures at atmospheric temperature and pressure. This is not necessarily true at temperatures and pressures that deviate significantly from ordinary atmospheric conditions, or for other gas-vapor mixtures.

## 11.6   The Psychrometric Chart

Properties of air-water vapor mixtures are given in graphical form on psychrometric charts. These are available in a number of different forms, and only the main features are considered here.

The basic psychrometric chart consists of a plot of dry-bulb temperature (abscissa) and humidity ratio (ordinate). If we fix the total pressure for which the chart is to be constructed (which in our chart is 1 bar, or

0.1 MPa), lines of constant relative humidity and wet-bulb temperature can be drawn on the chart, because for a given dry-bulb temperature, total pressure, and humidity ratio, the relative humidity and wet-bulb temperature are fixed. The partial pressure of the water vapor is fixed by the humidity ratio and total pressure, and therefore a second ordinate scale that indicates the partial pressure of the water vapor can be constructed.

Most psychrometric charts give the enthalpy of an air-vapor mixture per kg of dry air. The values given assume that the enthalpy of the dry air is zero at −20°C, and the enthalpy of the vapor is taken from the steam tables (which are based on the assumption that the enthalpy of saturated liquid is zero at 0°C). This procedure is satisfactory because we are usually concerned only with differences in enthalpy. The fact that lines of constant enthalpy are essentially parallel to lines of constant wet-bulb temperature is evident from the fact that the wet-bulb temperature is essentially equal to the adiabatic saturation temperature. Thus, in Fig. 11.8, if we neglect the enthalpy of the liquid entering the adiabatic saturator, the enthalpy of the air-vapor mixture entering is equal to the enthalpy of the saturated air-vapor mixture leaving, and a given adiabatic saturation temperature fixes the enthalpy of the mixture entering.

Some charts are available that give corrections for variation from standard atmospheric pressures. Before using a given chart one should fully understand the assumptions made in constructing it, and that it is applicable to the particular problem at hand.

## 11.7  Partial Molal Properties

We now turn our attention to some general considerations of mixtures and solutions. These considerations are not restricted to gaseous mixtures alone; they also are valid for solid, liquid, and gaseous solutions. In a later chapter the concepts introduced here will be extended to include equilibrium between phases, including mixtures that involve more than one phase. We begin by a consideration of partial molal properties.

The term partial property was first used in connection with the expression for the internal energy of a mixture, Eq. 11.3. In the general case, any extensive property $X$ is a function of the temperature and pressure of the mixture and the number of moles of each component. Thus, for a mixture of two components,

$$X = f(T, P, n_A, n_B)$$

Therefore,

$$dX_{T,P} = \left(\frac{\partial X}{\partial n_A}\right)_{T,P,n_B} dn_A + \left(\frac{\partial X}{\partial n_B}\right)_{T,P,n_A} dn_B \tag{11.22}$$

Since at constant temperature and pressure an extensive property is directly proportional to the mass, Eq. 11.22 can be integrated to give

$$X_{T,P} = \bar{X}_A n_A + \bar{X}_B n_B \tag{11.23}$$

where

$$\bar{X}_A = \left(\frac{\partial X}{\partial n_A}\right)_{T,P,n_B} ; \quad \bar{X}_B = \left(\frac{\partial X}{\partial n_B}\right)_{T,P,n_A} \tag{11.24}$$

$\bar{X}$ is defined as the partial molal property for a component in a mixture. It is particularly important to note that the partial molal property is defined under conditions of constant temperature and pressure. We also note that the general expression Eq. 11.23 is of the same form as Eq. 11.3.

The general extensive property $X$ discussed above can be any of the properties $V$, $H$, $U$, $S$, $A$, or $G$. As an example of the characteristics of partial properties, consider it to be the volume $V$. Then,

$$V = n_A \bar{V}_A + n_B \bar{V}_B \tag{11.25}$$

where, by definition,

$$\bar{V}_A \equiv \left(\frac{\partial V}{\partial n_A}\right)_{T,P,n_B} ; \quad \bar{V}_B \equiv \left(\frac{\partial V}{\partial n_B}\right)_{T,P,n_A} \tag{11.26}$$

For the special case of pure substance $A$,

$$\bar{V}_A = \left(\frac{\partial V}{\partial n_A}\right)_{T,P} = \bar{v}_A$$

and

$$V_A = \bar{v}_A\, n_A \tag{11.27}$$

Thus the partial volume of $A$ when no $B$ is present reduces to the specific volume of pure $A$ as would naturally be expected.

For the special case of an ideal gas mixture of $A$ and $B$,

$$\bar{V}_A = \left(\frac{\partial V}{\partial n_A}\right)_{T,P,n_B} = \frac{\bar{R}T}{P} = \bar{v}$$

and

$$\overline{V}_B = \left(\frac{\partial V}{\partial n_B}\right)_{T,P,n_A} = \frac{\overline{R}T}{P} = \overline{v}$$

Since $\overline{v}$, $\overline{v}_A$, $\overline{v}_B$ are the same per mole,

$$\overline{V}_A = \overline{V}_B = \overline{v} = \overline{v}_A = \overline{v}_B \qquad (11.28)$$

Therefore, for a mixture of ideal gases we conclude from Eq. 11.25 that

$$V = n_A\overline{v}_A + n_B\overline{v}_B = V_A + V_B$$

which is Amagat's rule of additive volumes.

Consider now the general case in which partial molal volumes can be determined from experimental data by the method of intercepts, as shown in Fig. 11.10. Curve $(DGI)$ represents the molal specific volume of a mixture, consisting of components $A$ and $B$, as a function of $y_A$ at constant $T$, $P$. It is desired to determine the partial volumes of $A$ and $B$ for a mixture at point $G$. Note the following lengths in Fig. 11.10.

$(CF) = (GH) = (KM) = \overline{v}$ (for the mixture)
$(DF) = \overline{v}_B$ (no $A$ present)
$(IM) = \overline{v}_A$ (no $B$ present)
$(CG) = (FH) = y_A$ (at the desired point $G$)
$(GK) = (HM) = y_B = (1-y_A)$

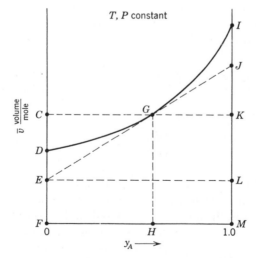

**Fig. 11.10**  Specific volume-composition diagram for a mixture of two components.

The partial volume of $A$ at point $G$ can be determined as follows:

$$V = n\bar{v} = (n_A + n_B)\bar{v}$$

$$\bar{V}_A = \left(\frac{\partial V}{\partial n_A}\right)_{T,P,n_B} = \bar{v} + (n_A + n_B)\left(\frac{\partial \bar{v}}{\partial n_A}\right)_{T,P,n_B}$$

But

$$y_A = \frac{n_A}{n_A + n_B}$$

and at constant, $T, P, n_B$,

$$\frac{dy_A}{dn_A} = \frac{n_B}{(n_A + n_B)^2}$$

or

$$\frac{n_A + n_B}{dn_A} = \left(\frac{n_B}{n_A + n_B}\right)\frac{1}{dy_A} = \frac{1 - y_A}{dy_A}$$

Therefore,

$$\bar{V}_A = \bar{v} + (1 - y_A)\left(\frac{\partial \bar{v}}{\partial y_A}\right)_{T,P,n_B}$$

or in terms of lengths in Fig. 11.10,

$$\bar{V}_A = (KM) + (GK) \times \frac{(JL)}{(EL)} = (KM) + (JK) = (JM)$$

That is, the partial molal volume of component $A$ at point $G$ is equal to the intercept at $y_A = 1.0$ of the tangent to the volume curve at point $G$.
  Similarly, for the partial volume of $B$,

$$\bar{V}_B = \bar{v} + (n_A + n_B)\left(\frac{\partial \bar{v}}{\partial n_B}\right)_{T,P,n_A} = \bar{v} - y_A\left(\frac{\partial \bar{v}}{\partial y_A}\right)_{T,P,n_A}$$

In terms of lengths,

$$\bar{V}_B = (CF) - (CG) \times \frac{(JL)}{(EL)} = (CF) - (CE) = (EF)$$

Thus the partial molal volume of component $B$ is equal to the intercept at $y_A = 0$ of the tangent to the volume curve at point $G$.
  The physical significance of the partial molal volume may be visualized by considering a reasonably large volume of a mixture of components $A$ and $B$. Let a very small amount of $A$ be added while the temperature, pressure and moles of $B$ remain constant. The composition of the mixture

will be essentially unchanged. The volume of the mixture will be increased by an amount equal to the specific volume of component $A$ in the mixture (the partial molal volume of $A$) multiplied by the number of moles of component $A$ added. Thus, in this case,

$$dV = \left(\frac{\partial V}{\partial n_A}\right)_{T,P,n_B} dn_A = \overline{V}_A \, dn_A$$

Similar analyses can of course be performed for the other extensive thermodynamic properties mentioned above. It can also be shown that the thermodynamic relations that apply to a mixture as a whole also hold for relations between partial molal properties for a given component. For example, consider the relation

$$dG = dH - T \, dS - S \, dT$$

Let us again consider a mixture of components $A$ and $B$, and let $T$, $P$, and $n_B$ be held constant, and let $n_A$ change by the amount $dn_A$. In this case, the preceding equation becomes,

$$\left(\frac{\partial G}{\partial n_A}\right)_{T,P,n_B} = \left(\frac{\partial H}{\partial n_A}\right)_{T,P,n_B} - T\left(\frac{\partial S}{\partial n_A}\right)_{T,P,n_B}$$

or

$$\overline{G}_A = \overline{H}_A - T\overline{S}_A \tag{11.29}$$

Similarly

$$\overline{H}_A = \overline{U}_A + P\overline{V}_A \tag{11.30}$$

$$\overline{A}_A = \overline{U}_A - T\overline{S}_A \tag{11.31}$$

## 11.8    Change in Properties upon Mixing

In this section, we will develop expressions for the changes in volume, enthalpy, and entropy upon mixing of two pure substances at constant temperature and pressure. We have previously seen that there is no volume change upon mixing of ideal gases at constant temperature and pressure. If two real gases each at $T$, $P$ were mixed, however, the volume of the mixture would not necessarily equal the sum of the volumes of the constituents. The same is true for solutions involving liquids and solids. The change of volume on mixing, $\Delta V_{\text{mix}}$, can be expressed as

$$\Delta V_{\text{mix}} = V_{\text{mixture}} - V_{\text{components}}$$

For a mixture of components $A$ and $B$

$$\Delta V_{\text{mix}} = (\overline{V}_A n_A + \overline{V}_B n_B) - (\overline{v}_A n_A + \overline{v}_B n_B)$$
$$= (\overline{V}_A - \overline{v}_A)n_A + (\overline{V}_B - \overline{v}_B)n_B \tag{11.32}$$

The significance of this can be shown graphically by reference to Fig. 11.10. For this mixture the change in volume on mixing would be

$$\Delta V_{\text{mix}} = (IJ)n_A + (DE)n_B$$

Similarly the enthalpy change of mixing is given by

$$\Delta H_{\text{mix}} = (\overline{H}_A - \overline{h}_A)n_A + (\overline{H}_B - \overline{h}_B)n_B \tag{11.33}$$

and the entropy change of mixing by

$$\Delta S_{\text{mix}} = (\overline{S}_A - \overline{s}_A)n_A + (\overline{S}_B - \overline{s}_B)n_B \tag{11.34}$$

For a mixture of ideal gases, the special case previously discussed, we concluded (Eq. 11.28) that

$$(\overline{V}_A - \overline{v}_A) = 0$$
$$(\overline{V}_B - \overline{v}_B) = 0$$

Therefore,

$$\Delta V_{\text{mix}} = 0 \tag{11.35}$$

Also, for a mixture of ideal gases

$$(\overline{H}_A - \overline{h}_A) = 0$$
$$(\overline{H}_B - \overline{h}_B) = 0$$

and therefore

$$\Delta H_{\text{mix}} = 0 \tag{11.36}$$

But, the entropy of mixing for ideal gases is given by the relation

$$(\overline{S}_A - \overline{s}_A) = -\overline{R} \ln y_A$$
$$(\overline{S}_B - \overline{s}_B) = -\overline{R} \ln y_B \tag{11.37}$$
$$\Delta S_{\text{mix}} = -\overline{R}(n_A \ln y_A + n_B \ln y_B)$$

in which $\bar{S}_A$ is the partial entropy of $A$ in the mixture at $T$, $P$, while $\bar{s}_A$ is that for pure $A$ at the same temperature and pressure, and similarly for component $B$. This is the expression that was derived in Example 11.2.

## 11.9   The Thermodynamic Property Relation for Variable Composition

In Section 10.2 we considered several forms of the property relation for systems of fixed mass, such as

$$dU = T\,dS - P\,dV$$

We note that in this equation temperature is the intensive property or potential function associated with entropy and pressure is the intensive property associated with volume. Now suppose we have a chemical reaction in which the amounts of components $A$ and $B$ change. How would we modify the property relation for this situation? Intuitively we might write the equation

$$dU = T\,dS - P\,dV + \mu_A dn_A + \mu_B dn_B \tag{11.38}$$

where $\mu_A$ is the intensive property or potential function associated with $n_A$, and similarly $\mu_B$ for $n_B$. This potential function is called the chemical potential.

To derive an expression for this chemical potential, we examine Eq. 11.38 and conclude that it is reasonable to write an expression for $U$ in the form

$$U = f(S, V, n_A, n_B)$$

Therefore,

$$dU = \left(\frac{\partial U}{\partial S}\right)_{V,n_A,n_B} dS + \left(\frac{\partial U}{\partial V}\right)_{S,n_A,n_B} dV + \left(\frac{\partial U}{\partial n_A}\right)_{S,V,n_B} dn_A + \left(\frac{\partial U}{\partial n_B}\right)_{S,V,n_A} dn_B$$

Since the expressions

$$\left(\frac{\partial U}{\partial S}\right)_{V,n_A,n_B} \quad \text{and} \quad \left(\frac{\partial U}{\partial V}\right)_{S,n_A,n_B}$$

imply constant composition, it follows from Eq. 10.15 that

$$\left(\frac{\partial U}{\partial S}\right)_{V,n_A,n_B} = T \quad \text{and} \quad \left(\frac{\partial U}{\partial V}\right)_{S,n_A\,n_B} = -P$$

Thus

$$dU = T\,dS - P\,dV + \left(\frac{\partial U}{\partial n_A}\right)_{S,V,n_B} dn_A + \left(\frac{\partial U}{\partial n_B}\right)_{S,V,n_A} dn_B \tag{11.39}$$

On comparing this equation with Eq. 11.38 it follows that the chemical potential can be defined by the relation

$$\mu_A = \left(\frac{\partial U}{\partial n_A}\right)_{S,V,n_B} \quad ; \quad \mu_B = \left(\frac{\partial U}{\partial n_B}\right)_{S,V,n_A} \tag{11.40}$$

We can also relate the chemical potential to the partial molal Gibbs function. To do so we proceed as follows.

$$G = U + PV - TS$$

$$dG = dU + P\,dV + V\,dP - T\,dS - S\,dT$$

Substituting Eq. 11.38 into this relation we have

$$dG = -S\,dT + V\,dP + \mu_A dn_A + \mu_B dn_B \tag{11.41}$$

This equation suggests that we write an expression for $G$ in the following form.

$$G = f(T, P, n_A, n_B)$$

Proceeding as we did in the case of a similar expression for internal energy, Eq. 11.39 we have

$$dG = \left(\frac{\partial G}{\partial T}\right)_{P,n_A,n_B} dT + \left(\frac{\partial G}{\partial P}\right)_{T,n_A,n_B} dP + \left(\frac{\partial G}{\partial n_A}\right)_{T,P,n_B} dn_A + \left(\frac{\partial G}{\partial n_B}\right)_{T,P,n_A} dn_B$$

$$= -S\,dT + V\,dP + \left(\frac{\partial G}{\partial n_A}\right)_{T,P,n_B} dn_A + \left(\frac{\partial G}{\partial n_B}\right)_{T,P,n_A} dn_B$$

Comparing this with Eq. 11.41 it follows that

$$\mu_A = \left(\frac{\partial G}{\partial n_A}\right)_{T,P,n_B} \quad ; \quad \mu_B = \left(\frac{\partial G}{\partial n_B}\right)_{T,P,n_A}$$

Note that, since partial molal properties are defined at constant temperature and pressure, the quantities $(\partial G/\partial n_A)_{T,P,n_B}$ and $(\partial G/\partial n_B)_{T,P,n_A}$ are the partial molal Gibbs functions for the two components. That is, the chemical potential is equal to the partial molal Gibbs function.

$$\mu_A = \overline{G}_A = \left(\frac{\partial G}{\partial n_A}\right)_{T,P,n_B} \quad ; \quad \mu_B = \overline{G}_B = \left(\frac{\partial G}{\partial n_B}\right)_{T,P,n_A} \tag{11.42}$$

In terms of the enthalpy

$$H = U + PV$$

Equation 11.38 becomes

$$dH = T \, dS + V \, dP + \mu_A dn_A + \mu_B dn_B \qquad (11.43)$$

Following the same procedure as before, we find that

$$\mu_A = \left(\frac{\partial H}{\partial n_A}\right)_{S,P,n_B} \qquad (11.44)$$

In terms of the Helmholtz function

$$A = U - TS$$

the property relation is

$$dA = -S \, dT - P \, dV + \mu_A dn_A + \mu_B dn_B \qquad (11.45)$$

As before, we find an expression for the chemical potential

$$\mu_A = \left(\frac{\partial A}{\partial n_A}\right)_{T,V,n_B} \qquad (11.46)$$

Thus, we have found four different expressions for the chemical potential in terms of other properties, Eqs. 11.40, 11.42, 11.44 and 11.46. Of these, only the one, Eq. 11.42, satisfies the definition of a partial property. The partial molal Gibbs function is an extremely important property in the thermodynamic analysis of chemical reactions because at constant temperature and pressure (the conditions under which many chemical reactions occur) it is a measure of the chemical potential or the driving force tending to cause a chemical reaction to take place.

Finally, we also realize that the surface, magnetic and other terms considered in Chapter 10 may be added to the property relation, Eq. 11.38, thereby giving a completely general form of the relation. This introduces a problem with identification of the chemical potential, and this matter is discussed in the following section.

## 11.10 A General Definition of Gibbs Function and Enthalpy

In the previous section we have extended the thermodynamic property relation to systems of variable composition, and found that the chemical potential can be identified with the partial molal Gibbs function according to Eq. 11.42. Let us now consider the general situation, in which we have a substance that is not a simple compressible substance, as in Eq. 10.17, in addition to allowing variable composition, as in Eq. 11.38. We can then write the general thermodynamic property relation

$$dU = T\,dS - P\,dV + \mathcal{T}\,dL + \mathcal{S}\,d\mathcal{A} + \mu_0 \mathcal{H}\,d(V\mathcal{M}) + \mathcal{E}\,dZ$$
$$+ \sum_i \mu_i\,dn_i + \dots \qquad (11.47)$$

The problem that arises in connection with this general relationship is that Eq. 11.42 is not valid if the Gibbs function is defined in the usual manner, according to

$$G = U + PV - TS \qquad (11.48)$$

To demonstrate that this is in fact the case, it is sufficient to consider only one of the additional effects. Let us analyze a system of two components $A$ and $B$ in which surface effects are of importance. For this case Eq. 11.47 reduces to

$$dU = T\,dS - P\,dV + \mathcal{S}\,d\mathcal{A} + \mu_A dn_A + \mu_B dn_B \qquad (11.49)$$

From Eqs. 11.48 and 11.49,

$$dG = -S\,dT + V\,dP + \mathcal{S}\,d\mathcal{A} + \mu_A dn_A + \mu_B dn_B \qquad (11.50)$$

which suggests a functional relationship of the form

$$G = f(T, P, \mathcal{A}, n_A, n_B)$$

Proceeding as in Section 11.9, we conclude that

$$\mu_A = \left(\frac{\partial G}{\partial n_A}\right)_{T,P,\mathcal{A},n_B} \qquad (11.51)$$

There is nothing incorrect about this result, but we do realize that it is not consistent with the concept of a partial molal property since the extensive parameter $\mathcal{A}$ is held constant. Instead, the corresponding intensive property $\mathcal{S}$ should be held constant as are the intensive properties $T$ and $P$ if this quantity is to be a partial molal property.

We can explain why this difficulty occurs by analogy to the situation of boundary movement work when the mass is not constant. In that case we recall a boundary movement work (flow work) associated with pushing mass across a control surface, this being in addition to the ordinary boundary movement work. In the property relation, that effect $P\bar{V}_A$ appears as part of the chemical potential. In the present case, in addition to the ordinary surface extension work $\mathscr{S}\,d\mathscr{A}$, there is then an additional work associated with the creation of new surface, $\mathscr{S}\bar{\mathscr{A}}_A dn_A$, as the mass of component $A$ changes. Thus, it is appropriate to treat the $\mathscr{S}\mathscr{A}$ term as analogous to the $PV$ term. Let us therefore define a general Gibbs function for this situation as

$$G = U + PV - TS - \mathscr{S}\mathscr{A} \tag{11.52}$$

Then,

$$dG = dU + P\,dV + V\,dP - T\,dS - S\,dT - \mathscr{S}\,d\mathscr{A} - \mathscr{A}\,d\mathscr{S}$$

and from Eq. 11.50,

$$dG = -S\,dT + V\,dP - \mathscr{A}\,d\mathscr{S} + \mu_A dn_A + \mu_B dn_B \tag{11.53}$$

We note that Eq. 11.53 suggests the relation

$$G = f(T, P, \mathscr{S}, n_A, n_B)$$

which, proceeding as before, leads to

$$\mu_A = \left(\frac{\partial G}{\partial n_A}\right)_{T,P,\mathscr{S},n_B} = \bar{G}_A \tag{11.54}$$

and we conclude that if $G$ is defined according to Eq. 11.52, then the chemical potential in Eqs. 11.49 and 11.53 is equal to the partial molal Gibbs function. It also follows for this case that

$$\mu_A = \bar{G}_A = \bar{U}_A + P\bar{V}_A - T\bar{S}_A - \mathscr{S}\bar{\mathscr{A}}_A \tag{11.55}$$

where the partial properties are all defined holding the intensive property $\mathscr{S}$ constant.

Having defined a general Gibbs function by Eq. 11.52 for the case of surface tension and variable mass, it is necessary to define enthalpy in a similar manner by

$$H = G + TS = U + PV - \mathscr{S}\mathscr{A} \tag{11.56}$$

and the property relation Eq. 11.49 can then be written in terms of $H$ in a consistent manner. It should also be noted that there is no effect on the

Helmholtz function, as that property does not include the energy terms associated with mass transfer modes of work.

It is not difficult to extend our analysis to the general case associated with Eq. 11.47. That expression contains the various quasiequilibrium work modes represented in Eq. 4.13, which can be written in general variables as

$$\delta W = -\sum_k F_k \, dX_k \tag{11.57}$$

In this equation, $F_k$ is the intensive property or driving force, and $X_k$ is the corresponding extensive property affected by $F_k$. A general definition of the Gibbs function is then

$$G = U - TS - \sum_k F_k X_k \tag{11.58}$$

and the enthalpy is

$$H = G + TS = U - \sum_k F_k X_k \tag{11.59}$$

The general thermodynamic property relation, Eq. 11.47, is

$$dU = T \, dS + \sum_k F_k \, dX_k + \sum_i \mu_i dn_i \tag{11.60}$$

or, in terms of $G$, from Eq. 11.58,

$$dG = -S \, dT - \sum_k X_k \, dF_k + \sum_i \mu_i dn_i \tag{11.61}$$

The chemical potential of component $i$ in Eqs. 11.60 and 11.61 is

$$\mu_i = \left( \frac{\partial G}{\partial n_i} \right)_{T, F_k, n_{j \neq i}} \tag{11.62}$$

The property relation can similarly be expressed in terms of enthalpy, using its definition Eq. 11.59, or in terms of the Helmholtz function.

### 11.11   Fugacity in a Mixture and Its Relation to Other Properties

Let us define the fugacity of component $A$ in a mixture by a similar procedure to that for a pure substance in Section 10.12. At constant temperature we define $\bar{f}_A$ by

$$(d\bar{G}_A)_T = \bar{R}T \, d(\ln \bar{f}_A)_T \tag{11.63}$$

along with the requirement that

$$\lim_{p \to 0} \left( \frac{\bar{f}_A}{y_A P} \right) = 1 \tag{11.64}$$

so that as pressure approaches zero, the fugacity of component $A$ approaches the ideal gas mixture partial pressure of component $A$. The fugacity of a component in a mixture as defined by these equations is not a true partial property according to the definition of Eq. 11.24. Nevertheless, we include the bar above the symbol as a reminder that the substance involved is a component of a mixture. Note that in the sense that $f$ is essentially a pseudopressure, $\bar{f}$ can be considered a pseudopartial pressure. The fugacity of component $B$ in the mixture is defined by a pair of equations analogous to Eqs. 11.63 and 11.64.

To determine the relation of fugacity of a component to other properties, consider first the property relation written in terms of the Gibbs function. For a mixture of components $A$ and $B$, this relation is given by Eq. 11.41.

At constant $T$ and $n_B$ this expression reduces to

$$dG_{T,n_B} = V\, dP_{T,n_B} + \bar{G}_A\, dn_{A\,T,n_B}$$

Taking a Maxwell cross-partial derivative we have

$$\left( \frac{\partial \bar{G}_A}{\partial P} \right)_{T,n_A,n_B} = \left( \frac{\partial V}{\partial n_A} \right)_{T,P,n_B} = \bar{V}_A \tag{11.65}$$

Therefore, from Eqs. 11.63 and 11.65, at constant temperature and composition,

$$(d\bar{G}_A)_{T,n_A,n_B} = \bar{R}T\, d(\ln \bar{f}_A)_{T,n_A,n_B} = \bar{V}_A\, dP_{T,n_A,n_B} \tag{11.66}$$

For pure substance $A$ at the same constant temperature,

$$(d\bar{g}_A)_T = \bar{R}T\, d(\ln f_A)_T = \bar{v}_A\, dP_T \tag{11.67}$$

In order to derive an expression that permits us to evaluate the fugacity of component $A$ in the mixture from measurable quantities, let us subtract Eq. 11.67 from Eq. 11.66, and integrate the resultant expression from $P^*$ to $P$ (where $P^*$ is a very low pressure) at constant temperature and composition.

$$\int_{\ln \bar{f}_A^*/f_A^*}^{\ln \bar{f}_A/f_A} \bar{R}T\, d\left( \ln \frac{\bar{f}_A}{f_A} \right) = \int_{P^* \to 0}^{P} (\bar{V}_A - \bar{v}_A)\, dP \tag{11.68}$$

At this low pressure $P^*$, $\bar{f}_A^* = y_A P^*$ (Eq. 11.64) and $f_A^* = P^*$. Therefore as $P^* \to 0$,

$$\left( \ln \frac{\bar{f}_A^*}{f_A^*} \right) \to \ln \left( \frac{y_A P^*}{P^*} \right) = \ln y_A$$

Therefore we can write

$$\int_{\ln y_A}^{\ln \bar{f}_A/f_A} \bar{R}T \, d\left( \ln \frac{\bar{f}_A}{f_A} \right) = \int_{P^* \to 0}^{P} (\bar{V}_A - \bar{v}_A) \, dP$$

This integration results in the expression

$$\bar{R}T \ln \left( \frac{\bar{f}_A}{y_A f_A} \right) = \int_0^P (\bar{V}_A - \bar{v}_A) \, dP \tag{11.69}$$

in which $\bar{f}_A$ is the fugacity of component $A$ in the mixture of given composition at the given temperature and pressure $P$, whereas $f_A$ is the fugacity of pure $A$ at the same temperature and pressure. Equation 11.69 expresses the relation between $\bar{f}_A$ and $f_A$ in terms of the difference between $\bar{V}_A$ and $\bar{v}_A$, a measurable quantity. It is also convenient to find a relation between $\bar{f}_A$ and $f_A$ in terms of the other quantities discussed in Sections 11.7 and 11.8, namely $\bar{H}_A - \bar{h}_A$ and $\bar{S}_A - \bar{s}_A$, for such relations will permit evaluation of one or more of these quantities in terms of $\bar{V}_A - \bar{v}_A$.

First we shall determine the partial molal Gibbs function of component $A$ in a mixture at $T$, $P$, with respect to a state at which the Gibbs function is known. Let us integrate Eq. 11.63 at constant $T$ from $P^*$ to $P$, where again $P^*$ is sufficiently low to assume ideal gas behavior,

$$\int_{\bar{G}_A^*}^{\bar{G}_A} d\bar{G}_A = \int_{\bar{f}_A^* \to y_A P^*}^{\bar{f}_A} \bar{R}T \, d(\ln \bar{f}_A)_T \tag{11.70}$$

This integration yields the result

$$\bar{G}_A = \bar{G}_A^* + \bar{R}T \ln \frac{\bar{f}_A}{y_A P^*} = \bar{G}_A^* - \bar{R}T \ln y_A + \bar{R}T \ln \frac{\bar{f}_A}{P^*}$$

We have noted, Eq. 11.29, that

$$\bar{G}_A = \bar{H}_A - T\bar{S}_A$$

Therefore,

$$\bar{G}_A = \bar{H}_A^* - T\bar{S}_A^* - \bar{R}T \ln y_A + \bar{R}T \ln \frac{\bar{f}_A}{P^*} \tag{11.71}$$

But, at low pressures,

$$\bar{H}_A^* = \bar{h}_A^*$$

Also, from Eq. 11.37,

$$\bar{S}_A^* + \bar{R} \ln y_A = \bar{s}_A^*$$

Therefore,

$$\bar{G}_A = \bar{h}_A^* - T\bar{s}_A^* + \bar{R}T \ln \frac{\bar{f}_A}{P^*}$$

$$\bar{G}_A = \bar{g}_A^* + \bar{R}T \ln \frac{\bar{f}_A}{P^*} \tag{11.72}$$

To determine the change in $\bar{G}_A$ with respect to temperature, we take another Maxwell cross-partial derivative of Eq. 11.41,

$$\left(\frac{\partial \bar{G}_A}{\partial T}\right)_{P,n_A,n_B} = -\left(\frac{\partial S}{\partial n_A}\right)_{T,P,n_B} = -\bar{S}_A \tag{11.73}$$

or, at constant pressure and composition,

$$d\bar{G}_A = -\bar{S}_A \, dT \tag{11.74}$$

But, from Eq. 11.29,

$$\bar{S}_A = \frac{\bar{H}_A - \bar{G}_A}{T}$$

Substituting,

$$(d\bar{G}_A)_{P,n_A,n_B} = -\frac{\bar{H}_A - \bar{G}_A}{T} \, dT_{P,n_A,n_B}$$

which may be rearranged to the form

$$\frac{T \, d\bar{G}_A}{T^2} \frac{\bar{G}_A \, dT}{T^2} = -\frac{\bar{H}_A}{T^2} \, dT_{P,n_A,n_B} \tag{11.75}$$

We note that Eq. 11.75 is, in effect,

$$d\left(\frac{\bar{G}_A}{T}\right)_{P,n_A,n_B} = -\frac{\bar{H}_A}{T^2} \, dT_{P,n_A,n_B} \tag{11.76}$$

By the same procedure for pure $A$ at constant pressure, we can write

$$d\left(\frac{\bar{g}_A}{T}\right)_P = -\frac{\bar{h}_A}{T^2} \, dT_P \tag{11.77}$$

Now, let us divide Eq. 11.72 by $T$ and differentiate at constant pressure and composition.

$$d\left(\frac{\bar{G}_A}{T}\right)_{P,n_A,n_B} = d\left(\frac{\bar{g}_A^*}{T}\right)_{P,n_A,n_B} + \bar{R}\,d(\ln \bar{f}_A)_{P,n_A,n_B} \tag{11.78}$$

Substituting Eqs. 11.76 and 11.77, we see that at constant pressure and composition,

$$(d\ln \bar{f}_A)_{P,n_A,n_B} = \frac{\bar{h}_A^* - \bar{H}_A}{\bar{R}T^2}\,dT_{P,n_A,n_B} \tag{11.79}$$

Again, by the same procedure for pure $A$ at constant pressure,

$$(d\ln f_A)_P = \frac{\bar{h}_A^* - \bar{h}_A}{\bar{R}T^2}\,dT_P \tag{11.80}$$

Combining Eqs. 11.79 and 11.80,

$$d\left(\ln\frac{\bar{f}_A}{f_A}\right)_{P,n_A,n_B} = \frac{\bar{h}_A - \bar{H}_A}{\bar{R}T^2}\,dT_{P,n_A,n_B} \tag{11.81}$$

which gives another relation between $\bar{f}_A$ and $f_A$ at a given $T$, $P$, in this case in terms of the difference between $(\bar{h}_A - \bar{H}_A)$.

Finally let us determine the relation between $\bar{f}_A$ and $f_A$ in terms of the difference $(\bar{S}_A - \bar{s}_A)$. For pure component $A$,

$$\bar{g}_A = \bar{h}_A - T\bar{s}_A$$

Also,

$$\bar{G}_A = \bar{H}_A - T\bar{S}_A$$

Combining these equations,

$$(\bar{S}_A - \bar{s}_A) = \left(\frac{\bar{H}_A - \bar{h}_A}{T}\right) - \left(\frac{\bar{G}_A - \bar{g}_A}{T}\right) \tag{11.82}$$

Substituting Eqs. 10.76 and 11.72, we have

$$(\bar{S}_A - \bar{s}_A) = \left(\frac{\bar{H}_A - \bar{h}_A}{T}\right) - \bar{R}\ln\left(\frac{\bar{f}_A}{f_A}\right) \tag{11.83}$$

which can also be written in the form

$$(\bar{S}_A - \bar{s}_A) = \left(\frac{\bar{H}_A - \bar{h}_A}{T}\right) - \bar{R}\ln\left(\frac{\bar{f}_A}{y_A f_A}\right) - \bar{R}\ln y_A \tag{11.84}$$

so that the second term on the right side of the equation is similar to the first term of Eq. 11.69.

## 11.12   The Ideal Solution

There are many mixtures and solutions for which the change of volume on mixing is negligibly small. These are referred to as ideal solutions. By this definition a mixture of ideal gases is an ideal solution, and this is quite acceptable terminology. However, ideal solutions also include certain solid solutions and liquid solutions as well as mixtures of nonideal gases. A number of significant simplifications result from this assumption.

We have noted (Eq. 11.28) that for a mixture of ideal gases consisting of components $A$ and $B$,

$$(\bar{V}_A - \bar{v}_A) = (\bar{V}_B - \bar{v}_B) = 0$$

Let us define an ideal solution as any solution or mixture for which

$$(\bar{V}_A - \bar{v}_A) = 0; \qquad (\bar{V}_B - \bar{v}_B) = 0 \tag{11.85}$$

at the pressure of the mixture and for all lower pressures as well. This definition therefore requires that

$$\Delta V_{\text{mix}} = 0$$

as in the case of mixtures of ideal gases.

Since, in accordance with Eq. 11.25, the volume of a mixture of components $A$ and $B$ is

$$V = \bar{V}_A n_A + \bar{V}_B n_B$$

the assumption of an ideal solution, Eq. 11.70 is equivalent to assuming Amagat's rule of additive volumes,

$$V = \bar{v}_A n_A + \bar{v}_B n_B$$

where $\bar{v}_A$ and $\bar{v}_B$ are the molal specific volumes of pure $A$ and pure $B$ each in the same phase as the mixture, and each at $P$ and $T$, the pressure and temperature of the mixture. It should be noted that in the case of an ideal solution, $\bar{v}_A$ and $\bar{v}_B$ are the actual specific volumes of these components, and no assumption of ideal gas behavior is made.

The fugacity of a component in an ideal solution is readily determined by reference to Eq. 11.69.

$$\bar{R}T \ln \left( \frac{\bar{f}_A}{y_A f_A} \right) = \int_0^P (\bar{V}_A - \bar{v}_A)\, dP$$

From the definition of an ideal solution, $(\bar{V}_A - \bar{v}_A) = 0$, it follows that

$$\bar{f}_A = y_A f_A \tag{11.86}$$

Similarly, for component $B$,

$$\bar{f}_B = y_B f_B \tag{11.87}$$

For the general case of an ideal solution we can write

$$\bar{f}_i = y_i f_i \tag{11.88}$$

Eq. 11.88 constitutes the Lewis-Randall rule, which holds for an ideal solution. Note the similar appearance of these equations and those for the partial pressures of ideal gas mixture components as discussed in Section 11.1. This is consistent with the concept that fugacity is a pseudo-pressure.

The enthalpy of mixing for a mixture of components $A$ and $B$ is given by Eq. 11.33.

$$\Delta H_{\mathrm{mix}} = (\bar{H}_A - \bar{h}_A)n_A + (\bar{H}_B - \bar{h}_B)n_B$$

The fact that the enthalpy of mixing is zero for an ideal solution can be demonstrated by consideration of Eq. 11.81, which was written for constant composition.

$$d\left(\ln \frac{\bar{f}_A}{f_A}\right)_{P,n_A,n_B} = \left(\frac{\bar{h}_A - \bar{H}_A}{RT^2}\right) dT_{P,n_A,n_B}$$

Substituting Eq. 11.69 it follows that for an ideal solution,

$$(\bar{h}_A - \bar{H}_A) = 0 \tag{11.89}$$

A similar expression can be written for component $B$, and therefore one concludes that the enthalpy of mixing for an ideal solution is equal to zero.

$$\Delta H_{\mathrm{mix}} = 0$$

We had reached the same conclusion for a mixture of ideal gases, but for an ideal solution $\bar{h}_A$ and $\bar{h}_B$ are actual enthalpies, and no assumption of ideal gas behavior is made.

The entropy of mixing for an ideal solution can be found by consideration of Eq. 11.84.

$$\overline{S}_A - \overline{s}_A = \left(\frac{\overline{H}_A - \overline{h}_A}{T}\right) - \overline{R} \ln \left(\frac{\overline{f}_A}{y_A f_A}\right) - \overline{R} \ln y_A$$

Since $\overline{H}_A - \overline{h}_A = 0$ and $\overline{f}_A = y_A f_A$, it follows that for an ideal solution

$$
\begin{aligned}
(\overline{S}_A - \overline{s}_A) &= -\overline{R} \ln y_A \\
(\overline{S}_B - \overline{s}_B) &= -\overline{R} \ln y_B
\end{aligned}
\tag{11.90}
$$

in which $\overline{s}_A$ and $\overline{s}_B$ are for pure $A$ and $B$ at $T$, $P$, the pressure and temperature of the mixture.

The entropy of mixing, Eq. 11.34, is

$$\Delta S_{\mathrm{mix}} = (\overline{S}_A - \overline{s}_A)n_A + (\overline{S}_B - \overline{s}_B)n_B$$

Therefore, for an ideal solution the entropy of mixing is given by the relation

$$\Delta S_{\mathrm{mix}} = -\overline{R}(n_A \ln y_A + n_B \ln y_B)$$

Note that the same expression for the entropy of mixing was obtained for a mixture of ideal gases, Eq. 11.14, although for an ideal solution $\overline{s}_A$ and $\overline{s}_B$ are actual entropies and not ideal gas entropies.

It also follows that since $\overline{V}_A - \overline{v}_A = 0$ and $\overline{V}_B - \overline{v}_B = 0$ for an ideal solution, the specific volume-composition diagram is a straight line for an ideal solution. The same would be true for an enthalpy-composition diagram. These are shown in Fig. 11.11.

**Example 11.8**

A gas mixture consisting of 75% $CH_4$ and 25% $C_2H_4$ on a mole basis is stored at 25°C, 8.25 MPa in a cylinder having a volume of 0.5 m³.

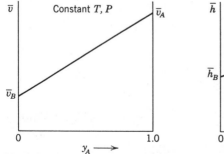
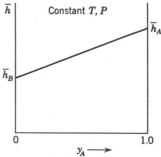

**Fig. 11.11** Typical specific volume-composition and enthalpy-composition diagrams for an ideal solution.

Determine the mass of gas in the tank assuming an ideal solution, and compare these results with those obtained from assuming a mixture of ideal gases.

From Eq. 11.25, which applies to all two component solutions,

$$V = n_A \overline{V}_A + n_B \overline{V}_B$$

Since we are assuming an ideal solution, we conclude from Eq. 11.85 that

$$\overline{V}_{CH_4} = \overline{v}_{CH_4}, \qquad \overline{V}_{C_2H_4} = \overline{v}_{C_2H_4}$$

and therefore

$$V = (n\overline{v})_{CH_4} + (n\overline{v})_{C_2H_4}$$

or

$$\overline{v} = (y\overline{v})_{CH_4} + (y\overline{v})_{C_2H_4}$$

which is Amagat's rule of additive volumes.

If we had a table of thermodynamic properties that gave the specific volume of $CH_4$ and $C_2H_4$ at 25°C and 8.25 MPa, we would simply find the gas phase molal specific volume $\overline{v}$ of each component at this temperature and pressure, and proceed with the solution. In the absence of these data we can use the generalized charts, which will yield fairly accurate results. For the $CH_4$ at 25°C and 8.25 MPa,

$$T_r = \frac{298.2}{191.1} = 1.56 \quad ; \quad P_r = \frac{8.25}{4.64} = 1.78$$

For the $C_2H_4$ at 25°C and 8.25 MPa,

$$T_r = \frac{298.2}{282.4} = 1.06 \quad ; \quad P_r = \frac{8.25}{5.12} = 1.61$$

From the generalized compressibility chart, Fig. A.5,

$$Z_{CH4} = 0.88 \quad ; \quad Z_{C2H4} = 0.35$$

Therefore,

$$\overline{v}_{CH_4} = \frac{Z\overline{R}T}{P} = \frac{0.88 \times 8.31434 \times 298.2}{8250} = 0.2645 \text{ m}^3/\text{kmol}$$

$$\overline{v}_{C_2H_4} = \frac{Z\overline{R}T}{P} = \frac{0.35 \times 8.31434 \times 298.2}{8250} = 0.1052 \text{ m}^3/\text{kmol}$$

$$\overline{v} = 0.75 (0.2645) + 0.25 (0.1052)$$

$$= 0.2247 \text{ m}^3/\text{kmol}$$

$$M = (yM)_{CH_4} + (yM)_{C_2H_4} = 0.75(16) + 0.25(28.0) = 19.0$$

$$v = \frac{\bar{v}}{M} = \frac{0.2247}{19} = 0.011\ 825\ m^3/kg$$

$$m = \frac{V}{v} = \frac{0.5}{0.011\ 825} = 42.3\ kg$$

Assuming this to be a mixture of ideal gases we can use either the partial pressure or additive volume approach. Let us use the former.

$$P_{CH_4} = y_{CH_4}P = 0.75\ (8.25) = 6.188\ MPa$$

$$m_{CH_4} = \frac{P_{CH_4}V}{RT} = \frac{6188 \times 0.5}{0.51835 \times 298.2} = 20.02\ kg$$

$$P_{C_2H_4} = y_{C_2H_4}P = 0.25\ (8.25) = 2.063\ MPa$$

$$m_{C_2H_4} = \frac{P_{C_2H_4}V}{RT} = \frac{2063 \times 0.5}{0.29637 \times 298.2} = 11.67\ kg$$

$$m = 20.02 + 11.67 = 31.69\ kg$$

This compares with 42.3 kg for the assumption of an ideal solution. We note two difficulties in particular from the results of this example. First, it is quite tedious to utilize the ideal solution model with the generalized charts, as it requires evaluating the entire problem separately for each of the components. Second, it required that we use the specific volume $\bar{v}_A$ of pure gaseous $A$ at the temperature and pressure of the gas mixture, and similarly for component $B$. However, there are certain pressures and temperatures at which the mixture would exist as a gas, but one of the components would, as a pure substance, be a solid or liquid. A familiar example is an air-water vapor mixture at ambient conditions. This poses the question of how to evaluate the properties of this component as a gas at the pressure and temperature of the mixture. There is no simple solution for this problem. In view of these difficulties, we may very well question the use of this model, even though the results are certainly more accurate than assuming ideal gas mixture behavior. The answer to this question is that there are other mixture models, described in Section 11.14, that are much simpler to use with the generalized charts, and which are as accurate as the ideal solution for $P$-$v$-$T$, enthalpy and entropy calculations. The ideal solution does find important application, however, in connection with phase and chemical equilibria correlations, as will be noted in Chapter 13. It should also be pointed out that examination of the ideal solution is beneficial from an educational standpoint, as it is the general model of mixtures of which the ideal gas mixture is a special case. Thus, we are able to develop an intuitive feeling for the validity of assuming ideal gas when analyzing gas mixture behavior.

## 11.13 Activity and Activity Coefficient

In our consideration of equilibrium in a following chapter we will find it convenient to make use of an additional property, the activity. It is appropriate, however, to introduce this property at this point.

We have noted that the partial molal Gibbs function for a component in a mixture can be expressed in terms of the low pressure ideal gas state, as was done in Eq. 11.72.

$$\bar{G}_A = \bar{g}_A^* + \bar{R}T \ln (\bar{f}_A/P^*)$$

A more general approach would be to express the partial Gibbs functions in terms of a state at which we consider the mixture to behave as an ideal solution instead of an ideal gas mixture. The temperature of this state is the same as that of the mixture, and the pressure is the standard-state pressure $P°$, a reference pressure at the system temperature. Standard-state values were mentioned briefly in Chapter 7 in connection with entropy, and will be discussed more generally at a later point. It should be pointed out that this state may very well be a hypothetical one. That is, the mixture at $T$ and $P°$ may not actually behave as an ideal solution, but as long as we consistently calculate the mixture properties and develop the equations as though it did, the fact that it is a hypothetical reference state will introduce no difficulties. The significance of this statement will become evident as we proceed further.

The procedure for expressing $\bar{G}_A$ is similar to that followed in the development of Eq. 11.72. In this case Eq. 11.63 is integrated at constant temperature and composition from the pressure $P°$ to the mixture pressure $P$,

$$\int_{\bar{G}_A^°}^{\bar{G}_A} (d\bar{G}_A)_T = \int_{\bar{f}_A^° = y_A f_A^°}^{\bar{f}_A} \bar{R}T \, d(\ln \bar{f}_A)_T$$

in which $f_A^°$ is the fugacity of pure $A$ at $T, P°$. Therefore,

$$\bar{G}_A = \bar{G}_A^° + \bar{R}T \ln \left(\frac{\bar{f}_A}{y_A f_A^°}\right)$$

$$= \bar{G}_A^° - \bar{R}T \ln y_A + \bar{R}T \ln \left(\frac{\bar{f}_A}{f_A^°}\right)$$

$$= \bar{H}_A^° - T\bar{S}_A^° - \bar{R}T \ln y_A + \bar{R}T \ln \left(\frac{\bar{f}_A}{f_A^°}\right) \qquad (11.91)$$

Since we assume an ideal solution at $P^\circ$, from Eq. 11.89,

$$\bar{H}_A^\circ = \bar{h}_A^\circ \tag{11.92}$$

and from Eq. 11.90

$$\bar{S}_A^\circ + \bar{R} \ln y_A = \bar{s}_A^\circ \tag{11.93}$$

Therefore,

$$\bar{G}_A = \bar{h}_A^\circ - T\bar{s}_A^\circ + \bar{R}T \ln\left(\frac{\bar{f}_A}{f_A^\circ}\right)$$

$$= \bar{g}_A^\circ + \bar{R}T \ln\left(\frac{\bar{f}_A}{f_A^\circ}\right) \tag{11.94}$$

Thus we have an expression for the partial molal Gibbs function in terms of known values. It is from this important equation that the equilibrium constant is defined in Chapter 13. Consequently Eq. 11.94 is a very powerful relation. The values $f_A^\circ$ and $\bar{g}_A^\circ$ are for pure substance $A$ at the temperature $T$ and pressure $P^\circ$, which is referred to as the standard state pressure. For gaseous mixtures, $P^\circ$ is commonly taken as 0.1 MPa. For liquid and vapor phases in two-phase systems (discussed in Chapter 13), the standard state for each component is usually taken as the pure substance in that phase at the pressure of the mixture.

Examining Eq. 11.94, we find it convenient to define a quantity called activity. The activity $a_A$ of component $A$ in a mixture at $T, P$, is defined as

$$a_A = \frac{\bar{f}_A}{f_A^\circ} \tag{11.95}$$

The activity of component $B$ is, of course, similarly defined. Equation 11.94 may now be written in the convenient form

$$\bar{G}_A = \bar{g}_A^\circ + \bar{R}T \ln a_A \tag{11.96}$$

Several special cases for evaluating $\bar{G}_A$ from this equation have already been considered. For example, in the case for which the mixture can be assumed an ideal gas mixture at the standard state pressure, the equation reduces to Eq. 11.72. If the mixture can be assumed an ideal solution at pressure $P$ as well as at $P^\circ$, then from Eq. 11.86,

$$a_A = \frac{y_A f_A}{f_A^\circ} \tag{11.97}$$

If conditions are such that an ideal gas mixture can be assumed at both

$P°$ and $P$, then from Eq. 11.64

$$a_A = \frac{y_A P}{P°} \tag{11.98}$$

Another parameter commonly used in the description of mixtures is the activity coefficient $\gamma$, which for any component $A$ is defined in terms of its activity and mole fraction as

$$\gamma_A = \frac{a_A}{y_A} \tag{11.99}$$

The activity coefficient is very useful for indicating the nonideality of a mixture and is particularly important for liquid or solid solutions. For example, if the standard state pressure $P°$ is taken as the pressure of the mixture, and if the mixture behaves as an ideal solution, then from Eqs. 11.97 and 11.99 the activity coefficient is unity. Departures of $\gamma$ from unity then indicate nonideal behavior of the mixture.

## 11.14 Equations of State and Pseudocritical State for Mixtures

Frequently it is desirable to have an equation of state for a mixture of gases. The question arises as to how an equation of state for a mixture can be developed from the equations of state for the pure components. For example, suppose the equations of state of the components are available in a given form, with given constants for each component. How should these constants be combined to provide an equation of state that accurately represents the $P$-$v$-$T$ behavior of the mixture?

An extensive discussion of this question is beyond the scope of this text. Instead, we shall simply note that various empirical combining rules have been proposed for the different equations of state and compared with experimental data for mixtures. As an example, for both the van der Waals equation, Eq. 10.53, and the Redlich-Kwong equation, Eq. 10.59, the two pure substance constants $a$ and $b$ are commonly combined according to the relations,

$$a_m = \left( \sum_i y_i a_i^{1/2} \right)^2, \qquad b_m = \sum_i y_i b_i \tag{11.100}$$

When limited $P$-$v$-$T$ data are available, the generalized compressibility chart can often be used with considerable accuracy. The mixture may be treated as a pseudo-pure substance, for which a set of critical constants are determined by some empirical combinations of the pure substance constants. W. B. Kay in 1936 first suggested a simple linear combination.

$$(P_c)_{mix} = \sum_i y_i P_{ci}$$

$$(T_c)_{mix} = \sum_i y_i T_{ci}$$

$$(11.101)$$

These relations give results of reasonable accuracy for mixtures. A number of other procedures for defining pseudocritical constants for mixtures have also been proposed. These procedures are in general more complicated to use than Kay's rule, but may yield somewhat more accurate results.

### Example 11.9

A mixture of 59.39% $CO_2$ and 40.61% $CH_4$ (mole basis) is maintained at 310.94 K, 86.19 bar, at which condition the specific volume has been measured as 0.2205 m³/kmol. Calculate the per cent deviation if the specific volume had been calculated by (a) Kay's Rule, (b) Ideal Solution, and (c) van der Waals' equation of state.

a.  For convenience, let

$$CO_2 = A, \qquad CH_4 = B.$$

Then

$$T_{c_A} = 304.2 \text{ K}, \qquad P_{cA} = 7.39 \text{ MPa}$$

$$T_{c_B} = 191.1 \text{ K}, \qquad P_{cB} = 4.64 \text{ MPa}$$

For Kay's Rule, Eq. 11.101,

$$T_{cm} = \sum_i y_i T_{ci} = y_A T_{c_A} + y_B T_{c_B}$$

$$= 0.5939(304.2) + 0.4061(191.1)$$

$$= 258.3 \text{ K}$$

$$P_{cm} = \sum_i y_i P_{ci} = y_A P_{c_A} + y_B P_{c_B}$$

$$= 0.5939 \ (7.39) + 0.4061 \ (4.64)$$

$$= 6.273 \text{ MPa}$$

Therefore, the pseudo-reduced properties of the mixture are

$$T_{rm} = \frac{T}{T_{cm}} = \frac{310.94}{258.3} = 1.204$$

$$P_{rm} = \frac{P}{P_{cm}} = \frac{8.619}{6.273} = 1.374$$

From the generalized chart,

$$Z_m = 0.705$$

and

$$\bar{v} = \frac{Z_m \bar{R} T}{P} = \frac{0.705 \times 8.31434 \times 310.94}{8619}$$

$$= 0.2115 \text{ m}^3/\text{kmol}$$

The per cent deviation from the experimental value is

$$\% \text{ Dev.} = \left(\frac{0.2205 - 0.2115}{0.2205}\right) \times 100 = 4.08\%$$

b.  For an ideal solution, using the same procedure as in Example 11.8,

$$T_{r_A} = \frac{310.94}{304.2} = 1.022, \qquad P_{r_A} = \frac{8.619}{7.39} = 1.166$$

$$Z_A = 0.35$$

$$\bar{v}_A = \frac{Z_A \bar{R} T}{P} = \frac{0.35 \times 8.31434 \times 310.94}{8619} = 0.105 \text{ m}^3/\text{kmol}$$

$$T_{r_B} = \frac{310.94}{191.1} = 1.63, \qquad P_{r_B} = \frac{8.619}{4.64} = 1.858$$

$$Z_B = 0.91$$

$$\bar{v}_B = \frac{Z_B \bar{R} T}{P} = \frac{0.91 \times 8.31434 \times 310.94}{8619} = 0.273 \text{ m}^3/\text{kmol}$$

Therefore,

$$\bar{v} = y_A \bar{v}_A + y_B \bar{v}_B$$

$$= 0.5939 \, (0.105) + 0.4061 \, (0.273)$$

$$= 0.1732 \text{ m}^3/\text{kmol}$$

and

$$\% \text{ Dev.} = \left(\frac{0.2205 - 0.1732}{0.2205}\right) \times 100 = 21.5\%$$

c. For van der Waals' equation, the pure substance constants are

$$a_A = \frac{27\bar{R}^2 T_{cA}^2}{64 P_{cA}} = 365.185 \ \frac{\text{kPa m}^6}{\text{kmol}^2}$$

$$b_A = \frac{\bar{R} T_{cA}}{8 P_{cA}} = .04278 \ \text{m}^3/\text{kmol}$$

and

$$a_B = \frac{27\bar{R}^2 T_{cB}^2}{64 P_{cB}} = 229.532 \ \frac{\text{kPa m}^6}{\text{kmol}^2}$$

$$b_B = \frac{\bar{R} T_{cB}}{8 P_{cB}} = .0428 \ \text{m}^3/\text{kmol}$$

Therefore, for the mixture, from Eq. 11.100,

$$a_m = (y_A\sqrt{a_A} + y_B\sqrt{a_B})^2$$

$$= (0.5939 \ \sqrt{365.185} + 0.4061 \ \sqrt{229.532})^2$$

$$= 306.315 \ \frac{\text{kPa m}^6}{\text{kmol}^2}$$

$$b_m = y_A b_A + y_B b_B$$

$$= (0.5939 \times .04278 + 0.4061 \times .0428)$$

$$= .04279 \ \text{m}^3/\text{kmol}$$

The equation of state for the mixture of this composition is

$$P = \frac{\bar{R}T}{\bar{v} - b_m} - \frac{a_m}{\bar{v}^2}$$

$$8619 = \frac{8.31434 \times 310.94}{\bar{v} - .04279} - \frac{306.315}{\bar{v}^2}$$

Solving for $\bar{v}$ by trial and error,

$$\bar{v} = 0.2061 \ \text{m}^3/\text{kmol}$$

$$\% \text{ Dev.} = \left(\frac{0.2205 - 0.2061}{.2205}\right) \times 100 = 6.5\%$$

As a point of interest from the ideal gas law, $\bar{v} = 0.300$ m³/kmol, which is a deviation of 36 per cent from the measured value. Also, if we use the Redlich-Kwong equation of state, and follow the same procedure as for the van der Waals equation, the calculated specific volume of the mixture is 0.2127 m³/kmol, which is in error by 3.5%.

We must be careful not to draw too general a conclusion from the results of this example. We have calculated per cent deviation in $v$ at only a single point for only one mixture. We do note, however, that the various methods used give quite different results. From a more general study of these models for a number of mixtures, we find that the results found here are fairly typical, at least qualitatively. Kay's rule is very useful because it is fairly accurate and yet relatively simple. The ideal solution model does not in general yield values as accurate for $P$-$v$-$T$ behavior of mixtures, but as mentioned earlier it is of considerable value for use in phase equilibrium predictions, which we will discuss in Chapter 13. The van der Waals equation is too simplified an expression to accurately represent $P$-$v$-$T$ behavior except at moderate densities, but it is useful to demonstrate the procedures followed in utilizing more complex analytical equations of state. The Redlich-Kwong equation is considerably better, and is still relatively simple to use.

The more sophisticated generalized behavior models and empirical equations of state will represent mixture $P$-$v$-$T$ behavior to within about one per cent over a wide range of density, but they are of course more difficult to use than the methods considered in Example 11.9. The generalized models have the advantage of being easier to use, and they are suitable for hand computations. Calculations with the complex empirical equations of state become very involved, but have the advantage of expressing the $P$-$v$-$T$-composition relations in analytical form, which is of great value when using a digital computer for such calculations.

The theoretical virial equation of state, Eq. 10.64, can be applied to mixtures, in which case the virial coefficients are dependent upon composition as well as upon temperature. For a mixture of $A$ and $B$, the second virial coefficient becomes

$$B(T,y) = y_A{}^2 B_A(T) + 2y_A y_B B_{AB}(T) + y_B{}^2 B_B(T) \qquad (11.101)$$

in which the interaction coefficient $B_{AB}(T)$ results from forces between unlike molecules. For the Lennard-Jones (6-12) potential, this can be found from the parameters

$$\begin{aligned} \epsilon_{AB} &= (\epsilon_A \epsilon_B)^{1/2} \\ b_{\circ AB} &= \tfrac{1}{8}(b_{\circ A}^{1/3} + b_{\circ B}^{1/3})^3 \end{aligned} \qquad (11.102)$$

and the values in Tables 10.1 and 10.2.


## PROBLEMS

11.1   An analysis of the products of a certain combustion process yields the following composition on a volumetric basis:

| Constituent | Percent by Volume |
|---|---|
| $N_2$ | 70 |
| $CO_2$ | 15 |
| $O_2$ | 11 |
| CO | 4 |

Determine:

(a) The composition of this mixture on a mass basis.

(b) The mass of $0.2$ m³ of this mixture at a condition of 100 kPa, 20°C.

(c) The heat transfer required to heat the mixture at constant volume from the initial state to 150°C.

11.2  The mixture of Problem 11.1 is compressed in a reversible adiabatic process in a cylinder from the initial state to a volume of 0.1 m³. Calculate the final temperature and the work for the process.

11.3  Two heavily insulated tanks are connected by a valve. Tank *A* has a volume of 5 m³ and initially contains oxygen at 15°C, 400 kPa. Tank *B* has a volume of 35 m³ and initially contains nitrogen at 35°C, 150 kPa. The valve is now opened and remains open until the resulting mixture comes to a uniform state. Determine

(a) The final pressure and temperature.

(b) The entropy change for the process.

11.4  The vacuum insulating space of the vessel shown in Fig. 11.12 contained carbon dioxide with a trace of helium at ambient conditions when the vessel was empty. When the vessel was then filled with liquid hydrogen, essentially all the $CO_2$ froze out on the cold inner wall, leaving only the helium gas in the vacuum space. If the gas initially contained 0.01% helium by volume, what is the pressure in the vacuum space, if the average temperature in this space is assumed to be 80 K?

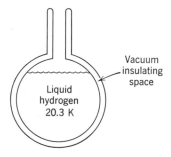

**Fig. 11.12**  Sketch for Problem 11.4.

11.5  A gas mixture containing 50% argon, 50% carbon dioxide by volume is contained in a cylinder at 300 kPa, 150°C. The gas is then expanded to 150 kPa in a reversible adiabatic process. Determine the final temperature, the work done per kmol of mixture, and the entropy changes of the Ar and the $CO_2$ during the process.

11.6 Hydrogen and nitrogen are mixed in an adiabatic, steady-flow process in the ratio 2 kmol of hydrogen per kmol of nitrogen. The hydrogen enters at 0.5 MPa, 20°C and the nitrogen enters at 0.5 MPa, 200°C. The pressure after mixing is 0.49 MPa. Determine the temperature of the mixture and the net entropy change per kmol of mixture.

11.7 Consider the compression of a fuel-air mixture which takes place in an internal combustion engine. Assume that the fuel is ethane and that the air-fuel ratio on a mass basis is 15 to 1. The engine has a compression ratio of 9 to 1 and before the compression stroke begins the pressure in the cylinder is 88 kPa and the temperature is 35°C. Determine the pressure and temperature after compression and the work of compression per kg of mixture.

11.8 An insulated 100 litre tank contains nitrogen gas at 200 kPa, 25°C. Carbon dioxide at 600 kPa, 90°C flows from a pipe into the tank until the pressure inside reaches 400 kPa, at which time the valve is closed. Calculate the final temperature inside the tank and the change in entropy for the process.

11.9 (a) 0.79 kmol of nitrogen at 100 kPa, 25°C are separated from 0.21 kmol of oxygen at 100 kPa, 25°C by a membrane. The membrane ruptures and the gases mix in an adiabatic process to form a uniform mixture. Determine the reversible work and irreversibility for this process.

(b) Determine the minimum work required to separate 1 kmol of air (assume composition to be 79% nitrogen and 21% oxygen by volume) at 100 kPa, 25°C into nitrogen and oxygen at 100 kPa, 25°C.

(c) How would you evaluate the performance of an air separation plant regarding work input?

11.10 A room of dimensions 4 m × 6 m × 2.4 m contains an air-water vapor mixture at a total pressure of 100 kPa and a temperature of 25°C. The partial pressure of the water vapor is 1.4 kPa. Calculate:
(a) The humidity ratio.
(b) The dew point.
(c) The total mass of water vapor in the room.

11.11 A certain apparatus involves a precise knowledge of the amount of water vapor in an electrical conductivity test cell. This is obtained by charging the bomb with a mixture of nitrogen and water vapor at 7 MPa pressure at a temperature of 25°C. The water vapor in the mixture is saturated. What is the humidity ratio of the mixture in the cell?

11.12 A mixture of ideal gases contains the following amounts of the components indicated:

| Component | Number of kmol |
|---|---|
| $CO_2$ | 8 |
| $H_2O$ | 9 |
| $O_2$ | 2.5 |
| $N_2$ | 56.4 |

Determine the temperature at which water will begin to condense if the mixture is cooled at constant pressure. Determine the mass of water that will have condensed if the mixture is cooled by 10°C below this condensation temperature. Assume that the total pressure remains at 100 kPa.

11.13 One method of removing moisture from atmospheric air is to cool the air so that the moisture condenses or freezes out. Suppose an experiment requires a humidity ratio of 0.0001. To what temperature must the air be cooled at a pressure of 0.1 MPa in order to achieve this humidity? To what temperature must it be cooled if the pressure is 10 MPa?

11.14 An air-water vapor mixture is contained in a vertical cylinder fitted with a frictionless piston. The initial volume is 50 litres, and the mixture is at 55°C, 150 kPa, 40% relative humidity. The system is then allowed to cool to the temperature of the surroundings, 15°C.
Calculate:
(a) Heat transfer during the process.
(b) Entropy change of the system.
(c) Irreversibility of the process.

11.15 An air-water vapor mixture at 25°C, 100 kPa, 50% relative humidity is compressed to 50°C, 300 kPa, and is then cooled at constant pressure. At what temperature will water begin to condense?

11.16 An air-water vapor mixture enters an air-conditioning unit at a pressure of 150 kPa, a temperature of 30°C, and a relative humidity of 80%. The mass of dry air entering is 1 kg/s. The air-vapor mixture leaves the air-conditioning unit at 125 kPa, 10°C, 100% relative humidity. The moisture condensed leaves at 10°C. Determine the heat transfer rate for the process.

11.17 An air-water vapor mixture enters a heater-humidifier unit at 5°C, 100 kPa, 50% relative humidity. The flow rate of dry air is 0.1 kg/s. Liquid water at 10°C is sprayed into the mixture at the rate of 0.0022 kg/s. The mixture leaves the unit at 30°C, 100 kPa. Calculate:
(a) The relative humidity at the outlet.
(b) The rate of heat transfer to the unit.

11.18 (a) Determine (by a first-law analysis) the humidity ratio and relative humidity of an air-water vapor mixture that has a dry-bulb temperature of 30°C, an adiabatic saturation temperature of 25°C, and a pressure of 100 kPa.
(b) By use of the psychrometric chart determine the humidity ratio and relative humidity of an air-water vapor mixture that has a dry-bulb temperature of 30°C, a wet-bulb temperature of 25°C, and a pressure of 100 kPa.

11.19 In areas where the temperature is high and the humidity is low, some measure of air conditioning can be achieved by evaporative cooling. This involves spraying water into the air, which subsequently evaporates with a resulting decrease in the temperature of the mixture. Such a scheme is shown in Fig. 11.13.

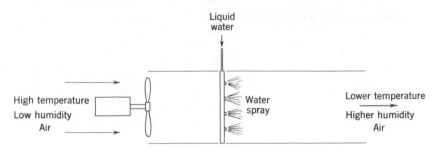

**Fig. 11.13**   Sketch for Problem 11.19.

Consider the case of atmospheric air at 40°C, 10% relative humidity, and 100 kPa. Cooling water at 10°C is sprayed into the air. If the air-water vapor mixture is to be cooled to 25°C, what will be the relative humidity? What are the disadvantages of this approach to air conditioning?

11.20  An air-water vapor mixture at 100 kPa, 35°C and 70% relative humidity is contained in a 0.5 m³ closed tank. The tank is cooled until the water just begins to condense. Determine the temperature at which condensation begins and the heat transfer for the process.

11.21  An uninsulated tank having a volume of 0.4 m³ contains an air-water vapor mixture with a relative humidity of 90% at 35°C, 110 kPa. Dry air at 35°C, 300 kPa flows from a pipe into the tank until the pressure reaches 300 kPa. Heat is transferred during this process such that the temperature of the material within the tank remains constant at 35°C. Determine the amount of heat transferred, as well as the relative humidity and humidity ratio existing at the completion of the process.

11.22  Air at 35°C; 10% relative humidity passes through a chamber into which cooling water at 15°C is sprayed in a steady-state, steady-flow adiabatic process. The amount of cooling water added is such that when it evaporates into the air stream the exiting air-water vapor mixture will be at 25°C. Determine the exiting relative humidity and the irreversibility for the process per kg of air. Assume a constant pressure of 100 kPa.

11.23  A combination air cooler and dehumidifier unit receives outside air at 35°C, 100 kPa, relative humidity of 90%. The air-water vapor mixture is first cooled to a low temperature to condense the proper amount of water, after which the air-vapor mixture is heated, leaving the unit at 20°C, 100 kPa, relative humidity of 30%. The volume flow rate of the air-vapor mixture at the outlet is 0.01 m³/s.

(a) Find the temperature to which the mixture is initially cooled, and the mass of water condensed per kg of dry air. Show the process the H₂O undergoes on a T-s diagram.

(b  If all the liquid condensed leaves the unit at the minimum temperature, calculate the heat transfer rate.

11.24  A 0.3 m³ vessel initially contains an air-water vapor mixture at 150 kPa, 40°C with a relative humidity of 10%. A supply line connected to the vessel by a valve carries steam at 600 kPa, 200°C. The valve is opened and steam flows into the vessel until the relative humidity of the resultant air-water vapor mixture is 90%, at which time the valve is closed. Heat is transferred from the vessel such that the temperature in the vessel remains at 40°C during this process. Determine the heat transfer for the process, the mass of steam entering the vessel, and the final pressure in the vessel.

11.25  The initial process in an air liquefaction and separation plant involves compressing ambient air (0.1 MPa, 25°C, 50% relative humidity) at the rate of 2.5 m³/s to a pressure of 20 MPa. At this point, the air is cooled to 30°C in an aftercooler and then fed to a heat exchanger in which the air will be cooled to −100°C. It may be assumed that any water entering the heat exchanger will freeze out on the tubes of the heat exchanger at such a low temperature. The designer of this plant is trying to decide whether to put a liquid trap between the aftercooler and the heat exchanger. What do you recommend? How much water will freeze on the heat exchanger tubes per hour with and without the liquid trap?

11.26  Air at 40°C, 300 kPa, with a relative humidity of 35% is to be expanded in a reversible adiabatic nozzle. To how low a pressure can the gas be expanded if no condensation is to take place? What is the exit velocity at this condition?

11.27  The process of injection cooling may be demonstrated by the following problem: Consider a column of water 3 m high which is insulated from the surroundings. Let dry air be bubbled through the water. During this process some of the water will evaporate and be carried away by the air and the temperature of the remaining liquid will decrease.

Assume that the water is initially at a temperature of 30°C, and that the dry air enters at 30°C and a pressure slightly above 100 kPa. Assume that the air vapor mixture leaves the surface of the water at 100 kPa pressure and the same temperature as the water and at a relative humidity of 100%.

(a) What is the heat transfer from the water for each kg of air entering under the initial conditions?

(b) What will be the ratio of the mass of air bubbled through the liquid to the mass of water cooled if the temperature of the water is decreased 5°C?

11.28  A problem of current interest is that of thermal pollution of riverwater by using the water for power plant condenser cooling water and then returning it to the river at a higher temperature. The use of a cooling tower presents a solution to this particular problem. Consider the case shown in Fig. 11.14, in which 2.5 kg/s of water at 35°C (from the condenser) enters the top of the cooling tower, and the cool water leaves the bottom at 15°C. The air-water vapor mixture enters the bottom of the cooling tower at 100 kPa, and has a dry-bulb temperature of 22°C and a wet-bulb temperature of 15°C. The air-water vapor mixture leaving the tower has a pressure of 95 kPa, a temperature of 30°C, and a relative humidity of 80%. Determine the kg of dry air per second that must be used and the

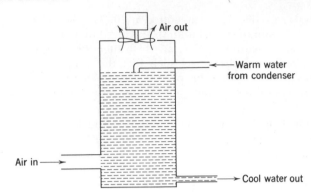

**Fig. 11.14** Sketch for Problem 11.28.

fraction of the incoming water that evaporates. Assume the process to be adiabatic.

11.29 An air-water vapor mixture at 25°C, 200 kPa, 50% relative humidity is to be mixed with nitrogen at 25°C, 200 kPa, and the resulting mixture expanded through an isentropic turbine to 100 kPa.

(*a*) What is the minimum ratio, kmol $N_2$/kmol air-vapor mixture, such that no condensation will occur in the turbine?

(*b*) Find the increase of entropy per kmol of initial air-vapor mixture for the overall process.

11.30 A manufacturing plant needs a supply of relatively dry air for purging equipment. This air requirement is 0.1 m³/s at conditions of 20°C, 100 kPa and 20% relative humidity. It is proposed to supply this air by compressing atmospheric air (20°C, 100 kPa and 17°C wet-bulb temperature) to a suitable pressure, cooling at constant pressure to 20°C, removing the water that condenses, and throttling the air down to atmospheric pressure, as shown in Fig. 11.15.

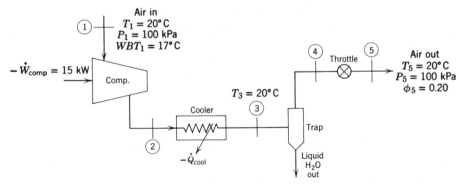

**Fig. 11.15** Sketch for Problem 11.30.

(a) What pressure is required at point 2 to achieve the desired condition at the exit, point 5?

(b) How many kg/s of condensate are removed by the trap?

(c) What is the required rate of heat rejection from the cooler?

11.31 Two streams of air, each saturated with water vapor, and at the pressure and temperature shown in Fig. 11.16 are mixed together in an adiabatic, steady-flow process. Determine the exit temperature, $T_3$.

$T_1 = 50°C$
$\phi_1 = 100\%$
$\dot{m}_1 = 0.2$ kg/s
$T_2 = -15°C$
$\phi_2 = 100\%$
$\dot{m}_2 = 0.1$ kg/s

$P_1 = P_2 = P_3 = 105$ kPa

**Fig. 11.16** Sketch for Problem 11.31.

11.32 One means for condensing steam during an emergency blowdown of a nuclear reactor is by use of a containment vessel as shown in Fig. 11.17. The vessel is insulated and has a total volume of 30 m³, with liquid $H_2O$ initially occupying 1 m³. The initial state in the vessel is 35°C, 100 kPa. Now, during a short period of time, 55 kg of $H_2O$ enters the vessel at an average state of $P_i = 700$ kPa, $x_i = 0.50$. Find the temperature and pressure in the vessel at the end of this time.

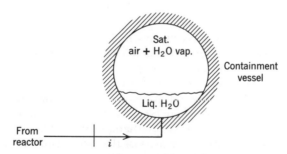

Sat.
air + $H_2O$ vap.

Containment
vessel

Liq. $H_2O$

From
reactor

$i$

**Fig. 11.17** Sketch for Problem 11.32.

11.33 The following data have been measured for carbon dioxide, methane, and their mixtures at 8.618 MPa, 37.8°C:

| $y_{CO_2}$ | $y_{CH_4}$ | $Z$ |
| --- | --- | --- |
| 1.0000 | 0.0000 | 0.3117 |
| 0.7961 | 0.2039 | 0.6262 |
| 0.5939 | 0.4061 | 0.7350 |
| 0.3944 | 0.6056 | 0.8084 |
| 0.1528 | 0.8472 | 0.8634 |
| 0.0000 | 1.0000 | 0.8892 |

(*a*) Determine the partial volumes of the components for a mixture of 50% $CO_2$, 50% $CH_4$ at this pressure and temperature.

(*b*) Evaluate the assumption of ideal solution for this system at the given temperature and pressure.

11.34  A gas mixture of 75% ethylene and 25% ethane on a mole basis enters a compressor at 35°C, 0.5 MPa pressure, at the rate of 12.5 litres/s. The power input to the compressor is 10 kW and the mixture exits at 65°C, 2 MPa. Assuming Kay's rule for the mixture, determine
(*a*) Mass flow rate.
(*b*) Heat transfer rate from the compressor.

11.35  A 40 litre tank containing air at 15 MPa, 300 K, is cooled to 190 K. Determine the final pressure in the tank and the heat transfer for the process, assuming the air to be
(*a*) An ideal gas.
(*b*) A pseudo-pure substance using Kay's rule.
(*c*) A pseudo-pure substance using the Beattie-Bridgeman equation of state.

11.36  One mole of mixture contains 50% $CO_2$, 50% $C_2H_6$ (mole basis) and is compressed in a cylinder in a reversible isothermal process. The initial state is 700 kPa, 35°C and the final pressure is 5.5 MPa. Calculate the heat transfer and work for the process, using
(*a*) The van der Waals equation of state.
(*b*) Kay's rule.
(*c*) Ideal gas behavior.

11.37  A mixture of 50% $N_2$, 50% $O_2$ on a mole basis flows through a long pipe. At the pipe inlet, the mixture is at 3.5 MPa, −100°C, with a velocity of 30 m/s. If the pressure drop in the pipe is estimated to be 500 kPa, determine the exit temperature and velocity. Assume Kay's rule.

11.38  A mixture of ethane (*A*) and propane (*B*) is contained in the cylinder as shown in Fig. 11.18. The initial state is 30°C (room temperature), 70 kPa, 0.6 m³, and a mole fraction of ethane of 0.20. The line contains pure ethane at 30°C, 7 MPa. The mass of the piston and weights is such that a cylinder pressure of 3.5 MPa would be required to float the piston.

The valve is now cracked open and the pin pulled from the piston. When the valve is closed, the temperature inside is 65°C and the mole fraction of ethane is 0.60.
Calculate
(*a*) The final volume in the cylinder.
(*b*) The net entropy change for the process.

11.39  It is frequently necessary to have a gas mixture of a certain composition and pressure. Such a mixture is prepared by initially charging one of the components into the tank at some pressure such that when the second component is charged until the final desired pressure is reached the composition will be of the proper proportions. It is desired to prepare a mix-

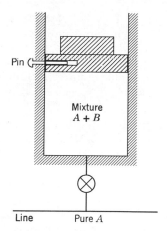

**Fig. 11.18**   Sketch for Problem 11.38.

ture of 80% $C_2H_4$, 20% $CO_2$ at 10 MPa, 25°C, in an uninsulated 50 litre tank (Fig. 11.19). The tank is initially to contain $CO_2$ at 25°C, and some pressure $P_1$. The valve is then opened slightly and $C_2H_4$ flows slowly into the tank from the line at 25°C, 10 MPa until the tank pressure reaches 10 MPa. Assume Kay's rule for the mixture.

(*a*) If the final composition is to be 80% $C_2H_4$, what must the initial pressure $P_1$ be?

(*b*) Determine the heat transfer and the net entropy change for the process of charging $C_2H_4$ into the tank.

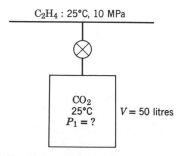

**Fig. 11.19**   Sketch for Problem 11.39.

11.40 A mixture of 50% methane and 50% nitrogen (mole basis) enters an insulated nozzle at −15°C, 20 MPa. The exit conditions are stated to be −85°C, 5.9 MPa.

(*a*) Calculate the exit velocity.

(*b*) Does this process violate the second law?

11.41 Two heavily insulated tanks, *A* and *B,* each of 20 litres volume, are connected by a valve. Tank *A* contains methane at 25°C, 24 MPa, and tank B contains carbon dioxide at 25°C, 1.4 MPa. The valve is opened and methane flows quickly from $A$ to $B$ until the pressure in $A$ drops to 17 MPa, at which time the valve is closed. Assuming that tank $B$ contains a homogeneous mixture at the final state, and tank $A$ contains only methane, determine the final temperature in $A$ and the final temperature, pressure, and composition in $B$. Specify any additional assumptions made in the analysis.

11.42   A mixture of 85% methane, 15% ethane on a mole basis is contained in an insulated 0.2 m³ tank at 3.5 MPa, 40°C. The valve is opened accidentally, and the pressure quickly drops to 2 MPa before the valve is closed.

(a) Using Kay's rule and assuming a homogeneous mixture, calculate the mass that escaped from the tank.

(b) Eventually the temperature inside the tank returns to that of the surroundings, 40°C. What is the tank pressure at this time?

11.43   An insulated vertical cylinder fitted with a piston is divided into two parts by a partition as indicated in Fig. 11.20. Side $A$ contains argon at 310 K,

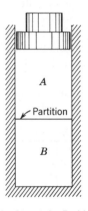

**Fig. 11.20**   Sketch for Problem 11.43.

14 MPa and side $B$ contains carbon dioxide at 310 K, 7 MPa. Each has a volume of 0.1 m³. A leak develops in the partition, and the two gases mix together.

(a) Assuming that the resulting mixture is represented by Kay's rule, find the pseudocritical pressure and temperature.

(b) Determine the final temperature in the system.

11.44   Capsule $A$ containing 1 litre of saturated liquid butane at 110°C is located inside Tank $B$ which has a total volume (including $A$) of 51 litres. Tank $B$ contains ethane at 1.4 MPa, and is at room temperature, 25°C.

The capsule now breaks and the two substances eventually form a uniform gas mixture at 25°C. Determine the final pressure and the irreversibility of the process.

11.45 A gas mixture of 80% ethylene and 20% propane enters a device at 30°C, 3.5 MPa, at the steady rate of 5 litres/s. Two separate streams leave the device; one is pure ethylene gas at 30°C, 2.8 MPa, and the other is pure liquid propane at 65°C, 3.5 MPa. Determine the minimum power input (kilowatts) required to operate this device.

11.46 Boil-off vapor from a liquid propane storage tank at 0°C is throttled into an adiabatic mixing chamber and mixed with nitrogen in the ratio 4 kmol $C_3H_8$/1 kmol $N_2$, as shown in Fig. 11.21. The mixture then enters a compressor. The exit state is 4.2 MPa, 60°C. It is claimed that the compression is adiabatic. Using the second law, show that this cannot be true.

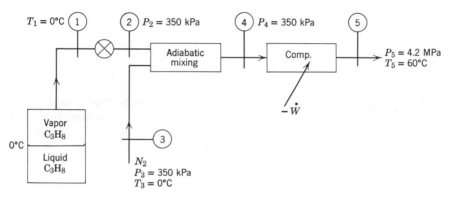

**Fig. 11.21** Sketch for Problem 11.44.

11.47 Repeat Problem 11.42 using the Redlich-Kwong equation of state, instead of Kay's rule and the generalized charts.

11.48 Consider a mixture of 50% nitrogen and 50% argon, on a mole basis, in a 1 litre volume at 180 K, 2 MPa. Calculate the mass of mixture, using the virial equation of state, and determine the percent error if ideal gas behavior's is assumed.

# 12

---

# Chemical Reactions

Many thermodynamic problems involve chemical reactions. Among the most familiar of these is the combustion of hydrocarbon fuels, for this process is utilized in most of our power generating devices. However, we can all think of a host of other processes involving chemical reactions, including those that occur in the human body.

It is our purpose in this chapter to consider a first and second law analysis of systems undergoing a chemical reaction. In many respects, this chapter is simply an extension of our previous consideration of the first and second laws. However, a number of new terms are introduced, and it will also be necessary to introduce the third law of thermodynamics.

In this chapter the combustion process is considered in detail. There are two reasons for this emphasis. The first is that the combustion process is of great significance in many problems and devices with which the engineer is concerned. This is especially true today in view of the many problems related to air pollution. The second is that the combustion process provides an excellent vehicle for teaching the basic principles of the thermodynamics of chemical reactions. The student should keep both of these objectives in mind as the study of this chapter progresses.

Chemical equilibrium will be considered in Chapter 13, and, therefore, the matter of dissociation will be deferred until then.

## 12.1  Fuels

A thermodynamics textbook is not the place for a detailed treatment of fuels. However, some knowledge of them is a prerequisite to a consideration of combustion, and this section is therefore devoted to a brief discussion of some of the hydrocarbon fuels. Most fuels fall into one of the three categories—coal, liquid hydrocarbons, or gaseous hydrocarbons.

Most liquid and gaseous hydrocarbon fuels are a mixture of many different hydrocarbons. For example, gasoline consists primarily of a mixture of about forty hydrocarbons, with many others present in very small quantities. In discussing hydrocarbon fuels, therefore, brief consideration should be given to the most important families of hydrocarbons, which are summarized in Table 12.1.

**Table 12.1**
Characteristics of Some of the Hydrocarbon Families

| Family | Formula | Structure | Saturated |
|---|---|---|---|
| Paraffin | $C_nH_{2n+2}$ | Chain | Yes |
| Olefin | $C_nH_{2n}$ | Chain | No |
| Diolefin | $C_nH_{2n-2}$ | Chain | No |
| Naphthene | $C_nH_{2n}$ | Ring | Yes |
| Aromatic | | | |
| Benzene | $C_nH_{2n-6}$ | Ring | No |
| Naphthalene | $C_nH_{2n-12}$ | Ring | No |

Three terms should be defined. The first pertains to the structure of the molecule. The important types are the ring and chain structures; the difference between the two is illustrated in Fig. 12.1. The same figure

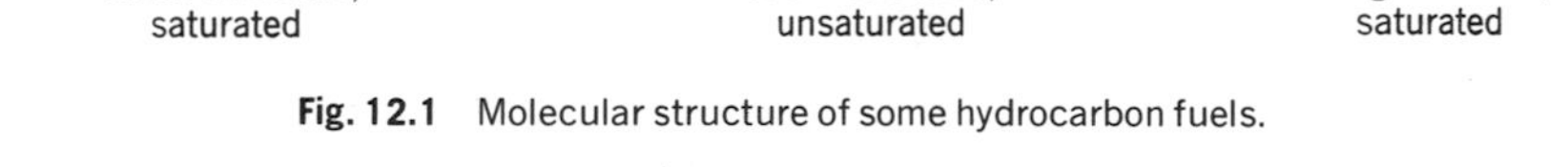

Chain structure, saturated

Chain structure, unsaturated

Ring structure, saturated

**Fig. 12.1**   Molecular structure of some hydrocarbon fuels.

illustrates the definition of saturated and unsaturated hydrocarbons. An unsaturated hydrocarbon has two or more adjacent carbon atoms joined by a double or triple bond, whereas in a saturated hydrocarbon all the carbon atoms are joined by a single bond. The third term to be defined is an isomer. Two hydrocarbons with the same number of carbon and hydrogen atoms and different structures are called isomers. Thus, there are several different octanes ($C_8H_{18}$) each having 8 carbon atoms and 18 hydrogen atoms, but each has a different structure.

The various hydrocarbon families are identified by a common suffix. The compounds comprising the paraffin family all end in "-ane" (as propane and octane). Similarly, the compounds comprising the olefin family end in "-ylene" or "-ene" (as propene and octene), and the diolefin family ends in "-diene" (as butadiene). The naphthene family has the same chemical formula as the olefin family, but has a ring rather than chain structure. The hydrocarbons in the naphthene family are named by adding the prefix "cyclo-" (as cyclopentane).

The aromatic family includes the benzene series ($C_nH_{2n-6}$) and the naphthalene series ($C_nH_{2n-12}$). The benzene series has a ring structure and is unsaturated.

Alcohols are sometimes used as a fuel in internal combustion engines. The characteristic feature of the alcohol family is that one of the hydrogen atoms is replaced by an OH radical. Thus methyl alcohol, also called methanol, is $CH_3OH$.

Most liquid hydrocarbon fuels are mixtures of hydrocarbons that are derived from crude oil through distillation and cracking processes. Thus, from a given crude oil a variety of different fuels can be produced, some of the common ones being gasoline, kerosene, diesel fuel, and fuel oil. Within each of these classifications there is a wide variety of grades, and each is made up of a large number of different hydrocarbons. The important distinction between these fuels is the distillation curve, Fig. 12.2. The distillation curve is obtained by slowly heating a sample of fuel so that it vaporizes. The vapor is then condensed and the amount measured. The more volatile hydrocarbons are vaporized first, and thus the temperature of the nonvaporized fraction increases during the process. The distillation curve, which is a plot of the temperature of the nonvaporized fraction vs. the amount of vapor condensed, is an indication of the volatility of the fuel.

In dealing with combustion of liquid fuels it is convenient to express the composition in terms of a single hydrocarbon, even though it is a mixture of many hydrocarbons. Thus gasoline is usually considered to be octane, $C_8H_{18}$, and diesel fuel is considered to be dodecane, $C_{12}H_{26}$. The composition of a hydrocarbon fuel may also be given in terms of percentage of carbon and hydrogen.

The two primary sources of gaseous hydrocarbon fuels are natural gas wells and certain chemical manufacturing processes. Table 12.2

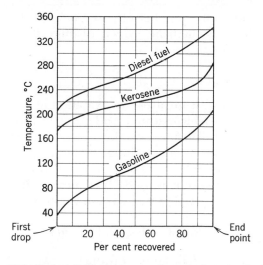

**Fig. 12.2**   Typical distillation curves of some hydrocarbon fuels.

## Table 12.2
### Volumetric Analyses of Some Typical Gaseous Fuels

| Constituent | Various Natural Gases | | | | Producer Gas from Bituminous Coal | Carbureted Water Gas | Coke-Oven Gas |
|---|---|---|---|---|---|---|---|
| | A | B | C | D | | | |
| Methane | 93.9 | 60.1 | 67.4 | 54.3 | 3.0 | 10.2 | 32.1 |
| Ethane | 3.6 | 14.8 | 16.8 | 16.3 | | | |
| Propane | 1.2 | 13.4 | 15.8 | 16.2 | | | |
| Butanes plus[a] | 1.3 | 4.2 | | 7.4 | | | |
| Ethene | | | | | | 6.1 | 3.5 |
| Benzene | | | | | | 2.8 | 0.5 |
| Hydrogen | | 7.5 | | | 14.0 | 40.5 | 46.5 |
| Nitrogen | | 7.5 | | 5.8 | 50.9 | 2.9 | 8.1 |
| Oxygen | | | | | 0.6 | 0.5 | 0.8 |
| Carbon monoxide | | | | | 27.0 | 34.0 | 6.3 |
| Carbon dioxide | | | | | 4.5 | 3.0 | 2.2 |

[a]This includes butane and all heavier hydrocarbons.

gives the composition of a number of gaseous fuels. The major constituent of natural gas is methane, which distinguishes it from manufactured gas.

## 12.2 The Combustion Process

The combustion process involves the oxidation of constituents in the fuel that are capable of being oxidized, and can therefore be represented by a chemical equation. During a combustion process the mass of each element remains the same. Thus, writing chemical equations and solving problems involving quantities of the various constituents basically involves the conservation of mass of each element. A brief review of this subject, particularly as it applies to the combustion process, is presented in this chapter.

Consider first the reaction of carbon with oxygen.

$$\text{Reactants} \quad \text{Products}$$

$$C + O_2 \rightarrow CO_2$$

This equation states that one kmol of carbon reacts with one kmol of oxygen to form one kmol of carbon dioxide. This also means that 12 kg of carbon react with 32 kg of oxygen to form 44 kg of carbon dioxide. All the initial substances that undergo the combustion process are called the reactants, and the substances that result from the combustion process are called the products.

When a hydrocarbon fuel is burned both the carbon and the hydrogen are oxidized. Consider the combustion of methane as an example.

$$CH_4 + 2O_2 \rightarrow CO_2 + 2H_2O \tag{12.1}$$

In this case the products of combustion include both carbon dioxide and water. The water may be in the vapor, liquid, or solid phases, depending on the temperature and pressure of the products of combustion.

It should be pointed out that in the combustion process there are many intermediate products formed during the chemical reaction. In this book we are concerned with the initial and final products and not with the intermediate products, but this aspect is very important in a detailed consideration of combustion.

In most combustion processes the oxygen is supplied as air rather than as pure oxygen. The composition of air on a molal basis is approximately 21 per cent oxygen, 78 per cent nitrogen, and 1 per cent argon. We will assume that the nitrogen and the argon do not undergo chemical reaction (except for dissociation which will be considered in Chapter 13).

They do leave at the same temperature as the other products, however, and therefore undergo a change of state if the products are at a temperature other than the original air temperature. It should be pointed out that at the high temperatures achieved in internal combustion engines, there is actually some reaction between the nitrogen and oxygen, and this gives rise to the air pollution problem associated with the oxides of nitrogen in the engine exhaust.

In combustion calculations involving air, the argon is usually neglected, and the air is considered to be composed of 21 per cent oxygen and 79 per cent nitrogen by volume. When this assumption is made, the nitrogen is sometimes referred to as "atmospheric nitrogen." Atmospheric nitrogen has a molecular weight of 28.16 (which takes the argon into account) as compared to 28.013 for pure nitrogen. This distinction will not be made in this text, and we will consider the 79 per cent nitrogen to be pure nitrogen.

The assumption that air is 21.0 per cent oxygen and 79.0 per cent nitrogen by volume leads to the conclusion that for each mole of oxygen, $79.0/21.0 = 3.76$ moles of nitrogen are involved. Therefore, when the oxygen for the combustion of methane is supplied as air, the reaction can be written

$$CH_4 + 2O_2 + 2(3.76)N_2 \rightarrow CO_2 + 2H_2O + 7.52N_2 \qquad (12.2)$$

The minimum amount of air that supplies sufficient oxygen for the complete combustion of all the carbon, hydrogen, and any other elements in the fuel that may oxidize is called the "theoretical air." When complete combustion is achieved with theoretical air, the products contain no oxygen. In practice, it is found that complete combustion is not likely to be achieved unless the amount of air supplied is somewhat greater than the theoretical amount. The amount of air actually supplied is expressed in terms of per cent theoretical air. Thus 150 per cent theoretical air means that the air actually supplied is 1.5 times the theoretical air. The complete combustion of methane with 150 per cent theoretical air is written

$$CH_4 + 2(1.5)O_2 + 2(3.76)(1.5)N_2 \rightarrow CO_2 + 2H_2O + O_2 + 11.28N_2$$

$$(12.3)$$

The amount of air actually supplied may also be expressed in terms of per cent excess air. The excess air is the amount of air supplied over and above the theoretical air. Thus 150 per cent theoretical air is equivalent to 50 per cent excess air. The terms theoretical air and excess air are both in current usage.

Two important parameters applied to combustion processes are the air-fuel ratio (designated $AF$) and its reciprocal, the fuel-air ratio (designated $FA$). The air-fuel ratio is usually expressed on a mass basis, but a mole basis is also used at times. The theoretical air-fuel ratio is the ratio of the mass (or moles) of theoretical air to the mass (or moles) of fuel.

When the amount of air supplied is less than the theoretical air required, the combustion is incomplete. If there is only a slight deficiency of air, the usual result is that some of the carbon unites with the oxygen to form carbon monoxide (CO) instead of carbon dioxide ($CO_2$). If the air supplied is considerably less than the theoretical air, there may also be some hydrocarbons in the products of combustion.

Even when some excess air is supplied there may be small amounts of carbon monoxide present, the exact amount depending on a number of factors including the mixing and turbulence during combustion. Thus, the combustion of methane with 110 per cent theoretical air might be as follows:

$$CH_4 + 2(1.1)O_2 + 2(1.1)3.76N_2 \rightarrow$$

$$+ 0.95CO_2 + 0.05CO + 2H_2O + 0.225O_2 + 8.27N_2 \qquad (12.4)$$

The material covered so far in this section is illustrated by the following examples.

### Example 12.1

Calculate the theoretical air-fuel ratio for the combustion of octane, $C_8H_{18}$.

The combustion equation is

$$C_8H_{18} + 12.5O_2 + 12.5(3.76)N_2 \rightarrow 8CO_2 + 9H_2O + 47.0N_2$$

The air-fuel ratio on a mole basis is

$$AF = \frac{12.5 + 47.0}{1} = 59.5 \text{ mol air/mol fuel}$$

The theoretical air-fuel ratio on a mass basis is found by introducing the molecular weight of the air and fuel.

$$AF = \frac{59.5(28.97)}{114.2} = 15.0 \text{ kg air/kg fuel}$$

### Example 12.2

Determine the molal analysis of the products of combustion when octane, $C_8H_{18}$, is burned with 200 per cent theoretical air, and determine the dew point of the products if the pressure is 0.1 MPa.

The equation for the combustion of octane with 200 per cent theoretical air is

$$C_8H_{18} + 12.5(2)O_2 + 12.5(2)(3.76)N_2 \rightarrow 8CO_2 + 9H_2O + 12.5O_2 + 94.0N_2$$

Total moles of product $= 8 + 9 + 12.5 + 94.0 = 123.5$
Molal analysis of products:

$$
\begin{array}{rll}
CO_2 = 8/123.5 & = & 6.47\% \\
H_2O = 9/123.5 & = & 7.29 \\
O_2 = 12.5/123.5 & = & 10.12 \\
N_2 = 94/123.5 & = & \underline{76.12} \\
& & 100.00\%
\end{array}
$$

The partial pressure of the $H_2O$ is $100 (0.0729) = 7.29$ kPa.

The saturation temperature corresponding to this pressure is 39.7°C, which is also the dew-point temperature.

The water condensed from the products of combustion usually contains some dissolved gases and therefore may be quite corrosive. For this reason the products of combustion are often kept above the dew point until discharged to the atmosphere.

## Example 12.3

Producer gas from bituminous coal (see Table 12.2) is burned with 20 per cent excess air. Calculate the air-fuel ratio on a volumetric basis and on a mass basis.

To calculate the theoretical air requirement, let us write the combustion equation for the combustible substances in 1 mole of fuel.

$$
\begin{array}{ll}
0.14H_2 + 0.070O_2 \rightarrow 0.14H_2O \\
0.27CO + 0.135O_2 \rightarrow 0.27CO_2 \\
0.03CH_4 + 0.06O_2 \rightarrow 0.03CO_2 + 0.06H_2O
\end{array}
$$

$$
\begin{array}{rl}
\underline{0.265} & = \text{moles oxygen required/mole fuel} \\
-0.006 & = \text{oxygen in fuel/mole fuel} \\
\overline{0.259} & = \text{moles oxygen required from air/mole fuel}
\end{array}
$$

Therefore, the complete combustion equation for 1 mole of fuel is

$$\overbrace{0.14H_2 + 0.27CO + 0.03CH_4 + 0.006O_2 + 0.509N_2 + 0.045CO_2}^{\text{fuel}}$$

$$+ \overbrace{0.259O_2 + 0.259(3.76)N_2}^{\text{air}} \rightarrow 0.20H_2O + 0.345CO_2 + 1.482N_2$$

$$\left(\frac{\text{moles air}}{\text{mole fuel}}\right)_{\text{theo}} = 0.259 \times \frac{1}{0.21} = 1.233$$

If the air and fuel are at the same pressure and temperature, this also represents the ratio of the volume of air to the volume of fuel.

For 20 per cent excess air, $\dfrac{\text{moles air}}{\text{mole fuel}} = 1.233 \times 1.200 = 1.48$

The air-fuel ratio on a mass basis is

$$AF = \frac{1.48(28.97)}{0.14(2) + 0.27(28) + 0.03(16) + 0.006(32) + 0.509(28) + 0.045(44)}$$

$$= \frac{1.48(28.97)}{24.74} = 1.73 \text{ kg air/kg fuel}$$

An analysis of the products of combustion affords a very simple method for calculating the actual amount of air supplied in a combustion process. There are various experimental methods by which such an analysis can be made. Some yield results on a "dry" basis; i.e., the fractional analysis of all the components, except for water vapor. Other experimental procedures give results that include the water vapor. In this presentation we are not concerned with the experimental devices and procedures, but rather with the use of such information in a thermodynamic analysis of the chemical reaction. The examples that follow illustrate how an analysis of the products can be used to determine the chemical reaction and the composition of the fuel.

In using the analysis of the products of combustion to obtain the actual fuel-air ratio, the basic principle is the conservation of the mass of each of the elements. Thus, in changing from reactants to products we can make a carbon balance, hydrogen balance, oxygen balance, and nitrogen balance (plus any other elements that may be involved). Further, we recognize that there is a definite ratio between the amounts of some of these elements. Thus, the ratio between the nitrogen and oxygen supplied in the air is fixed, as well as the ratio between carbon and hydrogen if the composition of a hydrocarbon fuel is known.

**Example 12.4**

Methane ($CH_4$) is burned with atmospheric air. The analysis of the products on a "dry" basis is as follows:

| | |
|---|---|
| $CO_2$ | 10.00% |
| $O_2$ | 2.37 |
| CO | 0.53 |
| $N_2$ | 87.10 |
| | 100.00% |

Calculate the air-fuel ratio, the per cent theoretical air, and determine the combustion equation.

The solution involves writing the combustion equation for 100 moles of dry products, introducing letter coefficients for the unknown quantities, and then solving for them.

From the analysis of the products, the following equation can be written, keeping in mind that this analysis is on a dry basis.

$$a\text{CH}_4 + b\text{O}_2 + c\text{N}_2 \rightarrow 10.0\text{CO}_2 + 0.53\text{CO} + 2.37\text{O}_2 + d\text{H}_2\text{O} + 87.1\text{N}_2$$

A balance for each of the elements involved will enable us to solve for all the unknown coefficients:

Nitrogen balance: $c = 87.1$

Since all the nitrogen comes from the air,

$$\frac{c}{b} = 3.76; \quad b = \frac{87.1}{3.76} = 23.16$$

Carbon balance:       $a = 10.00 + 0.53 = 10.53$
Hydrogen balance:     $d = 2a = 21.06$

Oxygen balance: All the unknown coefficients have been solved for, and in this case the oxygen balance provides a check on the accuracy. Thus, $b$ can also be determined by an oxygen balance.

$$b = 10.00 + \frac{0.53}{2} + 2.37 + \frac{21.06}{2} = 23.16$$

Substituting these values for $a$, $b$, $c$, $d$, and $e$ we have,

$$10.53\text{CH}_4 + 23.16\text{O}_2 + 87.1\text{N}_2 \rightarrow$$
$$10.0\text{CO}_2 + 0.53\text{CO} + 2.37\text{O}_2 + 21.06\text{H}_2\text{O} + 87.1\text{N}_2$$

Dividing through by 10.53 yields the combustion equation per mole of fuel.

$$\text{CH}_4 + 2.2\text{O}_2 + 8.27\text{N}_2 \rightarrow 0.95\text{CO}_2 + 0.05\text{CO} + 2\text{H}_2\text{O} + 0.225\text{O}_2 + 8.27\text{N}_2$$

The air-fuel ratio on a mole basis is

$$2.2 + 8.27 = 10.47 \text{ moles air/mole fuel}$$

The air-fuel ratio on a mass basis is found by introducing the molecular weights.

$$AF = \frac{10.47 \times 28.97}{16.0} = 18.97 \text{ kg air/kg fuel}$$

The theoretical air-fuel ratio is found by writing the combustion equation for theoretical air.

$$CH_4 + 2O_2 + 2(3.76)N_2 \rightarrow CO_2 + 2H_2O + 7.52N_2$$

$$AF_{\text{theo}} = \frac{(2 + 7.52)28.97}{16.0} = 17.23 \text{ kg air/kg fuel}$$

The per cent theoretical air is $\dfrac{18.97}{17.23} = 110\%$

## Example 12.5

The products of combustion of a hydrocarbon fuel of unknown composition have the following composition as measured on a dry basis:

|        |         |
|--------|---------|
| $CO_2$ | 8.0%    |
| CO     | 0.9     |
| $O_2$  | 8.8     |
| $N_2$  | 82.3    |
|        | 100.0%  |

Calculate (1) The air-fuel ratio, (2) The composition of the fuel on a mass basis, and (3) The per cent theoretical air on a mass basis.

Let us first write the combustion equation for 100 moles of dry product.

$$C_aH_b + dO_2 + cN_2 \rightarrow 8.0CO_2 + 0.9CO + 8.8O_2 + eH_2O + 82.3N_2$$

Nitrogen balance: $c = 82.3$

From composition of air, $\dfrac{c}{d} = 3.76$;   $d = \dfrac{82.3}{3.76} = 21.9$

Oxygen balance: $21.9 = 8.0 + \dfrac{0.9}{2} + 8.8 + \dfrac{e}{2}$;   $e = 9.3$

Carbon balance:   $a = 8.0 + 0.9 = 8.9$

Hydrogen balance: $b = 2e = 2 \times 9.3 = 18.6$

Thus the composition of the fuel could be written $C_{8.9}H_{18.6}$

The air-fuel ratio can now be found, using the molecular weights.

$$AF = \frac{(21.9 + 82.3)28.97}{8.9(12) + 18.6(1)} = 24.1 \text{ kg air/kg fuel}$$

On a mass basis the composition of fuel is:

$$\text{Carbon} = \frac{8.9(12)}{8.9(12) + 18.6(1)} = 85.2\%$$

$$\text{Hydrogen} = \frac{18.6(1)}{8.9(12) + 18.6(1)} = 14.8\%$$

The theoretical air requirement can be found by writing the theoretical combustion equation.

$$C_{8.9}H_{18.6} + 13.5O_2 + 13.5(3.76)N_2 \rightarrow 8.9CO_2 + 9.3H_2O + 50.8N_2$$

$$AF_{\text{theo}} = \frac{(13.5 + 50.8)28.97}{8.9(12) + 18.6(1)} = 14.9 \text{ kg air/kg fuel}$$

$$\% \text{ theoretical air} = \frac{24.1}{14.9} = 162\%$$

## 12.3   Enthalpy of Formation

In the first eleven chapters of this book the problems considered always involved a fixed chemical composition, and never involved a change of composition due to chemical reaction. Therefore, in dealing with a thermodynamic property we made use of tables of thermodynamic properties for the given substance, and in each of these tables the thermodynamic properties were given relative to some arbitrary base. In the steam tables, for example, the internal energy of saturated liquid at $0.01°C$ is assumed to be zero. This procedure is quite adequate when no change in composition is involved, because we are concerned with the changes in the properties of a given substance. However, it is evident that this procedure is quite inadequate when dealing with a chemical reaction, because the composition changes during the process.

To understand the procedure used in the case of chemical reactions, consider the simple SSSF combustion process shown in Fig. 12.3. This reaction involves the combustion of carbon and oxygen to form $CO_2$. The carbon and oxygen each enter the control volume at $25°C$ and $0.1$ MPa pressure, and the heat transfer is such that the $CO_2$ leaves at $25°C$,

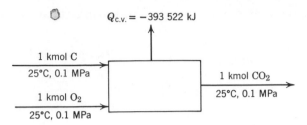

**Fig. 12.3**  Example of combustion process.

0.1 MPa pressure. If the heat transfer were accurately measured, it would be $-393\ 522$ kJ per kmol of $CO_2$ formed. This reaction can be written

$$C + O_2 \rightarrow CO_2$$

Applying the first law to this process we have

$$Q_{c.v.} + H_R = H_P \tag{12.5}$$

where the subscripts $R$ and $P$ refer to the reactants and products respectively. We will find it convenient to also write the first law for such a process in the form

$$Q_{c.v.} + \sum_R n_i \bar{h}_i = \sum_P n_e \bar{h}_e \tag{12.6}$$

where the summations refer respectively to all the reactants or all the products.

Thus a measurement of the heat transfer really gives us the difference between the enthalpy of the products and the reactants.

However, suppose we assign the value of zero to the enthalpy of all the elements at 25°C and 0.1 MPa pressure. In this case the enthalpy of the reactants is zero, and

$$Q_{c.v.} = H_P = -393\ 522 \text{ kJ/kmol}$$

The enthalpy of $CO_2$ at 25°C and 0.1 MPa pressure (with reference to this base in which the enthalpy of the elements is assumed to be zero) is called the enthalpy of formation. We designate this with the symbol $\bar{h}_f^\circ$. Thus, for $CO_2$,

$$\bar{h}_f^\circ = -393\ 522 \text{ kJ/kmol}$$

The enthalpy of $CO_2$ in any other state, relative to this base in which the enthalpy of the elements is zero, would be found by adding the change of enthalpy between 25°C, 0.1 MPa and the given state to the enthalpy of formation. This is, the enthalpy at any temperature and pressure, $\bar{h}_{T,P}$ is

$$\bar{h}_{T,P} = (\bar{h}_f^\circ)_{298,\,0.1\text{ MPa}} + (\Delta\bar{h})_{298,\,0.1\text{ MPa}\rightarrow T,P} \qquad (12.7)$$

where the term $(\Delta\bar{h})_{298,\,0.1\text{ MPa}\rightarrow T,P}$ represents the difference in enthalpy between any given state and the enthalpy at 298.15 K, 0.1 MPa. For convenience we usually drop the subscripts in the examples that follow.

The procedure which we have demonstrated for $CO_2$ can be applied to any compound.

Table 12.3 gives values of the enthalpy of formation for a number of substances in the units kJ/kmol.

Three further observations should be made in regard to enthalpy of formation.

1. We have demonstrated the concept of enthalpy of formation in terms of the measurement of the heat transfer in a chemical reaction in which a compound is formed from the elements. Actually, the enthalpy of formation is usually found by the application of statistical thermodynamics, using observed spectroscopic data.

2. The justification of this procedure of arbitrarily assigning the value of zero to the enthalpy of the elements at 25°C, 0.1 MPa rests on the fact that in the absence of nuclear reactions the mass of each element is conserved in a chemical reaction. No conflicts or ambiguities arise with this choice of reference state, and it proves to be very convenient in studying chemical reactions from a thermodynamic point of view.

3. In certain cases an element or compound can exist in more than one state at 25°C, 0.1 MPa. Carbon, for example, can be in the form of graphite or diamond. It is essential that the state to which a given value is related be clearly identified. Thus, in Table 12.3 the enthalpy of formation of graphite is given the value of zero, and the enthalpy of each substance that contains carbon is given relative to this base.

It will be noted from Table 12.3 that two values for $H_2O$ at 0.1 MPa, 25°C, are given; one is for liquid $H_2O$ and the other for gaseous $H_2O$. We know from the thermodynamic properties of water that under equilibrium conditions water exists as a slightly compressed liquid at 0.1 MPa, 25°C. However, in many cases, we find it convenient to consider a pseudo-equilibrium state, in which water is a vapor at 0.1 MPa, 25°C. In doing so we assume that the substance behaves as an ideal gas between

## Table 12.3
Enthalpy of Formation, Gibbs Function of Formation and Absolute Entropy of Various Substances at 25°C, 0.1 MPa Pressure

| Substance | Formula | M | State | $\bar{h}_f^\circ$ kJ/kmol | $\bar{g}_f^\circ$ kJ/kmol | $\bar{s}^\circ$ kJ/kmol K |
|---|---|---|---|---|---|---|
| Carbon monoxide[a] | CO | 28.011 | gas | −110 529 | −137 150 | 197.653 |
| Carbon dioxide[a] | CO$_2$ | 44.011 | gas | −393 522 | −394 374 | 213.795 |
| Water[a,b] | H$_2$O | 18.015 | gas | −241 827 | −228 583 | 188.833 |
| Water[b] | H$_2$O | 18.015 | liq. | −285 838 | −237 178 | 70.049 |
| Methane[a] | CH$_4$ | 16.043 | gas | − 74 873 | − 50 751 | 186.256 |
| Acetylene[a] | C$_2$H$_2$ | 26.038 | gas | +226 731 | +209 234 | 200.958 |
| Ethene[a] | C$_2$H$_4$ | 28.054 | gas | + 52 283 | + 68 207 | 219.548 |
| Ethane[c] | C$_2$H$_6$ | 30.070 | gas | − 84 667 | − 32 777 | 229.602 |
| Propane[c] | C$_3$H$_8$ | 44.097 | gas | −103 847 | − 23 316 | 270.019 |
| Butane[c] | C$_4$H$_{10}$ | 58.124 | gas | −126 148 | − 16 914 | 310.227 |
| Octane[c] | C$_8$H$_{18}$ | 114.23 | gas | −208 447 | + 16 859 | 466.835 |
| Octane[c] | C$_8$H$_{18}$ | 114.23 | liq. | −249 952 | + 6 940 | 360.896 |
| Carbon[a] (graphite) | C | 12.011 | solid | 0 | 0 | 5.795 |

[a] From JANAF *Thermochemical Data*, The Dow Chemical Company, Thermal Laboratory, Midland, Mich.
[b] From *Circular 500*, National Bureau of Standards.
[c] From F. D. Rossini et al., *API Research Project 44*.

saturated vapor at 25°C and this pseudo-equilibrium state at 0.1 MPa, 25°C. Thus, the enthalpy in the pseudo-equilibrium state is the same as that of saturated vapor at this same temperature, and the change in entropy between these two states would be found from the appropriate ideal gas relations. It also follows that the difference between the enthalpy of formation of $H_2O(l)$ and $H_2O(g)$ at 0.1 MPa, 25°C is $h_{fg}$, the enthalpy of evaporation at 25°C. (The correction due to the fact that $H_2O$ is a compressed liquid at 0.1 MPa, 25°C, is very small and may be neglected.) This same procedure is used in the case of other substances which have a vapor pressure of less than 0.1 MPa at 25°C.

Frequently students are bothered by the minus sign when the enthalpy of formation is negative. For example, the enthalpy of formation of the $CO_2$ is negative. This is quite evident because the heat transfer was negative during the steady-flow chemical reaction, and the enthalpy of the $CO_2$ must be less than the sum of enthalpy of the carbon and oxygen initially, both of which are assigned the value of zero. This is quite analogous to the situation we would have in the steam tables if we let the enthalpy of saturated vapor be zero at 0.1 MPa pressure, for in this case the enthalpy of the liquid would be negative, and we would simple use the negative value for the enthalpy of the liquid when solving problems.

## 12.4    First Law Analysis of Reacting Systems

The significance of the enthalpy of formation is that it is most convenient in performing a first law analysis of a reacting system, for the enthalpies of different substances can be added or subtracted, since they are all given relative to the same base.

In such problems we will write the first law for a steady-state, steady-flow process in the form

$$Q_{c.v.} + H_R = W_{c.v.} + H_P$$

or

$$Q_{c.v.} + \sum_R n_i \bar{h}_i = W_{c.v.} + \sum_P n_e \bar{h}_e$$

where $R$ and $P$ refer to the reactants and products respectively. In each problem it is necessary to choose one parameter as the basis of the solution. Usually this is taken as one kmol of fuel.

### Example 12.6

Consider the following reaction, which occurs in a steady-state, steady-flow process.

$$CH_4 + 2O_2 \rightarrow CO_2 + 2H_2O(l)$$

The reactants and products are each at a total pressure of 0.1 MPa and 25°C. Determine the heat transfer per kmol of fuel entering the combustion chamber.

Applying the first law,

$$Q_{c.v.} + \sum_R n_i \bar{h}_i = \sum_P n_e \bar{h}_e$$

From Table 12.3

$$\sum_R n_i \bar{h}_i = (\bar{h}_f^\circ)_{CH_4} = -74\ 874 \text{ kJ}$$

$$\sum_P n_e \bar{h}_e = (\bar{h}_f^\circ)_{CO_2} + 2(\bar{h}_f^\circ)_{H_2O(l)}$$

$$= -393\ 522 + 2\ (-285\ 838) = -965\ 198 \text{ kJ}$$

$$Q_{c.v.} = -965\ 198 - (-74\ 873) = -890\ 325 \text{ kJ}$$

In most cases however, the substances which comprise the reactants and products in a chemical reaction are not at a temperature of 25°C and a pressure of 0.1 MPa (the state at which the enthalpy of formation is given). Therefore the change of enthalpy between 25°C and 0.1 MPa and the given state must be known. In the case of a solid or liquid, this change of enthalpy can usually be found from a table of thermodynamic properties or from specific heat data. In the case of gases, this change of enthalpy can usually be found by one of the following procedures.

1. Assume ideal gas behavior between 25°C, 0.1 MPa, and the given state. In this case, the enthalpy is a function of the temperature only, and can be found by use of an equation of $C_{po}$ or from tabulated values of enthalpy as a function of temperature (which assumes ideal gas behavior). Table A.9 gives an equation for $\bar{C}_{po}$ for a number of substances and Table A.11 gives values of $(\bar{h}^\circ - \bar{h}_{298}^\circ)$, (that is, the $\Delta \bar{h}$ of Eq. 12.7) in kJ/kmol ($\bar{h}_{298}^\circ$ refers to 25°C or 298.15 K. For simplicity this is designated $\bar{h}_{298}^\circ$). The superscript $^\circ$ is used to designate that this is the enthalpy at 0.1 MPa pressure, based on ideal gas behavior, that is, the standard-state enthalpy.
2. If a table of thermodynamic properties is available, $\Delta \bar{h}$ can be found directly from these tables. This is necessary when the deviation from ideal gas behavior is significant.
3. If the deviation from ideal gas behavior is significant, but no tables of thermodynamic properties are available, the value for $\Delta \bar{h}$ can be found from the generalized charts and the values for $\bar{C}_{po}$ or $\Delta \bar{h}$ at 0.1 MPa pressure as indicated above.

Thus, in general, for applying the first law to a steady-state, steady-flow process involving a chemical reaction and negligible changes in kinetic and potential energy we can write

$$Q_{c.v.} + \sum_R n_i[\bar{h}_f^\circ + \Delta\bar{h}]_i = W_{c.v.} + \sum_P n_e[\bar{h}_f^\circ + \Delta\bar{h}]_e \qquad (12.8)$$

## Example 12.7

Calculate the enthalpy of $H_2O$ (on a kmol basis) at 3.5 MPa, 300°C, relative to the 25°C and 0.1 MPa base, using the following procedures.

1. Assume the steam to be an ideal gas with the value of $\bar{C}_{po}$ given in the Appendix, Table A.9.
2. Assume the steam to be an ideal gas with the value for $\Delta\bar{h}$ as given in the Appendix, Table A.11.
3. The steam tables.
4. The specific heat equations given in (1) above and the generalized charts.

For each of these procedures we can write,

$$\bar{h}_{T,P} = [\bar{h}_f^\circ + \Delta\bar{h}]$$

The only difference is in the procedure by which we calculate $\Delta\bar{h}$. From Table 12.3 we note that

$$(\bar{h}_f^\circ)_{H_2O(g)} = -241\ 827\ \text{kJ/kmol}$$

1. Using the specific heat equation for $H_2O(g)$ from Table A.9,

$$\bar{C}_{po} = 143.05 - 183.54\ \theta^{0.25} + 82.751\ \theta^{0.5} - 3.69890\ \text{kJ/kmol K}$$

where $\quad \theta = T/100$

Therefore,

$$\Delta\bar{h} = \int_{298.15}^{573.15} \bar{C}_{po}(T)\ dT$$

$$= \int_{2.9815}^{5.7315} \bar{C}_{po}(\theta)\ 100\ d\theta$$

$$= 9\ 517\ \text{kJ/kmol}$$

$$\bar{h}_{T,P} = -241\ 827 + 9\ 517 = -232\ 310\ \text{kJ/kmol}$$

**2.** Using Table A.11 for $H_2O(g)$,

$$\Delta \bar{h} = 9\ 537\ \text{kJ/kmol}$$

$$\bar{h}_{T,P} = -241\ 827 + 9\ 537 = -232\ 290\ \text{kJ/kmol}$$

**3.** Using the steam tables, either the liquid reference or the gaseous reference state may be used.

For the liquid,

$$\Delta \bar{h} = 18.015\ (2977.5 - 104.9) = 51\ 750\ \text{kJ/kmol}$$

$$\bar{h}_{T,P} = -285\ 838 + 51\ 750 = -234\ 088\ \text{kJ/kmol}$$

For the gas,

$$\Delta \bar{h} = 18.015\ (2977.5 - 2547.2) = 7\ 752\ \text{kJ/kmol}$$

$$\bar{h}_{T,P} = -241\ 827 + 7\ 752 = -234\ 075\ \text{kJ/kmol}$$

The very small difference results from using the enthalpy of saturated vapor at 25°C (which is almost but not exactly an ideal gas), in calculating the $\Delta h$.

**4.** When using the generalized charts we use the notation introduced in Chapter 10.

$$\bar{h}_{T,P} = \bar{h}_f^\circ - (\bar{h}_2^* - \bar{h}_2) + (\bar{h}_2^* - \bar{h}_1^*) + (\bar{h}_1^* - \bar{h}_1)$$

where subscript 2 refers to the state at 3.5 MPa, 300°C and state 1 refers to the state at 0.1 MPa, 25°C.

From part (2) $\bar{h}_2^* - \bar{h}_1^* = 9\ 537\ \text{kJ/kmol}$

$$\bar{h}_1^* - \bar{h}_1 = 0\ \text{(ideal gas reference)}$$

$$P_{r2} = \frac{3.5}{22.09} = 0.158\ ;\ T_{r2} = \frac{573.2}{647.3} = 0.886$$

From the generalized enthalpy chart,

$$\frac{\bar{h}_2^* - \bar{h}_2}{T_c} = 2.3\ ;\ \bar{h}_2^* - \bar{h}_2 = 2.3\ (647.3) = 1\ 489\ \text{kJ/kmol}$$

$$\bar{h}_{T,P} = -241\ 827 - 1\ 489 + 9\ 537 = -233\ 779 \text{ kJ/kmol}$$

The particular approach that is used in a given problem will depend on the data available for the given substance.

**Example 12.8**

A small gas turbine uses $C_8H_{18}(l)$ for fuel, and 400 per cent theoretical air. The air and fuel enter at 25°C and the products of combustion leave at 900 K. The output of the engine and the fuel consumption are measured and it is found that the specific fuel consumption is 0.25 kg/s of fuel per MW output. Determine the heat transfer from the engine per kmol of fuel. Assume complete combustion.
The combustion equation is

$$C_8H_{18}(l) + 4(12.5)O_2 + 4(12.5)(3.76)N_2$$
$$\rightarrow 8CO_2 + 9H_2O + 37.5O_2 + 188.0N_2$$

We consider that we have a steady-state, steady-flow process involving ideal gases (except for the $C_8H_{18}(l)$). Therefore we can write:

First law:   $Q_{c.v.} + \sum_R n_i[(\bar{h}_f^\circ + \Delta\bar{h})]_i = W_{c.v.} + \sum_P n_e[\bar{h}_f^\circ + \Delta\bar{h}]_e$

Property relation: Table A.11 for ideal gases and Table 12.3 for $C_8H_{18}(l)$. Since the air is composed of elements and enters at 25°C the enthalpy of the reactants is equal to that of the fuel,

$$\sum_R n_i[\bar{h}_f^\circ + \Delta\bar{h}]_i = (\bar{h}_f^\circ)_{C_8H_{18}(l)} = -249\ 952 \text{ kJ/kmol fuel}$$

Considering the products

$$\sum_P n_e[\bar{h}_f^\circ + \Delta\bar{h}]_e = n_{CO_2}[\bar{h}_f^\circ + \Delta\bar{h}]_{CO_2} + n_{H_2O}[\bar{h}_f^\circ + \Delta\bar{h}]_{H_2O} + n_{O_2}[\Delta\bar{h}]_{O_2}$$
$$+ n_{N_2}[\Delta\bar{h}]_{N_2}$$
$$= 8(-393\ 522 + 28\ 041) + 9(-241\ 827 + 21\ 924)$$
$$+ 37.5(19\ 246) + 188(18\ 221)$$
$$= -755\ 702 \text{ kJ/kmol fuel}$$

$$W_{c.v.} = \left(\frac{1000\ \frac{\text{kJ}}{\text{s}}}{0.25\ \frac{\text{kg}}{\text{s}}}\right) \times \frac{114.23\ \text{kg}}{\text{kmol}} = 456\ 920 \text{ kJ/kmol fuel}$$

Therefore, from the first law,

$$Q_{c.v.} = -755\ 702 + 456\ 920 - (-249\ 952)$$

$$= -48\ 830\ \text{kJ/kmol fuel}$$

## Example 12.9

A mixture of one kmol of gaseous ethene and three kmol of oxygen at 25°C react in a constant volume bomb. Heat is transferred until the products are cooled to 600 K. Determine the amount of heat transfer from the reactants.

The chemical reaction is

$$C_2H_4 + 3O_2 \rightarrow 2CO_2 + 2H_2O(g)$$

First law: $\quad Q + U_R = U_P$

$$Q + \sum_R n[\bar{h}_f^\circ + \Delta\bar{h} - \bar{R}T] = \sum_P n[\bar{h}_f^\circ + \Delta\bar{h} - \bar{R}T]$$

$$\sum_R n[\bar{h}_f^\circ + \Delta\bar{h} - \bar{R}T] = [\bar{h}_f^\circ - \bar{R}T]_{C_2H_4} - n_{O_2}[\bar{R}T]_{O_2} = (\bar{h}_f^\circ)_{C_2H_4} - 4\bar{R}T$$

$$= 52\ 283 - 4 \times 8.31434 \times 298.2 = 42\ 366\ \text{kJ}$$

$$\sum_P n[\bar{h}_f^\circ + \Delta\bar{h} - \bar{R}T] = 2[(\bar{h}_f^\circ)_{CO_2} + \Delta\bar{h}_{CO_2}] + 2[(\bar{h}_f^\circ)_{H_2O(g)} + \Delta\bar{h}_{H_2O(g)}] - 4\bar{R}T$$

$$= 2(-393\ 522 + 12\ 916) + 2(-241\ 827 + 10\ 498)$$

$$- 4 \times 8.31434 \times 600$$

$$= -1\ 243\ 824\ \text{kJ}$$

Therefore,

$$Q = -1\ 243\ 824 - 42\ 366 = -1\ 286\ 190\ \text{kJ}$$

## 12.5 Adiabatic Flame Temperature

Consider a given combustion process that takes place adiabatically and with no work or changes in kinetic or potential energy involved. For such a process the temperature of the products is referred to as the adiabatic flame temperature. With the assumptions of no work and no changes in kinetic or potential energy, this is the maximum temperature that can be achieved for the given reactants because any heat transfer from the reacting substances and any incomplete combustion would tend to lower the temperature of the products.

For a given fuel and given pressure and temperature of the reactants, the maximum adiabatic flame temperature that can be achieved is with a stoichiometric mixture. The adiabatic flame temperature can be controlled by the amount of excess air that is used. This is important, for example, in gas turbines, where the maximum permissible temperature is determined by metallurgical considerations in the turbine, and close control of the temperature of the products is essential.

Example 12.10 shows how the adiabatic flame temperature may be found. The dissociation that takes place in the combustion products, which has a significant effect on the adiabatic flame temperature, will be considered in the next chapter.

**Example 12.10**

Liquid octane at 25°C is burned with 400 per cent theoretical air at 25°C in a steady-flow process. Determine the adiabatic flame temperature.

The reaction

$$C_8H_{18}(l) + 4(12.5)O_2 + 4(12.5)(3.76)N_2 \rightarrow$$
$$8CO_2 + 9H_2O(g) + 37.5O_2 + 188.0N_2$$

We assume ideal gas behavior for all the constituents except the liquid octane.

In this case the first law reduces to

$$H_R = H_P$$
$$\sum_R n_i[h_f^\circ + \Delta h]_i = \sum_P n_e[h_f^\circ + \Delta h]_e$$

where $\Delta \bar{h}_e$ refers to each constituent in the products at the adiabatic flame temperature.

$$H_R = \sum_R n_i[\bar{h}_f^\circ + \Delta \bar{h}]_i = (\bar{h}_f^\circ)_{C_8H_{18}(l)} = -249\ 952 \text{ kJ/kmol fuel}$$

$$H_P = \sum_P n_e[\bar{h}_f^\circ + \Delta \bar{h}]_e = 8[-393\ 522 + \Delta \bar{h}_{CO_2}] + 9[-241\ 827 + \Delta \bar{h}_{H_2O}]$$
$$+ 37.5\Delta \bar{h}_{O_2} + 188.0\Delta \bar{h}_{N_2}$$

By trial-and-error solution, a temperature of the products is found that satisfies this equation. Assume

$$T_P = 900 \text{ K}$$

$$H_P = \sum_P n_e [\bar{h}_f^\circ + \Delta \bar{h}]_e$$

$$= 8\,(-393\ 522 + 28\ 041) + 9(-241\ 827 + 21\ 924)$$

$$+ 37.5(19\ 246) + 188(18\ 221)$$

$$= -755\ 702 \text{ kJ/kmol fuel}$$

Assume

$$T_P = 1000 \text{ K}$$

$$H_P = \sum_P n_e [\bar{h}_f^\circ + \Delta \bar{h}]_e$$

$$= 8(-393\ 522 + 33\ 405) + 9(-241\ 827 + 25\ 978)$$

$$+ 37.5(22\ 707) + 188(21\ 460)$$

$$= 62\ 416 \text{ kJ/kmol fuel}$$

Since $H_P = H_R = -249\ 952$ kJ, we find, by interpolation that the adiabatic flame temperature is 961.8 K.

## 12.6    Enthalpy and Internal Energy of Combustion; Heat of Reaction

The enthalpy of combustion, $h_{RP}$, is defined as the difference between the enthalpy of the products and the enthalpy of the reactants when complete combustion occurs at a given temperature and pressure. That is,

$$\bar{h}_{RP} = H_P - H_R$$

$$\bar{h}_{RP} = \sum_P n_e [\bar{h}_f^\circ + \Delta \bar{h}]_e - \sum_R n_i [\bar{h}_f^\circ + \Delta \bar{h}]_i \qquad (12.9)$$

The usual parameter for expressing the enthalpy of combustion is a unit mass of fuel, such as a kg ($h_{RP}$) or a kmol ($\bar{h}_{RP}$) of fuel.

The tabulated values of the enthalpy of combustion of fuels are usually given for a temperature of 25°C and a pressure of 0.1 MPa. The enthalpy of combustion for a number of hydrocarbon fuels at this temperature and pressure, which we designate $h_{RP_0}$, is given in Table 12.4.

The internal energy of combustion is defined in a similar manner.

$$\bar{u}_{RP} = U_P - U_R$$

$$= \sum_P n_e [\bar{h}_f^\circ + \Delta \bar{h} - P\bar{v}]_e - \sum_R n_i [\bar{h}_f^\circ + \Delta \bar{h} - P\bar{v}]_i \qquad (12.10)$$

When all the gaseous constituents can be considered as ideal gases, and the volume of the liquid and solid constituents is negligible compared to the value of the gaseous constituents, this relation for $\bar{u}_{RP}$ reduces to

$$\bar{u}_{RP} = \bar{h}_{RP} - \bar{R}T \, (n_{\text{gaseous products}} - n_{\text{gaseous reactants}}) \qquad (12.11)$$

Frequently the term "heating value" or "heat of reaction" is used. This represents the heat transferred from the chamber during combustion or reaction at constant temperature. In the case of a constant pressure or steady-flow process, we conclude from the first law of thermodynamics that this is equal to the negative of the enthalpy of combustion. For this reason this heat transfer is sometimes designated the constant-pressure heating value for combustion processes.

In the case of a constant volume process the heat transfer is equal to the negative of the internal energy of combustion. This is sometimes designated the constant-volume heating value in the case of combustion.

When the term heating value is used the terms "higher" and "lower" heating value are used. The higher heating value is the heat transfer with liquid $H_2O$ in the products, and the lower heating value is the heat transfer with vapor $H_2O$ in the products.

## Example 12.11

Calculate the enthalpy of combustion of propane at 25°C on both a kmol and kg basis under the following conditions.

1. Liquid propane with liquid $H_2O$ in the products.
2. Liquid propane with gaseous $H_2O$ in the products.
3. Gaseous propane with liquid $H_2O$ in the products.
4. Gaseous propane with gaseous $H_2O$ in the products.

This example is designed to show how enthalpy of combustion can be determined from enthalpies of formation.

The enthalpy of evaporation of propane is 370 kJ/kg.

The basic combustion equation is:

$$C_3H_8 + 5O_2 \rightarrow 3CO_2 + 4H_2O$$

From Table 12.3, $(\bar{h}_f^\circ)_{C_3H_8(g)} = -103\ 847$ kJ/kmol. Therefore,

$$(\bar{h}_f^\circ)_{C_3H_8(l)} = -103\ 847 - 44.097\ (370) = -120\ 163 \text{ kJ/kmol}$$

1. Liquid propane − liquid $H_2O$

$$\bar{h}_{RP_0} = 3(\bar{h}_f^\circ)_{CO_2} + 4(\bar{h}_f^\circ)_{H_2O(l)} - (\bar{h}_f^\circ)_{C_3H_8(l)}$$

$$= 3(-393\ 522) + 4(-285\ 838) - (-120\ 163)$$

$$= -2\ 203\ 755 \text{ kJ/kmol} = -\frac{2\ 203\ 755}{44.097} = -49\ 975 \text{ kJ/kg}$$

## Table 12.4
### Enthalpy of Combustion of Some Hydrocarbons at 25°C

| Hyrocarbon | Formula | Liquid $H_2O$ in Products (Negative of Higher Heating Value) | | Vapor $H_2O$ in Products (Negative of Lower Heating Value) | |
|---|---|---|---|---|---|
| | | Liquid Hydro-carbon, kJ/kg fuel | Gaseous Hydro-carbon, kJ/kg fuel | Liquid Hydro-carbon, kJ/kg fuel | Gaseous Hydro-carbon, kJ/kg fuel |
| *Paraffin Family* | | | | | |
| Methane | $CH_4$ | | −55 496 | | −50 010 |
| Ethane | $C_2H_6$ | | −51 875 | | −47 484 |
| Propane | $C_3H_8$ | −49 975 | −50 345 | −45 983 | −46 353 |
| Butane | $C_4H_{10}$ | −49 130 | −49 500 | −45 344 | −45 714 |
| Pentane | $C_5H_{12}$ | −48 643 | −49 011 | −44 983 | −45 351 |
| Hexane | $C_6H_{14}$ | −48 308 | −48 676 | −44 733 | −45 101 |
| Heptane | $C_7H_{16}$ | −48 071 | −48 436 | −44 557 | −44 922 |
| Octane | $C_8H_{18}$ | −47 893 | −48 256 | −44 425 | −44 788 |
| Decane | $C_{10}H_{22}$ | −47 641 | −48 000 | −44 239 | −44 598 |
| Dodecane | $C_{12}H_{26}$ | −47 470 | −47 828 | −44 109 | −44 467 |
| *Olefin Family* | | | | | |
| Ethene | $C_2H_4$ | | −50 296 | | −47 158 |
| Propene | $C_3H_6$ | | −48 917 | | −45 780 |
| Butene | $C_4H_8$ | | −48 453 | | −45 316 |
| Pentene | $C_5H_{10}$ | | −48 134 | | −44 996 |
| Hexene | $C_6H_{12}$ | | −47 937 | | −44 800 |
| Heptene | $C_7H_{14}$ | | −47 800 | | −44 662 |
| Octene | $C_8H_{16}$ | | −47 693 | | −44 556 |
| Nonene | $C_9H_{18}$ | | −47 612 | | −44 475 |
| Decene | $C_{10}H_{20}$ | | −47 547 | | −44 410 |
| *Alkylbenzene Family* | | | | | |
| Benzene | $C_6H_6$ | −41 831 | −42 266 | −40 141 | −40 576 |
| Methylbenzene | $C_7H_8$ | −42 437 | −42 847 | −40 527 | −40 937 |
| Ethylbenzene | $C_8H_{10}$ | −42 997 | −43 395 | −40 924 | −41 322 |
| Propylbenzene | $C_9H_{12}$ | −43 416 | −43 800 | −41 219 | −41 603 |
| Butylbenzene | $C_{10}H_{14}$ | −43 748 | −44 123 | −41 453 | −41 828 |

The higher heating value of liquid propane is therefore 49 975 kJ/kg.

**2.** Liquid propane – gaseous $H_2O$

$$\bar{h}_{RP_0} = 3(\bar{h}_f^\circ)_{CO_2} + 4(\bar{h}_f^\circ)_{H_2O(g)} - (\bar{h}_f^\circ)_{C_3H_8(l)}$$

$$= 3(-393\ 522) + 4(-241\ 827) - (-120\ 163)$$

$$= -2\ 027\ 711\ \text{kJ/kmol} = -\frac{2\ 027\ 711}{44.097} = -45\ 983\ \text{kJ/kg}$$

The lower heating value of liquid propane is 45 983 kJ/kg.

**3.** Gaseous propane – liquid $H_2O$

$$\bar{h}_{RP_0} = 3(\bar{h}_f^\circ)_{CO_2} + 4(\bar{h}_f^\circ)_{H_2O(l)} - (\bar{h}_f^\circ)_{C_3H_8(g)}$$

$$= 3(-393\ 522) + 4(-285\ 838) - (-103\ 847)$$

$$= -2\ 220\ 071\ \text{kJ/kmol} = -\frac{2\ 220\ 071}{44.097} = -50\ 345\ \text{kJ/kg}$$

The higher heating value of gaseous propane is 50 345 kJ/kg.

**4.** Gaseous propane – gaseous $H_2O$

$$\bar{h}_{RP_0} = 3(\bar{h}_f^\circ)_{CO_2} + 4(\bar{h}_f^\circ)_{H_2O(g)} - (\bar{h}_f^\circ)_{C_3H_8(g)}$$

$$= 3(-393\ 522) + 4(-241\ 827) - (-103\ 847)$$

$$= -2\ 044\ 027\ \text{kJ/kmol} = -\frac{2\ 044\ 027}{44.097} = -46\ 353\ \text{kJ/kg}$$

The lower heating value of gaseous propane is 46 353 kJ/kg. Each of the four values calculated in this example corresponds to the appropriate value given in Table 12.4.

**Example 12.12**

Calculate the enthalpy of combustion of gaseous propane at 500 K. (At this temperature all the water formed during combustion will be in the vapor phase.) This example will demonstrate how the enthalpy of combustion of propane varies with temperature. The average constant-pressure specific heat of propane between 25°C and 500 K is 2.1 kJ/kg K.

$$(\bar{h}_{RP})_T = \sum_P n_e[\bar{h}_f^\circ + \Delta\bar{h}]_e - \sum_R n_i[\bar{h}_f^\circ + \Delta\bar{h}]_i$$

$$C_3H_8(g) + 5O_2 \rightarrow 3CO_2 + 4H_2O(g)$$

$$\bar{h}_{R_{500}} = [\bar{h}_f^\circ + \bar{C}_{pav}(\Delta T)]_{C_3H_8(g)} + n_{O_2}(\Delta\bar{h})_{O_2}$$

$$= -103\ 847 + 2.1 \times 44.097\ (500 - 298.2) + 5(6\ 088)$$

$$= -54\ 720 \text{ kJ}$$

$$\bar{h}_{P_{500}} = n_{CO_2}[\bar{h}_f^\circ + \Delta\bar{h}]_{CO_2} + n_{H_2O}[\bar{h}_f^\circ + \Delta\bar{h}]_{H_2O}$$

$$= 3\,(-393\ 522 + 8\ 314) + 4(-241\ 827 + 6\ 920)$$

$$= -2\ 095\ 252 \text{ kJ}$$

$$\bar{h}_{RP_{500}} = -2\ 095\ 252 - (-54\ 720) = -2\ 040\ 532 \text{ kJ/kmol}$$

$$h_{RP_{500}} = \frac{-2\ 040\ 532}{44.097} = -46\ 274 \text{ kJ/kg}$$

This compares with a value of $-46\ 353$ at 25°C.

This problem could also have been solved using the given value of the enthalpy of combustion at 25°C by noting that

$$(\bar{h}_{RP})_{500} = (H_P)_{500} - (H_R)_{500}$$

$$= n_{CO_2}[\bar{h}_f^\circ + \Delta\bar{h}]_{CO_2} + n_{H_2O}[\bar{h}_f^\circ + \Delta\bar{h}]_{H_2O}$$

$$- [\bar{h}_f^\circ + \bar{C}_{pav}(\Delta T)]_{C_3H_8(g)} - n_{O_2}(\Delta\bar{h})_{O_2}$$

$$= \bar{h}_{RP_0} + n_{CO_2}\Delta\bar{h}_{CO_2} + n_{H_2O}\Delta\bar{h}_{H_2O}$$

$$- \bar{C}_{pav}(\Delta T)_{C_3H_8(g)} - n_{O_2}\Delta\bar{h}_{O_2}$$

$$(\bar{h}_{RP_0})_{500} = -46\ 353 \times 44.097 + 3(8\ 314) + 4(6\ 920)$$

$$- 2.1 \times 44.097\ (500 - 298.2) - 5(6\ 088)$$

$$= -2\ 040\ 532 \text{ kJ/kmol}$$

$$(h_{RP_0})_{500} = \frac{-2\ 040\ 532}{44.097} = -46\ 274 \text{ kJ/kg}$$

## 12.7   The Third Law of Thermodynamics and Absolute Entropy

As we consider a second law analysis of chemical reactions, we face the same problem we did in regard to the first law, namely, what base shall we use for the entropy of the various substances? This question leads directly to a consideration of the third law of thermodynamics. The third law of thermodynamics was formulated during the early part of the twentieth century. The initial work was done primarily by W. H. Nernst (1864–1941) and Max Planck (1858–1947). The third law deals with the entropy of substances at the absolute zero of temperature, and in essence states that the entropy of a perfect crystal is zero at the absolute zero of temperature. From a statistical point of view this means that the crystal structure is such that it has the maximum degree of order. Further, since the temperature is absolute zero, the thermal energy is minimum. It also follows that a substance that does not have a perfect crystalline structure at absolute zero, but instead has a degree of randomness, such as a solid solution or a glassy solid, has a finite value of entropy at absolute zero. The experimental evidence on which the third law rests is primarily data on chemical reactions at low temperatures and measurements of heat capacity at temperatures approaching absolute zero. It should be noted that in contrast to the first and second laws, which lead respectively to the properties of internal energy and entropy, the third law deals only with the question of entropy at absolute zero. However, the implications of the third law are quite profound, particularly in regard to chemical equilibrium.

For our consideration at this point, the particular relevance of the third law is that it provides an absolute base from which the entropy of each substance can be measured. The entropy relative to this base is termed the absolute entropy. The increase in entropy between absolute zero and any given state can be found either from calorimetric data or by procedures based on statistical thermodynamics. The calorimetric method involves precise measurements of specific heat data over the temperature range, as well as the energy associated with phase transformations. These results are in agreement with the calculations based on statistical thermodynamics and observed molecular data.

Table 12.3 gives the absolute entropy at 25°C and 0.1 MPa pressure for a number of substances. Table A.11 gives the absolute entropy for a number of gases at 0.1 MPa pressure and various temperatures. In the case of gases, ideal gas behavior is assumed at 0.1 MPa pressure in all these tables. If the saturation pressure at a given temperature is less than 0.1 MPa pressure, the ideal gas data at 0.1 MPa pressure and this given temperature refers to a state that does not exist under equilibrium conditions. However, by applying ideal gas relations to the data as given, the absolute entropy at pressures less than 0.1 MPa is readily found. From one of these states, the absolute entropy in any other is readily found by procedures that are outlined in a subsequent paragraph. The absolute entropy at 0.1 MPa pressure as given in these tables, is designated $\bar{s}^\circ$. The temperature is designated with a subscript such as $\bar{s}^\circ_{1000}$.

With the value of the absolute entropy known as a function of temperature at 0.1 MPa pressure, as given in Table A.11, how does one find the absolute entropy at any other state? Let us first consider the case in which the assumption of ideal gas behavior at 0.1 MPa pressure yields no significant error. Actually, this assumption is valid for most of the examples and problems given in this book, as well as most problems encountered in professional practice. In this case, we can write

$$\bar{s}_{T,P} = \bar{s}_T^\circ + (\Delta \bar{s})_{T,0.1 \text{ MPa} \to T,P} \tag{12.12}$$

where $\bar{s}_T^\circ$ refers to the absolute entropy at 0.1 MPa and temperature $T$, and $(\Delta \bar{s})_{T,0.1\text{MPa} \to T,P}$ refers to the change of entropy along the isotherm $T$ as the pressure changes from 0.1 MPa to pressure $P$. This is shown schematically in Fig. 12.4.

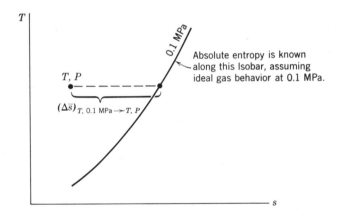

**Fig. 12.4**  Sketch showing the procedure for the calculation of absolute enthalpy when ideal gas assumption at 0.1 MPa pressure yields no significant error.

The quantity $(\Delta \bar{s})_{T,0.1\text{MPa} \to T,P}$ can be found in various ways. If the assumption of ideal gas behavior along this isotherm yields no significant error, this change of entropy can be found from the ideal gas relation.

$$\bar{s}_2 - \bar{s}_1 = -\bar{R}\ln \frac{P_2}{P_1} \tag{12.13}$$

If $P_2$ refers to the pressure in the given state $T,P$, and if this pressure is expressed in MPa, this relation reduces to

$$(\Delta \bar{s})_{T,\ 0.1\ \text{MPa} \longrightarrow T,P} = -\bar{R}\ln{(P/0.1)} \tag{12.14}$$

When the assumption of ideal gas behavior along the isotherm yields significant error, there are several procedures that can be used. If we have a table of thermodynamic properties that covers the change of state along the isotherm, this change of entropy can be found directly from the tables. If we have an equation of state that is valid in this range, we can follow the procedures outlined in Section 10.4 to find the change of entropy along the isotherm. One can also use the generalized charts, as presented in Section 10.13 to find this change of entropy.

If the assumption of ideal gas behavior is not valid at 0.1 MPa pressure, one must first apply ideal gas relations to find the absolute entropy at a pressure that is low enough to make the assumption of ideal gas behavior valid. From this state, one can use the same procedures outlined above to find the absolute entropy in any other state. This procedure is shown schematically in Fig. 12.5.

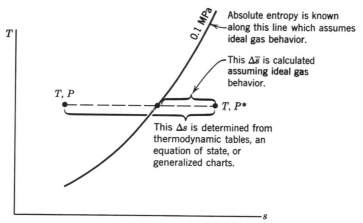

**Fig. 12.5**   Sketch showing procedure for calculation of absolute entropy when the assumption of ideal gas behavior at 0.1 MPa pressure is not valid.

If the absolute entropy is known at only one temperature, such as 25°C and 0.1 MPa, we would need appropriate specific heat data, as well as an equation of state to find the absolute entropy in other states.

In the case of mixtures of ideal gases, the absolute entropy of each component in the mixture can be found at the given temperature and pressure at which it exists in the mixture. Suppose, for example, that the mole fraction of oxygen in a mixture of ideal gases is 0.1, that the temperature of the mixture is 1000 K, and that the total pressure is 0.4 MPa. What is the absolute entropy of the oxygen as it exists in the mixture?

Since the mole fraction is 0.1 and the total pressure is 0.4 MPa, the partial pressure of the oxygen is 0.04 MPa. Therefore, from Eqs. 12.12 and 12.14,

$$\bar{s} = \bar{s}_T^{\circ} + (\Delta \bar{s})_{T,\, 0.1\ \text{MPa} \to T,P}$$
$$= \bar{s}_T^{\circ} - \bar{R} \ln P/0.1$$
$$= 243.585 - 8.31434 \ln 0.4 = 251.203 \ \text{kJ/kmol K}$$

In the case of a real gas mixture, the ideal solution model might be used, or more conveniently, a pseudocritical method such as Kay's method could be used to evaluate the non-ideal contribution to entropy and this value added to the ideal gas mixture entropy.

## 12.8   Second Law Analysis of Reacting Systems

The concepts of reversible work, irreversibility, and availability were introduced in Chapter 8. These concepts involved both the first and second laws of thermodynamics. At the conclusion of Chapter 8 the Gibbs function was introduced and the statement was made that this property would be particularly relevant in dealing with chemical reactions. We proceed now to develop this matter further, and we shall be particularly concerned with determining the maximum work (availability) that can be done through a combustion process and with examining the irreversibilities associated with such processes.

The reversible work, Eq. 8.9, for a steady-state, steady-flow process, in the absence of changes in kinetic and potential energy is,

$$W_{\text{rev}} = \sum m_i(h_i - T_0 s_i) - \sum m_e(h_e - T_0 s_e)$$

Applying this to an SSSF process involving a chemical reaction, and introducing the symbols we have been using in this chapter we have

$$W_{\text{rev}} = \sum_R n_i[\bar{h}_f^{\circ} + \Delta \bar{h} - T_0 \bar{s}]_i - \sum_P n_e[\bar{h}_f^{\circ} + \Delta \bar{h} - T_0 \bar{s}]_e \qquad (12.15)$$

Similarly, the irreversibility for such a process becomes

$$I = \sum_P n_e T_0 \bar{s}_e - \sum_R n_i T_0 \bar{s}_i - Q_{\text{c.v.}} \qquad (12.16)$$

The availability, $\Psi$, in the absence of kinetic and potential energy changes, for an SSSF process was defined, Eq. 8.16, as

$$\Psi = (h - T_0 s) - (h_0 - T_0 s_0)$$

It was also indicated in Chapter 8 that when an SSSF chemical reaction takes place in such a manner that both the reactants and products are in temperature equilibrium with the surroundings, the Gibbs function $(g = h - Ts)$ becomes a significant variable, Eq. 8.28. For such a process, in the absence of changes in kinetic and potential energy, the reversible work is given by the relation

$$W_{rev} = \sum_R n_i \bar{g}_i - \sum_P n_e \bar{g}_e \qquad (12.17)$$

Since the Gibbs function is so relevant for processes involving chemical reactions the Gibbs function of formation, $\bar{g}_f^\circ$, has been defined by a procedure similar to that used in defining the enthalpy of formation. That is, the Gibbs function of each of the elements at 25°C and 0.1 MPa pressure is assumed to be zero, and the Gibbs function of each substance is found relative to this base. Table 12.3 lists the Gibbs function of formation of a number of substances at 25°C and 0.1 MPa pressure. It is also evident that the Gibbs function can be found directly from data for $\bar{h}_f^\circ$ and $\bar{s}^\circ$ at the given temperature, as indicated in the following example.

## Example 12.13

Determine the Gibbs function of formation of $CO_2$.
Consider the reaction

$$C + O_2 \rightarrow CO_2$$

Assume that the carbon and oxygen are each initially at 25°C and 0.1 MPa pressure, and that the $CO_2$ is finally at 25°C and 0.1 MPa pressure. The change in Gibbs function for this reaction is found first.

$$G_P - G_R = (H_P - H_R) - T_0(S_P - S_R)$$

$$\sum_P n_e(\bar{g}_f^\circ)_e - \sum_R n_i(\bar{g}_f^\circ)_i$$

$$= \sum_P n_e(\bar{h}_f^\circ)_e - \sum_R n_i(\bar{h}_f^\circ)_i - T_0 \left[ \sum_P n_e(\bar{s}_{298}^\circ)_e - \sum_R n_i(\bar{s}_{298}^\circ)_i \right]$$

Therefore,

$$G_P - G_R = (\bar{h}_f^\circ)_{CO_2} - 298.15 [(\bar{s}_{298}^\circ)_{CO_2} - (\bar{s}_{298}^\circ)_C - (\bar{s}_{298}^\circ)_{O_2}]$$
$$= -393\ 522 - 298.15\ (213.795 - 5.795 - 205.142)$$
$$= -394\ 374 \text{ kJ/kmol}$$

Since the Gibbs function of the reactants, $G_R$, is zero (in accordance with the assumption that the Gibbs function of the elements is zero at 25°C and 0.1 MPa pressure), it follows that

$$G_P = (\bar{g}_f^\circ)_{CO_2} = -394\ 374 \text{ kJ/kmol}$$

This is the value given in Table 12.3.

Let us now consider the question of the maximum work that can be done during a chemical reaction. For example, consider one kmol of hydrocarbon fuel and the necessary air for complete combustion, each at 0.1 MPa pressure and 25°C, the pressure and temperature of the surroundings. What is the maximum work that can be done as this fuel reacts with the air? From our considerations in Chapter 8 we conclude that the maximum work would be done if this chemical reaction took place reversibly and the products were finally in pressure and temperature equilibrium with the surroundings. From Eq. 12.17 we conclude this reversible work could be calculated from the relation

$$W_{rev} = \sum_R n_i \bar{g}_i - \sum_P n_e \bar{g}_e$$

However, since the final state is in equilibrium with the surroundings we could consider this amount of work to be the availability of the fuel and air.

**Example 12.14**

Ethene ($g$) at 25°C and 0.1 MPa pressure is burned with 400 per cent theoretical air at 25°C and 0.1 MPa pressure. Assume that this reaction takes place reversibly at 25°C and that the products leave at 25°C and 0.1 MPa pressure. To further simplify this problem assume that the oxygen and nitrogen are separated before the reaction takes place (each at 0.1 MPa, 25°C), that the constituents in the products are separated, and that each is at 25°C and 0.1 MPa. Thus, the reaction takes place as shown in Fig. 12.6. For purposes of comparison between this and the two subsequent examples, we consider all of the $H_2O$ in the products to be a gas (a hypothetical situation in this example and Example 12.16).

Determine the reversible work for this process (i.e., the work that would be done if this chemical action took place reversibly and isothermally).

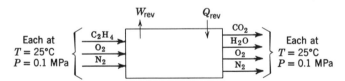

**Fig. 12.6**    Sketch for Example 12.14.

The equation for this chemical reaction is

$$C_2H_4(g) + 3(4)O_2 + 3(4)(3.76)N_2 \rightarrow 2CO_2 + 2H_2O(g) + 9O_2 + 45.1N_2$$

The reversible work for this process is equal to the decrease in Gibbs function during this reaction, Eq. 12.17. The values for the Gibbs function can be taken directly from Table 12.3, since each is at 25°C and 0.1 MPa pressure. From Eq. 12.17

$$W_{rev} = \sum_R n_i \bar{g}_i - \sum_P n_e \bar{g}_e$$

Since each of the reactants and each of the products are at 0.1 MPa pressure and 25°C this reduces to

$$\begin{aligned}
W_{rev} &= (\bar{g}_f^\circ)_{C_2H_4} - 2(\bar{g}_f^\circ)_{CO_2} - 2(\bar{g}_f^\circ)_{H_2O(g)} \\
&= 68\ 207 - 2(-394\ 374) - 2(-228\ 583) \\
&= 1\ 314\ 121 \text{ kJ/kmol } C_2H_4 \\
&= \frac{1\ 314\ 121}{28.054} = 46\ 843 \text{ kJ/kg}
\end{aligned}$$

Therefore we might say that when one kg of ethene is at 25°C, 0.1 MPa pressure, the temperature and pressure of the surroundings, it has an availability of 46 843 kJ.

Thus it would seem logical to rate the efficiency of a device designed to do work by utilization of a combustion process, such as an internal-combustion engine or a steam power plant, as the ratio of actual work to the decrease in Gibbs function for the given reaction, rather than to the heating value, as is current practice. However, as is evident from the preceding example, the difference between the decrease in Gibbs function and the heating value is small for hydrocarbon fuels and the efficiency defined in terms of heating value is essentially equal to that defined in terms of decrease in Gibbs function.

It is of particular interest to study the irreversibility that takes place during a combustion process. The following examples are presented to illustrate this matter, where we consider the same hydrocarbon fuel that was used in Example 12.14, namely Ethene (g) at 25°C and 0.1 MPa. We determined the availability and found that it was 46 843 kJ/kg. Now let us burn this fuel with 400 per cent theoretical air in a steady-state, steady-flow adiabatic process. We can determine the irreversibility of this process in two ways. The first is to calculate the increase in entropy for the process. Since the process is adiabatic, the increase in entropy is due entirely to the irreversibilities for the process and we can find the irreversibility from Eq. 12.16. We can also calculate the availability of the products of combustion at the adiabatic flame temperature, and note that they are less than the availability of the fuel and air before the combustion process, the difference being the irreversibility that occurs during the combustion process.

**Example 12.15**

Consider the same combustion process as in Example 12.14, but let it take place adiabatically, and assume that each constituent in the products is at 0.1 MPa pressure and at the adiabatic flame temperature. This combustion process is shown schematically in Fig. 12.7. The temperature of the surroundings is 25°C.

**Fig. 12.7**   Sketch for Example 12.15.

For this combustion process determine: (1) The increase in entropy during combustion. (2) The availability of the products of combustion.

The combustion equation is

$$C_2H_4 + 12O_2 + 12(3.76)N_2 \rightarrow 2CO_2 + 2H_2O(g) + 9O_2 + 45.1N_2$$

The adiabatic flame temperature is determined first.

$$H_R = H_P$$

$$\sum_R n_i(\bar{h}_f^\circ)_i = \sum_P n_e[\bar{h}_f^\circ + \Delta\bar{h}]_e$$

$$52\ 283 = 2[-393\ 522 + \Delta\bar{h}_{CO_2}] + 2[-241\ 827 + \Delta\bar{h}_{H_2O(g)}]$$

$$+ 9\Delta\bar{h}_{O_2} + 45.1\Delta\bar{h}_{N_2}$$

By a trial and error solution we find the adiabatic flame temperature to be 1016 K.

We now proceed to find the change in entropy during this adiabatic combustion process.

$$S_R = \sum_R (n_i\bar{s}_i^\circ)_{298} = (\bar{s}_{C_2H_4}^\circ + 12\bar{s}_{O_2}^\circ + 45.1\bar{s}_{N_2}^\circ)_{298}$$

$$= 219.548 + 12(205.142) + 45.1\ (191.611)$$

$$= 11\ 322.908 \text{ kJ/kmol (fuel) K}$$

$$S_P = \sum_P (n_e \bar{s}_e^\circ)_{1016} = (2\bar{s}_{CO_2}^\circ + 2\bar{s}_{H_2O(g)}^\circ + 9\bar{s}_{O_2}^\circ + 45.1\bar{s}_{N_2}^\circ)_{1016}$$

$$= 2(270.162) + 2(233.344) + 9(244.120) + 45.1(228.670)$$

$$= 13\ 517.109 \text{ kJ/kmol (fuel) K}$$

$$S_P - S_R = 2194.201 \text{ kJ/kmol (fuel) K}$$

Since this is an adiabatic process, the increase in entropy indicates the irreversibility of the adiabatic combustion process. This irreversibility can be found from Eq. 12.16.

$$I = T_0 \left[ \sum_P n_e \bar{s}_e - \sum_R n_i \bar{s}_i \right]$$

$$= 298.15 \times 2194.201 = 654\ 201 \text{ kJ/kmol}$$

$$= \frac{654\ 201}{28.054} = 23\ 319 \text{ kJ/kg fuel}$$

Therefore, the availability after the combustion process is

$$\Psi_P = 46\ 843 - 23\ 319 = 23\ 524 \text{ kJ/kg}$$

The availability of the products can also be found from the relation

$$\Psi_P = \sum_P \left[ (\bar{h}_e - T_0 \bar{s}_e) - (\bar{h}_0 - T_0 \bar{s}_0) \right]$$

Since in this problem the products are separated and each is at 0.1 MPa pressure and the adiabatic flame temperature of 1016 K, this reduces to

$$\Psi_P = \sum_P n_e [ (\bar{h}_e^\circ - \bar{h}_0^\circ) - T_0 (\bar{s}_e^\circ - \bar{s}_0^\circ) ]$$

$$= 2(34\ 283) + 2(26\ 648) + 9(23\ 269) + 45.1\ (21\ 988)$$

$$- 298.15\ [2(270.162 - 213.795) + 2(233.344 - 188.833)$$

$$+ 9(244.120 - 205.142) + 45.1\ (228.670 - 191.611)]$$

$$= 659\ 942 \text{ kJ/kmol} = 23\ 524 \text{ kJ/kg}$$

In other words, if every process after the adiabatic combustion process were reversible, the maximum amount of work that could be done is

23 524 kJ/kg fuel. This compares to a value of 46 843 kJ/kg for the reversible isothermal reaction. This means that if we had an engine that had the indicated adiabatic combustion process, and if all other processes were completely reversible, the efficiency would be about 50 per cent.

In the two prior examples we made the assumption, for purposes of simplifying the calculation, that the constituents in the reactants and products were separated, and each was at 0.1 MPa pressure. This of course is not a realistic problem, and in the following example, Example 12.14 is repeated with the assumption that the reactants and products each consist of a mixture at 0.1 MPa pressure.

### Example 12.16

Consider the same combustion process of Example 12.14, but assume that the reactants consist of a mixture at 0.1 MPa pressure and 25°C and that the products also consist of a mixture at 0.1 MPa and 25°C. Thus, the combustion process is as shown in Fig. 12.8. Assume each constituent to be an ideal gas.

Reactants — $P = 0.1$ MPa, $T = 25°C$

Products — $P = 0.1$ MPa, $T = 25°C$

**Fig. 12.8** Sketch for Example 12.16.

Determine the work that would be done if this combustion process took place reversibly and in pressure and temperature equilibrium with the surroundings.

The combustion equation, as noted previously, is

$$C_2H_4(g) + 3(4)O_2 + 3(4)(3.76)N_2 \rightarrow 2CO_2 + 2H_2O(g) + 9O_2 + 45.1N_2$$

In this case we must find the entropy of each substance as it exists in the mixture; i.e., at its partial pressure and the given temperature of 25°C. Since the absolute entropies given in Tables 12.3 and A.11 are at 0.1 MPa pressure and 25°C, the entropy of each constituent in the mixture can be found, using Eq. 12.12, from the relation

$$\bar{s} - \bar{s}° = -\bar{R} \ln y \frac{P}{P°}$$

where $\bar{s}$ = entropy of the constituent in the mixture

$\bar{s}°$ = absolute entropy at the same temperature and 0.1 MPa pressure

$P$ = pressure of the mixture

$P°$ = 0.1 MPa pressure

$y$ = mole fraction of the constituent

Since $P°$ and the pressure of the mixture are both 0.1 MPa pressure, the entropy of each constituent can be found by the relation

$$\bar{s} = \bar{s}° - \bar{R}\ln y = \bar{s}° + R\ln\frac{1}{y}$$

For the reactants:

| | $n$ | $1/y$ | $\bar{R}\ln 1/y$ | $\bar{s}°$ | $\bar{s}$ |
|---|---|---|---|---|---|
| $C_2H_4$ | 1 | 58.1 | 33.774 | 219.548 | 253.322 |
| $O_2$ | 12 | 4.842 | 13.114 | 205.142 | 218.256 |
| $N_2$ | 45.1 | 1.288 | 2.104 | 191.611 | 193.715 |
| | 58.1 | | | | |

For the products:

| | $n$ | $1/y$ | $\bar{R}\ln 1/y$ | $\bar{s}°$ | $\bar{s}$ |
|---|---|---|---|---|---|
| $CO_2$ | 2 | 29.05 | 28.011 | 213.795 | 241.806 |
| $H_2O$ | 2 | 29.05 | 28.011 | 188.833 | 216.844 |
| $O_2$ | 9 | 6.456 | 15.506 | 205.142 | 220.648 |
| $N_2$ | 45.1 | 1.288 | 2.104 | 191.611 | 193.715 |
| | 58.1 | | | | |

With the assumption of ideal gas behavior, the enthalpy of each constituent is equal to the enthalpy of formation at 25°C. The values of entropy are as calculated above. Therefore, from Eq. 12.15

$$W_{rev} = \sum_R n_i(\bar{h}_f°)_i - \sum_P n_e(\bar{h}_f°)_e - T_0\left[\sum_R n_i\bar{s}_i - \sum_P n_e\bar{s}_e\right]$$

$$= (\bar{h}_f°)_{C_2H_4} - 2(\bar{h}_f°)_{CO_2} - 2(\bar{h}_f°)_{H_2O(g)}$$

$$- 298.15\left[\bar{s}_{C_2H_4} + 12\bar{s}_{O_2} + 45.1\bar{s}_{N_2} - 2\bar{s}_{CO_2} - 2\bar{s}_{H_2O(g)} - 9\bar{s}_{O_2} - 45.1\bar{s}_{N_2}\right]$$

$$= 52\ 283 - 2(-393\ 522) - 2(-241\ 827)$$

$$-298.15\ [253.322 + 12(218.256) + 45.1(193.715)$$

$$-2(241.806) - 2(216.844) - 9(220.648) - 45.1(193.715)]$$

$$= 1\ 332\ 145 \text{ kJ/kmol}$$

$$= \frac{1\ 332\ 145}{28.054} = 47\ 485 \text{ kJ/kg}$$

Note that this is essentially the same value that was obtained in Example 12.14, when the reactants and products were each separated and at 0.1 MPa pressure.

These examples raise the question of the possibility of a reversible chemical reaction. Some reactions can be made to approach reversibility by having them take place in an electrolytic cell, as described in Chapter 1. When a potential exactly equal to the electromotive force of the cell is applied, no reaction takes place. When the applied potential is increased slightly, the reaction proceeds in one direction, and if the applied potential is decreased slightly, the reaction proceeds in the opposite direction. The work involved is the electrical energy supplied or delivered.

Consider a reversible reaction occuring at constant temperature equal to that of its environment. The work output of the fuel cell would then be

$$W = -\left[ \Sigma n_e \bar{g}_e - \Sigma n_i \bar{g}_i \right] = -\Delta G$$

where the $\Delta G$ is the change in Gibbs function for the overall chemical reaction. We also realize that the work is given in terms of the charged electrons flowing through an electrical potential $\mathscr{E}$ as

$$W = \mathscr{E} n_e N_0 e$$

in which $n_e$ is the number of moles of electrons flowing through the external circuit, and

$$N_0 e = 6.02217 \times 10^{26} \text{ elec/kmol} \times 1.6022 \times 10^{-22} \text{ kJ//elec V}$$

$$= 96\ 487 \text{ kJ/kmol V}$$

Thus, for a given reaction, the maximum (reversible reaction) electrical potential $\mathscr{E}^{\circ}$ of a fuel cell at a given temperature is

$$\mathscr{E}^{\circ} = \frac{-\Delta G}{96\ 487\ n_e} \tag{12.18}$$

**Example 12.17**

Calculate the reversible EMF at 25°C for the hydrogen-oxygen fuel cell described in Section 1.2.

The anode side reaction was stated to be

$$2H_2 \rightarrow 4H^+ + 4e^-$$

and that at the cathode is

$$4H^+ + 4e^- + O_2 \rightarrow 2H_2O$$

Therefore, the overall reaction is, in kmol,

$$2H_2 + O_2 \rightarrow 2H_2O$$

for which 4 kmol of electrons flow through the external circuit. Let us assume that each component is at its standard-state pressure of 0.1 MPa and that the water formed is liquid. Then

$$\Delta G_{25C} = 2\bar{g}^{\circ}_{f_{H_2O}} - 2\bar{g}^{\circ}_{f_{H_2}} - \bar{g}^{\circ}_{f_{O_2}}$$
$$= 2(-237\ 178) - 2(0) - 1(0)$$
$$= -474\ 356 \text{ kJ}$$

Therefore, from Eq. 12.18,

$$\mathscr{E}^{\circ} = \frac{-(-474\ 356)}{96\ 487 \times 4} = 1.229 \text{ V}$$

Much effort is being directed toward the development of fuel cells in which carbon, hydrogen, or hydrocarbons will react with oxygen and produce electricity directly. If a fuel cell can be developed that utilizes a hydrocarbon fuel and has a sufficiently high efficiency and capacity (for a given volume or weight) drastic changes will be possible in our techniques for generating electricity on a commercial scale. At the present time, however, fuel cells are not competitive with conventional power plants for the production of electricity on a large scale.

## 12.9   Evaluation of Actual Combustion Processes

In evaluating the performance of an actual combustion process a number of different parameters can be defined, depending on the nature of the process and the system considered. In the combustion chamber of a gas turbine, for example, the objective is to raise the temperature of the products to a given temperature (usually the maximum temperature the metals in the turbine can withstand). If we had a combustion process in which complete combustion was achieved and which was adiabatic, the temperature of the products would be the adiabatic flame temperature. Let us designate the fuel-air ratio needed to reach a given temperature under these conditions as the ideal fuel-air ratio. In the actual combustion chamber the combustion will be incomplete to some extent and there will be some heat transfer to the surroundings. Therefore more fuel will be required to reach the given temperature, and this we designate as the actual fuel-air ratio. In this case, the combustion efficiency, $\eta_{comb}$, is defined as

$$\eta_{comb} = \frac{FA_{ideal}}{FA_{actual}} \tag{12.19}$$

On the other hand, in the furnace of a steam generator (boiler) the purpose is to transfer the maximum possible amount of heat to the steam (water). In practice, the efficiency of a steam generator is defined as the ratio of the heat transferred to the steam to the higher heating value of the fuel. For a coal this is the heating value as measured in a bomb calorimeter, which is the constant-volume heating value, and it corresponds to the internal energy of combustion. One observes a minor inconsistency, since the boiler involves a flow process, and the change in enthalpy is the significant factor. However, in most cases the error thus introduced is less than the experimental error involved in measuring the heating value, and the efficiency of a steam generator is defined by the relation,

$$\eta_{steam\ generator} = \frac{\text{heat transferred to steam/kg fuel}}{\text{higher heating value of the fuel}} \tag{12.20}$$

In an internal combustion engine the purpose is to do work. The logical way to evaluate the performance of an internal-combustion engine would be to compare the actual work done to the maximum work which would be done by a reversible change of state from the reactants to the products. This, as we noted previously is equal to the decrease in Gibbs function $(h - Ts)$.

However, in practice the efficiency of an internal combustion engine is defined as the ratio of the actual work to the negative of the enthalpy of combustion of the fuel (i.e., the constant-pressure heating value). This is usually called the thermal efficiency, $\eta_{th}$

$$\eta_{th} = \frac{w}{(-h_{RP_0})} = \frac{w}{\text{heating value}} \tag{12.21}$$

The over-all efficiency of a gas turbine or steam power plant is defined in the same way. It should be pointed out that in an internal combustion engine or fuel-burning steam power plant the fact that the combustion is itself irreversible is a significant factor in the relatively low thermal efficiency of these devices.

One other factor should be pointed out regarding efficiency. We have noted that the enthalpy of combustion of a hydrocarbon fuel varies considerably with the phase of the water in the products (which leads to the concept of higher and lower heating values). Therefore, in considering the thermal efficiency of an engine, the heating value used to determine this efficiency must be borne in mind. Two engines made by different manufacturers may have identical performance, but if one manufacturer bases his efficiency on the higher heating value and the other on the lower heating value, the latter will be able to claim a higher thermal efficiency. This claim is not significant, of course, as the performance is the same, and a consideration of the way in which the efficiency was defined would reveal this.

The whole matter of efficiencies of devices involving combustion processes is treated in detail in textbooks dealing with particular applications, and our discussion is intended only as an introduction to the subject. However, two examples will be cited to illustrate these remarks.

### Example 12.18

The combustion chamber of a gas turbine uses a liquid hydrocarbon fuel which has an approximate composition of $C_8H_{18}$. On test the following data are obtained.

$$T_{air} = 400 \text{ K} \qquad T_{products} = 1100 \text{ K}$$

$$V_{air} = 100 \text{ m/s} \qquad V_{products} = 150 \text{ m/s}$$

$$T_{fuel} = 50°C \qquad FA_{actual} = 0.0211 \text{ kg fuel/kg air}$$

Calculate the combustion efficiency for this process.

For the ideal chemical reaction the heat transfer is zero. Therefore, writing the first law for a control volume that includes the combustion chamber we have,

$$H_R + KE_R = H_P + KE_P$$

$$H_R + KE_R = \sum_R n_i \left[ \bar{h}_f^\circ + \Delta \bar{h} + \frac{MV^2}{2} \right]_i$$

$$= [\bar{h}_f^\circ + \bar{C}_p (50 - 25)]_{C_8H_{18}(l)} + n_{O_2} \left[ \Delta \bar{h} + \frac{MV^2}{2} \right]_{O_2}$$

$$+ 3.76 n_{O_2} \left[ \Delta \bar{h} + \frac{MV^2}{2} \right]_{N_2}$$

$$= -249\ 952 + 1.7113 \times 114.23\ (50 - 25)$$

$$+ n_{02}\left[3\ 029 + \frac{32 \times (100)^2}{2 \times 1000}\right]$$

$$+ 3.76\ n_{02}\left[2\ 971 + \frac{28.02 \times (100)^2}{2 \times 1000}\right]$$

$$= -245\ 065 + 14\ 886\ n_{02}$$

$$H_P + \mathrm{KE}_P = \sum_P n_e\left[\bar{h}_f^\circ + \Delta\bar{h} + \frac{M\mathbf{V}^2}{2}\right]_e$$

$$= 8\left[\bar{h}_f^\circ + \Delta\bar{h} + \frac{M\mathbf{V}^2}{2}\right]_{CO_2} + 9\left[\bar{h}_f^\circ + \Delta\bar{h} + \frac{M\mathbf{V}^2}{2}\right]_{H_2O}$$

$$+ (n_{O_2} - 12.5)\left[\Delta\bar{h} + \frac{M\mathbf{V}^2}{2}\right]_{O_2} + 3.76 n_{O_2}\left[\Delta\bar{h} + \frac{M\mathbf{V}^2}{2}\right]_{N_2}$$

$$= 8\left[-393\ 522 + 38\ 894 + \frac{44.01 \times (150)^2}{2 \times 1000}\right]$$

$$+ 9\left[-241\ 827 + 30\ 167 + \frac{18.02 \times (150)^2}{2 \times 1000}\right]$$

$$+ (n_{02} - 12.5)\left[26\ 217 + \frac{32 \times (150)^2}{2 \times 1000}\right]$$

$$+ 3.76\ n_{02}\left[24\ 757 + \frac{28.02 \times (150)^2}{2 \times 1000}\right]$$

$$= -5\ 068\ 391 + 120\ 849\ n_{02}$$

Therefore,

$$-245\ 065 + 14\ 886\ n_{02} = -5\ 068\ 391 + 120\ 849\ n_{02}$$
$$n_{02} = 45.519\ \text{kmol O}_2/\text{kmol fuel}$$

kmol air/kmol fuel $= 4.76\ (45.519) = 216.67$

$$FA_{\text{Ideal}} = \frac{114.23}{216.67 \times 28.97} = 0.0182\ \text{kg fuel/kg air}$$

$$\eta_{\text{comb}} = \frac{0.0182}{0.0211} \times 100 = 86.2\%$$

## Example 12.19

In a certain steam power plant 325 000 kg of water per hour enter the boiler at a pressure of 12.5 MPa and a temperature of 200°C. Steam leaves the boiler at 9 MPa, 500°C. The power output of the turbine is 81 000 kW. Coal is used at the rate of 26 700 kg/hour, and has a higher heating value of 33 250 kJ/kg. Determine the efficiency of steam generator and over-all thermal efficiency of the plant.

In power plants the efficiency of both the boiler and the over-all efficiency of the plant are based on the higher heating value of the fuel.

The efficiency of the boiler is defined by Eq. 12.20 as

$$\eta_{\text{steam generator}} = \frac{\text{heat transferred to } H_2O/\text{kg fuel}}{\text{higher heating value}}$$

Therefore

$$\eta_{\text{steam generator}} = \frac{325\ 000\ (3386.1 - 857.1)}{26\ 700 \times 33\ 250} \times 100 = 92.6\%$$

The thermal efficiency is defined by Eq. 12.21.

$$\eta_{\text{th}} = \frac{w}{\text{heating value}} = \frac{81\ 000 \times 3600}{26\ 700 \times 33\ 250} \times 100 = 32.8\%$$

## PROBLEMS

12.1  Butane is burned with air and a volumetric analysis of the combustion products on a dry basis yields the following composition:

$$
\begin{array}{ll}
CO_2 & 7.8\% \\
CO & 1.1 \\
O_2 & 8.2 \\
N_2 & 82.9
\end{array}
$$

Determine the percent of theoretical air used in this combustion process.

12.2  A hydrocarbon fuel is burned with air and the following volumetric analysis on a dry basis is obtained from the products of combustion.

$$
\begin{array}{ll}
CO_2 & 10.5\% \\
O_2 & 5.3 \\
N_2 & 84.2
\end{array}
$$

Determine the composition of the fuel on a mass basis and the percent of theoretical air utilized in the combustion process.

12.3 Octane is burned with the theoretical air in a constant pressure process ($P = 100$ kPa) and the products are cooled to 25°C.

(a) How many kg of water are condensed per kg of fuel?

(b) Suppose the air used for combustion has a relative humidity of 90% and is at a temperature of 25°C and 100 kPa pressure. How many kg of water will be condensed per kg of fuel when the products are cooled to 25°C?

12.4 The field of coal gasification has become one of increased activity in recent years. Consider the combustion of producer gas (Table 12.2) with 130% theoretical air at 0.1 MPa pressure.

(a) Determine the dew point of the products.

(b) How many kg of water will be condensed per kg of fuel if the products are cooled 10°C below the dew point temperature?

12.5 The hot exhaust gas from an internal combustion engine is analyzed and found to have the following composition on a volumetric basis:

$$
\begin{array}{ll}
CO_2 & 10\% \\
H_2O & 13 \\
CO & 2 \\
O_2 & 3 \\
N_2 & 72
\end{array}
$$

This gas is to be fed to an exhaust gas reactor and mixed with a certain amount of air, as shown in Fig. 12.9, to eliminate the CO. It is decided that a mole fraction of $O_2$ of 10% in the mixture at point 3 will ensure that no CO remains. What must the ratio of flows be entering the reactor?

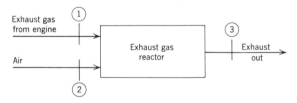

**Fig. 12.9** Sketch for Problem 12.5.

12.6 In a test of rocket propellant performance, liquid hydrazine ($N_2H_4$) at 25°C, 0.1 MPa, and oxygen gas at 25°C, 0.1 MPa, are fed to a combustion chamber in the ratio 0.5 kg $O_2$/kg $N_2H_4$. The heat transfer from the chamber to the surroundings is estimated to be 100 kJ/kg $N_2H_4$. Determine the temperature of the products, assuming only $H_2O$, $H_2$, and $N_2$ to be present. The enthalpy of formation of liquid $N_2H_4$ is $+ 50\,417$ kJ/kmol.

12.7 Repeat Problem 12.6, assuming that saturated liquid oxygen at 90 K is used instead of 25°C oxygen gas in the combustion process.

12.8 An internal combustion engine burns liquid octane ($C_8H_{18}$) and uses 125% theoretical air. The air and fuel enter at 25°C, the products leave the engine exhaust ports at 900 K. In the engine 85% of the carbon burns to $CO_2$ and the remainder burns to CO. The heat transfer from this engine is just equal to the work done by the engine. Determine:

(a) Power output of the engine if the engine burns 0.005 kg fuel/s.

(b) The composition and the dew point of the products of combustion.

12.9  A natural gas consisting of 80% methane and 20% ethane (on a volume basis) is burned with 150% theoretical air in a steady-state, steady-flow process. Heat is transferred from the products of combustion until the temperature reaches 700 K. The fuel enters the combustion chamber at 25°C and the air at 400 K.

Determine the heat transfer per kmol of fuel.

12.10  Liquid ethanol ($C_2H_5OH$) is burned with 150% theoretical oxygen in a steady-state, steady-flow process. The reactants enter the combustion chamber at 25°C, and the products are cooled and leave at 65°C, 0.1 MPa pressure. Calculate the heat transfer per kmol of ethanol. The enthalpy of formation of $C_2H_5OH(l)$ is −277 634 kJ/kmol.

12.11  It has been proposed to use hydrogen ($H_2$) and an oxidizer consisting of 50% oxygen ($O_2$) and 50% fluorine ($F_2$), on a mole basis, as a rocket propellent combination . In a test, the hydrogen and oxidizer both enter a combustion chamber at 25°C, 0.1 MPa, with excess $H_2$ being used to control the temperature of the products. It may be assumed that the products consist of $H_2O$, HF, and the excess $H_2$, and also that the heat transfer from the combustion chamber to the surroundings is 2300 kJ/ kmol of products. Determine the ratio of $H_2$ to oxidizer to be fed to the combustion chamber if the temperature of the products is 3000 K. For HF:

$$\bar{h}_f^\circ = -268\ 613 \text{ kJ/kmol}; \quad \Delta \bar{h}^\circ_{\underset{298 \to 3000 \text{ K}}{}} = 86\ 469 \text{ kJ/kmol}$$

12.12  A thermoelectric generator such as shown in Fig. 1.11 converts heat directly to electrical energy without moving parts. Such a device has been proposed for use in a portable power supply. For the purpose of thermodynamic analysis, the power supply unit may be represented as shown in Fig. 12.10.

Other data are as follows:

$$\eta_{\text{generator}} = \frac{W_{\text{elec}}}{Q_H} = 0.15$$

$$\eta_{\text{elec. motor}} = 0.95$$

Electric motor output = 2 kW

Fan requirement = 0.1 kW

Volumetric (dry) analysis of combustion products

| | |
|---|---|
| $CO_2$ | 9.0% |
| CO | 1.0 |
| $O_2$ | 7.0 |
| $N_2$ | 83.0 |

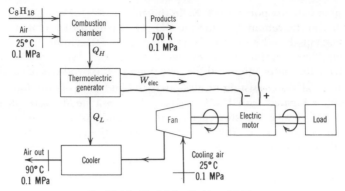

**Fig. 12.10** Sketch for Problem 12.12.

(*a*) Write the combustion equation per mole of octane, and determine the per cent theoretical air.

(*b*) Calculate the octane flow rate.

(*c*) Calculate the volume flow rate of cooling air required (inlet conditions).

12.13 A bomb is charged with 1 kmol of CO and 2 kmol of $O_2$. The total pressure is 0.1 MPa and the temperature is 298 K before combustion. Combustion then occurs and the products are cooled to 1300 K. Assume complete combustion. Determine:

(*a*) The final pressure. (Note that the number of moles changes during combustion.)

(*b*) The heat transfer.

12.14 Liquid octane enters the combustion chamber of a gas turbine engine at 25°C and air enters from the compressor at 500 K. It is determined that 98% of the carbon in the fuel burns to form $CO_2$ and the remaining 2% burns to form CO. What amount of excess air will be required if the temperature of the products is to be limited to 1100 K?

12.15 Hydrogen peroxide enters a gas generator at the rate of 0.1 kg/s, and is decomposed to steam and oxygen. The resulting mixture is expanded through a turbine to atmospheric pressure as shown in Fig. 12.11. Determine the power output of the turbine, and the heat transfer rate in the gas generator. The enthalpy of formation of liquid $H_2O_2$ is $-187\ 583$ kJ/kmol.

12.16 A jet engine is to be test run using liquid methane as a fuel with 300% theoretical air. The test conditions are as shown in Fig. 12.12. Determine the exit velocity of the products.

12.17 Gaseous propane at 25°C is mixed with air at 400 K and burned; 300% theoretical air is used. What is the adiabatic flame temperature?

12.18 Gaseous propane at 25°C is mixed with air at 400 K and burned. What percentage of theoretical air must be used if the temperature of the

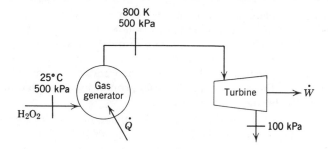

**Fig. 12.11**    Sketch for Problem 12.15.

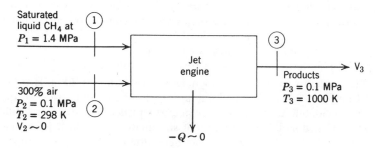

**Fig. 12.12**    Sketch for Problem 12.16.

products is to be 1200 K? Assume an adiabatic process and complete combustion.

12.19 How much error is introduced if the adiabatic flame temperature of Problem 12.17 is calculated assuming that all substances have constant specific heat, values from Table A.8?

12.20 The natural gas B from Table 12.2 is burned with 150% theoretical air. Calculate the adiabatic flame temperature for complete combustion, if the temperature of the reactants is 25°C.

12.21 Sulfur at 25°C is burned with 50% excess air, the air being at a temperature of 400 K. Assuming all the sulfur is burned to $SO_2$, calculate the adiabatic flame temperature. The enthalpy of formation of $SO_2$ is $-296\ 831$ kJ/kmol. The constant-pressure specific heat of $SO_2$ is given by the relation

$$\overline{C}_p = 32.238 + 0.0219\ T - 3.527 \times 10^{-6}\ T^2$$

where $\overline{C}_p = $ kJ/kmol K
    T = K

12.22 Acetylene gas at 25°C, 0.1 MPa pressure, is fed to the head of a cutting torch. Calculate the flame temperature if the acetylene is burned with
    (a) One hundred percent theoretical air at 25°C.
    (b) One hundred percent theoretical oxygen at 25°C.

12.23 Gaseous propane at 25°C is reacted with moist air (air containing water vapor) at 400 K in a steady-state, steady-flow process. The combustion process is adiabatic and the exiting temperature is observed to be 1200 K. A sample of the products is tested and found to have a dew point temperature of 70°C. Determine the percent theoretical air used and the relative humidity of this air. Assume complete combustion and that the pressure is 100 kPa throughout the process.

12.24 A mixture of 20% methane and 80% ethane on a mole basis is throttled from 65°C, 10 MPa to 100 kPa and fed to a combustion chamber where it undergoes complete combustion with air, which enters at 400 K, 100 kPa. The amount of air is such that the products of combustion leave at 1200 K, 100 kPa. Assuming the combustion process to be adiabatic and that all components behave as ideal gases except the fuel, which behaves according to Kay's rule, determine

(a) The percent theoretical air required

(b) The dew point temperature of the products.

12.25 The enthalpy of formation of magnesium oxide, MgO (s) is $-$ 601 827 kJ/kmol at 25°C. The melting point of magnesium oxide is approximately 3000 K, and the increase in enthalpy between 298 K and 3000 K is 128 449 kJ/kmol. The enthalpy of sublimation at 3000 K is estimated at 418 000 kJ/kmol, and the specific heat of magnesium-oxide vapor above 3000 K is estimated at 37.24 kJ/kmol K.

(a) Determine the enthalpy of combustion per kg of magnesium.

(b) Estimate the adiabatic flame temperature when magnesium is burned with theoretical oxygen.

12.26 Consider natural gas D listed in Table 12.2. Calculate the enthalpy of combustion of this gas at 25°C, considering the products to include:

(a) Vapor water.

(b) Liquid water.

12.27 Hydrogen peroxide is sometimes used as the oxidizer in special power plants such as torpedoes and rockets. Determine the enthalpy of combustion at 25°C per kg of reactants for the following combustion process:

$$4H_2O_2(l) + CH_4 \rightarrow 6H_2O + CO_2$$

The enthalpy of formation $H_2O_2(l)$ is $-187\ 583$ kJ/kmol.

12.28 (a) Determine the enthalpy of formation of liquid benzene at 25°C.

(b) Benzene (l) at 25°C is burned with air at 500 K in a steady-flow process. The products of combustion are cooled to 1400 K and have the following volumetric analysis on a dry basis:

| | % by volume |
|---|---|
| $CO_2$ | 10.7 |
| CO | 3.6 |
| $O_2$ | 5.3 |
| $N_2$ | 80.4 |

Calculate the heat transfer per kmol of fuel during the combustion process.

12.29 One kmol of carbon at 25°C, 0.1 MPa and two kmol of $O_2$ at 400 K, 0.1 MPa enter a combustion chamber, and the products leave at 1400 K, 0.1 MPa. Assuming complete combustion and a steady-state steady-flow process, determine:

(*a*) The heat transfer per kmol of fuel.

(*b*) The net entropy change for the process, if the heat is transferred to the surroundings at 25°C.

12.30 Methyl alcohol ($CH_3OH$) is burned with 120% theoretical air at 150 kPa, after which the products of combustion are passed through a heat exchanger and cooled to 50°C. Considering the process to be steady-flow, calculate the absolute entropy of the products leaving the heat exchanger per kmol of alcohol burned.

12.31 The following data are taken from the test of a gas turbine on a test stand.

Fuel — $C_4H_{10}$ (g) at 25°C, 0.1 MPa.
Air — 300% theoretical air at 25°C, 0.1 MPa
Velocity of inlet air = 70 m/s
Velocity of products at exit = 700 m/s
Temperature and pressure of products = 900 K, 0.1 MPa

Assuming complete combustion, determine
(*a*) The net heat transfer per kmol of fuel.
(*b*) The net increase of entropy per kmol of fuel.

12.32 Calculate the irreversibility for the process described in Problem 12.13.

12.33 An inventor claims to have built a device that will take 0.0013 kg/s of water from the faucet at 10°C, 100 kPa, and produce separate streams of hydrogen and oxygen gas, each at 400 K, 175 kPa. He says that his device will operate in a 25°C room on 15 kW electrical input. How do you evaluate this claim?

12.34 Consider the combustion of gaseous propane at 0.1 MPa, 25°C, with theoretical air at 0.1 MPa, 25°C in a steady-state, steady-flow process. Assume that combustion is complete and that the products leave at 25°C. Determine the decrease in Gibbs function for the two following cases:

(*a*) The reactants and products are both separated into their various constituents and each constituent is at 0.1 MPa pressure, 25°C. This is shown schematically in Fig. 12.13.

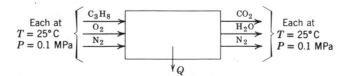

**Fig. 12.13** Sketch for Problem 12.34*a*.

(*b*) The reactants and products each consist of a mixture at a total pressure of 0.1 MPa and a temperature of 25°C. This is shown schematically in Fig. 12.14.

**Fig. 12.14** Sketch for Problem 12.34*b*.

12.35 A mixture of butane and 150% theoretical air enters a combustion chamber at 25°C, 150 kPa, and the products of combustion leave at 1000 K, 150 kPa. Assuming a complete combustion, determine the heat transfer from the combustion chamber and the irreversibility for the process, both per kmol of butane.

12.36 Ammonia is to be produced by the process indicated in Fig. 12.15. A mixture of nitrogen and hydrogen at 25°C, 0.1 MPa is compressed to the appropriate pressure, and enters the constant-pressure catalytic converter. Saturated liquid ammonia leaves at 40°C. Calculate the heat transfer and net entropy change for the overall process.
Data for ideal gas $NH_3$:

$$\bar{h}_f^0 = -46\ 191 \text{ kJ/kmol}$$
$$\bar{s}^0 = 192.548 \text{ kJ/kmol K}$$
$$\bar{g}_f^0 = -16\ 622 \text{ kJ/kmol}$$
$$\bar{C}_{p0} = 25.464 + 0.06647T - 20.416 \times 10^{-6}T^2 \text{ kJ/kmol K}$$

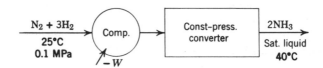

**Fig. 12.15** Sketch for Problem 12.36.

12.37 Consider one cylinder of a spark-ignition internal-combustion engine. Before the compression stroke the cylinder is filled with a mixture of air and ethane. Assume that 150% theoretical air has been used, and that the pressure is 0.1 MPa and the temperature 25°C before compression. The compression ratio of the engine is 9 to 1.

(*a*) Determine the pressure and temperature after compression assuming a reversible adiabatic compression. Assume that the ethane behaves as an ideal gas.

(*b*) Assume further that complete combustion is achieved while the piston remains at top dead center (i.e., after the reversible adiabatic compression), and that the combustion process is adiabatic. Determine the

temperature and pressure after combustion, and the increase in entropy
during the combustion process.

(d) What is the irreversibility for this process?

12.38 Consider the process described in Problem 12.24.

(a) Calculate the absolute entropy of the fuel mixture before it is
throttled into the combustion chamber.

(b) Calculate the irreversibility for the overall process.

12.39 In Example 12.17, a basic hydrogen-oxygen fuel cell reaction was analyzed
at 25°C, 0.1 MPa pressure. Repeat this calculation if the fuel cell operates
on air at 25°C, 0.1 MPa, instead of on pure oxygen at this state.

12.40 Consider the hydrogen-air fuel cell shown in Fig. 12.16.

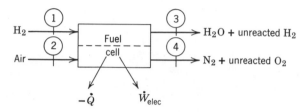

**Fig. 12.16** Sketch for Problem 12.40.

The cell temperature is maintained at 280°C by rejecting heat to the surroundings. The following data have been taken:

$$\dot{W}_{elec.} = 3 \text{ kilowatts}$$

| | | |
|---|---|---|
| $T_1 = 25°C$ | $P_1 = 700 \text{ kPa}$ | $\dot{V}_1 = 0.1 \text{ litre/s}$ |
| $T_2 = 25°C$ | $P_2 = 700 \text{ kPa}$ | $\dot{V}_2 = 0.22 \text{ litre/s}$ |
| $T_3 = 280°C$ | $P_3 = 630 \text{ kPa}$ | |
| $T_4 = 280°C$ | $P_4 = 630 \text{ kPa}$ | $(y_{O_2})_4 = 0.05$ |

(a) Write the overall reaction equation, on a basis of kmol/s for each
substance.

(b) Calculate the heat transfer rate from the cell.

(c) Calculate the entropy flow at point 4.

12.41 A small air-cooled gasoline engine is tested, and the output is found to
be 1.0 kW. The temperature of the products is measured and found to be
660 K. The products are analyzed with the following results, on a dry
volumetric basis:

$$
\begin{aligned}
CO_2 &\quad 11.4\% \\
O_2 &\quad 1.6 \\
CO &\quad 2.9 \\
N_2 &\quad 84.1
\end{aligned}
$$

The fuel used may be considered to be $C_8H_{18}$, and the air and fuel enter
the engine at 25°C. The rate at which fuel is used is 0.000 15 kg/s.

(*a*) Determine the rate of heat transfer from the engine.

(*b*) What is the efficiency of the engine?

12.42 Consider the combustion process of Problem 12.18, in which propane at 25°C is burned with air at 400 K and the temperature of the products is 1200 K. Given these same states, assume that the combustion efficiency is 90% and that 95% of the carbon in the propane burns to form $CO_2$ and 5% burns to form CO. What is the heat transfer from the combustion chamber for this process?

# 13

---

# Introduction to Phase and Chemical Equilibrium

In most of our considerations up to this point we have assumed that we are dealing either with systems that are in equilibrium or with those in which the deviation from equilibrium is infinitesimal, as in a quasi-equilibrium or reversible process. For irreversible processes we made no attempt to describe the state of the system during the process but dealt only with the initial and final states of the system, at which time we considered the system to be in equilibrium.

In this chapter we shall examine the criteria for equilibrium and from them derive certain relations which will enable us, under certain conditions, to determine the properties of a system when it is in equilibrium. The specific systems we shall consider are those involving equilibrium between phases and those involving chemical equilibrium in a single phase (homogeneous equilibrium) as well as certain related topics.

## 13.1 Requirements for Equilibrium

As a general requirement for equilibrium we postulate that a system is in equilibrium when there is no possibility that it can do any work when it is isolated from its surroundings. In applying this criterion to a system it is helpful to divide the system into two or more subsystems, and consider the possibility of doing work by any conceivable interaction between these two subsystems. For example, in Fig. 13.1 a system has been divided into two systems and an engine, of any conceivable variety, placed between these subsystems. A system may be so defined as to include the immediate surroundings. In this case we can let the immediate surroundings be a subsystem and thus consider the general case of the equilibrium between a system and its surroundings.

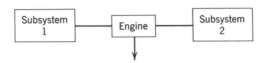

**Fig. 13.1**  Two subsystems that communicate through an engine.

The first requirement for equilibrium is that the two subsystems have the same temperature, for otherwise we could operate a heat engine between the two systems and do work. Thus we conclude that one requirement for equilibrium is that a system must be at a uniform temperature to be in equilibrium. It is also evident that there must be no unbalanced mechanical forces between the two systems, or else one could operate a turbine or piston engine between the two systems and do work.

However, we would like to establish general criteria for equilibrium which would apply to all simple compressible systems, including those that undergo chemical reactions. We will find that the Gibbs function is a particularly significant property in defining the criteria for equilibrium.

Let us first consider a qualitative example to illustrate this point. Consider a natural gas well that is one km deep, and let us assume that the temperature of the gas is constant throughout the gas well. Suppose we have analyzed the composition of the gas at the top of the well, and we would like to know the composition of the gas at the bottom of the well. Further, let us assume that equilibrium conditions prevail in the well. If this is true we would expect that an engine such as shown in Fig. 13.2 (which operates on the basis of the pressure and composition change with elevation and does not involve combustion) would not be capable of doing any work.

If we consider a steady-state, steady-flow process for a control volume around this engine, we could apply Eq. 8.28,

$$W_{\text{rev}} = m_i\left(g_i + \frac{V_i^2}{2} + gZ_i\right) - m_e\left(g_e + \frac{V_e^2}{2} + gZ_e\right)$$

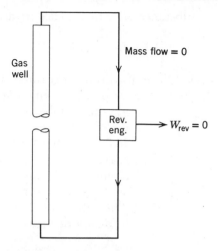

**Fig. 13.2**   Illustration showing the relation between reversible work and the criteria for equilibrium.

However,

$$W_{\text{rev}} = 0, \quad m_i = m_e, \quad \text{and} \quad \frac{V_i^2}{2} = \frac{V_e^2}{2}$$

Then we can write,

$$g_i + gZ_i = g_e + gZ_e$$

and the requirement for equilibrium in the well between two levels that are a distance $dZ$ apart would be

$$dg_T + g\,dZ_T = 0$$

In contrast to a deep gas well, most of the systems that we consider are of such size that $\Delta Z$ is negligibly small, and therefore we consider the pressure in the system to be uniform.

This leads to the general statement of equilibrium that applies to simple compressible systems that may undergo a change in chemical composition, namely, that at equilibrium

$$dG_{T,P} = 0 \tag{13.1}$$

In the case of a chemical reaction, it is helpful to think of the equilibrium state as the state in which the Gibbs function is a minimum. For example, consider a system consisting initially of $n_A$ moles of substance

$A$ and $n_B$ moles of substance $B$, which react in accordance with the relation

$$\nu_A A + \nu_B B \rightleftharpoons \nu_C C + \nu_D D$$

Let the reaction take place at constant pressure and temperature. If

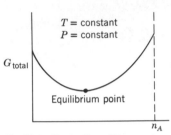

we plot $G$ for this system as a function of $n_A$, the number of moles of $A$ present, we would have a curve as shown in Fig. 13.3. At the minimum point on the curve, $dG_{T,P} = 0$, and this will be the equilibrium composition for this system at the given temperature and pressure. The subject of chemical equilibrium will be developed further in Section 13.6.

**Fig. 13.3**  Illustration of the requirement for chemical equilibrium.

## 13.2  Equilibrium Between Two Phases of a Pure Substance

As another example of this requirement for equilibrium, let us consider the equilibrium between two phases of a pure substance. Consider a system consisting of two phases of a pure substance at equilibrium. We know that under these conditions the two phases are at the same pressure and temperature. Consider the change of state associated with a transfer of $dn$ moles from phase 1 to phase 2 while the temperature and pressure remain constant. That is,

$$dn^1 = -dn^2$$

The Gibbs function of this system is given by

$$G = f(T, P, n^1, n^2)$$

where $n^1$ and $n^2$ designate the number of moles in each phase. Therefore,

$$dG = \left(\frac{\partial G}{\partial T}\right)_{P,n^1,n^2} dT + \left(\frac{\partial G}{\partial P}\right)_{T,n^1,n^2} dP$$
$$+ \left(\frac{\partial G}{\partial n^1}\right)_{T,P,n^2} dn^1 + \left(\frac{\partial G}{\partial n^2}\right)_{T,P,n^1} dn^2$$

By definition,

$$\left(\frac{\partial G}{\partial n^1}\right)_{T,P,n^2} = \bar{g}^1; \quad \left(\frac{\partial G}{\partial n^2}\right)_{T,P,n^1} = \bar{g}^2$$

Therefore, at constant temperature and pressure,

$$dG = \bar{g}^1 \, dn^1 + \bar{g}^2 \, dn^2 = dn^1(\bar{g}^1 - \bar{g}^2)$$

Now at equilibrium (Eq. 13.1)

$$dG_{T,P} = 0$$

Therefore, at equilibrium, we have

$$\bar{g}^1 = \bar{g}^2 \tag{13.2}$$

That is, under equilibrium conditions, the Gibbs function of each phase of a pure substance is equal. Let us check this by determining the Gibbs function of saturated liquid (water) and saturated vapor (steam) at 300 kPa. From the steam tables:
For the liquid:

$$g_f = h_f - T s_f = 561.47 - 406.7 \times 1.6718 = -118.4 \text{ kJ/kg}$$

For the vapor:

$$g_g = h_g - T s_g = 2725.3 - 406.7 \times 6.9919 = -118.4 \text{ kJ/kg}$$

Equation 13.2 can also be derived by applying the relation

$$T \, ds = dh - v \, dP$$

to the change of phase that takes place at constant pressure and temperature. For this process this relation can be integrated as follows:

$$\int_f^g T \, ds = \int_f^g dh$$

$$T(s_g - s_f) = (h_g - h_f)$$

$$h_f - T s_f = h_g - T s_g$$

$$g_f = g_g$$

The Clapeyron equation, which was derived in Section 10.3 can be derived by an alternate method by considering the fact that the Gibbs functions of two phases in equilibrium are equal. In Chapter 10 we considered the relation (Eq. 10.9) for a simple compressible substance,

$$dg = v \, dP - s \, dT$$

Consider a system that consists of a saturated liquid and a saturated vapor in equilibrium, and let this system undergo a change of pressure $dP$. The corresponding change in temperature, as determined from the vapor-pressure curve, is $dT$. Both phases will undergo the change in Gibbs function, $dg$, but since the phases always have the same value of the Gibbs function when they are in equilibrium, it follows that

$$dg_f = dg_g$$

But, from Eq. 10.9,

$$dg = v\,dP - s\,dT$$

it follows that

$$dg_f = v_f\,dP - s_f\,dT$$

$$dg_g = v_g\,dP - s_g\,dT$$

Since

$$dg_f = dg_g$$

it follows that

$$v_f\,dP - s_f\,dT = v_g\,dP - s_g\,dT$$

$$dP(v_g - v_f) = dT(s_g - s_f) \tag{13.3}$$

$$\frac{dP}{dT} = \frac{s_{fg}}{v_{fg}} = \frac{h_{fg}}{Tv_{fg}}$$

In summary, when different phases of a pure substance are in equilibrium, each phase has the same value of the Gibbs function per unit mass. This fact is relevant to different solid phases of a pure substance and is important in metallurgical applications of thermodynamics. Example 13.1 illustrates this principle.

**Example 13.1**

What pressure is required to make diamonds from graphite at a temperature of 25°C? The following data are given for a temperature of 25°C and a pressure of 0.1 MPa.

| Graphite | | Diamond |
|---|---|---|
| $\overline{g}$ | 0 | 2867.8 kJ/kmol |
| $v$ | 0.000 444 m³/kg | 0.000 284 m³/kg |
| $\beta_T$ | 0.304 × 10⁻⁶ 1/MPa | 0.016 × 10⁻⁶ 1/MPa |

The basic principle in the solution is that graphite and diamond can exist in equilibrium when they have the same value of the Gibbs function. At 0.1 MPa pressure the Gibbs function of the diamond is greater than that of the graphite. However, the rate of increase in Gibbs function with pressure is greater for the graphite than the diamond, and therefore, at some pressure they can exist in equilibrium, and our problem is to find this pressure.

We have already considered the relation

$$dg = v\,dP - s\,dT$$

Since we are considering a process that takes place at constant temperature this reduces to

$$dg_T = v\,dP_T \tag{a}$$

Now at any pressure $P$ and the given temperature the specific volume can be found from the following relation, which utilizes the isothermal compressibility factor.

$$v = v^\circ + \int_{P=1}^{P} \left(\frac{\partial v}{\partial P}\right)_T dP = v^\circ + \int_{P=1}^{P} \frac{v}{v}\left(\frac{\partial v}{\partial P}\right)_T dP$$

$$= v^\circ - \int_{P=1}^{P} v\beta_T\,dP \tag{b}$$

The superscript $^\circ$ will be used in this example to indicate the properties at a pressure of 0.1 MPa and a temperature of 25°C.

The specific volume changes only slightly with pressure, so that $v \approx v^\circ$. Also, we assume that $\beta_T$ is constant and that we are considering a very high pressure. With these assumptions this equation can be integrated to give

$$v = v^\circ - v^\circ \beta_T P = v^\circ(1 - \beta_T P) \tag{c}$$

We can now substitute this into Eq. $a$ to give the relation

$$dg_T = [v^\circ(1 - \beta_T P)]\,dP_T$$

$$g - g^\circ = v^\circ(P - P^\circ) - v^\circ \beta_T \frac{(P^2 - P^{\circ 2})}{2} \tag{d}$$

If we assume that $P^\circ \ll P$ this reduces to

$$g - g^\circ = v^\circ \left[ P - \frac{\beta_T P^2}{2} \right] \tag{e}$$

For the graphite, $g° = 0$ and we can write

$$g_G = v_G° \left[ P - (\beta_T)_G \frac{P^2}{2} \right]$$

For the diamond, $g°$ has a definite value and we have

$$g_D = g_D° + v_D° \left[ P - (\beta_T)_D \frac{P^2}{2} \right]$$

But, at equilibrium the Gibbs function of the graphite and diamond are equal.

$$g_G = g_D$$

Therefore,

$$v_G° \left[ P - (\beta_T)_G \frac{P^2}{2} \right] = g_D° + v_D° \left[ P - (\beta_T)_D \frac{P^2}{2} \right]$$

$$(v_G° - v_D°)P - [v_G°(\beta_T)_G - v_D°(\beta_T)_D] \frac{P^2}{2} = g_D°$$

$$(0.000\ 444 - 0.000\ 284)P$$

$$-(0.000\ 444 \times 0.304 \times 10^{-6} - 0.000\ 284 \times 0.016 \times 10^{-6})P^2/2$$

$$= \frac{2867.8}{12.011 \times 1000}$$

Solving this for $P$ we find

$$P = 1493 \text{ MPa}$$

That is, at 1493 MPa, 25°C, graphite and diamond can exist in equilibrium, and the possibility exists for conversion from graphite to diamonds.

The preceding example could also have been evaluated in terms of the fugacities instead of the Gibbs functions for the two forms of carbon. Similarly, the equilibrium requirement for liquid and vapor water discussed earlier could be expressed in terms of the fugacities of the two phases. That is,

$$f^L = f^V = f^{\text{sat}}$$

where $f^{\text{sat}}$ is the fugacity of the substance determined at the given temperature and its saturation pressure $P^{\text{sat}}$.

The fugacity of a pure compressed liquid or solid phase at pressures moderately greater than the saturation pressure can be calculated with considerable accuracy as follows. For a pure substance at constant temperature,

$$dg_T = RT(d \ln f)_T = v \, dP_T$$

Integrating at constant temperature between the saturation state and the pressure $P$, we have

$$\int_{f^{sat}}^{f} RT(d \ln f)_T = \int_{P^{sat}}^{P} v \, dP_T$$

$$RT \ln \frac{f}{f^{sat}} = \int_{P^{sat}}^{P} v \, dP_T$$

If we assume that $v$ is a constant, which often holds with considerable accuracy for liquids and solids, we have

$$RT \ln \frac{f}{f^{sat}} \approx v(P - P^{sat}) \tag{13.4}$$

Equation 13.4 is used to determine the fugacity coefficients of compressed liquid presented in the generalized chart given in Appendix Fig. A.7.

Since $v$ is small for the liquid and solid phases, the quantity $v(P - P^{sat})$ is small for moderate changes of pressure. Therefore, for liquids and solids at moderate pressures we conclude that

$$f^L \approx f^{sat} \quad \text{and} \quad f^S \approx f^{sat} \tag{13.5}$$

## 13.3  Equilibrium of a Multicomponent, Multiphase System

As an introduction to a consideration of equilibrium in multicomponent, multiphase systems, let us consider a system consisting of two components and two phases. To show the general characteristics of such a system, consider a mixture of oxygen and nitrogen at a pressure of 0.1 megapascal and in the range of temperature where both the liquid and vapor phases are present.

Figure 13.4 shows the composition of the liquid and vapor phases which are in equilibrium as a function of temperature at a pressure of 0.1 MPa. The upper line, marked "vapor line" gives the composition of the vapor phase, and the lower line gives the composition of the liquid phase. If we have pure nitrogen, the boiling point is 77.3 K. If we have pure oxygen, the boiling point is 90.2 K. If we have a mixture of nitrogen and oxygen with both phases present at a temperature of 84 K, the vapor

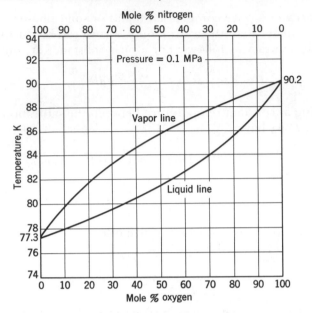

**Fig. 13.4**  Equilibrium diagram for liquid-vapor phases of the nitrogen-oxygen system at a pressure of 0.1 MPa.

will have a composition of 64 per cent $N_2$ and 36 per cent $O_2$. The liquid will have a composition of 30 per cent $N_2$ and 70 per cent $O_2$.

Consider a mixture consisting of 21 per cent oxygen and 79 per cent nitrogen (approximately the composition of air) and at a pressure of 0.1 MPa and an initial temperature of 74 K. At this state the mixture will be in the liquid phase. Let this liquid be slowly heated while the pressure remains constant. By referring to Fig. 13.4 we note that when the temperature reaches 78.8 K, the first bubble will form. This vapor will have a composition of approximately 6 per cent $O_2$ and 94 per cent $N_2$. As more heat is added the temperature increases and the mole fraction of oxygen in the liquid increases. When the last drop of liquid remains, the vapor will have a composition of 21 per cent $O_2$ and 79 per cent $N_2$, the temperature will be 81.9 K, and the liquid composition will be 54 per cent $O_2$ and 46 per cent $N_2$.

To understand the phase behavior of a system such as described here, it may be helpful to briefly discuss the more general phase diagram. We recall that the phase behavior for a pure substance is dictated by the saturation line on $P–T$ coordinates, such as in Fig. 3.2. For a two-component system $A$ and $B$, overall composition is now an additional independent property, and phase behavior is represented in terms of surfaces on a pressure-composition-temperature diagram, as shown in Fig. 13.5a. This diagram includes three constant-temperature planes to indicate that there is a lower vapor surface and an upper liquid surface that together form an equilibrium envelope between the two pure substance saturation lines. Only within this envelope is it possible to have two

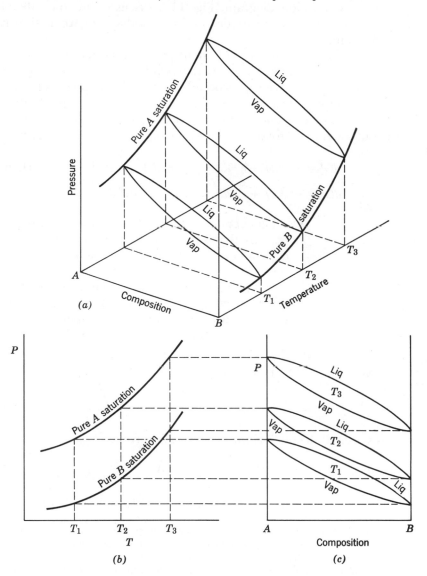

**Fig. 13.5** Phase diagram for a two-component system.

phases present in equilibrium. Beneath the envelope the two-component system is all superheated vapor and above the envelope the system is all compressed liquid.

We realize now that a constant-pressure plane (0.1 MPa) of the three-dimensional phase diagram, as viewed from above, results in the temperature-composition diagram, Fig. 13.4, discussed in some detail previously. The other two projections of such a diagram are as shown in Figs. 13.5b and 13.5c.

In studying the characteristics of such a two-component system, we must first ask the question, "What is the requirement for phase equilibrium in this system?" We can proceed in a manner analogous to that followed in Section 13.2 for a pure substance, with superscripts 1 and 2 denoting the two phases, and in this case subscripts $A$ and $B$ referring to the two components. It follows from applying Eq. 11.41 to each phase that[1]

$$dG^1 = -S^1\, dT + V^1\, dP + \mu_A{}^1\, dn_A{}^1 + \mu_B{}^1\, dn_B{}^1 \tag{13.6}$$

$$dG^2 = -S^2\, dT + V^2\, dP + \mu_A{}^2\, dn_A{}^2 + \mu_B{}^2\, dn_B{}^2$$

$T$ and $P$
remain constant

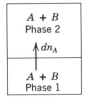

**Fig. 13.6**  An equilibrium mixture involving two components and two phases.

Let us consider this two-phase, two-component mixture that is in equilibrium as a system, and let each phase be considered as a subsystem. One possible change of state that might occur is for a very small amount of component $A$ to be transferred from phase 1 to phase 2 while the moles of $B$ in each phase and the temperature and pressure remain constant. This is shown schematically in Fig. 13.6.

For this change of state

$$dn_A{}^2 = -dn_A{}^1 \tag{13.7}$$

Since this system is in equilibrium, $dG = 0$. Therefore,

$$dG = dG^1 + dG^2 = 0 \tag{13.8}$$

Since $T$, $P$, $n_B{}^1$ and $n_B{}^2$ are all constant, it follows from Eqs. 13.6 and 13.8

---

[1]For those students who have not studied the behavior of real mixtures in Sections 11.7–11.12, the variable $\mu_A$ in Eq. 13.6 is the chemical potential of component $A$, the intensive thermodynamic property that drives a mass transfer $dn_A$. It is identical to $\bar{G}_A$, the partial molal Gibbs function of component $A$; that is, the value of the Gibbs function of $A$ as it exists in the mixture. See also the discussion related to Eq. 11.3.

that

$$dG = \mu_A{}^1 \, dn_A{}^1 + \mu_A{}^2 \, dn_A{}^2$$
$$= \mu_A{}^1 \, dn_A{}^1 - \mu_A{}^2 \, dn_A{}^1 - dn_A{}^1(\mu_A{}^1 - \mu_A{}^2)$$
$$= 0$$

Therefore, at equilibrium

$$\mu_A{}^1 = \mu_A{}^2 \tag{13.9}$$

Thus the requirement for equilibrium is that the chemical potential of each component is the same in all phases. If the chemical potential is not the same in all phases there will be a tendency for mass to pass from one phase to the other. When the chemical potential of a component is the same in both phases, there is no tendency for a net transfer of mass from one phase to the other.

This requirement for equilibrium is readily extended to multicomponent, multiphase systems, for the chemical potential of each component must be the same in all phases.

$$\mu_A{}^1 = \mu_A{}^2 = \mu_A{}^3 = \cdots \text{for all phases}$$
$$\mu_B{}^1 = \mu_B{}^2 = \mu_B{}^3 = \cdots \text{for all phases}$$
$$\underline{\quad} \quad \underline{\quad} \quad \underline{\quad}$$
$$\underline{\quad} \quad \underline{\quad} \quad \underline{\quad} \tag{13.10}$$
$$\underline{\quad} \quad \underline{\quad} \quad \underline{\quad}$$

for all components

Let us now consider alternate means of expressing the requirement for phase equilibrium at a given temperature and pressure. We found in Section 11.9, Eq. 11.42, that the chemical potential of component $A$ is identical to the partial Gibbs function $\bar{G}_A$ of that component in the mixture. Thus, the requirement for equilibrium, Eq. 13.9, can also be expressed as

$$\bar{G}_A{}^1 = \bar{G}_A{}^2 \tag{13.11}$$

That is, at equilibrium the partial Gibbs function of a given component is the same in each phase. In this form the result is analogous to that for pure substance phase equilibrium, Eq. 13.2. It also follows from the definition of fugacity of a component in a mixture, Eqs. 11.63 and 11.64,

$$(d\bar{G}_A)_T = \bar{R}T d(\ln \bar{f}_A)_T$$
$$\lim_{P \to 0} (\bar{f}_A/y_A P) = 1$$

that the requirement for phase equilibrium at a given temperature and pressure, Eq. 13.9 can be expressed in terms of fugacity as

$$\bar{f}_A{}^1 = \bar{f}_A{}^2 \tag{13.12}$$

That is, at equilibrium the fugacity of a given component is the same in each phase. This is perhaps the most useful formulation of the requirement for equilibrium between phases at a given temperature and pressure, as we have previous experience in evaluating fugacities. The most general concepts, however, are those of activity and activity coefficient, which were defined and discussed in Section 11.13. Thus, the definition of activity, Eq. 11.95, could be substituted into Eq. 13.12 for the fugacity in each phase, or the activity could be eliminated in favor of the activity coefficient using Eq. 11.99.

We find that the requirement for equilibrium can be expressed in numerous ways, in terms of different thermodynamic variables. The question of interest to us at this point is whether we can use such expressions to predict the equilibrium compositions of the different phases from the known properties of the pure substances that make up the system. In order to be able to do this, we need to utilize a model for each phase. In this section we will consider two models, namely the Ideal Solution model and the Raoult's Rule-Ideal Gas model. For each model, we will restrict our discussion to a two-phase (liquid and vapor), two-component system as shown in Fig. 13.6. The results of this analysis are then readily extended to multiphase or multicomponent systems. In the analysis, we will distinguish liquid phase mole fractions from those for the vapor phase by using the symbol $x$ for the liquid phase.

## Ideal Solution

A frequently used mixture model is the Ideal Solution, in which the fugacity of component $A$ in a mixture is expressed as the product of the mole fraction of $A$ and the fugacity of pure $A$ in the same phase as the mixture and at the pressure and temperature of the mixture (and similarly for component $B$). These specified conditions for pure $A$ may very well result in a hypothetical state. The Ideal Solution has been discussed in detail in Section 11.12, but for our purposes here we need only be familiar with the fugacity equation described above, Eq. 11.86.

Let us assume that both the liquid and vapor phases of a two-component system behave according to the Ideal Solution model. The requirement for equilibrium for component $A$ is, from Eqs. 13.12 and 11.86

$$x_A f_A{}^L = y_A f_A{}^V \tag{13.13}$$

Similarly, for component $B$,

$$x_B f_B{}^L = y_B f_B{}^V \tag{13.14}$$

We also realize that for the liquid phase,

$$x_A + x_B = 1 \tag{13.15}$$

and for the vapor phase

$$y_A + y_B = 1 \tag{13.16}$$

Now, at a given temperature and pressure, the four pure substance fugacities in Eqs. 13.13 and 13.14 are fixed values. Therefore, Eqs. 13.13 through 13.16 comprise a set of four equations in four unknowns, which can be solved for the equilibrium composition of the liquid and vapor phases.

The problem that arises in evaluating the Ideal Solution model is that connected with hypothetical states. This is most easily recognized by referring to the phase diagram, Fig. 13.5$b$. For an equilibrium two-phase mixture of $A$ and $B$, the system pressure $P$ is less than $P_A{}^{sat}$ and greater than $P_B{}^{sat}$ at the given temperature $T$. We realize from examining Fig. 13.5$b$ that at the conditions of $P$ and $T$, pure $A$ is a superheated vapor and pure $B$ is a compressed liquid. As a result, the fugacities $f_A{}^V$ and $f_B{}^L$ are readily found by conventional methods, as for example from Fig. A.7. However, the fugacities $f_A{}^L$ and $f_B{}^V$ cannot be found, as pure liquid $A$ and pure vapor $B$ do not exist at the conditions of $P$ and $T$. These are the hypothetical states. In order to determine the fugacities for these hypothetical states, vapor-phase fugacities have been extrapolated into the liquid region and liquid-phase fugacities extrapolated into the vapor region in a manner so as to give reasonable generalized correlations from Eqs. 13.13 and 13.14 with experimental phase equilibrium data. These extrapolated hypothetical values are presented in Fig. 13.7.

The following example should help to clarify the determination of fugacities in the Ideal Solution model.

**Example 13.2**

Calculate the phase compositions in the liquid-vapor system methane-carbon monoxide at $-100°C$, 4 MPa, assuming Ideal Solution in both phases. Let CO be component $A$ and $CH_4$ be component $B$. Using critical point properties from Table A.7,

$$T_{rA} = \frac{173.2}{133.0} = 1.302 \quad , \quad P_{rA} = \frac{4}{3.5} = 1.143$$

$$T_{rB} = \frac{173.2}{191.1} = 0.906 \quad , \quad P_{rB} = \frac{4}{4.64} = 0.862$$

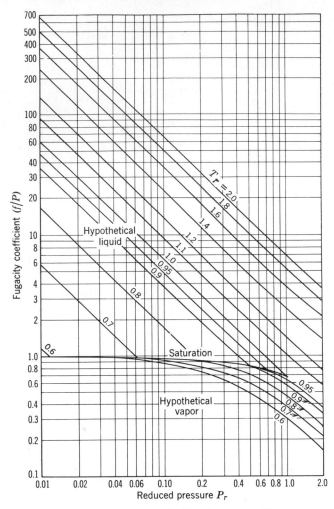

**Fig. 13.7** Hypothetical liquid and vapor fugacities.

From Fig. A.7, it is seen that pure $A$ is a gas and pure $B$ is a liquid at the given state. Therefore, from Fig. A.7,

$$f_A^V = 0.87 \times 4 = 3.48 \text{ MPa}$$
$$f_B^L = 0.49 \times 4 = 1.96$$

Pure liquid $A$ and pure gaseous $B$ are the hypothetical states. From Fig. 13.7,

$$f_A^L = 1.8 \times 4 = 7.2 \text{ MPa}$$
$$f_B^V = 0.65 \times 4 = 2.6$$

Substituting these values into the equilibrium equations, Eqs. 13.13 and 13.14, and solving,

$$x_A = \frac{3.48}{7.2}\, y_A = 0.4833\, y_A$$

$$x_B = \frac{2.6}{1.96}\, y_B = 1.3265\, y_B$$

$$x_A + x_B = 0.4833\, y_A + 1.3265\,(1 - y_A) = 1$$

$$y_A = 0.387 \quad , \quad x_A = 0.187$$

Although in this text we do not consider a more general phase equilibrium model than the Ideal Solution, a few brief comments are appropriate here. From the definition of activity, Eq. 11.95, it follows that since for a multicomponent, multiphase mixture the fugacities of each component are the same in all phases, the activity of each component depends upon the selection of the standard state and the fugacities of the pure components in that state. The activity coefficient as defined by Eq. 11.99 then similarly depends upon the standard states. In Chapter 11 it was stated that for phase equilibrium calculations, standard states are often selected as the pure substance in that phase at the pressure and temperature of the mixture. For such a choice, the activity coefficient of each component is unity for an ideal solution, and departures from unity indicate nonideality of the system. Thus, the activity coefficient becomes a very useful parameter in correlating and evaluating phase equilibrium data.

### Raoult's Rule – Ideal Gas

This model is a special case of the Ideal Solution, and while not as accurate, it is nevertheless quite reasonable for the behavior of many systems at low pressure. For the liquid phase, we make two simplifying assumptions:

1. The fugacity of pure liquid $A$ at $T$ and $P$ of the system is equal to the fugacity of saturated $A$ (liquid or vapor) at the same $T$, and its corresponding saturation pressure $P_A^{sat}$, or

$$f_A^L = f_A^{sat} \tag{13.17}$$

This is equivalent to assuming that the $\int v\, dP$ correction of Eq. 13.4 is negligibly small.

2. Pure saturated vapor $A$ at $T$ and $P_A^{sat}$ behaves as an ideal gas, or

$$f_A^{sat} = P_A^{sat} \tag{13.18}$$

Combining Eqs. 13.17 and 13.18, we obtain, for the two components $A$ and $B$,

$$f_A^L = P_A^{\text{sat}}, \qquad f_B^L = P_B^{\text{sat}} \tag{13.19}$$

which, when combined with the Ideal Solution model, Eq. 11.86, is termed Raoult's Rule.

For the vapor phase, we assume that pure gas $A$ and pure gas $B$ behave as ideal gases at $T$ and $P$, so that

$$f_A^V = P, \qquad f_B^V = P \tag{13.20}$$

Now, substituting Eqs. 13.19 and 13.20 for each component into the result for the Ideal Solution model, Eqs. 13.13 and 13.14, we obtain the result for the Raoult's Rule-Ideal Gas model:

$$x_A P_A^{\text{sat}} = y_A P \tag{13.21}$$

$$x_B P_B^{\text{sat}} = y_B P \tag{13.22}$$

which together with

$$x_A + x_B = 1 \tag{13.23}$$

$$y_A + y_B = 1 \tag{13.24}$$

forms a set of four equations in four unknowns at a given $T$ and $P$ ($P_A^{\text{sat}}$ and $P_B^{\text{sat}}$ depend only on $T$). This set of equations can then be solved to determine the equilibrium compositions in each phase.

The three additional assumptions made in this model should clearly indicate its limitations as compared with the Ideal Solution, which is the more accurate, especially at higher pressures.

**Example 13.3**

Air (assumed to be 21% $O_2$, 79% $N_2$) is cooled to 80 K, 0.1 MPa pressure. Calculate the composition of the liquid and vapor phases at this condition, assuming the Raoult's Rule–Ideal Gas model, and compare the results with Fig. 13.4. At 80 K, from Tables A.4 and A.5,

$$P_{N_2}^{\text{sat}} = 0.1370 \text{ MPa} \qquad P_{O_2}^{\text{sat}} = 0.030 \ 06 \text{ MPa}$$

For convenience let $N_2 = A$, and $O_2 = B$ in the calculations. Using Eqs. 13.21 and 13.22,

$$x_A P_A^{\text{sat}} = y_A P$$

$$x_B P_B^{\text{sat}} = y_B P$$

Therefore, at 80 K, 0.1 MPa pressure,

$$x_A \, (0.137) = y_A \, (0.1)$$

$$x_B \, (0.030 \; 06) = y_B \, (0.1)$$

But,

$$x_A + x_B = 1$$

$$y_A + y_B = 1$$

Substituting for $y_A$, $y_B$, in the last equation,

$$\frac{0.137}{0.1} \, x_A + \frac{0.030 \; 06}{0.1} x_B = 1.37 \, x_A + 0.3006 \, (1 - x_A) = 1$$

$$x_A = 0.654 \qquad y_A = 0.896$$

From Fig. 13.4, which is based on experimental data, we find that at 80 K, 0.1 MPa.

$$x_A = 0.66, \quad y_A = 0.89$$

Thus, we conclude that for this system at this temperature and pressure, Raoult's Rule gives quite accurate results.

## 13.4   The Gibbs Phase Rule (without Chemical Reaction)

The Gibbs phase rule, which was derived by Professor J. Willard Gibbs of Yale University in 1875, ranks among the truly significant contributions to physical science. In this section we consider the Gibbs phase rule for a system that does not involve a chemical reaction. For such a system the Gibbs phase rule is

$$\mathscr{P} + \mathscr{V} = \mathscr{C} + 2 \tag{13.25}$$

where $\mathscr{P}$ is the number of phases present, $\mathscr{V}$ is the variance, and $\mathscr{C}$ the number of components present. The term variance designates the number of intensive properties that must be specified in order to completely fix the state of the system. For example, in Example 13.3 we considered a two phase mixture of oxygen and nitrogen. For this system the number of phases $\mathscr{P}$ equals 2, the number of components $\mathscr{C}$ is 2, and therefore the variance $\mathscr{V}$ is 2. This is evident from Fig. 13.4, for this diagram is for a fixed pressure (0.1 MPa), and the fixing of one additional intensive property, such as temperature, mole fraction of a given component in the liquid phase, or mole fraction of a given component in the vapor phase, will completely determine all other intensive properties and thus fix the state of the system (although not the relative amounts of the two phases).

Let us consider further the application of the Gibbs phase rule to a pure substance. In this case $\mathscr{C} = 1$. When we have one phase present, such as superheated vapor, $\mathscr{P} = 1$, and we conclude that $\mathscr{V} = \mathscr{C} + 2 - 1 = 2$. That is, two intensive properties must be specified in order to fix the state. We are already familiar with the superheated vapor tables for a number of substances and recognize that these tables are presented with pressure and temperature as the two independent properties. Such a system is referred to as a bivariant system.

Suppose we have two phases of a pure substance in equilibrium, such as saturated liquid and saturated vapor. In this case $\mathscr{C} = 1$, $\mathscr{P} = 2$, and $\mathscr{V} = 1 + 2 - 2 = 1$. That is, one intensive property fixes the value of all other intensive properties and thus determines the state of each phase. Again, from our familiarity with table of thermodynamic properties we recall that either pressure or temperature is used as the independent intensive property for tabulating liquid-vapor equilibrium data for a pure substance. Such a system is known as monovariant.

Consider also the triple point of a pure substance. In this case $\mathscr{C} = 1$ and $\mathscr{P} = 3$. Therefore $\mathscr{V} = 1 + 2 - 3 = 0$. That is, at the triple point all intensive properties are fixed, and if any intensive property is varied, we are no longer at the triple point. This is known as an invariant system.

It might be well at this point to again emphasize that a pure substance can have several phases in the solid state. For example, Fig. 3.6 shows the various phases of water. Note that there are several states at which three phases exist in equilibrium, and each of these is a triple point.

The application of the phase rule to a two-component, two-phase system has already been made. As a final application let us consider a two-component, three-phase system. For such a system $\mathscr{V} = 2 + 2 - 3 = 1$. That is, it is a monovariant system, and specifying one property, such as pressure or temperature fixes the state of the system.

The validity of the Gibbs phase rule can be outlined by considering a system consisting of $\mathscr{C}$ components and $\mathscr{P}$ phases in equilibrium at a given temperature and pressure. Assuming that each component exists in each phase, the state of the system could be completely specified if the concentration of each component in each phase and the temperature and pressure were specified. This would be a total of $\mathscr{C}\mathscr{P}+2$ intensive properties.

We know, however, that these are not all independent intensive properties. If we determine the number of equations we have between these intensive properties, we can subtract this from the $\mathscr{C}\mathscr{P}+2$ intensive properties and find the number of independent intensive properties, or, as it has been defined, the variance. The fact that at equilibrium the chemical potential of each component is the same in all phases gives rise to a total of $\mathscr{C}(\mathscr{P}-1)$ equations between the intensive properties of the mixture.

Further, the fact that the sum of the mole fractions equals unity in each of the $\mathscr{P}$ phases gives $\mathscr{P}$ additional equations. Therefore the variance is

$$\mathscr{V} = \mathscr{C}\mathscr{P}+2-\mathscr{P}-\mathscr{C}(\mathscr{P}-1) = \mathscr{C}+2-\mathscr{P}$$

## 13.5  Metastable Equilibrium

Although the limited scope of this book precludes an extensive treatment of metastable equilibrium, a brief introduction to the subject is presented in this section. Let us first consider an example of metastable equilibrium.

Consider a slightly superheated vapor, such as steam, expanding in a convergent-divergent nozzle, as shown in Fig. 13.8. Assuming the process is reversible and adiabatic, the steam will follow path 1-$a$ on the $T$-$s$ diagram, and at point $a$ we would expect condensation to occur. However, if point $a$ is reached in the divergent section of the nozzle, it is observed that no condensation occurs until point $b$ is reached, and at this point the condensation occurs very abruptly in what is referred to as a condensation shock. Between points $a$ and $b$ the steam exists as a vapor, but the temperature is below the saturation temperature for the given pressure. This is known as a metastable state. The possibility of a metastable state exists with any phase transformation. The dotted lines on the equilibrium diagram shown in Fig. 13.9 represent possible metastable states for solid-liquid-vapor equilibrium.

The nature of a metastable state is often pictured schematically by the kind of diagram shown in Fig. 13.10. The ball is in a stable position (the "metastable state") for small displacements, but with a large displacement it moves to a new equilibrium position. The steam expanding in the nozzle is in a metastable state between $a$ and $b$. This means that droplets

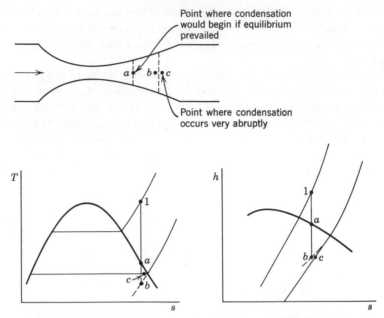

Point where condensation
would begin if equilibrium
prevailed

Point where condensation
occurs very abruptly

**Fig. 13.8**    Illustration of the phenomenon of supersaturation in a nozzle

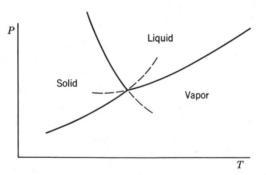

**Fig. 13.9**    Metastable states for solid-liquid-vapor equilibrium.

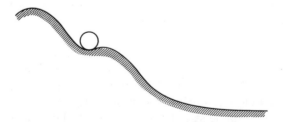

**Fig. 13.10**    Schematic diagram illustrating a metastable state.

smaller than a certain critical size will re-evaporate, and only when droplets of larger than this critical size have formed (this corresponds to moving the ball out of the depression) will the new equilibrium state appear.

## 13.6   Chemical Equilibrium

We now turn our attention to chemical equilibrium and consider first a chemical reaction involving only one phase. This is referred to as a homogeneous chemical reaction. It may be helpful to visualize this as a gaseous phase, but the basic considerations apply to any phase.

Consider a vessel, Fig. 13.11, that contains four compounds, $A$, $B$, $C$, and $D$, which are in chemical equilibrium at a given pressure and temperature. For example, these might consist of $CO_2$, $H_2$, $CO$, and $H_2O$ in equilibrium. Let the number of moles of each component be designated $n_A$, $n_B$, $n_C$, and $n_D$. Further, let the chemical reaction which takes place between these four constituents be

$$\nu_A A + \nu_B B \rightleftharpoons \nu_C C + \nu_D D \qquad (13.26)$$

where the $\nu$'s are the stoichiometric coefficients. It should be emphasized that there is a very definite relation between the $\nu$'s (the stoichiometric coefficients) whereas the $n$'s (the number of moles present) for any constituent can be varied simply by varying the amount of that component in the reaction vessel.

Let us now consider how the requirement for equilibrium, namely, that $dG_{T,P} = 0$ at equilibrium, applies to a homogeneous chemical reaction. When we considered phase equilibrium (Section 13.3) we proceeded by assuming that the two phases were in equilibrium at a given temperature and pressure, and then let a small quantity of one component be transferred from one phase to the other. In a similar manner, let us assume that the four components are in chemical equilibrium and then assume that from this equilibrium state, while the temperature and pressure remain constant, the reaction proceeds an infinitesimal amount toward the right as Eq. 13.26 is written. This results in a decrease in the moles of $A$ and $B$ and an increase in the moles of $C$ and $D$. Let us designate the degree of reaction by $\epsilon$, and define the degree of reaction

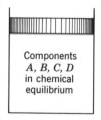

**Fig. 13.11** Schematic diagram for consideration of chemical equilibrium.

by the relations

$$dn_A = -\nu_A \, d\epsilon$$
$$dn_B = -\nu_B \, d\epsilon$$
$$dn_C = +\nu_C \, d\epsilon \quad\quad (13.27)$$
$$dn_D = +\nu_D \, d\epsilon$$

That is, the change in the number of moles of any component during a chemical reaction is given by the product of the stoichiometric coefficients (the $\nu$'s) and the degree of reaction.

Let us evaluate the change in the Gibbs function associated with this chemical reaction that proceeds to the right in the amount $d\epsilon$. In doing so we use, as would be expected, the Gibbs function of each component in the mixture, namely, the partial molal Gibbs function (or its equivalent, the chemical potential)[2]

$$dG_{T,P} = \bar{G}_C \, dn_C + \bar{G}_D \, dn_D + \bar{G}_A \, dn_A + \bar{G}_B \, dn_B$$

Substituting Eq. 13.27, we have

$$dG_{T,P} = (\nu_C \bar{G}_C + \nu_D \bar{G}_D - \nu_A \bar{G}_A - \nu_B \bar{G}_B) \, d\epsilon \quad\quad (13.28)$$

We have already considered (Section 11.13) how to evaluate the partial molal Gibbs function of a given component in terms of the Gibbs function of the pure component at the standard state and the activity[3] of that component. For the component $i$, we can write (Eq. 11.96)

$$\bar{G}_i = \bar{g}_i^\circ + \bar{R}T \ln a_i$$

Substituting this relation into Eq. 13.28 we have

$$dG_{T,P} = [\nu_C(\bar{g}_C^\circ + \bar{R}T \ln a_C) + \nu_D(\bar{g}_D^\circ + \bar{R}T \ln a_D)$$
$$- \nu_A(\bar{g}_A^\circ + \bar{R}T \ln a_A) - \nu_B(\bar{g}_B^\circ + \bar{R}T \ln a_B)] \, d\epsilon \quad (13.29)$$

Let us define $\Delta G^\circ$ as follows,

$$\Delta G^\circ = \nu_C \bar{g}_C^\circ + \nu_D \bar{g}_D^\circ - \nu_A \bar{g}_A^\circ - \nu_B \bar{g}_B^\circ \quad\quad (13.30)$$

---

[2] See the footnote in Section 13.3.

[3] For students who have not studied real mixture behavior, the activity was defined by Eq. 11.95 following the development to express the partial Gibbs function in terms of a pure substance reference value and the fugacity of that component in the mixture.

That is, $\Delta G°$ is the change in the Gibbs function that would occur if the chemical reaction given by Eq. 13.26 (which involves the stoichiometric amounts of each component) proceeded completely from left to right, with the reactants $A$ and $B$ initially separated and at temperature $T$ and the standard state pressure and the products $C$ and $D$ finally separated and at temperature $T$ and the standard state pressure. Note also that $\Delta G°$ for a given reaction is a function of only the temperature. This will be most important to bear in mind as we proceed with our developments of homogeneous chemical equilibrium. Let us now digress from our development to consider an example involving the calculation of $\Delta G°$.

**Example 13.4**

Determine the value of $\Delta G°$ for the reaction $H_2O \rightleftharpoons H_2 + \frac{1}{2}O_2$ at 298 K and at 2000 K, with the $H_2O$ in the gaseous phase.

We take 0.1 MPa as our standard state pressure, and recall that we assume further that $\bar{g}$ for all the elements is zero at 0.1 MPa pressure and 298 K. Therefore, at 298 K,

$$(\bar{g}_f°)_{H_2} = 0; \quad (\bar{g}_f°)_{O_2} = 0$$

From Table 12.3, at 298 K,

$$(\bar{g}_f°)_{H_2O} = -228\,583 \text{ kJ/kmol}$$
$$\Delta G° = (\bar{g}_f°)_{H_2} + \tfrac{1}{2}(\bar{g}_f°)_{O_2} - (\bar{g}_f°)_{H_2O}$$

$$= 0 + 0 - (-228\,583) = 228\,583 \text{ kJ}$$

At 2000 K, from Table A.11,

$$\bar{g}_{H_2}° = \bar{g}_{2000}° - \bar{g}_{298}° = (\bar{h}_{2000}° - \bar{h}_{298}°) - (2000\bar{s}_{2000}° - 298.15\bar{s}_{298}°)$$

$$= 52\,932 - (2000 \times 188.406 - 298.15 \times 130.684)$$

$$= -284\,917 \text{ kJ/kmol}$$

Similarly, for the $O_2$ at 2000 K,

$$\bar{g}_{O_2}° = (\bar{h}_{2000}° - \bar{h}_{298}°) - (2000\bar{s}_{2000}° - 298.15\bar{s}_{298}°)$$

$$= 59\,199 - (2000 \times 268.764 - 298.15 \times 205.142)$$

$$= -417\,166 \text{ kJ/kmol}$$

For the $H_2O$,

$$\bar{g}^\circ_{H_2O} = \bar{g}^\circ_f + \bar{g}^\circ_{2000} - \bar{g}^\circ_{298}$$

$$= -228\ 583 + 72\ 689 - (2000 \times 264.681 - 298.15 \times 188.833)$$

$$= -628\ 955\ \text{kJ/kmol}$$

Therefore, at 2000 K

$$\Delta G^\circ = -284\ 917 + \tfrac{1}{2}(-417\ 166) - (-628\ 955) = 135\ 455\ \text{kJ}$$

The value for $\Delta G^\circ$ at 2000 K can alternately be evaluated by using the fact that at constant $T$,

$$\Delta G^\circ = \Delta H^\circ - T\Delta S^\circ$$

At 2000 K,

$$\Delta H^\circ = (\bar{h}^\circ_{2000} - \bar{h}^\circ_{298})_{H_2} + \tfrac{1}{2}(\bar{h}^\circ_{2000} - \bar{h}^\circ_{298})_{O_2} - (\bar{h}^\circ_f + \bar{h}^\circ_{2000} - \bar{h}^\circ_{298})_{H_2O}$$

$$= 52\ 932 + \tfrac{1}{2}(59\ 199) - (-241\ 827 + 72\ 689)$$

$$= 251\ 670\ \text{kJ}$$

$$\Delta S^\circ = (\bar{s}^\circ_{2000})_{H_2} + \tfrac{1}{2}(\bar{s}^\circ_{2000})_{O_2} - (\bar{s}^\circ_{2000})_{H_2O}$$

$$= 188.406 + \tfrac{1}{2}(268.764) - (264.681)$$

$$= 58.107\ \text{kJ/K}$$

Therefore,

$$\Delta G^\circ = 251\ 670 - 2000 \times 58.107 = 135\ 455\ \text{kJ}$$

Note that while the two methods are identical in result, the second requires fewer calculations, and is therefore simpler to use.

Returning now to our development, substituting Eq. 13.30 into Eq. 13.29 and rearranging we can write

$$dG_{T,P} = \left\{ \Delta G^\circ + \bar{R}T \ln \left[ \frac{a_C^{\nu_C} \, a_D^{\nu_D}}{a_A^{\nu_A} \, a_B^{\nu_B}} \right] \right\} d\epsilon \tag{13.31}$$

At equilibrium $dG_{T,P} = 0$. Therefore, since $d\epsilon$ is arbitrary,

$$\ln \left[ \frac{a_C^{\nu_C} \, a_D^{\nu_D}}{a_A^{\nu_A} \, a_B^{\nu_B}} \right] = -\frac{\Delta G^\circ}{\bar{R}T} \tag{13.32}$$

For convenience, we define the equilibrium constant $K$ as

$$\ln K = -\frac{\Delta G°}{\bar{R}T} \tag{13.33}$$

which we note must be a function of temperature only for a given reaction, since $\Delta G°$ is given by Eq. 13.30 in terms of the properties of the pure substances at a given temperature and the standard-state pressure.

Combining Eqs. 13.32 and 13.33, we have

$$K = \frac{a_C{}^{\nu_C} a_D{}^{\nu_D}}{a_A{}^{\nu_A} a_B{}^{\nu_B}} \tag{13.34}$$

which is the chemical equilibrium equation corresponding to the reaction equation, Eq. 13.26.

### Example 13.5

Determine the equilibrium constant $K$, expressed as $\ln K$, for the reaction $H_2O \rightleftharpoons H_2 + \frac{1}{2} O_2$ at 298 K and at 2000 K.

We have already found, in Example 13.4, $\Delta G°$ for this reaction at these two temperatures. Therefore at 298 K,

$$(\ln K)_{298} = -\frac{\Delta G°_{298}}{\bar{R}T} = \frac{-228\ 583}{8.31434 \times 298.15} = -92.21$$

At 2000 K we have,

$$(\ln K)_{2000} = \frac{-\Delta G°_{2000}}{\bar{R}T} = \frac{-135\ 455}{8.31434 \times 2000} = -8.146$$

Table A.12 gives the values of the equilibrium constant for a number of reactions. Note that for each reaction the value of the equilibrium constant is determined from the properties of each of the pure constituents at the standard state pressure and is a function of temperature only.

For other reaction equations, the chemical equilibrium constant can be calculated as in Example 13.5 or can be determined analytically in the following manner. Consider the general reaction Eq. 13.26. The standard-state Gibbs function for each constituent can be expressed by writing Eq. 11.77 at the pressure $P°$. For component $A$

$$d\left(\frac{\bar{g}_A°}{T}\right)_{P°} = -\frac{\bar{h}_A°}{T^2} dT_{P°} \tag{13.35}$$

Therefore, for the reaction given by Eq. 13.26,

$$d\left(\frac{\Delta G^\circ}{T}\right)_{P^\circ} = -\frac{\Delta H^\circ}{T^2}\, dT_{P^\circ} \tag{13.36}$$

where $\Delta G^\circ$ is defined by Eq. 13.30 and $\Delta H^\circ$ similarly by

$$\Delta H^\circ = \nu_C \bar{h}_C^\circ + \nu_D \bar{h}_D^\circ - \nu_A \bar{h}_A^\circ - \nu_B \bar{h}_B^\circ \tag{13.37}$$

Now, substituting the definition of the equilibrium constant Eq. 13.33 into Eq. 13.36,

$$d \ln K = \frac{\Delta H^\circ}{\bar{R} T^2}\, dT_{P^\circ} \tag{13.38}$$

which is termed the van't Hoff equation. In integrating this equation, we must be careful to note that $\Delta H^\circ$ as defined by Eq. 13.37 is a function of temperature.

In applying the concept of the equilibrium constant to the determination of the equilibrium composition for a chemical reaction, we must be able to determine the activity of the various constituents in the mixture. The most general model that we will utilize in this text is the Ideal Solution. For this model, the activity of each component is, from Eq. 11.97,

$$a_i = \frac{y_i f_i}{f_i^\circ}$$

If we further assume that all of the standard-state values are for gases, then the standard-state fugacity $f_i^\circ$ is equal to the standard-state pressure $P^\circ$. The chemical equilibrium equation, Eq. 13.34, then can be rewritten for an ideal solution as

$$K = \frac{a_C^{\nu_C} a_D^{\nu_D}}{a_A^{\nu_A} a_B^{\nu_B}} = \frac{y_C^{\nu_C} y_D^{\nu_D}}{y_A^{\nu_A} y_B^{\nu_B}} \left(\frac{P}{P^\circ}\right)^{\nu_C + \nu_D - \nu_A - \nu_B} \left[\frac{(f/P)_C^{\nu_C} (f/P)_D^{\nu_D}}{(f/P)_A^{\nu_A} (f/P)_B^{\nu_B}}\right] \tag{13.39}$$

We have written the equation in this form because it demonstrates quite clearly the influence of various factors on the equilibrium composition (the $y$'s). That is, we know that temperature and pressure both influence the equilibrium composition. From Eq. 13.39 it can be seen that the influence of temperature enters through the value of $K$ (which is a function of temperature only) and the influence of pressure through the term $(P/P^\circ)^{\nu_C + \nu_D - \nu_A - \nu_B}$ (both temperature and pressure influence the various fugacity coefficients $f/P$).

Let us now consider a special case of the ideal solution, the ideal gas model, which will be found to be appropriate for most of our examples and applications. If each component behaves as an ideal gas, then each fugacity coefficient $f/P$ in Eq. 13.39 is unity, and the chemical equilibrium equation reduces to the form,

$$K = \frac{y_C^{\nu_C} y_D^{\nu_D}}{y_A^{\nu_A} y_B^{\nu_B}} \left(\frac{P}{P^\circ}\right)^{\nu_C + \nu_D - \nu_A - \nu_B} \tag{13.40}$$

Let us now consider a number of examples that illustrate the procedure for determining the equilibrium composition for a homogeneous reaction and the influence of certain variables on the equilibrium composition.

**Example 13.6**

One kmol of carbon at 25°C and 0.1 MPa pressure reacts with one kmol of oxygen at 25°C and 0.1 MPa pressure to form an equilibrium mixture of $CO_2$, $CO$, and $O_2$ at 3000 K, 0.1 MPa pressure, in a steady-flow process. Determine the equilibrium composition and the heat transfer for this process.

It is convenient to view the over-all process as though it occurred in two separate steps, a combustion process followed by a heating and dissociation of the combustion product $CO_2$, as indicated in Fig. 13.12.

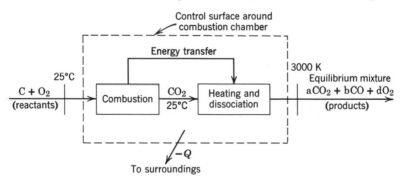

**Fig. 13.12**   Sketch for Example 13.6

This two-step process is represented as follows:

combustion: $\qquad\qquad\qquad$ $C + O_2 \rightarrow CO_2$
dissociation reaction: $\qquad$ $CO_2 \rightleftharpoons CO + \tfrac{1}{2}O_2$

That is, the energy released by the combustion of C and $O_2$ heats the $CO_2$ formed to high temperature, resulting in the dissociation of part of the $CO_2$ to $CO$ and $O_2$. Thus, the over-all reaction can be written

$$C + O_2 \rightarrow aCO_2 + bCO + dO_2$$

where the unknown coefficients $a$, $b$, and $d$ must be found by solution of the equilibrium equation associated with the dissociation reaction. Once this is accomplished, we can write the first law for a control volume around the combustion chamber to calculate the heat transfer.

From the combustion equation we find that the initial composition for the dissociation reaction is 1 kmol $CO_2$. Therefore, letting $z$ be the number of kmol of $CO_2$ dissociated, we find

|  | $CO_2 \rightleftharpoons CO + \frac{1}{2}O_2$ | | |
|---|---|---|---|
| Initial: | 1 | 0 | 0 |
| Change: | $-z$ | $+z$ | $+z/2$ |
| At equilibrium: | $(1-z)$ | $z$ | $z/2$ |

Therefore, the over-all reaction is

$$C + O_2 \rightarrow (1-z)CO_2 + z\,CO + \frac{z}{2}O_2$$

and the total number of kmol at equilibrium is

$$n = (1-z) + z + z/2 = 1 + z/2$$

The equilibrium mole fractions are

$$y_{CO_2} = \frac{1-z}{1+z/2}, \quad y_{CO} = \frac{z}{1+z/2}, \quad y_{O_2} = \frac{z/2}{1+z/2}$$

From Table A.12 we find that the value of the equilibrium constant at 3000 K for the dissociation reaction considered here is

$$\ln K = -1.111 \quad , \quad K = 0.329$$

Substituting these quantities along with $P = 0.1$ MPa into Eq. 13.40, we have the equilibrium equation,

$$K = 0.329 = \frac{y_{CO}y_{O_2}^{1/2}}{y_{CO_2}}\left(\frac{P}{P°}\right)^{1+1/2-1}$$

$$= \frac{\left(\dfrac{z}{1+z/2}\right)\left(\dfrac{z/2}{1+z/2}\right)^{1/2}(1)^{1/2}}{\left(\dfrac{1-z}{1+z/2}\right)}$$

or, in more convenient form,

$$\frac{K^2}{(P/P^\circ)} = \frac{(0.329)^2}{1} = \left(\frac{z}{1-z}\right)^2 \left(\frac{z}{2+z}\right)$$

In order to obtain the physically meaningful root of this mathematical relation, we note that the number of moles of each component must be greater than zero. Thus, the root of interest to us must lie in the range

$$0 \leqslant z \leqslant 1$$

Solving the equilibrium equation by trial and error, we find

$$z = 0.4372$$

Therefore, the over-all process is

$$C + O_2 \rightarrow 0.5628CO_2 + 0.4372CO + 0.2186O_2$$

where the equilibrium mole fractions are

$$y_{CO_2} = \frac{0.5628}{1.2186} = 0.4618$$

$$y_{CO} = \frac{0.4372}{1.2186} = 0.3588$$

$$y_{O_2} = \frac{0.2186}{1.2186} = 0.1794$$

The heat transfer from the combustion chamber to the surroundings can be calculated using the enthalpies of formation and Table A.11. For this process,

$$H_R = (\bar{h}_f^\circ)_C + (\bar{h}_f^\circ)_{O_2} = 0 + 0 = 0$$

The equilibrium products leave the chamber at 3000 K. Therefore,

$$\begin{aligned}
H_p &= n_{CO_2} (\bar{h}_f^\circ + \bar{h}_{3000}^\circ - \bar{h}_{298}^\circ)_{CO_2} \\
&+ n_{CO} (\bar{h}_f^\circ + \bar{h}_{3000}^\circ - \bar{h}_{298}^\circ)_{CO} \\
&+ n_{O_2} (\bar{h}_f^\circ + \bar{h}_{3000}^\circ - \bar{h}_{298}^\circ)_{O_2} \\
&= 0.5628 \, (-393 \, 522 + 152 \, 862) \\
&+ 0.4372 \, (-110 \, 529 + 93 \, 542) \\
&+ 0.2186 \, (98 \, 098) \\
&= -121 \, 426 \text{ kJ}
\end{aligned}$$

Substituting into the first law,

$$Q_{\text{c.v.}} = H_P - H_R$$

$$= -121\ 426\ \text{kJ/kmol C burned}$$

## Example 13.7

One kmol of carbon at 25°C reacts with two kmol of oxygen at 25°C to form an equilibrium mixture of $CO_2$, CO, and $O_2$ at 3000 K, 0.1 MPa pressure. Determine the equilibrium composition.

The over-all process can be imagined to occur in two steps as in the previous example. The combustion process is

$$C + 2O_2 \rightarrow CO_2 + O_2$$

and the subsequent dissociation reaction is

|  | $CO_2$ | $\rightleftharpoons$ CO | $+\tfrac{1}{2}O_2$ |
|---|---|---|---|
| Initial: | 1 | 0 | 1 |
| Change: | $-z$ | $+z$ | $+z/2$ |
| At equilibrium: | $(1-z)$ | $z$ | $(1+z/2)$ |

We find that in this case the over-all process is

$$C + 2O_2 \rightarrow (1-z)CO_2 + zCO + (1+z/2)O_2$$

and the total number of kmol at equilibrium is

$$n = (1-z) + z + (1+z/2) = 2 + z/2$$

The mole fractions are

$$y_{CO_2} = \frac{1-z}{2+z/2}, \quad y_{CO} = \frac{z}{2+z/2}, \quad y_{O_2} = \frac{1+z/2}{2+z/2}$$

The equilibrium constant for the reaction $CO_2 \rightleftharpoons CO + \tfrac{1}{2}O_2$ at 3000 K was found in Example 13.6 to be 0.329. Therefore, the equilibrium equation is, substituting these quantities and $P = 0.1$ MPa,

$$K = 0.329 = \frac{y_{CO}y_{O_2}^{1/2}}{y_{CO_2}}\left(\frac{P}{P^\circ}\right)^{1+1/2-1}$$

$$= \frac{\left(\dfrac{z}{2+z/2}\right)\left(\dfrac{1+z/2}{2+z/2}\right)^{1/2}}{\left(\dfrac{1-z}{2+z/2}\right)}(1)^{1/2}$$

or,

$$\frac{K^2}{(P/P^\circ)} = \frac{(0.329)^2}{1} = \left(\frac{z}{1-z}\right)^2 \left(\frac{2+z}{4+z}\right)$$

We note that in order for the number of kmol of each component to be greater than zero,

$$0 \leq z \leq 1$$

Solving the equilibrium equation for $z$, we find

$$z = 0.310$$

so that the over-all process is

$$C + 2O_2 \rightarrow 0.69\,CO_2 + 0.31\,CO + 1.155\,O_2$$

The mole fractions of the components in the equilibrium mixture are

$$y_{CO_2} = \frac{0.69}{2.155} = 0.320$$

$$y_{CO} = \frac{0.31}{2.155} = 0.144$$

$$y_{O_2} = \frac{1.155}{2.155} = 0.536$$

The heat transferred from the chamber in this process could be found by the same procedure followed in Example 13.6, considering the over-all process as expressed above.

## 13.7  Simultaneous Reactions

In developing the equilibrium equation and equilibrium constant expressions of Section 13.6, it was assumed that there was only a single chemical reaction equation relating the substances present in the system. To demonstrate the more general situation in which there is more than one chemical reaction, we will now analyze a system involving two simultaneous reactions by a procedure analogous to that followed in Section 13.6. These results are then readily extended to systems involving several simultaneous reactions.

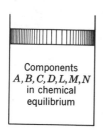

Components
$A, B, C, D, L, M, N$
in chemical
equilibrium

**Fig. 13.13** Sketch demonstrating simultaneous reactions.

Consider a mixture of substances $A$, $B$, $C$, $D$, $L$, $M$, and $N$ as indicated in Fig. 13.13. These substances are assumed to exist at a condition of chemical equilibrium at temperature $T$ and pressure $P$, and are related by the two independent reactions

$$(1) \quad \nu_{A_1}A + \nu_B B \rightleftharpoons \nu_C C + \nu_D D \qquad\qquad (13.41)$$

$$(2) \quad \nu_{A_2}A + \nu_L L \rightleftharpoons \nu_M M + \nu_N N \qquad\qquad (13.42)$$

We have considered the situation where one of the components (substance $A$) is involved in each of the reactions in order to demonstrate the effect of this condition on the resulting equations. As in the previous section, the changes in amounts of the components are related by the various stoichiometric coefficients (which are not the same as the number of moles of each substance present in the vessel). We also realize that the coefficients $\nu_{A_1}$, and $\nu_{A_2}$ are not necessarily the same. That is, substance $A$ does not in general take part in each of the reactions to the same extent.

Development of the requirement for equilibrium is completely analogous to that of Section 13.6. We consider that each reaction proceeds an infinitesimal amount toward the right side. This results in a decrease in the number of moles of $A$, $B$, and $L$, and an increase in the moles of $C$, $D$, $M$, and $N$. Letting the degrees of reaction be $\epsilon_1$ and $\epsilon_2$ for reactions 1 and 2, respectively, the changes in the number of moles are, for infinitesimal shifts from the equilibrium composition,

$$dn_A = -\nu_{A_1}d\epsilon_1 - \nu_{A_2}d\epsilon_2$$

$$dn_B = -\nu_B d\epsilon_1$$

$$dn_L = -\nu_L d\epsilon_2$$

$$dn_C = +\nu_C d\epsilon_1 \qquad\qquad (13.43)$$

$$dn_D = +\nu_D d\epsilon_1$$

$$dn_M = +\nu_M d\epsilon_2$$

$$dn_N = +\nu_N d\epsilon_2$$

The change in Gibbs function for the mixture in the vessel at constant temperature and pressure is

$$dG_{T,P} = \bar{G}_A dn_A + \bar{G}_B dn_B + \bar{G}_C dn_C + \bar{G}_D dn_D + \bar{G}_L dn_L + \bar{G}_M dn_M + \bar{G}_N dn_N$$

Substituting the expressions of Eq. 13.43 and collecting terms,

$$dG_{T,P} = (\nu_C \bar{G}_C + \nu_D \bar{G}_D - \nu_{A_1}\bar{G}_A - \nu_B \bar{G}_B)d\epsilon_1 \\ + (\nu_M \bar{G}_M + \nu_N \bar{G}_N - \nu_{A_2}\bar{G}_A - \nu_L \bar{G}_L)d\epsilon_2 \quad (13.44)$$

It is convenient to again express each of the partial molal Gibbs functions in terms of the activity as

$$\bar{G}_i = \bar{g}_i^\circ + \bar{R}T \ln a_i$$

Equation 13.44 written in this form becomes

$$dG_{T,P} = \left[ \Delta G_1^\circ + \bar{R}T \ln \frac{a_C^{\nu_C} a_D^{\nu_D}}{a_A^{\nu_{A_1}} a_B^{\nu_B}} \right] d\epsilon_1$$

$$+ \left[ \Delta G_2^\circ + \bar{R}T \ln \frac{a_M^{\nu_M} a_N^{\nu_N}}{a_A^{\nu_{A_2}} a_L^{\nu_L}} \right] d\epsilon_2 \qquad (13.45)$$

In this equation the standard-state change in Gibbs function for each reaction is defined as

$$\Delta G_1^\circ = \nu_C \bar{g}_C^\circ + \nu_D \bar{g}_D^\circ - \nu_{A_1} \bar{g}_A^\circ - \nu_B \bar{g}_B^\circ \qquad (13.46)$$

$$\Delta G_2^\circ = \nu_M \bar{g}_M^\circ + \nu_N \bar{g}_N^\circ - \nu_{A_2} \bar{g}_A^\circ - \nu_L \bar{g}_L^\circ \qquad (13.47)$$

Equation 13.45 expresses the change in Gibbs function of the system at constant $T$, $P$, for infinitesimal degrees of reaction of both reactions 1 and 2, Eqs. 13.41 and 13.42. The requirement for equilibrium is that $dG_{T,P} = 0$. Therefore, since reactions 1 and 2 are independent, $d\epsilon_1$ and $d\epsilon_2$ can be independently varied. It follows that at equilibrium each of the bracketed terms of Eq. 13.45 must be zero. Defining equilibrium constants for the two reactions by

$$\ln K_1 = -\frac{\Delta G_1^\circ}{\bar{R}T} \qquad (13.48)$$

and

$$\ln K_2 = -\frac{\Delta G_2^\circ}{\bar{R}T} \qquad (13.49)$$

we find that at equilibrium

$$K_1 = \frac{a_C^{\nu_C} a_D^{\nu_D}}{a_A^{\nu_{A_1}} a_B^{\nu_B}} \qquad (13.50)$$

and

$$K_2 = \frac{a_M^{\nu_M} a_N^{\nu_N}}{a_A^{\nu_{A_2}} a_L^{\nu_L}} \qquad (13.51)$$

can be expressed in terms of the mole fractions for the appropriate model of the mixture, after which these expressions must be solved simultaneously for the equilibrium composition of the mixture. The following example is presented to demonstrate and clarify this procedure.

**Example 13.8**

One kmol of $H_2O$ vapor is heated to 3000 K, 0.1 MPa pressure. Determine the equilibrium composition, assuming that $H_2O$, $H_2$, $O_2$, and OH are present.

There are two independent reactions relating the four components of the mixture at equilibrium. These can be written as:

$$(1) \ H_2O \rightleftharpoons H_2 + \tfrac{1}{2}O_2$$

$$(2) \ H_2O \rightleftharpoons \tfrac{1}{2}H_2 + OH$$

Let $a$ be the number of kmol of $H_2O$ dissociating according to reaction (1) during the heating, and $b$ the number of kmol of $H_2O$ dissociating according to reaction (2). Since the initial composition is 1 kmol $H_2O$, the changes according to the two reactions are

Change:
$$(1) \ H_2O \rightleftharpoons H_2 + \tfrac{1}{2}O_2$$
$$-a \qquad +a + \tfrac{1}{2}a$$

Change:
$$(2) \ H_2O \rightleftharpoons \tfrac{1}{2}H_2 + OH$$
$$-b \qquad +\tfrac{1}{2}b \ \ +b$$

Therefore, the number of kmol of each component at equilibrium is its initial number plus the change, so that at equilibrium

$$
\begin{aligned}
n_{H_2O} &= 1 - a - b \\
n_{H_2} &= a + \tfrac{1}{2}b \\
n_{O_2} &= \tfrac{1}{2}a \\
n_{OH} &= b \\
\hline
n &= 1 + \tfrac{1}{2}a + \tfrac{1}{2}b
\end{aligned}
$$

The overall chemical reaction that occurs during the heating process can be written

$$H_2O \rightarrow (1 - a - b)H_2O + (a + \tfrac{1}{2}b)H_2 + \tfrac{1}{2}aO_2 + bOH$$

The right-hand side of this expression is the equilibrium composition of the system. Since the number of kmol of each substance must necessarily be greater than zero, we find that the possible values of $a$ and $b$ are restricted to

$$a \geqslant 0$$
$$b \geqslant 0$$
$$(a + b) \leqslant 1$$

The two equilibrium equations are, assuming the mixture to behave as an ideal gas

$$K_1 = \frac{y_{H_2} y_{O_2}^{1/2}}{y_{H_2O}} \left(\frac{P}{P^\circ}\right)^{1+1/2-1}$$

$$K_2 = \frac{y_{H_2}^{1/2} y_{OH}}{y_{H_2O}} \left(\frac{P}{P^\circ}\right)^{1/2+1-1}$$

Since the mole fraction of each component is the ratio of the number of kmol of the component to the total number of kmol of the mixture, these equations can be written in the form,

$$K_1 = \frac{\left(\dfrac{a+\frac{1}{2}b}{1+\frac{1}{2}a+\frac{1}{2}b}\right)\left(\dfrac{\frac{1}{2}a}{1+\frac{1}{2}a+\frac{1}{2}b}\right)^{1/2}}{\left(\dfrac{1-a-b}{1+\frac{1}{2}a+\frac{1}{2}b}\right)} \left(\frac{P}{P^\circ}\right)^{1/2}$$

$$= \left(\frac{a+\frac{1}{2}b}{1-a-b}\right)\left(\frac{\frac{1}{2}a}{1+\frac{1}{2}a+\frac{1}{2}b}\right)^{1/2}\left(\frac{P}{P^\circ}\right)^{1/2}$$

and

$$K_2 = \frac{\left(\dfrac{a+\frac{1}{2}b}{1+\frac{1}{2}a+\frac{1}{2}b}\right)^{1/2}\left(\dfrac{b}{1+\frac{1}{2}a+\frac{1}{2}b}\right)}{\left(\dfrac{1-a-b}{1+\frac{1}{2}a+\frac{1}{2}b}\right)} \left(\frac{P}{P^\circ}\right)^{1/2}$$

$$= \left(\frac{a+\frac{1}{2}b}{1+\frac{1}{2}a+\frac{1}{2}b}\right)^{1/2}\left(\frac{b}{1-a-b}\right)\left(\frac{P}{P^\circ}\right)^{1/2}$$

giving two equations in the two unknowns $a$ and $b$, since $P = 0.1$ MPa and the values of $K_1$, $K_2$ are known. From Table A.12 at 3000 K, we find

$$K_1 = 0.0457 \quad , \quad K_2 = 0.0530$$

Therefore, the equations can be solved simultaneously for $a$ and $b$. The values satisfying the equations are

$$a = 0.1080 \quad , \quad b = 0.1086$$

Substituting these values into the expressions for the number of kmol of each component and of the mixture, we find the equilibrium mole fractions to be

$$y_{H_2O} = 0.7069$$
$$y_{H_2} = 0.1464$$
$$y_{O_2} = 0.0487$$
$$y_{OH} = 0.0980$$

The methods used in this section can readily be extended to equilibrium systems involving more than two independent reactions. In each case, the number of simultaneous equilibrium equations is equal to the number of independent reactions. Solution of a large set of nonlinear simultaneous equations naturally becomes quite difficult, however, and is not easily accomplished by hand calculations. Instead, solution of these problems is normally made using iterative procedures on a digital computer.

### 13.8 Ionization

In the final section of this chapter, we shall consider the equilibrium of systems involving ionized gases, or plasmas, a field that has found increasing application and study in recent years. In previous sections we have discussed chemical equilibrium, with a particular emphasis on molecular dissociation, as for example the reaction

$$N_2 \rightleftharpoons 2N$$

which occurs to an appreciable extent for most molecules only at high temperature (of the order of magnitude 3000–10 000 K). At still higher temperatures, such as found in electric arcs, the gas becomes ionized. That is, some of the atoms lose an electron, according to the reaction

$$N \rightleftharpoons N^+ + e^-$$

where $N^+$ denotes a singly-ionized nitrogen atom (one that has lost one electron and consequently has a positive charge) and $e^-$ represents the free electron. As temperature is increased still further, many of the ionized atoms lose another electron, according to the reaction

$$N^+ \rightleftharpoons N^{++} + e^-$$

and thus become doubly-ionized. As the temperature is increased further, the process continues until a temperature is reached at which all the electrons have been stripped from the nucleus.

Ionization generally is appreciable only at high temperature. However, dissociation and ionization both tend to occur to greater extents at low pressure, and consequently dissociation and ionization may be appreciable in such environments as the upper atmosphere, even at moderate temperature. Other effects such as radiation will also cause ionization, but these effects are not considered here.

The problems involved with analyzing the composition in a plasma become much more difficult than for an ordinary chemical reaction, since in an electric field the free electrons in the mixture do not exchange energy with the positive ions and neutral atoms at the same rate as they do with the field. Consequently, in a plasma in an electric field, the electron gas is not at exactly the same temperature as the heavy particles. However, for moderate fields, it is a reasonable approximation to assume a condition of thermal equilibrium in the plasma, at least for preliminary calculations. Under this condition, we can treat the ionization equilibrium in exactly the same manner as an ordinary chemical equilibrium analysis.

At these extremely high temperatures, we may assume that the plasma behaves as an ideal gas mixture of neutral atoms, positive ions, and electron gas. Thus, for the ionization of some atomic species $A$,

$$A \rightleftharpoons A^+ + e^- \tag{13.52}$$

we may write the ionization equilibrium equation in the form

$$K = \frac{y_{A^+} y_{e^-}}{y_A} \left(\frac{P}{P^\circ}\right)^{1+1-1} \tag{13.53}$$

The ionization-equilibrium constant $K$ is defined in the ordinary manner

$$\ln K = -\frac{\Delta G^\circ}{\bar{R}T} \tag{13.54}$$

and is a function of temperature only. The standard-state Gibbs function change for reaction 13.52 is found from

$$\Delta G^\circ = \bar{g}^\circ_{A^+} + \bar{g}^\circ_{e^-} - \bar{g}^\circ_A \tag{13.55}$$

The standard-state Gibbs function for each component at the given plasma temperature can be calculated using the procedures of statistical thermodynamics, so that ionization-equilibrium constants can be tabulated as functions of temperature.

Solution of the ionization-equilibrium equation, Eq. 13.53, is then accomplished in the same manner as for an ordinary chemical-reaction equilibrium.

### Example 13.9

Calculate the equilibrium composition if argon gas is heated in an arc to 10 000 K, 1 kPa, assuming the plasma to consist of Ar, Ar$^+$, $e^-$. The ionization-equilibrium constant for the reaction

$$Ar \rightleftharpoons Ar^+ + e^-$$

at this temperature is 0.000 42.

Consider an initial composition of 1 kmol neutral argon, and let $z$ be the number of kmol ionized during the heating process. Therefore,

$$
\begin{array}{lccc}
 & \text{Ar} \rightleftharpoons & \text{Ar}^+ + & e^- \\
\text{Initial:} & 1 & 0 & 0 \\
\text{Change:} & -z & +z & +z \\
\hline
\text{Equilibrium:} & (1-z) & z & z
\end{array}
$$

and

$$n = (1-z) + z + z = 1 + z$$

Since the number of kmol of each component must be positive, the variable $z$ is restricted to the range

$$0 \leqslant z \leqslant 1$$

The equilibrium mole fractions are

$$y_{Ar} = \frac{n_{Ar}}{n} = \frac{1-z}{1+z}$$

$$y_{Ar^+} = \frac{n_{Ar^+}}{n} = \frac{z}{1+z}$$

$$y_{e^-} = \frac{n_{e^-}}{n} = \frac{z}{1+z}$$

The equilibrium equation is

$$K = \frac{y_{Ar^+} y_{e^-}}{y_{Ar}} \left(\frac{P}{P^\circ}\right)^{1+1-1} = \frac{\left(\dfrac{z}{1+z}\right)\left(\dfrac{z}{1+z}\right)}{\left(\dfrac{1-z}{1+z}\right)} \left(\frac{P}{P^\circ}\right)$$

so that, at 10 000 K, 1 kPa,

$$0.000\ 42 = \left(\frac{z^2}{1-z^2}\right)(0.01)$$

Solving,

$$z = 0.2008$$

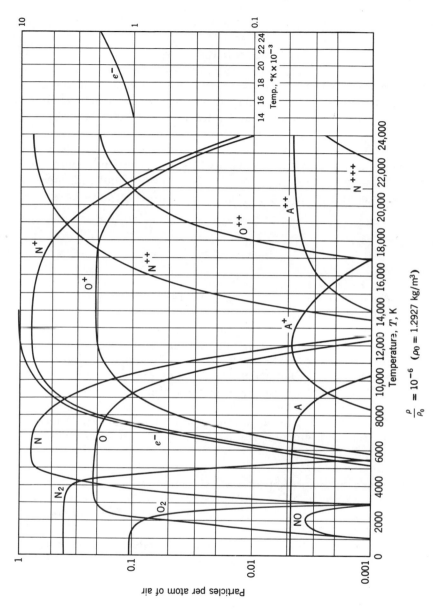

Fig. 13.14 Equilibrium composition of air (W. E. Moeckel & K. C. Weston, NACA TN 4265 (1958)).

$$\frac{\rho}{\rho_0} = 10^{-6} \quad (\rho_0 = 1.2927 \text{ kg/m}^3)$$

and the composition is found to be

$$y_{Ar} = 0.6656$$
$$y_{Ar^+} = 0.1672$$
$$y_{e^-} = 0.1672$$

Simultaneous reactions, such as simultaneous molecular dissociation and ionization reactions or multiple ionization reactions, can be analyzed in the same manner as the ordinary simultaneous chemical reactions of Section 13.7. In doing so, we again make the assumption of thermal equilibrium in the plasma, which as mentioned before is, in many cases, a reasonable approximation. Figure 13.14 shows the equilibrium composition of air at high temperature and low density, and indicates the overlapping regions of the various dissociation and ionization processes.

## PROBLEMS

13.1  Show that at the triple point of water all three phases have the same value of the Gibbs function.

13.2  Consider a gas well 2000 m deep filled with pure ethane. The temperature may be assumed to be uniform throughout the well at 35°C. The pressure is measured at the top of the well and found to be 5 MPa. What is the pressure at the bottom of the well?

13.3  Consider a gas well containing a mixture of methane and ethane at a uniform temperature of 30°C. The pressure at the top of the well is 14 MPa, and the molar composition is 90% $CH_4$, 10% $C_2H_6$. Determine the pressure and composition at a depth of 2000 metres, assuming
  (a) ideal gas mixture.
  (b) ideal solution.

13.4  Determine the phase compositions in the liquid-vapor equilibrium system nitrogen-oxygen at 110 K, 1 MPa pressure, and compare the results with the experimental values (50.0% $N_2$ in the liquid phase, 68.5% $N_2$ in the vapor phase).

13.5  Ethylene ($C_2H_4$) is burned with air and the products exit at 120°C, 3.5 MPa. Assuming complete combustion, what is the minimum percent theoretical air if there is to be no condensation of the products?

13.6  It is desired to maintain a liquid-vapor equilibrium mixture of argon and nitrogen at 100 K and a pressure at which the vapor will contain 50 mole percent of each component. What is this pressure? The saturation pressure for pure argon at 100 K is 315 kPa.

13.7 A spacecraft breathing mixture of oxygen and helium is to be prepared by the process shown in Fig. 13.15. Helium gas is bubbled through a liquid oxygen storage tank such that a saturated gas mixture of He and $O_2$ is in equilibrium with the pure liquid $O_2$ at 110 K and pressure $P$. The gas is withdrawn, throttled to 70 kPa and heated to 15°C. Assume that the tank temperature is uniform and constant at 110 K. What are the required tank pressure $P$ and the heat transfer in the heater?

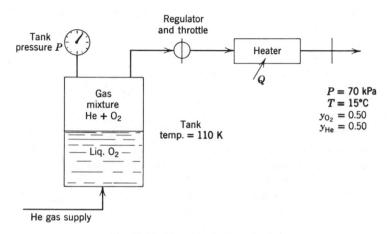

**Fig. 13.15**   Sketch for Problem 13.7.

13.8 Calculate the equilibrium composition of the liquid and vapor phases in the binary system carbon dioxide-sulfur dioxide at 30°C, 4 MPa, assuming that both phases behave as ideal solutions. Compare the results with the experimental values $x_{CO_2} = 0.46$, $y_{CO_2} = 0.88$, and evaluate the assumption of ideal solution for the two phases at this point.

13.9 An air-water vapor mixture is in equilibrium with liquid water (assuming the mole fraction of air in the liquid to be negligible) at 50°C, 3.5 MPa. Calculate the mole fraction of $H_2O$ in the gas phase, assuming
   (a) Raoult's rule — Ideal gas (the method used for saturated air-$H_2O$ mixtures in Chapter 11).
   (b) Ideal solution.

13.10 Consider the high-pressure combustion process shown in Fig. 13.16.
   (a) What is the phase of the products (all gas, gas + liquid, etc.)?
   (b) Calculate the heat transfer per mole of acetylene.
For $C_2H_2$: $T_c = 309$ K, $P_c = 6.24$ MPa

13.11 Consider the effect of gaseous nitrogen on the equilibrium of solid-vapor carbon dioxide at 190 K and a pressure of 10 MPa. Assume that the mole fraction of nitrogen in the solid phase is negligible.

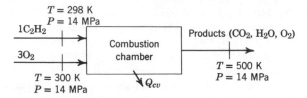

**Fig. 13.16**   Sketch for Problem 13.10.

Determine the mole fraction of $CO_2$ in the vapor by the following methods:

(a) Raoult's rule — ideal gas

(b) Ideal solution.

(c) Van der Waals' equation of state

(The experimental value for this pressure and temperature is $y_{CO_2} = 0.0324$). For carbon dioxide at 190 K,

$$P_{\text{sublimation}} = 68.8 \text{ kPa} \qquad \bar{v}_{\text{solid}} = 0.0280 \frac{\text{litre}}{\text{mol}} \quad .$$

13.12   Consider the hydrogen-oxygen fuel cell described in Example 12.17, in which it was assumed that pure liquid water exits at 25°C, 0.1 MPa.

(a) Determine the maximum EMF of the fuel cell operating at 25°C, assuming pure ideal gas water exits at 25°C, 0.1 MPa. What is the significance of this result?

(b) Repeat part a, considering the fuel cell to operate at a temperature of 600 K. The hydrogen and oxygen each enter at 600 K, 0.1 MPa, and ideal gas water exits at 600 K, 0.1 MPa.

13.13   Calculate the equilibrium constant for the reaction

$$N_2 \rightleftharpoons 2N$$

at temperatures of 298 K and 5000 K.

13.14   Plot to scale the values of $\ln K$ vs. $1/T$ for the reaction $CO_2 \rightleftharpoons CO + \frac{1}{2}O_2$. Write an equation for the $\ln K$ as a function of temperature.

13.15     (a) One mole of nitrogen at 25°C, 0.1 MPa pressure is heated to 5000 K in a constant-pressure process. Determine the equilibrium composition at 5000 K, 0.1 MPa pressure, assuming that $N_2$ and $N$ are the constituents present.

(b) Repeat part a for a pressure of 1 kPa.

13.16   Oxygen is heated from room temperature in a steady-flow process at a constant pressure of 1 MPa. At what temperature will the mole fraction of $O_2$ in the mixture of $O_2$ and $O$ be 40%? (That is, the mole fraction of the $O_2$ is 40% and of the $O$, 60%.)

13.17   Catalytic gas generators are frequently used to decompose a liquid, providing a desired gas mixture (spacecraft control systems, fuel cell gas supply, and so forth). Consider pure hydrazine $N_2H_4$ fed to a gas generator which yields an equilibrium mixture of $N_2$, $H_2$, $NH_3$ at 100°C, 350 kPa. Calculate the per cent composition of the mixture. For the reaction:

$$\tfrac{1}{2}N_2 + \tfrac{3}{2}H_2 \rightleftharpoons NH_3$$

$$\ln K = 3.491 \text{ at } 100°C$$

13.18 A mixture of one kmol CO, one kmol $N_2$ and one kmol $O_2$ at 298 K, 150 kPa is heated in a constant-pressure steady flow process. The exit mixture can be assumed in chemical equilibrium ($CO_2$, CO, $O_2$, $N_2$ present) and the mole fraction of $CO_2$ at this point is 0.176. Calculate the heat transfer for the process.

13.19 A mixture of 1 kmol $CO_2$, $\tfrac{1}{2}$ kmol CO, and 2 kmol $O_2$ at 25°C and 150 kPa pressure is heated in a constant pressure steady-flow process to 3000 K. Determine the equilibrium composition at 3000 K on a per cent basis, assuming the reaction equation to be

$$CO_2 \rightleftharpoons CO + \tfrac{1}{2}O_2$$

13.20 Repeat Problem 13.19 for the case where the initial mixture also includes 2 kmol $N_2$, which does not dissociate during the process.

13.21 Calculate the equilibrium constant at 3000 K for the reaction

$$2CO_2 \rightleftharpoons 2CO + O_2.$$

Repeat Problem 13.19 using this reaction equation instead of the one specified in Problem 13.19

13.22 Hydrogen fuel cells are in operation at the present time, for certain special applications. For most purposes, however, it would be more practical to be able to use a hydrocarbon fuel. One means of doing this is to "reform" the hydrocarbon to obtain $H_2$, which is then fed to the fuel cell. As a preliminary step in the analysis of such a procedure, consider the reforming reaction

$$CH_4 + H_2O \rightleftharpoons 3H_2 + CO$$

(a) Determine the equilibrium constant for this reaction at 800 K.

(b) One kmol $CH_4$ and 4 kmol $H_2O$ are fed to a catalytic reformer. Assume that an equilibrium mixture of $CH_4$, $H_2O$, $H_2$, and CO leaves at 800 K, 0.1 MPa. Find the equilibrium composition of this mixture.

13.23 Diatomic fluorine gas ($F_2$) enters a heat exchanger at 25°C, 700 kPa at the rate of 0.1 kmol/s. The gas is heated to 1600 K, during which process some of the fluorine dissociates according to the reaction

$$F_2 \rightleftharpoons 2F$$

Calculate:

(a) The equilibrium constant for the reaction at 1600 K.

(b) The composition of the mixture leaving the heat exchanger at 1600 K, 700 kPa.

(c) The heat transfer rate.

| | $(\bar{h}_f^\circ)_{298} \left(\dfrac{\text{kJ}}{\text{kmol}}\right)$ | $\bar{s}_{298}^\circ \left(\dfrac{\text{kJ}}{\text{kmol K}}\right)$ | $\bar{C}_{po} \left(\dfrac{\text{kJ}}{\text{kmol K}}\right)$ |
|---|---|---|---|
| $F_2$ | 0 | 202.715 | $27.196 + 4.142 \times 10^{-3}\, T$ |
| $F$ | 79 078 | 158.657 | 20.786 |

13.24 The presence of even small amounts of the various oxides of nitrogen in combustion products is an important factor in the air pollution problem. Consider a mixture comprised of the following basic products of combustion:

| | |
|---|---|
| $CO_2$ | 11 volume % |
| $H_2O$ | 12 |
| $O_2$ | 4 |
| $N_2$ | 73 |

At a condition of 2000 K, 0.2 MPa pressure, it is believed that NO (but not $NO_2$) will also be present in the mixture. Determine the percent composition of the mixture at this state without using Table A.12. The following data are known at 2000 K:

$$\tfrac{1}{2}N_2 + O_2 \rightleftharpoons NO_2, \; \Delta G^\circ = 142\ 549 \text{ kJ}$$

$$NO + \tfrac{1}{2}O_2 \rightleftharpoons NO_2, \; \Delta G^\circ = 77\ 195 \text{ kJ}$$

13.25 A mixture of 40% $CO_2$, 40% $O_2$, 20% $N_2$ (by volume) at 25°C, 175 kPa, is fed to a steady-flow heater at the rate of 0.1 kmol/s. An equilibrium mixture of $CO_2$, CO, $O_2$, $N_2$ leaves the heater at 2800 K, 175 kPa. Determine

  (a) The composition of the mixture leaving the heater.

  (b) Heat transfer to the heater.

13.26 One kmol of carbon monoxide at 0.1 MPa pressure, 1000 K, reacts with one kmol of oxygen at 0.1 MPa pressure, 1000 K to form an equilibrium mixture at 2800 K and a total pressure of 0.1 MPa.

  (a) Determine the heat transfer for this process, assuming that the reaction takes place in a steady-state, steady-flow process.

  (b) Calculate the irreversibility for this process.

13.27 Many combustion processes involve products of combustion at temperatures not greater than 1100°C. A gas turbine is a typical example. Usually the effects of dissociation are ignored at these temperatures. Is this justified for $CO_2$ and $H_2O$?

13.28 Methane is to be burned with oxygen in an adiabatic steady-flow process. Both enter the combustion chamber at 298 K, and the products leave at 70 kPa. What per cent excess $O_2$ should be used if the flame temperature is to be 2800 K? Assume that some of the $CO_2$ formed dissociates to CO and $O_2$, such that the products leaving the chamber consist of $CO_2$, CO, $H_2O$, and $O_2$ at equilibrium.

13.29 Methane at 25°C is burned with 200% theoretical oxygen at 400 K in a steady-state, steady-flow, adiabatic process. Assume that the only significant dissociation reaction in the products is that of $CO_2$ going to $CO$ and $O_2$. Determine the composition of the products and the exiting temperature of the products. Assume the pressure to be 100 kPa in the process.

13.30 One kmol of carbon (solid) and $b$ kmol of $O_2$ are contained in a closed tank. The initial conditions are $P_1 = 175$ kPa, $T_1 = 25°C$. The carbon is then ignited, and the final state is an equilibrium mixture of $CO_2$, $O_2$, and $CO$ at $T_2 = 3200$ K, and the mole fraction of $CO_2$ is 0.20. Calculate the initial number of kmol of $O_2$, $b$, and the final pressure, $P_2$.

13.31 A mixture of 2 kmol $CO_2$, 3 kmol $O_2$, and $b$ kmol $N_2$ at room temperature and 350 kPa is heated to 2200 K in a steady-flow process. Consider that at the outlet the mixture consists of $CO_2$, $O_2$, $N_2$, $CO$ in equilibrium. Determine the minimum value for $b$ if the mole fraction of $CO$ at the outlet is to be kept below 0.001.

13.32 An unknown hydrocarbon fuel $C_aH_b$ is burned with air. The resulting products are a chemical equilibrium mixture of $CO_2$, $CO$, $H_2O$, $O_2$ and $N_2$ at 3200 K, 175 kPa. The mole fraction of $O_2$ in this mixture is 0.0742 and the mole fraction of $H_2O$ is 0.0647. What is the fuel, and what percent theoretical air was used?

13.33 An important step in the manufacture of chemical fertilizer is the production of ammonia, according to the reaction

$$N_2 + 3H_2 \rightleftharpoons 2NH_3$$

   (a) Using the data given below, calculate the equilibrium constant for this reaction at 150°C.
   (b) For an initial composition of 25% $N_2$, 75% $H_2$, calculate the equilibrium composition at 150°C, 5 MPa pressure.

For $NH_3$: $(\bar{h}_f^\circ)_{298} = -46\ 191$ kJ/kmol; $(\bar{g}_f^\circ)_{298} = -16\ 622$ kJ/kmol

$$\bar{C}_{po} = 25.464 + 0.0664\,T - 20.416 \times 10^{-6}T^2 \text{ kJ/kmol K}$$

13.34 Repeat Problem 13.33 for a system pressure of 15 MPa, assuming the equilibrium to behave as an ideal solution.

13.35 A gas mixture at 25°C, 100 kPa, which consists of 50% $CO_2$ and 50% $CO$ (by volume) is fed to a chemical reactor at the rate of 1 kmol/s. Steam from a line at 200°C, 1.4 MPa is also fed to the reactor at the rate of 1 kmol/s. In the reactor the gases are heated in the presence of a catalyst such that the mixture leaving the reactor at 1000 K, 100 kPa may be assumed to be in equilibrium according to the reaction

$$CO + H_2O \rightleftharpoons CO_2 + H_2$$

Assume that all the constituents are ideal gases except for the inlet steam. Calculate:

*a*) The equilibrium constant for this reaction at 1000 K using the values from Table A.12.

(*b*) Enthalpy of the inlet steam relative to the elements at 25°C, 0.1 MPa.

(*c*) Composition (on a mole basis) of the mixture leaving the reactor.

(*d*) Heat transfer to the reactor.

13.36 One kmol of liquid hydrazine ($N_2H_4$) at 25°C enters an adiabatic combustor with $x$ kmol of $O_2$ at 25°C (where $x < 1$). If the products consist of an equilibrium mixture of $H_2O$, $H_2$, OH, $N_2$ at 3000 K, 150 kPa, determine the value of $x$. For liquid hydrazine, $\overline{h}_f^0 = +50\ 417$ kJ/kmol.

13.37 A problem of current interest is the production of methyl alcohol (to be used as a fuel replacing petroleum) from coal. Consider the following simplification of this process: a gas mixture of 50% CO, 50% $H_2$ leaves a coal gasifier at 400 K, 1 MPa, and enters a catalytic converter. A chemical equilibrium mixture of $CH_3OH$, CO and $H_2$ leaves the converter at the same temperature and pressure.

(*a*) Calculate the equilibrium constant for an appropriate reaction equation, and determine the equilibrium composition.

(*b*) Would it be more desirable to operate the converter at ambient pressure?

(*c*) Would it be more desirable to operate the converter at ambient temperature?

For ideal gas $CH_3OH$ at 25°C:

$$\overline{h}_f^0 = -201\ 167 \text{ kJ/kmol}$$
$$\overline{s}^0 = 239.70 \text{ kJ/kmol K}$$
$$\overline{C}_{po} = 45.02 \text{ kJ/kmol K}$$

13.38 A mixture of one kmol $H_2O$ and one kmol $O_2$ at 400 K is heated to 3000 K, 200 kPa pressure, in a steady-state, steady-flow process. Determine the equilibrium composition at the outlet of the heat exchanger, assuming that $H_2O$, $H_2$, $O_2$, and OH are present.

13.39 Butane is burned with 200% theoretical air, and the products, an equilibrium mixture containing $CO_2$, $O_2$, $H_2O$, $N_2$, NO and $NO_2$ only, exit at 1400 K, 2 MPa. Determine the equilibrium composition at this state.

13.40 One kmol of air (assume 78% $N_2$, 21% $O_2$, 1% Ar) at room temperature is heated to 4000 K, 200 kPa. Find the equilibrium composition at this temperature and pressure, assuming that only $N_2$, $O_2$, NO, O, and Ar are present.

13.41 Methane is burned with theoretical oxygen in a steady-state, steady-flow process. The products leave the combustion chamber at 3200 K, 700 kPa, and because of the high temperature some of the $CO_2$ and $H_2O$ will have dissociated. Calculate the equilibrium composition, assuming that $CO_2$, $H_2O$, CO, $H_2$, $O_2$, and OH are present.

13.42 One kmol of water vapor at 0.1 MPa pressure and 400 K is heated to 3000 K in a steady-flow constant-pressure process. Determine the final composition, assuming that $H_2O$, $H_2$, $H$, $O_2$, and $OH$ are present at equilibrium.

13.43 Consider a coal gasifier in which $\frac{1}{2}$ kmol of $O_2$ and $x$ kmols of $H_2O$ enter for each kmol of coal (C). Leaving the gasifier is a homogeneous chemical equilibrium ideal gas mixture at $T$ and $P$. This mixture contains $CH_4$, $H_2$, $H_2O$, $CO$ and $CO_2$.

(a) Determine the number of independent reaction equations for this system.

(b) Set up the equilibrium equations and specify the procedure followed to solve for the composition.

(c) Comment on any restrictions on $x$, and on whether it is desirable to have $x$ as small or as large as possible.

13.44 Operation of an MHD converter requires an electrically conducting gas. It is proposed to utilize helium gas "seeded" with 1.0 mole percent cesium, as shown in Fig. 13.17. The cesium is partly ionized ($Cs \rightleftharpoons Cs^+ + e^-$) by heating the mixture to 2500 K, 1 MPa in a nuclear reactor, in order to provide free electrons. No helium is ionized in this process, so that the mixture entering the converter consists of He, Cs, $Cs^+$, $e^-$. In order to analyze the converter process, it is necessary to know precisely the mole fraction of electrons in the mixture. Determine this fraction. At 2500 K, $1 \ln K = -13.4$ for the reaction given above.

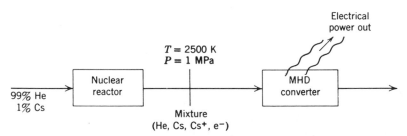

Fig. 13.17   Sketch for Problem 13.44.

13.45 One kmol of argon at room temperature is heated to 20 000 K, 0.1 MPa pressure. Assume that the plasma at this condition consists of an equilibrium mixture of Ar, $Ar^+$, $A^{++}$, $e^-$, according to the simultaneous reactions

(1) $Ar \rightleftharpoons Ar^+ + e^-$

(2) $Ar^+ \rightleftharpoons Ar^{++} + e^-$

The ionization-equilibrium constants for these reactions at 20 000 K have been calculated from spectroscopic data as: $\ln K_1 = 3.11$; $\ln K_2 = -4.92$. Determine the equilibrium composition of the plasma.

13.46 Plot the equilibrium composition of nitrogen at 10 kPa pressure (assume $N_2$, N, $N^+$, $e^-$ present) from 5000 K to 15 000 K. For the reaction

$$N \rightleftharpoons N^+ + e^-$$

the ionization-equilibrium constant $K$ has been calculated from spectroscopic data as

| $T(K)$ | $K$ |
|--------|-----|
| 10 000 | $6.26 \times 10^{-4}$ |
| 12 000 | $1.51 \times 10^{-2}$ |
| 14 000 | 0.151 |
| 16 000 | 0.92 |

13.47 Write a computer program to solve the following problem. One kmol of carbon at 25°C is to be burned with $b$ kmol of oxygen in a constant-pressure, adiabatic process. The products consist of an equilibrium mixture of $CO_2$, CO, and $O_2$. It is desired to determine the flame temperature for various combinations of $b$ and the pressure $P$, assuming variable specific heat for the constituents.

# 14

# Flow Through Nozzles and Blade Passages[1]

This chapter deals with the thermodynamic aspects of one-dimensional flow through nozzles and blade passages. In addition, the momentum equation for the control volume is developed and applied to these same problems. The sonic velocity is defined in terms of thermodynamic properties, and the importance of the Mach number as a variable in compressible flow is noted.

## 14.1 Stagnation Properties

In dealing with problems involving flow, many discussions and equations can be simplified by introducing the concept of the isentropic stagnation state and the properties associated with it. The isentropic stagnation state is the state a flowing fluid would attain if it underwent a reversible adiabatic deceleration to zero velocity. This state is designated in this chapter with the subscript 0. From the first law for a steady-state,

[1]As indicated in the preface, the basic idea of the Thermal and Transport Sciences Series is to present basic texts in thermodynamics, fluid mechanics, heat transfer, and statistical thermodynamics, followed by certain applications of these subjects. As the first book in this series, the original intention had been to limit the coverage in this book to thermodynamics, and to encourage the use of the next volume in this series, *Fluid Mechanics*, by A. G. Hansen, for a thorough coverage of fluid mechanics. However, this final chapter has been included in order to make the book complete in itself and perhaps even to contribute to the continuity of the series. It should be emphasized however, that a complete analysis of fluid flow problems involves not only thermodynamics, but many other topics, such as viscosity, turbulence, similarity laws, and boundary layer theory.

steady-flow process we conclude that

$$h + \frac{V^2}{2} = h_0 \qquad (14.1)$$

The actual and the isentropic stagnation states for a typical gas or vapor are shown on the $h$-$s$ diagram of Fig. 14.1. Sometimes it is advantageous to make a distinction between the actual and the isentropic stagnation states. The actual stagnation state is the state achieved after

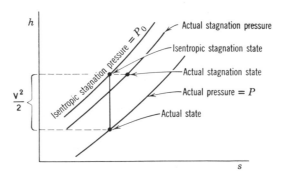

**Fig. 14.1**   Enthalpy-entropy diagram illustrating the definition of stagnation state.

an actual deceleration to zero velocity (as at the nose of a body placed in a fluid stream), and there may be irreversibilities associated with the deceleration process. Therefore, the term stagnation property is sometimes reserved for the properties associated with the actual state, and the term total property is used for the isentropic stagnation state.

It is evident from Fig. 14.1 that the enthalpy is the same for both the actual and isentropic stagnation states (assuming that the actual process is adiabatic). Therefore, for an ideal gas, the actual stagnation temperature is the same as the isentropic stagnation temperature. However, the actual stagnation pressure may be less than the isentropic stagnation pressure and for this reason the term total pressure (meaning isentropic stagnation pressure) has particular meaning compared to the actual stagnation pressure.

**Example 14.1**

Air flows in a duct at a pressure of 150 kPa with a velocity of 200 m/s. The temperature of the air is 300 K. Determine the isentropic stagnation pressure and temperature.

If we assume that the air is an ideal gas with constant specific heat as given in Table A.8, the calculation is as follows. From Eq. 14.1

$$\frac{V^2}{2} = h_0 - h = C_{po}(T_0 - T)$$

$$\frac{(200)^2}{2 \times 1000} = 1.0035\ (T_0 - 300)$$

$$T_0 = 319.9\ \text{K}$$

The stagnation pressure can be found from the relation

$$\frac{T_0}{T} = \left(\frac{P_0}{P}\right)^{(k-1)/k}$$

$$\frac{319.9}{300} = \left(\frac{P_0}{150}\right)^{0.286}$$

$$P_0 = 187.8\ \text{kPa}$$

The Air Tables, Table A.10, which are abridged from Keenan and Kayes' *Gas Tables* could also have been used, and then the variation of specific heat with temperature would have been taken into account. Since the actual and stagnation states have the same entropy, we proceed as follows: Using Table 1 of the *Gas Tables* (Appendix Table A.10),

$$T = 300\ \text{K} \qquad h = 300.19 \qquad P_r = 1.3860$$

$$h_0 = h + \frac{V^2}{2} = 300.19 + \frac{(200)^2}{2 \times 1000} = 320.19$$

$$T_0 = 319.9\ \text{K} \qquad P_{ro} = 1.7357$$

$$P_0 = 150 \times \frac{1.7357}{1.3860} = 187.8\ \text{kPa}$$

## 14.2 The Momentum Equation for the Control Volume

Before proceeding it will be advantageous to develop the momentum equation for the control volume. Newton's second law states that the sum of the external forces acting on a body in a given direction is proportional to the rate of change of momentum in the given direction.

Writing this in equation form for the $x$-direction we have

$$\sum F_x \propto \frac{d(mV_x)}{dt}$$

For the system of units used in this book, this proportionality can be written directly as an equality.

$$\sum F_x = \frac{d(mV_x)}{dt} \tag{14.2}$$

Equation 14.2 has been written for a body of fixed mass, or in thermo-dynamic parlance, for a system. We now proceed to write the momentum equation for a control volume, and follow a procedure similar to that used in writing the continuity equation and the first and second laws of thermodynamics for a control volume.

Consider the system and control volume shown in Fig. 14.2. Let the control volume be fixed relative to its coordinate frame. During the time interval $\delta t$, the mass $\delta m_i$ enters the control volume with a velocity $(V_r)_i$ and velocity components $(V_x)_i$, $(V_y)_i$, and $(V_z)_i$. During this same time interval the mass $\delta m_e$ leaves the control volume, with velocity $(V_r)_e$ and velocity components $(V_x)_e$, $(V_y)_e$, and $(V_z)_e$.

If we write the $x$-momentum equation for the system during this time interval we have

$$\left(\sum F_x\right)_{av} = \frac{\Delta(mV_x)}{\delta t} = \frac{(mV_x)_2 - (mV_x)_1}{\delta t} \tag{14.3}$$

Let $(mV_x)_t = x$-momentum in the control volume at time $t$.
$(mV_x)_{t+\delta t} = x$-momentum in the control volume at time $t + \delta t$.

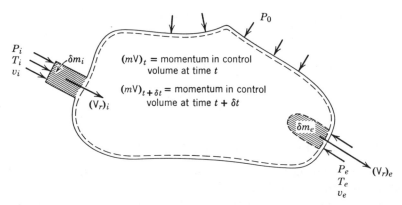

**Fig. 14.2** Schematic diagram for the development of the momentum equation for a control volume.

Then

$$(m\mathbf{V}_x)_1 = (m\mathbf{V}_x)_t + (\mathbf{V}_x)_i \, \delta m_i = x\text{-momentum of the system at time } t.$$

$$(m\mathbf{V}_x)_2 = (m\mathbf{V}_x)_{t+\delta t} + (\mathbf{V}_x)_e \, \delta m_e = x\text{-momentum of the system at time}$$
$$t + \delta t.$$

It follows that

$$(m\mathbf{V}_x)_2 - (m\mathbf{V}_x)_1 = [(m\mathbf{V}_x)_{t+\delta t} - (m\mathbf{V}_x)_t] + [(\mathbf{V}_x)_e \, \delta m_e - (\mathbf{V}_x)_i \, \delta m_i]$$

$$(14.4)$$

The first bracketed term on the right side of Eq. 14.4 represents the change of $x$-momentum within the control volume during the time interval $\delta t$. This can be written as a volume integral in the form

$$(m\mathbf{V}_x)_{t+\delta t} - (m\mathbf{V}_x)_t = \delta \int_V \mathbf{V}_x \rho \, dV \qquad (14.5)$$

Dividing by $\delta t$, we have

$$\frac{(m\mathbf{V}_x)_{t+\delta t} - (m\mathbf{V}_x)_t}{\delta t} = \frac{\delta}{\delta t} \int_V \mathbf{V}_x \rho \, dV \qquad (14.6)$$

The second bracketed term in Eq. 14.4 represents the momentum flow across the control surface during $\delta t$, and can be written as a surface integral.

$$(\mathbf{V}_x)_e \, \delta m_e - (\mathbf{V}_x)_i \, \delta m_i = \left[ \iint_{\mathscr{A}} \mathbf{V}_x \rho \mathbf{V}_{rn} \, d\mathscr{A} \right]_{av} \delta t \qquad (14.7)$$

Dividing by $\delta t$ we have the average rate at which momentum crosses the control surface during $\delta t$.

$$\frac{(\mathbf{V}_x)_e \, \delta m_e - (\mathbf{V}_x)_i \, \delta m_i}{\delta t} = \left[ \iint_{\mathscr{A}} \mathbf{V}_x \rho \mathbf{V}_{rn} \, d\mathscr{A} \right]_{av} \qquad (14.8)$$

Substituting Eqs. 14.14, 14.6, and 14.8 into Eq. 14.3 we have

$$\left( \sum F_x \right)_{av} = \frac{\delta}{\delta t} \int_V \mathbf{V}_x \rho \, dV + \left[ \iint_{\mathscr{A}} \mathbf{V}_x \rho \mathbf{V}_{rn} \, d\mathscr{A} \right]_{av} \qquad (14.9)$$

We now establish the limit for each of these terms as $\delta t \to 0$.

$$\lim_{\delta t \to 0} \left( \sum F_x \right)_{av} = \sum F_x$$

$$\lim_{\delta t \to 0} \frac{\delta}{\delta t} \int_V \mathbf{V}_x \rho \, dV = \frac{d}{dt} \int_V \mathbf{V}_x \rho \, dV \qquad (14.10)$$

$$\lim_{\delta t \to 0} \left[ \iint_{\mathscr{A}} \mathbf{V}_x \rho \mathbf{V}_{rn} \, d\mathscr{A} \right]_{av} = \int \mathbf{V}_x \rho \mathbf{V}_{rn} \, d\mathscr{A}$$

Thus, as $\delta t \rightarrow 0$ we have a rate form of the momentum equation for the control volume.

$$\sum F_x = \frac{d}{dt} \int_V \mathbf{V}_x \rho \, dV + \int_{\mathcal{A}} \mathbf{V}_x \rho \mathbf{V}_{rn} \, d\mathcal{A} \qquad (14.11)$$

Similar equations can also be written for $y$ and $z$ directions.

$$\sum F_y = \frac{d}{dt} \int_V \mathbf{V}_y \rho \, dV + \int_{\mathcal{A}} \mathbf{V}_y \rho \mathbf{V}_{rn} \, d\mathcal{A} \qquad (14.12)$$

$$\sum F_z = \frac{d}{dt} \int_V \mathbf{V}_z \rho \, dV + \int_{\mathcal{A}} \mathbf{V}_z \rho \mathbf{V}_{rn} \, d\mathcal{A} \qquad (14.13)$$

In this chapter we will be concerned primarily with steady-state, steady-flow processes in which there is a single flow with uniform properties into the control surface, and a single flow with uniform properties out of the control surface. The assumption of steady-state, steady-flow means that the volume integrals in Eqs. 14.11, 14.12, and 14.13 are equal to zero. That is

$$\frac{d}{dt} \int_V \mathbf{V}_x \rho \, dV = 0; \qquad \frac{d}{dt} \int_V \mathbf{V}_y \rho \, dV = 0; \qquad \frac{d}{dt} \int_V \mathbf{V}_z \rho \, dV = 0$$

The assumption of uniform properties both into and out of the control volume means that we can write

$$\int_{\mathcal{A}} \mathbf{V}_x \rho \mathbf{V}_{rn} \, d\mathcal{A} = \sum \dot{m}_e (\mathbf{V}_e)_x - \sum \dot{m}_i (\mathbf{V}_i)_x$$
$$\int_{\mathcal{A}} \mathbf{V}_y \rho \mathbf{V}_{rn} \, d\mathcal{A} = \sum \dot{m}_e (\mathbf{V}_e)_y - \sum \dot{m}_i (\mathbf{V}_i)_y \qquad (14.14)$$
$$\int_{\mathcal{A}} \mathbf{V}_z \rho \mathbf{V}_{rn} \, d\mathcal{A} = \sum \dot{m}_e (\mathbf{V}_e)_y - \sum \dot{m}_i (\mathbf{V}_i)_z$$

Therefore, for such a process the momentum equation for the control volume reduces to

$$\sum F_x = \sum \dot{m}_e (\mathbf{V}_e)_x - \sum \dot{m}_i (\mathbf{V}_i)_x$$

$$\sum F_y = \sum \dot{m}_e (\mathbf{V}_e)_y - \sum \dot{m}_i (\mathbf{V}_i)_y \qquad (14.15)$$

$$\sum F_z = \sum \dot{m}_e (\mathbf{V}_e)_z - \sum \dot{m}_i (\mathbf{V}_i)_z$$

If with these same assumptions, there is a single flow into and out of the control volume these equations reduce to

$$\sum F_x = \dot{m}\,[(V_e)_x - (V_i)_x]$$

$$\sum F_y = \dot{m}\,[(V_e)_y - (V_i)_y] \tag{14.16}$$

$$\sum F_z = \dot{m}\,[(V_e)_z - (V_i)_z]$$

**Example 14.2**

On a level floor a man is pushing a wheelbarrow (Fig. 14.3) into which sand is falling at the rate of 1 kg/s. The man is walking at the rate of 1 m/s and the sand has a velocity of 10 m/s as it falls into the wheelbarrow. Determine the force the man must exert on the wheelbarrow and the force the floor exerts on the wheelbarrow due to the falling sand.

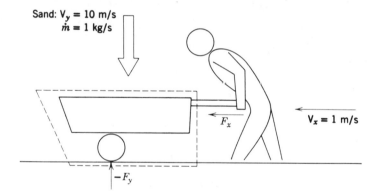

**Fig. 14.3**  Sketch for Example 14.2.

Consider a control surface around the wheelbarrow. Consider first the x-direction. From Eq. 14.11

$$\sum F_x = F_x = \frac{d}{dt}\int_V V_x \rho \, dV + \int_{\mathcal{A}} V_x \rho V_{rn}\, d\mathcal{A}$$

Let us analyze this problem from the point of view of an observer riding on the wheelbarrow. For this observer, $V_x$ of the material in the wheelbarrow is zero and therefore,

$$\frac{d}{dt}\int_V V_x \rho \, dV = 0$$

However, for this observer the sand crossing the control surface has an x-component velocity of $-1$ m/s, and $\int_{\mathscr{A}} \rho V_{rn} d\mathscr{A}$, the mass flow out of the control volume is $-1$ kg/s. Therefore,

$$F_x = (1 \text{ kg/s}) \times (1 \text{ m/s}) = 1\text{N}$$

If one considers this from the point of view of an observer who is stationary on the earth's surface we conclude that $V_x$ of the falling sand is zero and therefore

$$\int_{\mathscr{A}} V_x \rho V_{rn} d\mathscr{A} = 0$$

However for this observer there is a change of momentum within the control volume, namely,

$$F_x = \frac{d}{dt} \int_V V_x \rho \, dV = (1 \text{ m/s}) \times (1 \text{ kg/s}) = 1\text{N}$$

Next consider the vertical ($y$) direction.

$$\sum F_y = F_y = \int_V V_y \rho \, dV + \int_{\mathscr{A}} V_y \rho V_{rn} d\mathscr{A}$$

For both the stationary and moving observer the first term drops out because $V_y$ of the mass within the control volume is zero. However, for the mass crossing the control surface, $V_y = 10$ m/s and

$$\int_{\mathscr{A}} \rho V_{rn} d\mathscr{A} = -1 \text{ kg/s}$$

Therefore,

$$F_y = (10 \text{ m/s}) \times (-1 \text{ kg/s}) = -10\text{N}$$

The minus sign indicates that the force is in the opposite direction to $V_y$.

## 14.3   Forces Acting on a Control Surface

In the last section we considered the momentum equation for the control volume. We now wish to evaluate the net force on a control surface which causes this change in momentum. Let us do this by considering the system shown in Fig. 14.4, which involves a pipe bend. The

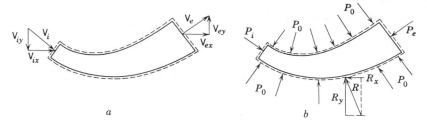

**Fig. 14.4**  Forces acting on a control surface.

control surface is designated by the dotted lines, and is so chosen that at the point where the fluid crosses the system boundary the flow is perpendicular to the control surface. The shear forces at the section where the fluid crosses the boundary of the system are assumed to be negligible. Figure 14.4a shows the velocities and Fig. 14.4b shows the forces involved. The force $R$ is the result of all external forces on the system, except for the pressure of the surroundings. The pressure of the surroundings, $P_0$, acts on the entire boundary except at $\mathscr{A}_i$ and $\mathscr{A}_e$, where the fluid crosses the control surface. $P_i$ and $P_e$ represent the absolute pressures at these points.

The net forces acting on the system in the $x$ and $y$ directions, $F_x$ and $F_y$, are the sum of the pressure forces and the external force $R$ in their respective directions. The influence of the pressure of the surroundings, $P_0$, is most easily taken into account by noting that it acts over the entire system boundary except at $\mathscr{A}_i$ and $\mathscr{A}_e$. Therefore, we can write

$$\sum F_x = (P_i \mathscr{A}_i)_x - (P_0 \mathscr{A}_i)_x + (P_e \mathscr{A}_e)_x - (P_0 \mathscr{A}_e)_x + R_x$$
$$\sum F_y = (P_i \mathscr{A}_i)_y - (P_0 \mathscr{A}_i)_y + (P_e \mathscr{A}_e)_y - (P_0 \mathscr{A}_e)_y + R_y$$

This equation may be simplified by combining the pressure terms.

$$\sum F_x = [(P_i - P_0)\mathscr{A}_i]_x + [(P_e - P_0)\mathscr{A}_e]_x + R_x$$
$$\sum F_y = [(P_i - P_0)\mathscr{A}_i]_y + [(P_e - P_0)\mathscr{A}_e]_y + R_y \tag{14.17}$$

The proper sign for each pressure and force must of course be used in all calculations.

Equations 14.15 and 14.17 may be combined to give

$$\sum F_x = \sum \dot{m}_e (V_e)_x - \sum \dot{m}_i (V_i)_x = \sum [(P_i - P_0)\mathscr{A}_i]_x$$
$$+ \sum [(P_e - P_0)\mathscr{A}_e]_x + R_x \tag{14.18}$$

$$\sum F_y = \sum \dot{m}_e(V_e)_y - \sum \dot{m}_i(V_i)_y = \sum \left[(P_i - P_0)\mathscr{A}_i\right]_y$$
$$+ \sum \left[(P_e - P_0)\mathscr{A}_e\right]_y + R_y$$

If there is a single flow across the control surface, Eqs. 14.16 and 14.17 can be combined to give

$$\sum F_x = \dot{m}\,(V_e - V_i)_x = \left[(P_i - P_0)\mathscr{A}_i\right]_x + \left[(P_e - P_0)\mathscr{A}_e\right]_x + R_x$$

$$\sum F_y = \dot{m}\,(V_e - V_i)_y = \left[(P_i - P_0)\mathscr{A}_i\right]_y + \left[(P_e - P_0)\mathscr{A}_e\right]_y + R_y$$

(14.19)

A similar equation could be written for the z-direction. These equations are very useful in analyzing the forces involved in a control volume analysis.

**Example 14.3**

A jet engine is being tested on a test stand (Fig. 14.5). The inlet area to the compressor is 0.2 m² and air enters the compressor at 95 kPa, 100 m/s. The pressure of the atmosphere is 100 kPa. The exit area

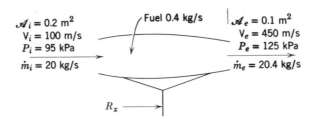

**Fig. 14.5**   Sketch for Example 14.3.

of the engine is 0.1 m², and the products of combustion leave the exit plane at a pressure of 125 kPa and a velocity of 450 m/s. The air-fuel ratio is 50 kg air/kg fuel, and the fuel enters with a low velocity. The rate of air flow entering the engine is 20 kg/s. Determine the thrust on the engine.

In the solution that follows it is assumed that forces and velocities to the right are positive.

Using Eq. 14.19

$$R_x + \left[(P_i - P_0)\mathscr{A}_i\right]_x + \left[(P_e - P_0)\mathscr{A}_e\right]_x = (\dot{m}_e V_e - \dot{m}_i V_i)_x$$

$$R_x + [(95 - 100) \times 0.2] - [(125 - 100) \times 0.1]$$

$$= \frac{20.4 \times 450 - 20 \times 100}{1000}$$

$$R_x = 10.68 \text{ kN}$$

(Note that the momentum of the fuel entering has been neglected.)

## 14.4   Adiabatic, One-Dimensional, Steady-State, Steady Flow of an Incompressible Fluid through a Nozzle

A nozzle is a device in which the kinetic energy of a fluid is increased in an adiabatic process. This increase involves a decrease in pressure and is accomplished by the proper change in flow area. A diffuser is a device that has the opposite function, namely, to increase the pressure by decelerating the fluid. In this section we discuss both nozzles and diffusers, but to minimize words we shall use only the term nozzle.

Consider the nozzle shown in Fig. 14.6, and assume an adiabatic, one-dimensional, steady-state, steady-flow process of an incompressible fluid. From the continuity equation we conclude that

$$\dot{m}_e = \dot{m}_i = \rho \mathcal{A}_i V_i = \rho \mathcal{A}_e V_e$$

or

$$\frac{\mathcal{A}_i}{\mathcal{A}_e} = \frac{V_e}{V_i} \tag{14.20}$$

The first law for this process is

$$h_e - h_i + \frac{V_e^2 - V_i^2}{2} + (Z_e - Z_i)g = 0 \tag{14.21}$$

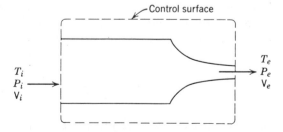

**Fig. 14.6**   Schematic sketch of a nozzle.

From the second law we conclude that $s_e \geq s_i$, where the equality holds for a reversible process. Therefore, from the relation

$$T\,ds = dh - v\,dP$$

we conclude that for the reversible process

$$h_e - h_i = \int_i^e v\,dP \tag{14.22}$$

If we assume that the fluid is incompressible, Eq. 14.22 can be integrated to give

$$h_e - h_i = v(P_e - P_i) \tag{14.23}$$

Substituting this in Eq. 14.21 we have

$$v(P_e - P_i) + \frac{V_e^2 - V_i^2}{2} + (Z_e - Z_i)g = 0 \tag{14.24}$$

This is of course the Bernouilli Equation which was derived in Section 7.13, Eq. 7.64, and for the reversible, adiabatic one-dimensional steady-state, steady-flow of an incompressible fluid through a nozzle the Bernoulli Equation represents a combined statement of the first and second laws of thermodynamics.

**Example 14.4**

Water enters the diffuser in a pump casing with a velocity of 30 m/s, a pressure of 350 kPa, and a temperature of 25°C. It leaves the diffuser with a velocity of 7 m/s and a pressure of 600 kPa. Determine the exit pressure for a reversible diffuser with these inlet conditions and exit velocity. Determine the increase in enthalpy, internal energy, and entropy for the actual diffuser.

Consider first a control surface around a reversible diffuser with the given inlet conditions and exit velocity. Equation 14.24, the Bernoulli equation, is a statement of the first and second laws of thermodynamics for this process. Since there is no change in elevation this equation reduces to

$$v[(P_e)_s - P_i] + \frac{V_e^2 - V_i^2}{2} = 0$$

where $(P_e)_s$ represents the exit pressure for the reversible diffuser. From the steam tables, $v = 0.001\,003$ m³/kg.

$$P_{es} - P_i = \frac{(30)^2 - (7)^2}{0.001\ 003 \times 2 \times 1000} = 424 \text{ kPa}$$

$$P_{es} = 774 \text{ kPa}$$

Next consider a control surface around the actual diffuser. The change in enthalpy can be found from the first law for this process, Eq. 14.21.

$$h_e - h_i = \frac{V_i^2 - V_e^2}{2} = \frac{(30)^2 - (7)^2}{2 \times 1000} = 0.4255 \text{ kJ/kg}$$

The change in internal energy can be found from the definition of enthalpy, $h_e - h_i = (u_e - u_i) + (P_e v_e - P_i v_i)$.

Thus, for an incompressible fluid

$$u_e - u_i = h_e - h_i - v(P_e - P_i)$$

$$= 0.4255 - 0.001\ 003\ (600 - 350)$$

$$= 0.174\ 75 \text{ kJ/kg}$$

The change of entropy can be approximated from the familiar relation

$$T\ ds = du + P\ dv$$

by assuming that the temperature is constant (which is approximately true in this case) and noting that for an incompressible fluid $dv = 0$. With these assumptions

$$s_e - s_i = \frac{u_e - u_i}{T} = \frac{0.174\ 75}{298.2} = 0.000\ 586 \text{ kJ/kg K}$$

Since this is an irreversible adiabatic process, the entropy will increase, as the above calculation indicates.

## 14.5   Velocity of Sound in an Ideal Gas

When a pressure disturbance occurs in a compressible fluid, the disturbance travels with a velocity that depends on the state of the fluid. A sound wave is a very small pressure disturbance; the velocity of sound, also called the sonic velocity, is an important parameter in compressible-fluid flow. We proceed now to determine an expression for the sonic velocity of an ideal gas in terms of the properties of the gas.

Let a disturbance be set up by the movement of the piston at the end of the tube, Fig. 14.7a. A wave travels down the tube with a velocity $c$, which is the sonic velocity. Assume that after the wave has passed the properties of the gas have changed an infinitesimal amount and that the gas is moving with the velocity $dV$ toward the wave front.

In Fig. 14.7b this process is shown from the point of view of an observer who travels with the wave front. Consider the control surface shown in Fig. 14.7b. From the first law for this steady-state, steady-flow process we can write,

$$h + \frac{c^2}{2} = (h + dh) + \frac{(c - dV)^2}{2} \qquad (14.25)$$

$$dh - c\,dV = 0$$

From the continuity equation we can write

$$\rho \mathscr{A} c = (\rho + d\rho)\mathscr{A}(c - dV)$$
$$c\,d\rho - \rho\,dV = 0 \qquad (14.26)$$

Consider also the relation between properties

$$T\,ds = dh - \frac{dP}{\rho}$$

If the process is isentropic, $ds = 0$, and this equation can be combined with Eq. 14.25 to give the relation

$$\frac{dP}{\rho} - c\,dV = 0 \qquad (14.27)$$

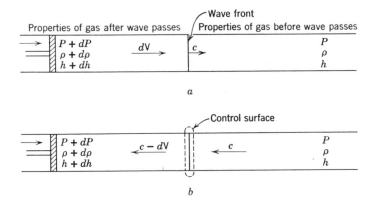

**Fig. 14.7**  Diagram illustrating sonic velocity. (a) Stationary observer. (b) Observer traveling with wave front.

This can be combined with Eq. 14.26 to give the relation

$$\frac{dP}{d\rho} = c^2$$

Since we have assumed the process to be isentropic this is better written as a partial derivative.

$$\left(\frac{\partial P}{\partial \rho}\right)_s = c^2 \tag{14.28}$$

An alternate derivation is to introduce the momentum equation. For the control volume of Fig. 14.7b the momentum equation is

$$P\mathscr{A} - (P+dP)\mathscr{A} = \dot{m}\ (c-dV-c) = \rho\mathscr{A}c\,(c-dV-c)$$

$$dV = \rho c\, dP \tag{14.29}$$

On combining this with Eq. 14.26 we obtain Eq. 14.28.

$$\left(\frac{\partial P}{\partial \rho}\right)_s = c^2$$

It will be of particular advantage to solve Eq. 14.28 for the velocity of sound in an ideal gas.

When an ideal gas undergoes an isentropic change of state, we found in Chapter 7 that, for this process, assuming constant specific heat,

$$\frac{dP}{P} - k\frac{d\rho}{\rho} = 0$$

or

$$\left(\frac{\partial P}{\partial \rho}\right)_s = \frac{kP}{\rho}$$

Substituting this equation in Eq. 14.28 we have an equation for the velocity of sound in an ideal gas,

$$c^2 = \frac{kP}{\rho} \tag{14.30}$$

Since for an ideal gas

$$\frac{P}{\rho} = RT$$

this equation may also be written

$$c^2 = kRT \qquad (14.31)$$

**Example 14.5**

Determine the velocity of sound in air at 300 K and at 1000 K. Using Eq. 14.31

$$c = \sqrt{kRT}$$

$$= \sqrt{1.4 \times 0.287 \times 300 \times 1000} = 347.2 \text{ m/s}$$

Similarly, at 1000 K, using $k = 1.4$

$$c = \sqrt{1.4 \times 0.287 \times 1000 \times 1000} = 633.9 \text{ m/s}$$

Note the significant increase in sonic velocity as the temperature increases.

The Mach number, $M$, is defined as the ratio of the actual velocity $V$ to the sonic velocity $c$.

$$M = \frac{V}{c} \qquad (14.32)$$

When $M > 1$ the flow is supersonic; when $M < 1$ the flow is subsonic: and when $M = 1$ the flow is sonic. The importance of the Mach number as a parameter in fluid-flow problems will be evident in the paragraphs which follow.

## 14.6   Reversible, Adiabatic, One-Dimensional, Steady Flow of an Ideal Gas through a Nozzle

A nozzle or diffuser with both a converging and diverging section is shown in Fig. 14.8. The minimum cross-sectional area is called the throat.

Our first consideration concerns the conditions that determine whether a nozzle or diffuser should be converging or diverging, and the conditions that prevail at the throat. For the control volume shown the following relations can be written.

First law:

$$dh + V\, dV = 0 \qquad (14.33)$$

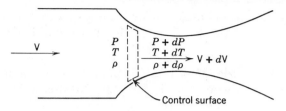

**Fig. 14.8**   One-dimensional reversible adiabatic steady flow through a nozzle.

Property relation:

$$T\, ds = dh - \frac{dP}{\rho} = 0 \qquad (14.34)$$

Continuity equation:

$$\rho \mathscr{A} V = \dot{m} = \text{constant}$$

$$\frac{d\rho}{\rho} + \frac{d\mathscr{A}}{\mathscr{A}} + \frac{dV}{V} = 0 \qquad (14.35)$$

Combining Eqs. 14.33 and 14.34 we have

$$dh = \frac{dP}{\rho} = -V\, dV$$

$$dV = -\frac{1}{\rho V}\, dP$$

Substituting this in Eq. 14.35

$$\frac{d\mathscr{A}}{\mathscr{A}} = \left(-\frac{d\rho}{\rho} - \frac{dV}{V}\right) = -\frac{d\rho}{\rho}\left(\frac{dP}{dP}\right) + \frac{1}{\rho V^2}\, dP$$

$$= \frac{-dP}{\rho}\left(\frac{d\rho}{dP} - \frac{1}{V^2}\right) = \frac{dP}{\rho}\left(-\frac{1}{(dP/d\rho)} + \frac{1}{V^2}\right)$$

Since the flow is isentropic

$$\frac{dP}{d\rho} = c^2 = \frac{V^2}{M^2}$$

and therefore,

$$\frac{d\mathscr{A}}{\mathscr{A}} = \frac{dP}{\rho V^2}\,(1 - M^2) \qquad (14.36)$$

This is a very significant equation, for from it we can draw the following conclusions about the proper shape for nozzles and diffusers:

For a nozzle, $dP < 0$. Therefore,
   for a subsonic nozzle, $M < 1$, $d\mathscr{A} < 0$, and the nozzle is converging.
   for a supersonic nozzle, $M > 1$, $d\mathscr{A} > 0$, and the nozzle is diverging.
For a diffuser, $dP > 0$. Therefore,
   for a subsonic diffuser, $M < 1$, $d\mathscr{A} > 0$, and the diffuser is diverging.
   for a supersonic diffuser, $M > 1$, $d\mathscr{A} < 0$, and the diffuser is converging.

When $M = 1$, $d\mathscr{A} = 0$, which means that sonic velocity can be achieved only at the throat of a nozzle or diffuser. These conclusions are summarized in Fig. 14.9.

We will now develop a number of relations between the actual properties, stagnation properties, and Mach number. These relations are very useful in dealing with isentropic flow of an ideal gas in a nozzle.

Equation 14.1 gives the relation between enthalpy, stagnation enthalpy, and kinetic energy.

$$h + \frac{V^2}{2} = h_0$$

For an ideal gas with constant specific heat Eq. 14.1 can be written

$$V^2 = 2C_{po}\,(T_0 - T) = 2\frac{kRT}{k-1}\left(\frac{T_0}{T} - 1\right)$$

Since

$$c^2 = k\,RT$$

$$V^2 = \frac{2c^2}{k-1}\left(\frac{T_0}{T} - 1\right)$$

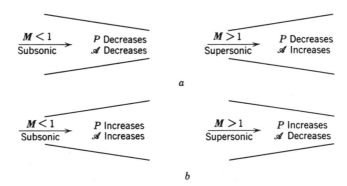

| | | | |
|---|---|---|---|
| $M < 1$ Subsonic → | P Decreases $\mathscr{A}$ Decreases | $M > 1$ Supersonic → | P Decreases $\mathscr{A}$ Increases |

*a*

| | | | |
|---|---|---|---|
| $M < 1$ Subsonic → | P Increases $\mathscr{A}$ Increases | $M > 1$ Supersonic → | P Increases $\mathscr{A}$ Decreases |

*b*

**Fig. 14.9**   Required area changes for (*a*) nozzles, and (*b*) diffusers.

$$\frac{V^2}{c^2} = M^2 = \frac{2}{k-1}\left(\frac{T_0}{T} - 1\right)$$

$$\frac{T_0}{T} = 1 + \frac{(k-1)}{2}M^2 \qquad (14.37)$$

For an isentropic process,

$$\left(\frac{T_0}{T}\right)^{k/(k-1)} = \frac{P_0}{P} \qquad \left(\frac{T_0}{T}\right)^{1/(k-1)} = \frac{\rho_0}{\rho}$$

Therefore,

$$\frac{P_0}{P} = \left[1 + \frac{(k-1)}{2}M^2\right]^{k/(k-1)} \qquad (14.38)$$

$$\frac{\rho_0}{\rho} = \left[1 + \frac{(k-1)}{2}M^2\right]^{1/(k-1)} \qquad (14.39)$$

Values of $P/P_0$, $\rho/\rho_0$, and $T/T_0$ are given as a function of $M$ in Tables 30 to 35 of the *Gas Tables*, each table being for a given value of $k$. Table A.13 of the Appendix has been abstracted from Table 30 of the *Gas Tables*, and applies to an ideal gas with $k = 1.4$.

The conditions at the throat of the nozzle can be found by noting that $M = 1$ at the throat. The properties at the throat are denoted by an asterisk.* Therefore,

$$\frac{T^*}{T_0} = \frac{2}{k+1} \qquad (14.40)$$

$$\frac{P^*}{P_0} = \left(\frac{2}{k+1}\right)^{k/(k-1)} \qquad (14.41)$$

$$\frac{\rho^*}{\rho_0} = \left(\frac{2}{k+1}\right)^{1/(k-1)} \qquad (14.42)$$

These properties at the throat of a nozzle when $M = 1$ are frequently referred to as critical pressure, critical temperature, and critical density and the ratios given by Eqs. 14.40, 14.41, and 14.42 are referred to as the critical-temperature ratio, critical-pressure ratio, and critical-density ratio. Table 14.1 gives these ratios for various values of $k$.

## 14.7 Mass Rate of Flow of an Ideal Gas through an Isentropic Nozzle

We now turn our attention to a consideration of the mass rate of flow per unit area, $\dot{m}/\mathscr{A}$, in a nozzle. From the continuity equation we proceed

**Table 14.1**
Critical Pressure, Density, and Temperature Ratios for Isentropic Flow of an Ideal Gas

|            | $k = 1.1$ | $k = 1.2$ | $k = 1.3$ | $k = 1.4$ | $k = 1.67$ |
|------------|-----------|-----------|-----------|-----------|------------|
| $P^*/P_0$  | 0.5847    | 0.5644    | 0.5457    | 0.5283    | 0.4867     |
| $\rho^*/\rho_0$ | 0.6139 | 0.6209   | 0.6276    | 0.6340    | 0.6497     |
| $T^*/T_0$  | 0.9524    | 0.9091    | 0.8696    | 0.8333    | 0.7491     |

as follows:

$$\frac{\dot{m}}{\mathscr{A}} = \rho V = \frac{PV}{RT} \sqrt{\frac{kT_0}{kT_0}}$$

$$= \frac{PV}{\sqrt{kRT}} \sqrt{\frac{k}{R}} \sqrt{\frac{T_0}{T}} \sqrt{\frac{1}{T_0}}$$

$$= \frac{PM}{\sqrt{T_0}} \sqrt{\frac{k}{R}} \sqrt{1 + \frac{k-1}{2} M^2} \tag{14.43}$$

By substituting Eq. 14.38 into Eq. 14.42 the flow per unit area can be expressed in terms of stagnation pressure, stagnation temperature, Mach number, and gas properties.

$$\frac{\dot{m}}{\mathscr{A}} = \frac{P_0}{\sqrt{T_0}} \sqrt{\frac{k}{R}} \times \frac{M}{\left(1 + \dfrac{k-1}{2} M^2\right)^{(k+1)/2(k-1)}} \tag{14.44}$$

At the throat, $M = 1$, and therefore the flow per unit area at the throat, $\dot{m}/\mathscr{A}^*$, can be found by setting $M = 1$ in Eq. 14.44.

$$\frac{\dot{m}}{\mathscr{A}^*} = \frac{P_0}{\sqrt{T_0}} \sqrt{\frac{k}{R}} \times \frac{1}{\left(\dfrac{k+1}{2}\right)^{(k+1)/2(k-1)}} \tag{14.45}$$

The area ratio $\mathscr{A}/\mathscr{A}^*$ can be obtained by dividing Eq. 14.45 by Eq. 14.44.

$$\frac{\mathscr{A}}{\mathscr{A}^*} = \frac{1}{M} \left[ \left(\frac{2}{k+1}\right) \left(1 + \frac{k-1}{2} M^2\right) \right]^{(k+1)/2(k-1)} \tag{14.46}$$

The area ratio $\mathscr{A}/\mathscr{A}^*$ is the ratio of the area at the point where the Mach number is $M$ to the throat area, and values of $\mathscr{A}/\mathscr{A}^*$ as a function of Mach number are given in Tables 30 through 35 of the *Gas Tables* and in Table A.13 in the Appendix. Figure 14.10 shows a plot of $\mathscr{A}/\mathscr{A}^*$ vs. $M$, which is in accordance with our previous conclusion that a subsonic nozzle is converging and a supersonic nozzle is diverging.

The final point to be made regarding the isentropic flow of an ideal gas through a nozzle involves the effect of varying the back pressure (the pressure outside the nozzle exit) on the mass rate of flow.

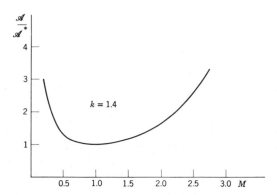

**Fig. 14.10** Area ratio as a function of Mach number for a reversible adiabatic nozzle.

Consider first a convergent nozzle as shown in Fig. 14.11, which also shows the pressure ratio $P/P_0$ along the length of the nozzle. The conditions upstream are the stagnation conditions, which are assumed to be constant. The pressure at the exit plane of the nozzle is designated $P_E$, and the back pressure $P_B$. Let us consider how the mass rate of flow $\dot{m}$ and the exit plane pressure $P_E/P_0$, vary as the back pressure $P_B$ is decreased. These quantities are plotted in Fig. 14.12.

When $P_B/P_0 = 1$ there is of course no flow, and $P_E/P_0 = 1$ as designated by point $a$. Next let the back pressure $P_B$ be lowered to that designated by

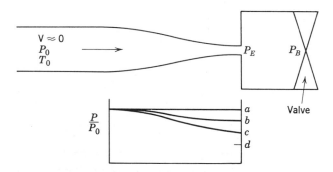

**Fig. 14.11** Pressure ratio as a function of back pressure for a convergent nozzle.

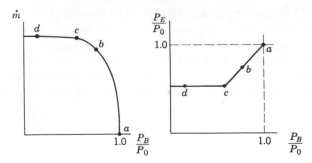

**Fig. 14.12**  Mass rate of flow and exit pressure as a function of back pressure for a convergent nozzle.

point $b$, so that $P_B/P_0$ is greater than the critical-pressure ratio. The mass rate of flow has a certain value and $P_E = P_B$. The exit Mach number is less than 1. Next let the back pressure be lowered to the critical pressure, designated by point $c$. The Mach number at the exit is now unity, and $P_E$ is equal to $P_B$. When $P_B$ is decreased below the critical pressure, designated by point $d$, there is no further increase in the mass rate of flow, and $P_E$ remains constant at a value equal to the critical pressure, and the exit Mach number is unity. The drop in pressure. from $P_E$ to $P_B$ takes place outside the nozzle exit. Under these conditions the nozzle is said to be choked, which means that for given stagnation conditions the nozzle is passing the maximum possible mass flow.

Consider next a convergent-divergent nozzle in a similar arrangement, Fig. 14.13. Point $a$ designates the condition when $P_B = P_0$ and there is no flow. When $P_B$ is descreased to the pressure indicated by point $b$, so that $P_B/P_0$ is less than 1 but considerably greater than the critical-pressure

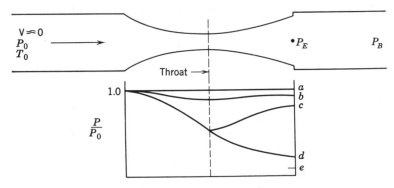

**Fig. 14.13**  Nozzle pressure ratio as a function of back pressure for a reversible convergent-divergent nozzle.

ratio, the velocity increases in the convergent section, but $M < 1$ at the throat. Therefore, the diverging section acts as a subsonic diffuser in which the pressure increases and velocity decreases. Point $c$ designates the back pressure at which $M = 1$ at the throat, but the diverging section acts as a subsonic diffuser (with $M = 1$ at the inlet) in which the pressure increases and velocity decreases. Point $d$ designates one other back pressure that permits isentropic flow, and in this case the diverging section acts as a supersonic nozzle, with a decrease in pressure and an increase in velocity. Between the back pressures designated by points $c$ and $d$, an isentropic solution is not possible, and shock waves will be present. This matter is discussed in the section that follows. When the back pressure is decreased below that designated by point $d$, the exit-plane pressure $P_E$ remains constant, and the drop in pressure from $P_E$ to $P_B$ takes place outside the nozzle. This is designated by point $e$.

### Example 14.6

A convergent nozzle has an exit area of 500 mm². Air enters the nozzle with a stagnation pressure of 1000 kPa and a stagnation temperature of 360 K. Determine the mass rate of flow for back pressures of 800 kPa, 528 kPa, and 300 kPa, assuming isentropic flow.

For air $k = 1.4$ and Table 30 of the *Gas Tables* (or Table A.13 in the Appendix) may be used. The critical-pressure ratio, $P^*/P_0$, is 0.528. Therefore, for a back pressure of 528 kPa, $M = 1$ at the nozzle exit and the nozzle is choked. Decreasing the back pressure below 528 kPa will not increase the flow.

For a back pressure of 528 kPa,

$$\frac{T^*}{T} - 0.8333 \quad , \quad T^* = 300 \text{ K}$$

At the exit

$$V = c = \sqrt{kRT}$$

$$= \sqrt{1.4 \times 0.287 \times 300 \times 1000} = 347.2 \text{ m/s}$$

$$\rho^* = \frac{P^*}{RT^*} = \frac{528}{0.287 \times 300} = 6.1324 \text{ kg/m}^3$$

$$\dot{m} = \rho \mathscr{A} V$$

Applying this relation to the throat section,

$$\dot{m} = 6.1324 \times 500 \times 10^{-6} \times 347.2 = 1.0646 \text{ kg/s}$$

For a back pressure of 800 kPa, $P_E/P_0 = 0.8$ (subscript $E$ designates the properties in the exit plane). From Table A.13,

$$M_E = 0.573 \qquad T_E/T_0 = 0.9381$$

$$T_E = 337.7 \text{ K}$$

$$c_E = \sqrt{kRT_E} = \sqrt{1.4 \times 0.287 \times 337.7 \times 1000} = 368.4 \text{ m/s}$$

$$V_E = M_E c_E = 211.1 \text{ m/s}$$

$$\rho_E = \frac{P_E}{RT_E} = \frac{800}{0.287 \times 337.7} = 8.2542 \text{ kg/m}^3$$

$$\dot{m} = \rho \mathscr{A} V$$

Applying this relation to the exit section,

$$\dot{m} = 8.2542 \times 500 \times 10^{-6} \times 211.1 = 0.8712 \text{ kg/s}$$

For a back pressure less than the critical pressure, which in this case is 528 kPa, the nozzle is choked and the mass rate of flow is the same as that for the critical pressure. Therefore, for an exhaust pressure of 300 kPa, the mass rate of flow is 1.0646 kg/s.

### Example 14.7

A converging-diverging nozzle has an exit area to throat area ratio of 2. Air enters this nozzle with a stagnation pressure of 1000 kPa and a stagnation temperature of 360 K. The throat area is 500 mm². Determine the mass rate of flow, exit pressure, exit temperature, exit Mach number, and exit velocity for the following conditions:

(a) Sonic velocity at the throat, diverging section acting as a nozzle. (Corresponds to point $d$ in Fig. 14.13.)

(b) Sonic velocity at the throat, diverging section acting as a diffuser. (Corresponds to point $c$ in Fig. 14.13.)

(a) In Table A.13 of the Appendix we find that there are two Mach numbers listed for $\mathscr{A}/\mathscr{A}^* = 2$. One of these is greater than unity and one is less than unity. When the diverging section acts as a supersonic nozzle we use the value for $M > 1$. The following are from Table A.13.

$$\frac{\mathscr{A}_E}{\mathscr{A}^*} = 2.0 \qquad M_E = 2.197 \qquad \frac{P_E}{P_0} = 0.0939 \qquad \frac{T_E}{T_0} = 0.5089$$

Therefore,

$$P_E = 0.0939 \ (1000) = 93.9 \ \text{kPa}$$

$$T_E = 0.5089 \ (360) = 183.2 \ \text{K}$$

$$c_E = \sqrt{kRT_E} = \sqrt{1.4 \times 0.287 \times 183.2 \times 1000} = 271.3 \ \text{m/s}$$

$$V_E = M_E c_E = 2.197 \ (271.3) = 596.1 \ \text{m/s}$$

The mass rate of flow can be determined by considering either the throat section or the exit section. However, in general it is preferable to determine the mass rate of flow from conditions at the throat. Since in this case $M = 1$ at the throat, the calculation is identical to the calculation for the flow in the convergent nozzle of Example 14.6 when it is choked.

(b) The following are from Table A.13

$$\frac{\mathscr{A}_E}{\mathscr{A}^*} = 2.0 \qquad M = 0.308 \qquad \frac{P_E}{P_0} = 0.936 \qquad \frac{T_E}{T_0} = 0.9812$$

$$P_E = 0.936 \ (1000) = 936 \ \text{kPa}$$

$$T_E = 0.9812 \ (360) = 353.3 \ \text{K}$$

$$c_E = \sqrt{kRT_E} = \sqrt{1.4 \times 0.287 \times 353.4 \times 1000} = 376.8 \ \text{m/s}$$

$$V_E = M_E c_E = 0.308 \ (376.8) = 116 \ \text{m/s}$$

Since $M = 1$ at the throat, the mass rate of flow is the same as in (a), which is also equal to the flow in the convergent nozzle of Example 14.6 when it is choked.

In the example above a solution assuming isentropic flow is not possible if the back pressure is between 936 kPa and 93.9 kPa. If the back pressure is in this range there will either be a normal shock in the nozzle or oblique shock waves outside the nozzle. The matter of normal shock waves is considered in the following section.

## 14.8   Normal Shock in an Ideal Gas Flowing through a Nozzle

A shock wave involves an extremely rapid and abrupt change of state. In a normal shock this change of state takes place across a plane normal to the direction of the flow. Figure 14.14 shows a control surface that includes such a normal shock. We can now determine the relations that govern the flow. Assuming steady-state, steady-flow we can write the following relations, where subscripts $x$ and $y$ denote the conditions

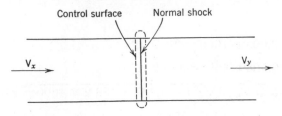

**Fig. 14.14**  One-dimensional normal shock.

upstream and downstream of the shock respectively. Note that no heat and work cross the control surface.

First law:

$$h_x + \frac{V_x^2}{2} = h_y + \frac{V_y^2}{2} = h_{0x} = h_{0y} \tag{14.47}$$

Continuity equation:

$$\frac{\dot{m}}{\mathscr{A}} = \rho_x V_x = \rho_y V_y \tag{14.48}$$

Momentum equation:

$$\mathscr{A}(P_x - P_y) = \dot{m}\,(V_y - V_x) \tag{14.49}$$

Second law: Since the process is adiabatic

$$s_y - s_x \geqslant 0 \tag{14.50}$$

The energy and continuity equations can be combined to give an equation which when plotted on the $h$-$s$ diagram is called the Fanno line. Similarly, the momentum and continuity equations can be combined to give an equation, the plot of which on the $h$-$s$ diagram is known as the Rayleigh line. Both of these lines are shown on the $h$-$s$ diagram of Fig. 14.15. It can be shown that the point of maximum entropy on each line, points $a$ and $b$, correspond to $M = 1$. The lower part of each line corresponds to supersonic velocities, and the upper part to subsonic velocities.

The two points where all three equations are satisfied are points $x$ and $y$, $x$ being in the supersonic region and $y$ in the subsonic region. Since the second law requires that $s_y - s_x \geqslant 0$ in an adiabatic process, we conclude that the normal shock can proceed only from $x$ to $y$. This means that the velocity changes from supersonic ($M > 1$) before the shock to subsonic ($M < 1$) after the shock.

The equations governing normal shock waves will now be developed. If we assume constant specific heats we conclude from Eq. 14.47, the

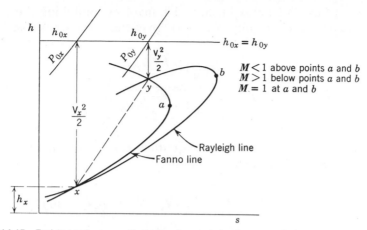

**Fig. 14.15**   End states for a one-dimensional normal shock on an enthalpy-entropy diagram.

energy equation, that

$$T_{0x} = T_{0y} \qquad (14.51)$$

That is, there is no change in stagnation temperature across a normal shock. Introducing Eq. 14.37

$$\frac{T_{0x}}{T_x} = 1 + \frac{k-1}{2} M_x^2 \qquad \frac{T_{0y}}{T_y} = 1 + \frac{k-1}{2} M_y^2$$

and substituting into Eq. 14.51 we have,

$$\frac{T_y}{T_x} = \frac{1 + \dfrac{k-1}{2} M_x^2}{1 + \dfrac{k-1}{2} M_y^2} \qquad (14.52)$$

The equation of state, the definition of Mach number, and the relation $c = \sqrt{kRT}$ can be introduced into the continuity equation as follows:

$$\rho_x V_x = \rho_y V_y$$

But

$$\rho_x = \frac{P_x}{RT_x} \qquad \rho_y = \frac{P_y}{RT_y}$$

$$\frac{T_y}{T_x} = \frac{P_y V_y}{P_x V_x} = \frac{P_y M_y c_y}{P_x M_x c_x} = \frac{P_y M_y \sqrt{T_y}}{P_x M_x \sqrt{T_x}}$$

$$= \left(\frac{P_y}{P_x}\right)^2 \left(\frac{M_y}{M_x}\right)^2 \qquad (14.53)$$

Combining Eqs. 14.52 and 14.53, which involves combining the energy equation and the continuity equation, gives the equation of the Fanno line.

$$\frac{P_y}{P_x} = \frac{M_x \sqrt{1 + \frac{k-1}{2} M_x^2}}{M_y \sqrt{1 + \frac{k-1}{2} M_y^2}} \tag{14.54}$$

The momentum and continuity equations can be combined as follows to give the equation of the Rayleigh line.

$$P_x - P_y = \frac{\dot{m}}{\mathscr{A}}(V_y - V_x) = \rho_y V_y^2 - \rho_x V_x^2$$

$$P_x + \rho_x V_x^2 = P_y + \rho_y V_y^2$$

$$P_x + \rho_x M_x^2 c_x^2 = P_y + \rho_y M_y^2 c_y^2$$

$$P_x + \frac{P_x M_x^2}{RT_x}(kRT_x) = P_y + \frac{P_y M_y^2}{RT_y}(kRT_y)$$

$$P_x(1 + kM_x^2) = P_y(1 + kM_y^2)$$

$$\frac{P_y}{P_x} = \frac{1 + kM_x^2}{1 + kM_y^2} \tag{14.55}$$

Equations 14.54 and 14.55 can be combined to give the following equation relating $M_x$ and $M_y$.

$$M_y^2 = \frac{M_x^2 + \frac{2}{k-1}}{\frac{2k}{k-1}M_x^2 - 1} \tag{14.56}$$

Tables 48 through 53 of the *Gas Tables* give the normal shock functions, which include $M_y$ as a function of $M_x$. Table A.14 of the Appendix has been abstracted from Table 48 of the *Gas Tables*, and applies to an ideal gas with $k = 1.4$. Note that $M_x$ is always supersonic and $M_y$ is always subsonic, which agrees with the previous statement that in a normal shock the velocity changes from supersonic to subsonic. These tables also give the pressure, density, temperature, and stagnation pressure ratios across a normal shock as a function of $M_x$. These are found from Eqs. 14.52, 14.53, and the equation of state. Note that there is always a drop in stagnation pressure across a normal shock and an increase in the static pressure.

**Example 14.8**

Consider the convergent-divergent nozzle of Example 14.7 in which the diverging section acts as a supersonic nozzle (Fig. 14.16). Assume that a normal shock stands in the exit plane of the nozzle. Determine the static pressure and temperature and the stagnation pressure just downstream of the normal shock.

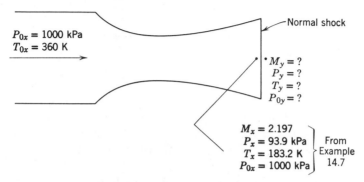

**Fig. 14.16**   Sketch for Example 14.8.

From Table 48 of the *Gas Tables* (Table A.14 of the Appendix)

$$M_x = 2.197 \qquad M_y = 0.547 \qquad \frac{P_y}{P_x} = 5.46 \qquad \frac{T_y}{T_x} = 1.854 \qquad \frac{P_{0y}}{P_{0x}} = 0.630$$

$$P_y = 5.46 \times P_x = 5.46\,(93.9) = 512.7 \text{ kPa}$$

$$T_y = 1.854 \times T_x = 1.854\,(183.2) = 339.7 \text{ K}$$

$$P_{0y} = 0.631 \times P_{0x} = 0.630\,(1000) = 630 \text{ kPa}$$

In the light of this example we can conclude the discussion concerning the flow through a convergent-divergent nozzle. Figure 14.13 is repeated here as Fig. 14.17 for convenience, except that points $f$, $g$, and $h$ have been added. Consider point $d$. We have already noted that with this back pressure the exit plane pressure $P_E$ is just equal to the back pressure $P_B$, and isentropic flow is maintained in the nozzle. Let the back pressure be raised to that designated by point $f$. The exit-plane pressure $P_E$ is not influenced by this increase in back pressure, and the increase in pressure from $P_E$ to $P_B$ takes place outside the nozzle. Let the back pressure be raised to that designated by point $g$, which is just sufficient to cause a normal shock to stand in the exit plane of the nozzle. The exit-plane pressure $P_E$ (downstream of the shock) is equal to the back pressure $P_B$, and $M < 1$ leaving the nozzle. This is the case in Example 14.8. Now let the

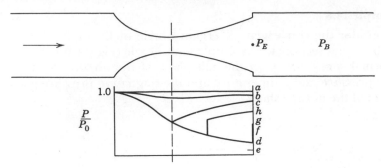

**Fig. 14.17**    Nozzle pressure ratio as a function of back pressure for a convergent-divergent nozzle.

back pressure be raised to that corresponding to point $h$. As the back pressure is raised from $g$ to $h$ the normal shock moves into the nozzle as indicated. Since $M < 1$ downstream of the normal shock, the diverging part of the nozzle which is downstream of the shock acts as a subsonic diffuser. As the back pressure is increased from $h$ to $c$ the shock moves further upstream and disappears at the nozzle throat where the back pressure corresponds to $c$. This is reasonable since there are no supersonic velocities involved when the back pressure corresponds to $c$, and hence no shock waves are possible.

**Example 14.9**

Consider the convergent-divergent nozzle of Examples 14.7 and 14.8. Assume that there is a normal shock wave standing at the point where $M = 1.5$. Determine the exit-plane pressure, temperature, and Mach number. Assume isentropic flow except for the normal shock (Fig. 14.18).

The properties at point $x$ can be determined from Table A.13, because the flow is isentropic to point $x$.

$$M_x = 1.5 \qquad \frac{P_x}{P_{0x}} = 0.2724 \qquad \frac{T_x}{T_{0x}} = 0.6897 \qquad \frac{\mathscr{A}_x}{\mathscr{A}_x^*} = 1.1762$$

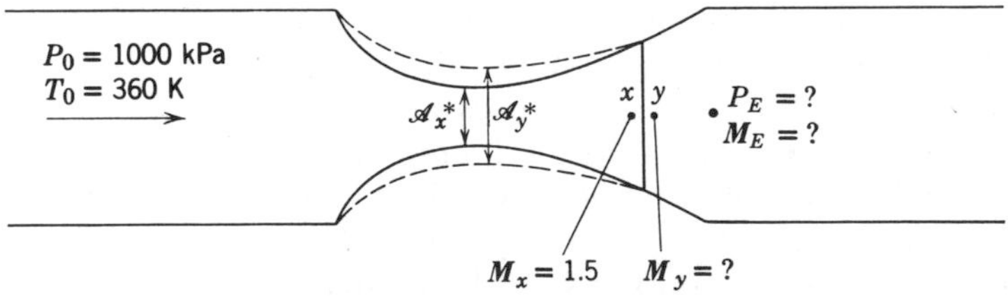

**Fig. 14.18**    Sketch for Example 14.9.

Therefore,

$$P_x = 0.2724\,(1000) = 272.4 \text{ kPa}$$

$$T_x = 0.6897\,(360) = 248.3 \text{ K}$$

The properties at point $y$ can be determined from the normal shock functions, Table A.14.

$$M_y = 0.7011 \quad \frac{P_y}{P_x} = 2.4583 \quad \frac{T_y}{T_x} = 1.320 \quad \frac{P_{0y}}{P_{0x}} = 0.9298$$

$$P_y = 2.4583P_x = 2.4583\,(272.4) = 669.6 \text{ kPa}$$

$$T_y = 1.320T_x = 1.320\,(248.3) = 327.8 \text{ K}$$

$$P_{0y} = 0.9298P_{0x} = 0.9298\,(1000) = 929.8 \text{ kPa}$$

Since there is no change in stagnation temperature across a normal shock,

$$T_{0x} = T_{0y} = 360 \text{ K}$$

From $y$ to $E$ the diverging section acts as a subsonic diffuser. In solving this problem it is convenient to think of the flow at $y$ as having come from an isentropic nozzle having a throat area $\mathscr{A}_y^*$. Such a hypothetical nozzle is shown by the dotted line. From the table of isentropic flow functions, Table A.13, we find the following for $M_y = 0.7011$

$$M_y = 0.7011 \quad \frac{\mathscr{A}_y}{\mathscr{A}_y^*} = 1.0938 \quad \frac{P_y}{P_{0y}} = 0.7202 \quad \frac{T_y}{T_{0y}} = 0.9105$$

From the statement of the problem

$$\frac{\mathscr{A}_E}{\mathscr{A}_x^*} = 2.0$$

Also, since the flow from $y$ to $E$ is isentropic,

$$\frac{\mathscr{A}_E}{\mathscr{A}_E^*} = \frac{\mathscr{A}_E}{\mathscr{A}_y^*} = \frac{\mathscr{A}_E}{\mathscr{A}_x^*} \times \frac{\mathscr{A}_x^*}{\mathscr{A}_x} \times \frac{\mathscr{A}_x}{\mathscr{A}_y} \times \frac{\mathscr{A}_y}{\mathscr{A}_y^*}$$

$$= \frac{\mathscr{A}_E}{\mathscr{A}_y^*} = 2.0 \times \frac{1}{1.1762} \times 1 \times 1.0938 = 1.860$$

From the table of isentropic flow functions for $\mathscr{A}/\mathscr{A}^* = 1.860$ and $M < 1$

$$M_E = 0.339 \qquad \frac{P_E}{P_{0E}} = 0.9222 \qquad \frac{T_E}{T_{0E}} = 0.9771$$

$$\frac{P_E}{P_{0E}} = \frac{P_E}{P_{0y}} = 0.9222$$

$$P_E = 0.9222\,(P_{0y}) = 0.9222\,(929.8) = 857.5 \text{ kPa}$$

$$T_E = 0.9771\,(T_{0E}) = 0.9771\,(360) = 351.7 \text{ K}$$

In conclusion it should be pointed out that in considering the normal shock we have ignored the effect of viscosity and thermal conductivity, which are certain to be present. The actual shock wave will occur over some finite thickness. However, the development as given here gives a very good qualitative picture of normal shocks, and also provides a basis for fairly accurate quantitive results.

## 14.9 Flow of a Vapor through a Nozzle

In this section we consider a vapor to be a substance that is in the gaseous phase but with limited superheat. Therefore, the vapor will probably deviate significantly from the ideal-gas relations, and the possibility of condensation must be considered. The most familiar example is the flow of steam through the nozzles of a steam turbine.

The principles that have been developed for the isentropic flow of an ideal gas apply also to the isentropic flow of a vapor. However, because the vapor deviates from the ideal-gas relationships, the appropriate tables of thermodynamic properties must be used. Further, the possibility of condensation must be borne in mind, as indicated in Fig. 14.19. If steam expands isentropically from state 1 to state 2, and if equilibrium is maintained throughout the nozzle, condensation would begin at point

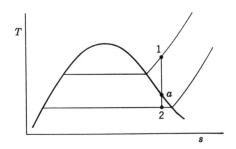

**Fig. 14.19** Reversible adiabatic expansion of a vapor.

$a$, and at pressures below this a mixture of liquid droplets and vapor would be present. In an actual nozzle the formation of the liquid tends to be delayed due to an effect known as supersaturation. This will be considered in a later paragraph.

Let us consider first the isentropic flow of steam in a nozzle without condensation. The value of the specific-heat ratio $k$ for steam varies, but $k = 1.3$ is a good approximation over a considerable range. Therefore, the critical pressure ratio, $P*/P_0$, can be found by the relation

$$\frac{P*}{P_0} = \left(\frac{2}{k+1}\right)^{k/(k-1)} = 0.545 \qquad (14.57)$$

This is also the value given in Table 14.1. Knowing, therefore, the critical-pressure ratio, the throat area for a given flow can be calculated. The exit area of the nozzle can be calculated in a similar manner. The example below illustrates this procedure.

**Example 14.10**

Steam at stagnation pressure of 800 kPa and a stagnation temperature of 350°C expands in a nozzle to 200 kPa. Determine the throat area and exit area required for a flow of 3 kg/s, assuming reversible adiabatic flow.

First consider the properties and flow at the throat.

$$\frac{P*}{P_0} = 0.545 \qquad\qquad P* = 436 \text{ kPa}$$

$$s* - s_0 - 7.4089 \text{ kJ/kg K} \qquad h_0 = 3161.7 \text{ kJ/kg}$$

$$T* = 268.7°C \qquad\qquad h* = 3001.4 \text{ kJ/kg}$$

$$h_0 = h* + \frac{V*^2}{2} \qquad\qquad V* = \sqrt{2\,(h_0 - h*)}$$

$$V* = \sqrt{2 \times 1000\,(3161.7 - 3001.4)} = 566.2 \text{ m/s}$$

$$v* = 0.5724 \text{ m}^3/\text{kg (from the steam tables)}$$

$$\dot{m}v* = \mathscr{A}*V*$$

$$\mathscr{A}* = \frac{3 \times 0.5724}{566.2} = 3.033 \times 10^{-3}\text{m}^2$$

At the nozzle exit

$$P_E = 200 \text{ kPa} \qquad s_E = s_0 = 7.4089 \text{ kJ/kg K}$$

Therefore,

$$T_E = 178.5°C$$

$$h_E = 2826.7 \text{ kJ/kg} \qquad v_e = 1.0284 \text{ m}^3/\text{kg}$$

$$V_E = \sqrt{2 \ (h_0 - h_E)} = \sqrt{2 \times 1000 \ (3161.7 - 2826.7)} = 818.5 \text{ m/s}$$

$$\dot{m}v_E = \mathscr{A}_E V_E$$

$$\mathscr{A}_E = \frac{3 \times 1.0284}{818.5} = 3.769 \times 10^{-3} \text{m}^2$$

For steam initially saturated the critical-pressure ratio is usually taken as

$$\frac{P*}{P_0} = 0.577 \tag{14.58}$$

If the flow with initially saturated steam is assumed to be isentropic there will be a certain amount of liquid entrained with the vapor. Calculation of throat and exit areas under most conditions is similar to the example above.

When the steam flowing through a nozzle is initially superheated, there will be a certain point in the nozzle where the steam becomes saturated. However, as noted in Section 13.5 in connection with a discussion of metastable equilibrium, if the point at which condensation would occur under equilibrium conditions (point $a$ in Fig. 13.8) occurs in the divergent section of the nozzle, a condition of metastable equilibrium exists. That is, the formation of droplets is delayed, and the vapor temperature is less than the saturation temperature for the given pressure. This is frequently referred to as supersaturation.

This phenomenon of supersaturation is observed only in the diverging portions of the nozzle. In a nozzle in which supersaturation occurs the rate of mass flow might be slightly greater than that obtained in an isentropic flow without supersaturation.

Supersaturation may also be present in a supersonic wind tunnel. An abrupt condensation of the moisture in the air would disturb the flow pattern, and for this reason most of the moisture is removed from the air to be used in the tunnel.

A similar phenomenon may occur in the flow of a saturated liquid through a nozzle or valve. When a saturated liquid flows through a nozzle or valve some of it will vaporize as the pressure is reduced. In practice the formation of the vapor is delayed, because a metastable state exists.

## 14.10  Nozzle and Diffuser Coefficients

Up to this point we have considered only isentropic flow and normal shocks. As was pointed out in Chapter 7, isentropic flow through a nozzle provides a standard to which the performance of an actual nozzle can be compared. For nozzles, the three important parameters by which actual flow can be compared to the ideal flow are nozzle efficiency, velocity coefficient, and discharge coefficient. These are defined as follows:

The nozzle efficiency $\eta_N$ is defined as

$$\eta_N = \frac{\text{Actual kinetic energy at nozzle exit}}{\text{Kinetic energy at nozzle exit with isentropic flow to same exit pressure}} \quad (14.59)$$

The efficiency can be defined in terms of properties. On the $h$-$s$ diagram of Fig. 14.20 state $0i$ represents the stagnation state of the fluid entering the nozzle; state $e$ represents the actual state at the nozzle exit; and state $s$ represents the state that would have been achieved at the nozzle exit if the flow had been reversible and adiabatic to the same exit pressure. Therefore, in terms of these states the nozzle efficiency is

$$\eta_N = \frac{h_{0i} - h_e}{h_{0i} - h_s}$$

Nozzle efficiencies vary in general from 90 to 99 per cent. Large nozzles usually have higher efficiencies than small nozzles, and nozzles with

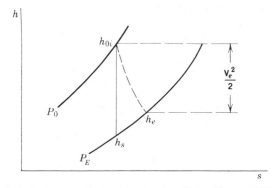

**Fig. 14.20**  Enthalpy-entropy diagram showing the effects of irreversibility in a nozzle.

straight axes have higher efficiencies than nozzles with curved axes. The irreversibilities, which cause the departure from isentropic flow, are primarily due to frictional effects, and are confined largely to the boundary layer. The rate of change of cross-sectional area along the nozzle axis (i.e., the nozzle contour) is an important parameter in the design of an efficient nozzle, particularly in the divergent section. Detailed consideration of this matter is beyond the scope of this text, and the reader is referred to standard references on the subject.

The velocity coefficient $C_V$ is defined as

$$C_V = \frac{\text{Actual velocity at nozzle exit}}{\text{Velocity at nozzle exit with isentropic flow and same exit pressure}} \tag{14.60}$$

It follows that the velocity coefficient is equal to the square root of the nozzle efficiency

$$C_V = \sqrt{\eta_N} \tag{14.61}$$

The coefficient of discharge $C_D$ is defined by the relation

$$C_D = \frac{\text{Actual mass rate of flow}}{\text{Mass rate of flow with isentropic flow}}$$

In determining the mass rate of flow with isentropic conditions, the actual back pressure is used if the nozzle is not choked. If the nozzle is choked, the isentropic mass rate of flow is based on isentropic flow and sonic velocity at the minimum section (i.e., sonic velocity at the exit of a convergent nozzle and at the throat of a convergent-divergent nozzle).

The performance of a diffuser is usually given in terms of diffuser efficiency, which is best defined with the aid of an $h$-$s$ diagram. On the $h$-$s$ diagram of Fig. 14.21 states 1 and 01 are the actual and stagnation states of the fluid entering the diffuser. States 2 and 02 are the actual and stagnation states of the fluid leaving the diffuser. State 3 is not attained in the diffuser, but it is the state that has the same entropy as the initial state and the pressure of the isentropic stagnation state leaving the diffuser. The efficiency of the diffuser $\eta_D$ is defined as

$$\eta_D = \frac{\Delta h_s}{V_1^2/2} = \frac{h_3 - h_1}{h_{01} - h_1} = \frac{h_3 - h_1}{h_{02} - h_1} \tag{14.62}$$

If we assume an ideal gas with constant specific heat this reduces to

$$\eta_D = \frac{T_3 - T_1}{T_{02} - T_1} = \frac{\dfrac{(T_3 - T_1)}{T_1} T_1}{\dfrac{V_1^2}{2 C_{p0}}}$$

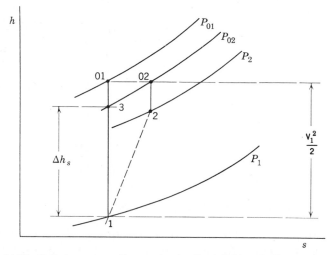

**Fig. 14.21**  Enthalpy-entropy diagram showing the definition of diffuser efficiency.

$$C_{po} = \frac{kR}{k-1} \qquad T_1 = \frac{c_1^2}{kR} \qquad V_1^2 = M_1^2 c_1^2 \qquad \frac{T_3}{T_1} = \left(\frac{P_{02}}{P_1}\right)^{(k-1)/k}$$

Therefore,

$$\eta_D = \frac{\left(\dfrac{P_{02}}{P_1}\right)^{(k-1)/k} - 1}{\dfrac{k-1}{2} M_1^2}$$

$$\left(\frac{P_{02}}{P_1}\right)^{(k-1)/k} = \left(\frac{P_{01}}{P_1}\right)^{(k-1)/k} \times \left(\frac{P_{02}}{P_{01}}\right)^{(k-1)/k}$$

$$= \left(1 + \frac{k-1}{2} M_1^2\right)\left(\frac{P_{02}}{P_{01}}\right)^{(k-1)/k}$$

$$\eta_D = \frac{\left(1 + \dfrac{k-1}{2} M_1^2\right)\left(\dfrac{P_{02}}{P_{01}}\right)^{(k-1)/k} - 1}{\dfrac{k-1}{2} M_1^2} \qquad (14.63)$$

## 14.11  Nozzles and Orifices as Flow-Measuring Devices

The mass rate of flow of a fluid flowing in a pipe is frequently determined by measuring the pressure drop across a nozzle or orifice in the line, as shown in Fig. 14.22. The ideal process for such a nozzle or orifice is assumed to be isentropic flow through a nozzle which has the measured

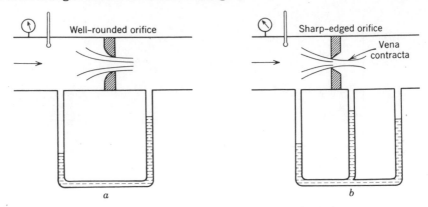

**Fig. 14.22**   Nozzles and orifices as flow-measuring devices.

pressure drop from inlet to exit and a minimum cross-sectional area equal to the minimum area of the nozzle or orifice. The actual flow is related to the ideal flow by the coefficient of discharge which is defined by Eq. 14.62.

The pressure difference measured across an orifice depends upon the location of the pressure taps as indicated in Fig. 14.22. Since the ideal flow is based on the measured pressure difference, it follows that the coefficient of discharge depends on the locations of the pressure taps. Also, the coefficient of discharge for a sharp-edged orifice is considerably less than that for a well-rounded nozzle, primarily due to a contraction of the stream, known as the vena contracta, as it flows through a sharp-edged orifice.

There are two approaches to determining the discharge coefficient of a nozzle or orifice. One is to follow a standard design procedure, such as the ones established by the American Society of Mechanical Engineers,[2] and use the coefficient of discharge given for a particular design. A more accurate method is to calibrate a given nozzle or orifice, and determine the discharge coefficient for a given installation by accurately measuring the actual mass rate of flow. The procedure to be followed will depend on the accuracy desired and other factors involved (such as time, expense, availability of calibration facilities) in a given situation.

For incompressible fluids flowing through an orifice the ideal flow for a given pressure drop can be found by the procedure outlined in Section 14.4. Actually, it is advantageous to combine Eqs. 14.24 and 14.20 to give the following relation, which is valid for reversible flow.

$$v(P_2 - P_1) + \frac{V_2^2 - V_1^2}{2} = v(P_2 - P_1) + \frac{V_2^2 - (\mathscr{A}_2/\mathscr{A}_1)^2 V_2^2}{2} = 0 \qquad (14.64)$$

[2] *Fluid Meters, Their Theory and Application*, ASME, 1959. *Flow Measurement*, ASME, 1959.

or

$$v(P_2 - P_1) + \frac{V_2^2}{2}\left[1 - \left(\frac{\mathscr{A}_2}{\mathscr{A}_1}\right)^2\right] = 0$$

$$V_2 = \sqrt{\frac{2v(P_1 - P_2)}{[1 - (\mathscr{A}_2/\mathscr{A}_1)^2]}} \qquad (14.65)$$

For an ideal gas it is frequently advantageous to use the following simplified procedure when the pressure drop across an orifice or nozzle is small. Consider the nozzle shown in Fig. 14.23. From the first law we conclude that

$$h_i + \frac{V_i^2}{2} = h_e + \frac{V_e^2}{2}$$

Assuming constant specific heat, this reduces to

$$\frac{V_e^2 - V_i^2}{2} = h_i - h_e = C_{po}(T_i - T_e)$$

Let $\Delta P$ and $\Delta T$ be the decrease in pressure and temperature across the nozzle. Since we are considering reversible adiabatic flow we note that

$$\frac{T_e}{T_i} = \left(\frac{P_e}{P_i}\right)^{(k-1)/k}$$

or

$$\frac{T_i - \Delta T}{T_i} = \left(\frac{P_i - \Delta P}{P_i}\right)^{(k-1)/k}$$

$$1 - \frac{\Delta T}{T_i} = \left(1 - \frac{\Delta P}{P_i}\right)^{(k-1)/k}$$

Using the binomial expansion on the right side of the equation we have

$$1 - \frac{\Delta T}{T_i} = 1 - \frac{k-1}{k}\frac{\Delta P}{P_i} - \frac{k-1}{2k^2}\frac{\Delta P^2}{P_i^2}\cdots$$

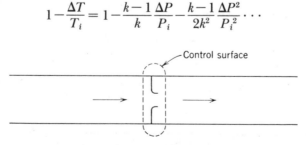

**Fig. 14.23**  Analysis of a nozzle as a flow-measuring device.

If $\Delta P/P_i$ is small this reduces to

$$\frac{\Delta T}{T_i} = \frac{k-1}{k}\frac{\Delta P}{P_i}$$

Substituting this into the first-law equation we have

$$\frac{V_e^2 - V_i^2}{2} = C_{po}\frac{k-1}{k}\Delta P\frac{T_i}{P_i}$$

But for an ideal gas

$$C_{po} = \frac{kR}{k-1} \quad \text{and} \quad v_i = R\frac{T_i}{P_i}$$

Therefore,

$$\frac{V_e^2 - V_i^2}{2} = v_i\Delta P$$

which is the same as Eq. 14.64, which was developed for incompressible flow. Therefore, when the pressure drop across a nozzle or orifice is small, the flow can be calculated with high accuracy by assuming incompressible flow.

The Pitot tube, Fig. 14.24, is an important instrument for measuring the velocity of a fluid. In calculating the flow with a Pitot tube it is assumed that the fluid is decelerated isentropically in front of the Pitot tube, and therefore the stagnation pressure of the free stream can be measured.

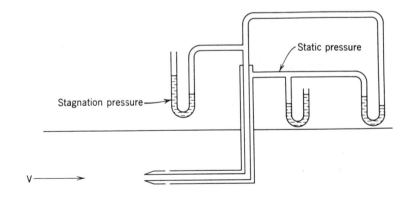

**Fig. 14.24**   Schematic arrangement of a Pitot tube.

Applying the first law to this process we have

$$h + \frac{V^2}{2} = h_0$$

If we assume incompressible flow for this isentropic process, the first law reduces to (because $T\,ds = dh - v\,dp$)

$$\frac{V^2}{2} = h_0 - h = v(P_0 - P)$$

or

$$V = \sqrt{2v(P_0 - P)} \qquad (14.66)$$

If we consider the compressible flow of an ideal gas with constant specific heat, the velocity can be found from the relations

$$\frac{V^2}{2} = h_0 - h = C_{po}(T_0 - T) = C_{po}T\left(\frac{T_0}{T} - 1\right)$$

$$= C_{po}T\left[\left(\frac{P_0}{P}\right)^{(k-1)/k} - 1\right] \qquad (14.67)$$

It is of interest to know the error introduced by assuming incompressible flow when using the Pitot tube to measure the velocity of an ideal gas. In order to do so we introduce Eq. 14.38 and rearrange it as follows:

$$\frac{P_0}{P} = \left(1 + \frac{k-1}{2}M^2\right)^{k/(k-1)} = \left[1 + \left(\frac{k-1}{2}\right)\left(\frac{V^2}{c^2}\right)\right]^{k/(k-1)} \qquad (14.68)$$

But

$$\frac{V^2}{2} + C_{po}T = C_{po}T_0$$

$$\frac{V^2}{2} + \frac{kRc^2}{(k-1)kR} = \frac{kRc_0^2}{(k-1)kR}$$

$$1 + \frac{2c^2}{(k-1)V^2} = \frac{2c_0^2}{(k-1)V^2}, \quad \text{where } c_0 = \sqrt{kRT_0}$$

$$\frac{c^2}{V^2} = \frac{k-1}{2}\left[\left(\frac{2}{k-1}\right)\left(\frac{c_0^2}{V^2}\right) - 1\right] = \frac{c_0^2}{V^2} - \frac{k-1}{2}$$

or

$$\frac{c^2}{V^2} = \frac{c_0^2}{V^2} - \frac{k-1}{2} \qquad (14.69)$$

Substituting this into Eq. 14.68 and rearranging

$$\frac{P}{P_0} = \left[1 - \frac{k-1}{2}\left(\frac{V}{c_0}\right)^2\right]^{k/(k-1)} \tag{14.70}$$

Expanding this equation by the binomial theorem, and including terms through $(V/c_0)^4$ we have

$$\frac{P}{P_0} = 1 - \frac{k}{2}\left(\frac{V}{c_0}\right)^2 + \frac{k}{8}\left(\frac{V}{c_0}\right)^4$$

On rearranging this we have

$$\frac{P_0 - P}{\rho_0 V^2/2} = 1 - \frac{1}{4}\left(\frac{V}{c_0}\right)^2 \tag{14.71}$$

For incompressible flow the corresponding equation is

$$\frac{P_0 - P}{\rho_0 V^2/2} = 1$$

Therefore, the second term on the right side of Eq. 14.71 represents the error involved if incompressible flow is assumed. The error in pressure for a given velocity, and the error in velocity for a given pressure which would result from assuming incompressible flow are given in Table 14.2

**Table 14.2**

| $V/c_0$ | Approximate Room-Temperature Velocity, m/s | Error in Pressure for a Given Velocity, % | Error in Velocity for a Given Pressure, % |
|---------|---------|---------|---------|
| 0.0 | 0 | 0 | 0 |
| 0.1 | 35 | 0.25 | −0.13 |
| 0.2 | 70 | 1.0 | −0.5 |
| 0.3 | 105 | 2.25 | −1.2 |
| 0.4 | 140 | 4.0 | −2.1 |
| 0.5 | 175 | 6.25 | −3.3 |

## 14.12   Flow through Blade Passages

We will now apply the principles we have been considering in this chapter to the flow of fluids through blade passages. We will limit our consideration to uniform, parallel, one-dimensional flow.

There are a number of different devices that involve flow through blade passages. Some of the familiar ones are steam and gas turbines, axial-flow compressors, and certain types of blowers, pumps, and torque converters. Some of these devices, such as the steam turbine, are made up of a number of stages. One stage consists of a nozzle (or blade passages which act as a nozzle) and the moving row of blades that follows. Sometimes the blades are also called buckets, but in this text the term blade will be used.

In the analysis of such problems it is necessary to discuss both absolute and relative velocities. By absolute velocity we mean the velocity a stationary observer would detect. The relative velocity is the velocity an observer traveling with the blade would detect, which is designated with a subscript $R$. The velocity of the blade is designated $V_B$.

It is advantageous to show these velocities on a vector diagram. Figure 14.25 shows typical vector diagrams for both a turbine and a compressor. In each case the first sketch shows the velocity vectors in relation to the blades, and the second sketch shows how these vector diagrams are usually drawn. In these diagrams $V_1$ represents the velocity of the fluid

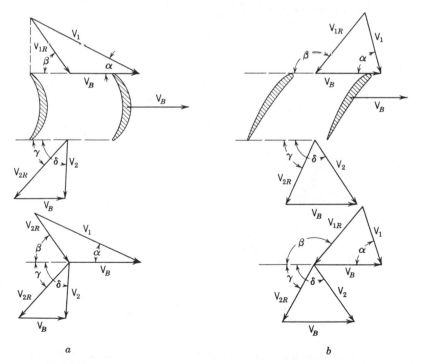

**Fig. 14.25**   Velocity vector diagrams for (*a*) a turbine, and (*b*) a compressor.

entering the blade passage, and $\alpha$ designates the angle at which it enters. $V_{1R}$ represents the relative velocity of the fluid entering the blade passage and $\beta$ the angle at which it enters. Similarly, $V_2$ and $V_{2R}$ represent the velocity and relative velocity of the fluid leaving at the angles $\delta$ and $\gamma$, respectively.

In analyzing the flow through a blade passage it is convenient to consider the control surface shown in Fig. 14.26, which is shown both from a stationary and a moving observer's point of view.

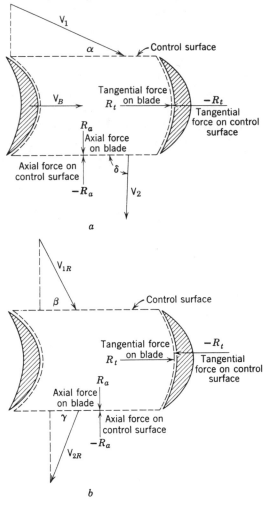

**Fig. 14.26** Analysis of forces on a turbine blade. (a) Stationary observer. (b) Observer moving with the blade.

Let us begin the analysis by writing the first law for both cases, assuming steady, adiabatic flow through the blade passage. The stationary observer is aware that work is being done on the blade, and the steady-flow equation for the first law is

$$h_1 + \frac{V_1^2}{2} = h_2 + \frac{V_2^2}{2} + w \tag{14.72}$$

The moving observer is not aware of any work being done on the blade, and in this case the first law is

$$h_1 + \frac{V_{1R}^2}{2} = h_2 + \frac{V_{2R}^2}{2} \tag{14.73}$$

By combining Eqs. 14.72 and 14.73, we can derive the relation

$$w = \frac{(V_1^2 - V_{1R}^2) - (V_2^2 - V_{2R}^2)}{2} \tag{14.74}$$

Applying the second law of thermodynamics to this process we conclude, since the process is adiabatic, that

$$s_2 \geqslant s_1$$

We will next consider the momentum equation, and evaluate both the tangential force $R_t$ and the axial force $R_a$ on the blade. For our sign convention we assume that tangential velocities and forces in the direction of the movement of the blade are positive and axial velocities and forces in the direction of the axial component of the velocity are positive.

Let us first apply the momentum equation in the tangential direction to the control volume as viewed by the stationary observer. In doing so we make the reasonable assumption that the pressure forces in the tangential direction balance. Then, using Eq. 14.19

$$-R_t = \dot{m} \left( -V_2 \cos \delta - V_1 \cos \alpha \right)$$

or

$$R_t = \dot{m} \left( V_1 \cos \alpha + V_2 \cos \delta \right) \tag{14.75}$$

Similarly, for the control volume as viewed by the moving observer we have

$$-R_t = \dot{m} \left( -V_{2R} \cos \gamma - V_{1R} \cos \beta \right)$$

or

$$R_t = \dot{m} \left( V_{1R} \cos \beta + V_{2R} \cos \gamma \right) \tag{14.76}$$

Equation 14.76 could have been developed from Eq. 14.75 and from the geometry of the vector diagram by noting that

$$V_1 \cos \alpha - V_B = V_{1R} \cos \beta$$

and

$$V_2 \cos \delta + V_B = V_{2R} \cos \gamma$$

Therefore,

$$V_1 \cos \alpha + V_2 \cos \delta = V_{1R} \cos \beta + V_{2R} \cos \gamma \qquad (14.77)$$

The work done by the control volume on the blade in time $dt$ as the blade moves a distance $dx$ is

$$\delta W = R_t \, dx$$

The rate at which work is done, or the power, is

$$\frac{\delta W}{dt} = R_t \frac{dx}{dt} = R_t V_B$$

Substituting for $R_t$ from Eqs. 14.75 and 14.76

$$\frac{\delta W}{dt} = \dot{W}_{\text{c.v.}} = \dot{m} \, V_B (V_1 \cos \alpha + V_2 \cos \delta)$$

$$\frac{\delta W}{\delta t} = \dot{W}_{\text{c.v.}} = \dot{m} \, V_B (V_{1R} \cos \beta + V_{2R} \cos \gamma) \qquad (14.78)$$

The work $w$ per kilogram of fluid flowing is therefore

$$\dot{W}_{\text{c.v.}}/\dot{m} = w = V_B (V_1 \cos \alpha + V_2 \cos \delta) \qquad (14.79)$$

or

$$\dot{W}_{\text{c.v.}}/\dot{m} = w = V_B (V_{1R} \cos \beta + V_{2R} \cos \gamma) \qquad (14.80)$$

The axial thrust on the blade can be found by applying the momentum equation in the axial direction. We must recognize that the pressure at the inlet section may be different from the pressure at the exit section. Consider first the control volume as viewed by the stationary observer. Again we use Eq. 14.19.

$$R_a = \dot{m} \, (V_2 \sin \delta - V_1 \sin \alpha) - (P_1 \mathscr{A}_1 - P_2 \mathscr{A}_2) \qquad (14.81)$$

As viewed by the moving observer we have

$$R_a = \dot{m} \ (V_{2R} \sin \gamma - V_{1R} \sin \beta) - (P_1 \mathscr{A}_1 - P_2 \mathscr{A}_2) \qquad (14.82)$$

Since there is no motion in the axial direction there is no work involved. However, adequate thrust bearings must be provided to handle the axial force. It is important to note in Eqs. 14.81 and 14.82 that if the pressure changes across the moving blade, the axial forces will be significantly influenced. This will be discussed more fully in the next section when the difference between an impulse and a reaction stage is considered.

### Example 14.11

Steam enters the blade passage of one stage of a steam turbine with a velocity of 550 m/s and at an angle ($\alpha$) of 20°. The steam leaves the blade (as seen by the moving observer) at an angle ($\gamma$) of 50°. There is no pressure change across the blade and there are no irreversibilities (which means that there is no change in the magnitude of the relative velocity during the flow through the blade passage). Determine the work per kilogram of steam and the axial thrust. The blade speed is 250 m/s.

We first draw the vector diagram as shown in Fig. 14.27.

$$V_{1R} \sin \beta = V_1 \sin \alpha = 550 \sin 20° = 550 \ (0.342) = 188.1 \text{ m/s}$$

$$V_{1R} \cos \beta + V_B = V_1 \cos \alpha = 550 \cos 20° = 550 \ (0.9397) = 516.8 \text{ m/s}$$

$$V_{1R} \cos \beta = 516.8 - 250 = 266.8 \text{ m/s}$$

$$\tan \beta = \frac{V_{1R} \sin \beta}{V_{1R} \cos \beta} = \frac{188.1}{266.8} = 0.7050$$

$$\beta = 35.185°$$

$$V_{1R} \sin \beta = 188.1$$

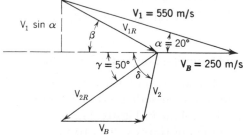

**Fig. 14.27**   Sketch for Example 14.11.

$$V_{1R} = \frac{188.1}{\sin 35.185°} = \frac{188.1}{0.5762} = 326.4 \text{ m/s}$$

$$V_{2R} = V_{1R} = 326.4 \text{ m/s}$$

$$V_2 \sin \delta = V_{2R} \sin \gamma = 326.4 \sin 50° = 326.4 \,(0.7660) = 250.1 \text{ m/s}$$

$$V_2 \cos \delta + V_B = V_{2R} \cos \gamma = 326.4 \cos 50° = 326.4 \,(0.6428) = 209.8 \text{ m/s}$$

$$V_2 \cos \delta = 209.8 - 250 = -40.2 \text{ m/s}$$

This implies that we drew our vector diagram slightly wrong and it should have been as shown in Fig. 14.28

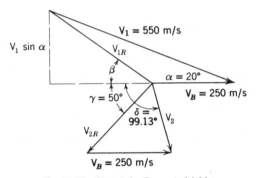

**Fig. 14.28**   Sketch for Example 14.11.

$$\cot \delta = \frac{V_2 \cos \delta}{V_2 \sin \delta} = \frac{-40.2}{250.1} = -0.1607$$

$$\delta = 99.13°$$

$$V_2 = \sqrt{(V_2 \sin \delta)^2 + (V_2 \cos \delta)^2}$$

$$= \sqrt{(250.1)^2 + (40.2)^2} = 253.3 \text{ m/s}$$

## 14.13   Impulse and Reaction Stages for Turbines

Two terms used in connection with turbines are impulse stage and reaction stage. A turbine stage is defined as the fixed nozzle or blade and the moving blades that follow it. Figure 14.29a shows an impulse stage. In an impulse stage the entire pressure drop takes place in a stationary nozzle, and the pressure remains constant as the fluid flows through the

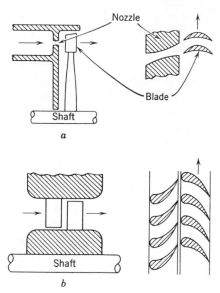

**Fig. 14.29**  Schematic arrangement for (*a*) an impulse stage, and (*b*) a reaction stage.

blade passage. There is a decrease in the kinetic energy of the fluid as it flows through the blade passage, and the enthalpy will increase due to irreversibilities associated with the fluid flow.

In the pure reaction stage the entire pressure drop occurs as the fluid flows through the moving blades. Thus, the moving blade acts as a nozzle, and the blade passage must have the proper contour for a nozzle (converging if the exit pressure is greater than the critical pressure and converging-diverging if the exit pressure is less than the critical pressure). In the pure reaction stage the only purpose of the stationary blade is to direct the fluid into the moving blade at the proper angle and velocity.

The pure reaction stage is used relatively infrequently. Rather, most turbines that are classified as reaction turbines have a pressure and enthalpy drop in both the fixed and moving blades. The degree of reaction is defined as the fraction of the enthalpy drop that occurs in the moving blades. Thus, in the 50 per cent reaction stage, which is very commonly used, half of the enthalpy drop across the stage occurs in the fixed blade and the other half in the moving blade.

A few comparisons between the impulse and reaction blade may be noted here. Since there is a pressure drop across both the fixed and moving blades in the reaction stage, there is a tendency for leakage to occur across the tip of the blade. Therefore, very close clearances must be maintained at the blade tips. Further, the pressure drop across the moving blade of the reaction stage gives rise to axial forces that must be balanced in order to prevent axial motion in reaction turbines. In an impulse stage it is possible to utilize only part of the periphery for admission of steam. This is usually termed partial admission. In fact, in turbines that utilize an impulse stage for the first stage the power output can be controlled by opening and closing nozzles. The main advantage of the reaction stage is that lower fluid velocities are utilized and higher efficiencies can be attained at lower velocities. This will be discussed more fully in the following sections.

## 14.14 Some Further Considerations of Impulse Stages

Figure 14.30 shows a typical velocity diagram for an impulse stage. The blade velocity coefficient $k_B$ is defined as the ratio of the relative velocity leaving the blade to the relative velocity entering the blade. That is,

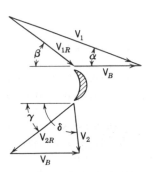

$$k_B = \frac{V_{2R}}{V_{1R}} \qquad (14.83)$$

In the ideal impulse stage $V_{2R} = V_{1R}$ and $k_B = 1$, but in the actual impulse stage $V_{2R}$ will be less than $V_{1R}$ due to irreversibilities in the fluid as it flows through the blade passage.

The blade efficiency $\eta_B$ is defined as the ratio of the actual work per kilogram of fluid flowing to the kinetic energy of the fluid entering the blade passage.

**Fig. 14.30** A velocity vector diagram for an impulse blade.

$$\eta_B = \frac{w}{V_1^2/2} \qquad (14.84)$$

Thus, a blade efficiency of 100 per cent means that the work is exactly equal to the kinetic energy of the fluid entering the blade, and the kinetic energy of the fluid leaving the blade is zero.

Quite obviously, the fluid must have some axial velocity if it is to flow out the blade passage. However, it is of interest to consider a turbine having a zero axial velocity and to determine the blade-speed ratio that will give an efficiency of 100 per cent. The blade-speed ratio is the ratio of the blade speed $V_B$ to the velocity of the fluid leaving the nozzle $V_1$ and the ratio is designated $V_B/V_1$.

Figure 14.31a shows a reversible impulse stage that has a very small entrance and blade exit angle, and a blade-speed ratio of 0.5. It is evident that as the angles $\alpha$ and $\gamma$ (the inlet and blade exit angles) approach zero that the exit velocity $V_2$ will also approach zero. Such a stage would have

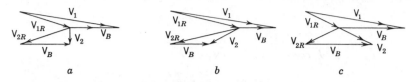

**Fig. 14.31**    A velocity vector diagram illustrating optimum blade-speed ratio.

an efficiency of 100 per cent. Figures 14.31*b* and *c* show reversible impulse stages with essentially zero angles, but with blade-speed ratios considerably less and considerably greater than 0.5. In both of these the exit velocity is large, and therefore the blade efficiency is considerably less than 100 per cent. We conclude, therefore, that for a reversible, zero-angle impulse turbine the optimum blade-speed ratio is 0.5. The curve marked impulse stage on Fig. 14.34 shows the variation of blade efficiency with blade-speed ratio for a reversible stage.

An expression for the blade efficiency as a function of the angles $\alpha$ and $\gamma$ (the inlet angle and blade exit angles), the blade-speed ratio, and the blade velocity coefficient can be derived.

From Eq. 14.79

$$w = V_B (V_1 \cos \alpha + V_2 \cos \delta)$$

$$V_{2R} = k_B V_{1R}$$

From the vector diagram we note that

$$V_2 \cos \delta = V_{2R} \cos \gamma - V_B = k_B V_{1R} \cos \gamma - V_B$$

$$V_{1R} = \frac{V_1 \cos \alpha - V_B}{\cos \beta}$$

Therefore,

$$w = V_B \left( V_1 \cos \alpha + k_B \cos \gamma \frac{V_1 \cos \alpha - V_B}{\cos \beta} - V_B \right)$$

$$= V_B \left[ (V_1 \cos \alpha - V_B) \left( 1 + \frac{k_B \cos \gamma}{\cos \beta} \right) \right]$$

$$\eta_B = \frac{w}{V_1^2/2} = \frac{2 V_B}{V_1^2} (V_1 \cos \alpha - V_B) \left( 1 + k_B \frac{\cos \gamma}{\cos \beta} \right)$$

$$= \frac{2 V_B^2}{V_1^2} \left( \frac{V_1}{V_B} \cos \alpha - 1 \right) \left( 1 + k_B \frac{\cos \gamma}{\cos \beta} \right)$$

But

$$\cos \beta = \frac{V_1 \cos \alpha - V_B}{\sqrt{(V_1 \cos \alpha - V_B)^2 + (V_1 \sin \alpha)^2}}$$

$$= \frac{\cos \alpha - V_B/V_1}{\sqrt{(\cos \alpha - V_B/V_1)^2 + \sin^2 \alpha}}$$

$$\eta_B = 2 \frac{V_B}{V_1} \left(\cos \alpha - \frac{V_B}{V_1}\right) \left[1 + \frac{k_B \cos \gamma}{\dfrac{\cos \alpha - V_B/V_1}{\sqrt{(\cos \alpha - V_B/V_1)^2 + \sin^2 \alpha}}}\right] \qquad (14.85)$$

This relation gives the blade efficiency as a function of the blade speed-ratio $V_B/V_1$, the angles $\alpha$ and $\gamma$, and the blade velocity coefficient $k_B$. From this relation it can be shown that for the usual values of angles and blade velocity coefficient, the optimum efficiency is obtained with blade velocity coefficient of approximately 0.5, which was the value obtained for the reversible zero-angle impulse stage.

This leads to consideration of the velocity-compounded turbine. Suppose one was attempting to build a turbine having one impulse stage, and that it was to receive steam at 700 kPa, 250°C and exhaust at a pressure of 15 kPa. The velocity of the steam leaving the nozzle would be about 1100 m/s. In order to have maximum efficiency a blade-speed ratio of 0.5 must be used, and this would require a blade speed of 550 m/s. Blade speeds of this magnitude result in very high stresses due to centrifugal force, and further, the irreversibilities associated with fluid flow increase as the fluid velocity increases.

For these reasons velocity compounding is used, and a schematic diagram and a vector diagram for a velocity-compounded stage are shown in Fig. 14.32. The stationary blade serves to change the direction of the fluid so that it will enter the second row of moving blades properly. It can be shown that for a reversible zero-angle turbine having two moving rows the blade-speed ratio for optimum efficiency is 1/4. Sometimes velocity-compounded turbines are built with as many as three moving rows.

Velocity-compounded turbines have a lower efficiency than comparable turbines that have a number of impulse or reaction stages. However, they are relatively inexpensive and are therefore frequently used in small sizes where initial cost is more important than efficiency. A velocity-compounded stage is also frequently used as the first stage in large steam turbines in order to have a large pressure and temperature drop before the steam enters the moving blades.

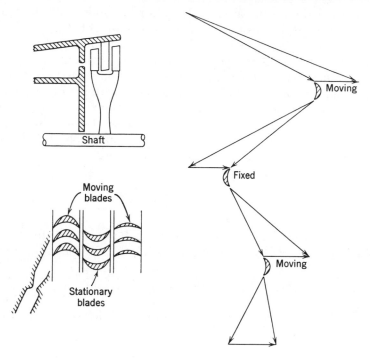

**Fig. 14.32**   A velocity-compounded impulse stage.

## 14.15   Some Further Considerations of Reaction Stages

First a remark should be made regarding the blade efficiency of a reaction stage. Both the fixed blades and moving blades act as a nozzle. Therefore, the efficiency of the moving blades can be defined in the same way as the efficiency of a nozzle if the relative velocities are used.

The second matter to consider is the blade-speed ratio for maximum efficiency of a reaction stage. Let us consider a reversible reaction stage in which the isentropic enthalpy drops across the fixed and moving blades are equal (i.e., a 50 per cent reaction stage). For a zero-angle turbine it is evident from Fig. 14.33 that for zero kinetic energy leaving the blade, a blade-speed ratio of 1.0 is required. Thus, maximum efficiency in a reaction stage is obtained for a blade-speed ratio of 1.0.

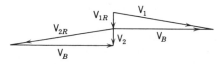

**Fig. 14.33**   A velocity vector diagram for a reaction stage.

This leads us to a comparison of the impulse stage with the reaction stage regarding the blade-speed ratio for maximum efficiency.

If we compare reversible impulse and reaction stages for a given velocity from the stationary nozzle or blade, it is evident that for maximum efficiency the blade velocity for the reaction stage would be twice the blade velocity of the impulse stage. However, this is not a good comparison because the reaction stage would have a greater enthalpy drop (due to the enthalpy drop in the moving row) than the impulse stage.

A better comparison can be made by considering equal enthalpy drops across the stage. Let this enthalpy drop be denoted $\Delta h_s$, and let the velocity $V_0$ be defined by the relation

$$V_0 = \sqrt{2\Delta h_s}$$

That is, $V_0$ would be the velocity leaving the nozzle of the reversible impulse stage that has this enthalpy drop. That is, for the impulse stage

$$V_1 = V_0$$

For the reversible 50 per cent reaction stage the velocity leaving the fixed blades would be

$$V_1 = \sqrt{\tfrac{1}{2} \times 2\Delta h_s} = \frac{V_0}{\sqrt{2}}$$

For maximum efficiency in the impulse stage, $V_B/V_1 = 0.5$. Therefore, for the impulse stage

$$\frac{V_B}{V_1} = \frac{V_B}{V_0} = 0.5$$

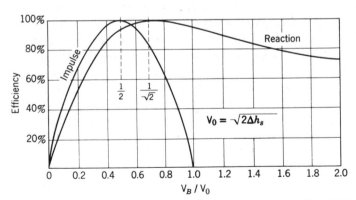

**Fig. 14.34**   Efficiency of an ideal impulse and reaction stage as a function of $V_B/V_0$.

For maximum efficiency in the reaction stage, $V_B/V_1 = 1.0$. Therefore, for the reaction stage

$$\frac{V_B}{V_1} = \frac{V_B \sqrt{2}}{V_0} \qquad \frac{V_B}{V_0} = \frac{1}{\sqrt{2}}$$

Thus, for a given enthalpy drop per stage, the reaction stage requires a higher blade speed for maximum efficiency than the impulse stage. Conversely, for a given blade speed the reaction turbine requires more stages than an impulse stage, and in this case the fluid velocity is lower in the reaction stage than in the impulse stage.

Figure 14.34 shows the blade efficiency as a function of $V_B/V_0$ for an impulse stage and reaction stage.

## PROBLEMS

14.1    Air leaves the compressor of a jet engine at a temperature of 150°C, a pressure of 300 kPa, and a velocity of 120 m/s. Determine the isentropic stagnation temperature and pressure.

14.2    The products of combustion of a jet engine leave the engine with a velocity relative to the plane of 400 m/s, a temperature of 480°C, and a pressure of 85 kPa. Assuming that $k = 1.34$, $C_p = 1.13$ kJ/kg K for the products, determine the stagnation pressure and temperature of the products (relative to the airplane).

14.3    Steam leaves a nozzle with a velocity of 250 m/s. The stagnation pressure is 800 kPa, and the stagnation temperature is 250°C. What is the static pressure and temperature?

14.4    Water is pumped from a lake and discharged through a nozzle as shown in Fig. 14.35 (See p. 640.) At the pump discharge the pressure is 700 kPa. and the temperature is 25°C. The nozzle is located 10 m above the pump. Assume 100 kPa atmospheric pressure and reversible flow throughout. Determine the velocity of the water leaving the nozzle.

14.5    In a water turbine that utilizes nozzles, the water enters the nozzle at 350 kPa, 25°C and leaves at atmospheric pressure. If the flow through the nozzle is reversible and adiabatic, determine the velocity and kinetic energy per kg of water leaving the nozzle.

14.6    Air is expanded in a nozzle from 2 MPa, 200°C, to 0.2 MPa. The mass rate of flow through the nozzle is 5 kg/s. Assume the flow to be reversible and adiabatic.

   (a) Determine specific volume, velocity, Mach number, and cross-sectional area for each 0.2 MPa decrease of pressure, and plot these as a function of pressure.

   (b) Determine the throat and exit areas for the nozzle.

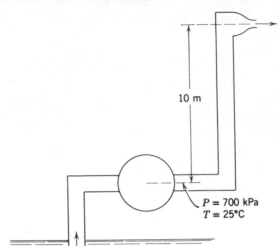

10 m

$P = 700$ kPa
$T = 25°C$

**Fig. 14.35** Sketch for Problem 14.4.

14.7 Consider the nozzle of Problem 14.6 and determine what back pressure will cause a normal shock to stand in the exit plane of the nozzle. What is the mass rate of flow under these conditions?

14.8 At what Mach number will the normal shock occur in the nozzle of Problem 14.6 if the back pressure is 1.4 MPa?

14.9 Consider the nozzle of Problem 14.6. What back pressure will be required to cause subsonic flow throughout the entire nozzle with $M = 1$ at the throat?

14.10 Determine the mass rate of flow through the nozzle of Problem 14.6 for a back pressure of 1.9 MPa.

14.11 A nozzle is designed assuming reversible adiabatic flow with an exit Mach number of 2.6 while flowing air having a stagnation pressure and temperature of 2 MPa and 150°C, respectively. The mass rate of flow is 5 kg/s, and $k$ may be assumed to be 1.40 and constant.

(a) Determine the exit pressure, temperature, and area, and the throat area.

(b) Suppose that the back pressure at the nozzle exit is raised to 1.35 MPa, and that the flow remains isentropic except for a normal shock wave. Determine the exit Mach number and temperature, and the mass flow through the nozzle.

14.12 A jet plane travels through the air with a speed of 1000 km per hour at an altitude of 6000 m, where the pressure is 40 kPa and the temperature is −12°C. Consider the diffuser of the engine. The air leaves the diffuser with a velocity of 100 m/s. Determine the pressure and temperature leaving the diffuser, and the ratio of inlet to exit area of the diffuser, assuming the flow to be reversible and adiabatic.

14.13 The products of combustion enter the nozzle of a jet engine at a total pressure of 125 kPa, and a total temperature of 650°C. The atmospheric pressure is  45 kPa. The nozzle is convergent, and the rate of flow is 25 kg/s. Assume the flow to be reversible and adiabatic. Determine the exit area of the nozzle.

14.14 Air is expanded in a nozzle from 700 kPa, 200°C, to 150 kPa in a nozzle having an efficiency of 90%. The mass rate of flow is 4 kg/s. Determine the exit area of the nozzle, the exit velocity, and the increase of entropy per kg of fluid. Compare these results with those of the reversible adiabatic nozzle of Problem 14.6.

14.15 Repeat Problem 14.12 assuming a diffuser efficiency of 80%.

14.16 Consider the diffuser of a supersonic aircraft flying at $M = 1.4$ at such an altitude that the temperature is $-20$°C, and the atmosphere pressure is 50 kPa. Consider two possible ways in which this diffuser might operate, and for each case calculate the throat area required for a flow of 50 kg/s.

  (a) The diffuser operates as a reversible adiabatic diffusers with subsonic exit velocity.

  (b) A normal shock stands at the entrance to the diffuser. Except for the normal shock the flow is reversible and adiabatic, and the exit velocity is subsonic. This is shown in Fig. 14.36. Assume a convergent-divergent diffuser with $M = 1$ at the throat.

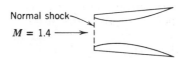

**Fig. 14.36**   Sketch for Problem 14.16.

14.17 Air enters a diffuser with a velocity of 200 m/s, a static pressure of 70 kPa, and a temperature of $-10$°C. The velocity leaving the diffuser is 65 m/s and the static pressure at the diffuser exit is 82 kPa. Determine the static temperature at the diffuser exit and the diffuser efficiency. Compare the stagnation pressures at the diffuser inlet and exit.

14.18 Steam at a pressure of 600 kPa and a temperature of 300°C expands to a pressure of 150 kPa in a nozzle having an efficiency of 90%. The mass rate of flow is 10 kg/s. Determine the nozzle exit area.

14.19 Steam at an initial pressure of 600 kPa and a temperature of 300°C flows through a convergent-divergent nozzle having a throat area of 300 mm². The pressure at the exit plane is 150 kPa and the exit velocity is 780 m/s. The flow from the nozzle entrance to the throat is reversible and adiabatic. Determine the exit area of the nozzle, the over-all nozzle efficiency, and the increase in entropy per kg of fluid.

14.20 A sharp-edged orifice is used to measure the flow of air in a pipe. The pipe diameter is 100 mm and the diameter of the orifice is 25 mm. Upstream of the orifice, the absolute pressure is 150 kPa and the temperature is 35°C. The pressure drop across the orifice is 12 kPa, and the coefficient of discharge is 0.62. Determine the mass rate of flow in the pipeline.

14.21 A steam turbine utilizes convergent nozzles. An estimate of the rate of steam flow is to be made from the pressure drop across the nozzles of one stage. The inlet conditions to these nozzles are 600 kPa, 250°C. The exit pressure is 400 kPa. The coefficient of discharge is estimated to be 0.94. The total exit area of the nozzles in this stage is 0.005 m². Determine the rate of flow under these conditions.

14.22 The coefficient of discharge of a sharp-edged orifice is determined at one set of conditions by use of an accurately calibrated gasometer. The orifice has a diameter of 20 mm and the pipe diameter is 50 mm. The absolute upstream pressure is 200 kPa and the pressure drop across the orifice is 10 kPa. The temperature of the air entering the orifice is 25°C. The mass rate of flow as measured by the gasometer is 0.04 kg/s. What is the coefficient of discharge of the orifice under these conditions?

14.23 Consider the flow of air through an impulse blade passage. The air enters the blade passage at an angle of 18° with a velocity of 460 m/s, a pressure of 110 kPa, and a temperature of 90°C. The blade velocity is 250 m/s and the air leaves the blade passage at an angle of 45° relative to the blade. The mass rate of flow is 10 kg/s, and it is assumed that the flow is reversible and adiabatic.

 (a) Draw the velocity diagram to scale.

 (b) What is the power output of the turbine?

 (c) If the blade wheel to which the blades are attached has a diameter of 0.3 m, what is the length of each blade?

14.24 Steam enters an impulse turbine in which all processes are assumed to be reversible and adiabatic. The inlet pressure is 700 kPa and the inlet temperature is 400°C. The exhaust pressure is 100 kPa. The steam leaves the nozzle and enters the turbine at an angle of 20°. The blade-speed ratio is 0.5 and the blade exit angle is 50°. Determine the blade efficiency of this turbine.

14.25 Consider an impulse turbine having the same inlet conditions as Problem 14.24. The nozzle efficiency is 92% and the blade velocity coefficient is 0.96. The blade-speed ratio is 0.5. Determine the work per kg of steam and draw the velocity diagram to scale.

14.26 Consider a two stage, reversible, velocity-compounded impulse turbine that has the same inlet conditions as Problem 14.24. Assume a blade-velocity coefficient of 0.25, and an inlet angle of 20°. Both the moving and fixed blades have equal inlet and exit angles (i.e., the angle at which the steam enters relative to the blade is equal to the relative angle at which it leaves). Determine the blade efficiency of the turbine and draw the velocity diagram to scale.

14.27  Consider a single-stage reaction turbine having equal enthalpy drops across the fixed and moving blades. Consider the same inlet conditions and exit pressure as Problem 14.24. All processes are reversible and adiabatic. The blade inlet angle is 20° and the blade-speed ratio is 0.9.

    (a) What is the pressure at the exit of the fixed blades?

    (b) Draw the velocity vector diagram to scale.

    (c) What is the net work per kg of steam flowing through the turbine?

14.28  A single-stage air turbine having 50% reaction operates with inlet pressure and temperature of 350 kPa, 1000 K, and exhaust pressure of 100 kPa. The fixed-blade exit angle is 18°, and exhaust from the turbine is in the axial direction. The blade-speed ratio is 0.9, and all processes are reversible and adiabatic. Determine:

    (a) All velocities and angles, and draw the velocity vector diagram.

    (b) The work/kg of air flowing through the turbine.

# Appendix

# Figures, Tables, and Charts

# Table A.1
## Thermodynamic Properties of Steam[a]
# Table A.1.1
## Saturated Steam: Temperature Table

| Temp. °C $T$ | Press. kPa $P$ | Specific Volume | | Internal Energy | | | Enthalpy | | | Entropy | | |
|---|---|---|---|---|---|---|---|---|---|---|---|---|
| | | Sat. Liquid $v_f$ | Sat. Vapor $v_g$ | Sat. Liquid $u_f$ | Evap. $u_{fg}$ | Sat. Vapor $u_g$ | Sat. Liquid $h_f$ | Evap. $h_{fg}$ | Sat. Vapor $h_g$ | Sat. Liquid $s_f$ | Evap. $s_{fg}$ | Sat. Vapor $s_g$ |
| 0.01 | 0.6113 | 0.001 000 | 206.14 | .00 | 2375.3 | 2375.3 | .01 | 2501.3 | 2501.4 | .0000 | 9.1562 | 9.1562 |
| 5 | 0.8721 | 0.001 000 | 147.12 | 20.97 | 2361.3 | 2382.3 | 20.98 | 2489.6 | 2510.6 | .0761 | 8.9496 | 9.0257 |
| 10 | 1.2276 | 0.001 000 | 106.38 | 42.00 | 2347.2 | 2389.2 | 42.01 | 2477.7 | 2519.8 | .1510 | 8.7498 | 8.9008 |
| 15 | 1.7051 | 0.001 001 | 77.93 | 62.99 | 2333.1 | 2396.1 | 62.99 | 2465.9 | 2528.9 | .2245 | 8.5569 | 8.7814 |
| 20 | 2.339 | 0.001 002 | 57.79 | 83.95 | 2319.0 | 2402.9 | 83.96 | 2454.1 | 2538.1 | .2966 | 8.3706 | 8.6672 |
| 25 | 3.169 | 0.001 003 | 43.36 | 104.88 | 2304.9 | 2409.8 | 104.89 | 2442.3 | 2547.2 | .3674 | 8.1905 | 8.5580 |
| 30 | 4.246 | 0.001 004 | 32.89 | 125.78 | 2290.8 | 2416.6 | 125.79 | 2430.5 | 2556.3 | .4369 | 8.0164 | 8.4533 |
| 35 | 5.628 | 0.001 006 | 25.22 | 146.67 | 2276.7 | 2423.4 | 146.68 | 2418.6 | 2565.3 | .5053 | 7.8478 | 8.3531 |
| 40 | 7.384 | 0.001 008 | 19.52 | 167.56 | 2262.6 | 2430.1 | 167.57 | 2406.7 | 2574.3 | .5725 | 7.6845 | 8.2570 |
| 45 | 9.593 | 0.001 010 | 15.26 | 188.44 | 2248.4 | 2436.8 | 188.45 | 2394.8 | 2583.2 | .6387 | 7.5261 | 8.1648 |
| 50 | 12.349 | 0.001 012 | 12.03 | 209.32 | 2234.2 | 2443.5 | 209.33 | 2382.7 | 2592.1 | .7038 | 7.3725 | 8.0763 |
| 55 | 15.758 | 0.001 015 | 9.568 | 230.21 | 2219.9 | 2450.1 | 230.23 | 2370.7 | 2600.9 | .7679 | 7.2234 | 7.9913 |
| 60 | 19.940 | 0.001 017 | 7.671 | 251.11 | 2205.5 | 2456.6 | 251.13 | 2358.5 | 2609.6 | .8312 | 7.0784 | 7.9096 |
| 65 | 25.03 | 0.001 020 | 6.197 | 272.02 | 2191.1 | 2463.1 | 272.06 | 2346.2 | 2618.3 | .8935 | 6.9375 | 7.8310 |
| 70 | 31.19 | 0.001 023 | 5.042 | 292.95 | 2176.6 | 2469.6 | 292.98 | 2333.8 | 2626.8 | .9549 | 6.8004 | 7.7553 |
| 75 | 38.58 | 0.001 026 | 4.131 | 313.90 | 2162.0 | 2475.9 | 313.93 | 2321.4 | 2635.3 | 1.0155 | 6.6669 | 7.6824 |
| 80 | 47.39 | 0.001 029 | 3.407 | 334.86 | 2147.4 | 2482.2 | 334.91 | 2308.8 | 2643.7 | 1.0753 | 6.5369 | 7.6122 |
| 85 | 57.83 | 0.001 033 | 2.828 | 355.84 | 2132.6 | 2488.4 | 355.90 | 2296.0 | 2651.9 | 1.1343 | 6.4102 | 7.5445 |
| 90 | 70.14 | 0.001 036 | 2.361 | 376.85 | 2117.7 | 2494.5 | 376.92 | 2283.2 | 2660.1 | 1.1925 | 6.2866 | 7.4791 |
| 95 | 84.55 | 0.001 040 | 1.982 | 397.88 | 2102.7 | 2500.6 | 397.96 | 2270.2 | 2668.1 | 1.2500 | 6.1659 | 7.4159 |

[a] Adapted from Joseph H. Keenan, Frederick G. Keyes, Philip G. Hill, and Joan G. Moore, *Steam Tables*, (New York: John Wiley & Sons, Inc., 1969).

## Table A.1.1 (Continued)
## Saturated Steam: Temperature Table

| Temp. °C T | Press. MPa P | Specific Volume | | Internal Energy | | | Enthalpy | | | Entropy | | |
|---|---|---|---|---|---|---|---|---|---|---|---|---|
| | | Sat. Liquid $v_f$ | Sat. Vapor $v_g$ | Sat. Liquid $u_f$ | Evap. $u_{fg}$ | Sat. Vapor $u_g$ | Sat. Liquid $h_f$ | Evap. $h_{fg}$ | Sat. Vapor $h_g$ | Sat. Liquid $s_f$ | Evap. $s_{fg}$ | Sat. Vapor $s_g$ |
| 100 | 0.101 35 | 0.001 044 | 1.6729 | 418.94 | 2087.6 | 2506.5 | 419.04 | 2257.0 | 2676.1 | 1.3069 | 6.0480 | 7.3549 |
| 105 | 0.120 82 | 0.001 048 | 1.4194 | 440.02 | 2072.3 | 2512.4 | 440.15 | 2243.7 | 2683.8 | 1.3630 | 5.9328 | 7.2958 |
| 110 | 0.143 27 | 0.001 052 | 1.2102 | 461.14 | 2057.0 | 2518.1 | 461.30 | 2230.2 | 2691.5 | 1.4185 | 5.8202 | 7.2387 |
| 115 | 0.169 06 | 0.001 056 | 1.0366 | 482.30 | 2041.4 | 2523.7 | 482.48 | 2216.5 | 2699.0 | 1.4734 | 5.7100 | 7.1833 |
| 120 | 0.198 53 | 0.001 060 | 0.8919 | 503.50 | 2025.8 | 2529.3 | 503.71 | 2202.6 | 2706.3 | 1.5276 | 5.6020 | 7.1296 |
| 125 | 0.2321 | 0.001 065 | 0.7706 | 524.74 | 2009.9 | 2534.6 | 524.99 | 2188.5 | 2713.5 | 1.5813 | 5.4962 | 7.0775 |
| 130 | 0.2701 | 0.001 070 | 0.6685 | 546.02 | 1993.9 | 2539.9 | 546.31 | 2174.2 | 2720.5 | 1.6344 | 5.3925 | 7.0269 |
| 135 | 0.3130 | 0.001 075 | 0.5822 | 567.35 | 1977.7 | 2545.0 | 567.69 | 2159.6 | 2727.3 | 1.6870 | 5.2907 | 6.9777 |
| 140 | 0.3613 | 0.001 080 | 0.5089 | 588.74 | 1961.3 | 2550.0 | 589.13 | 2144.7 | 2733.9 | 1.7391 | 5.1908 | 6.9299 |
| 145 | 0.4154 | 0.001 085 | 0.4463 | 610.18 | 1944.7 | 2554.9 | 610.63 | 2129.6 | 2740.3 | 1.7907 | 5.0926 | 6.8833 |
| 150 | 0.4758 | 0.001 091 | 0.3928 | 631.68 | 1927.9 | 2559.5 | 632.20 | 2114.3 | 2746.5 | 1.8418 | 4.9960 | 6.8379 |
| 155 | 0.5431 | 0.001 096 | 0.3468 | 653.24 | 1910.8 | 2564.1 | 653.84 | 2098.6 | 2752.4 | 1.8925 | 4.9010 | 6.7935 |
| 160 | 0.6178 | 0.001 102 | 0.3071 | 674.87 | 1893.5 | 2568.4 | 675.55 | 2082.6 | 2758.1 | 1.9427 | 4.8075 | 6.7502 |
| 165 | 0.7005 | 0.001 108 | 0.2727 | 696.56 | 1876.0 | 2572.5 | 697.34 | 2066.2 | 2763.5 | 1.9925 | 4.7153 | 6.7078 |
| 170 | 0.7917 | 0.001 114 | 0.2428 | 718.33 | 1858.1 | 2576.5 | 719.21 | 2049.5 | 2768.7 | 2.0419 | 4.6244 | 6.6663 |
| 175 | 0.8920 | 0.001 121 | 0.2168 | 740.17 | 1840.0 | 2580.2 | 741.17 | 2032.4 | 2773.6 | 2.0909 | 4.5347 | 6.6256 |
| 180 | 1.0021 | 0.001 127 | 0.194 05 | 762.09 | 1821.6 | 2583.7 | 763.22 | 2015.0 | 2778.2 | 2.1396 | 4.4461 | 6.5857 |
| 185 | 1.1227 | 0.001 134 | 0.174 09 | 784.10 | 1802.9 | 2587.0 | 785.37 | 1997.1 | 2782.4 | 2.1879 | 4.3586 | 6.5465 |
| 190 | 1.2544 | 0.001 141 | 0.156 54 | 806.19 | 1783.8 | 2590.0 | 807.62 | 1978.8 | 2786.4 | 2.2359 | 4.2720 | 6.5079 |
| 195 | 1.3978 | 0.001 149 | 0.141 05 | 828.37 | 1764.4 | 2592.8 | 829.98 | 1960.0 | 2790.0 | 2.2835 | 4.1863 | 6.4698 |
| 200 | 1.5538 | 0.001 157 | 0.127 36 | 850.65 | 1744.7 | 2595.3 | 852.45 | 1940.7 | 2793.2 | 2.3309 | 4.1014 | 6.4323 |
| 205 | 1.7230 | 0.001 164 | 0.115 21 | 873.04 | 1724.5 | 2597.5 | 875.04 | 1921.0 | 2796.0 | 2.3780 | 4.0172 | 6.3952 |
| 210 | 1.9062 | 0.001 173 | 0.104 41 | 895.53 | 1703.9 | 2599.5 | 897.76 | 1900.7 | 2798.5 | 2.4248 | 3.9337 | 6.3585 |

| | | | | | | | | | | | | |
|---|---|---|---|---|---|---|---|---|---|---|---|---|
| 215 | 2.104 | 0.001 181 | 0.094 79 | 918.14 | 1682.9 | 2601.1 | 920.62 | 1879.9 | 2800.5 | 2.4714 | 3.8507 | 6.3221 |
| 220 | 2.318 | 0.001 190 | 0.086 19 | 940.87 | 1661.5 | 2602.4 | 943.62 | 1858.5 | 2802.1 | 2.5178 | 3.7683 | 6.2861 |
| 225 | 2.548 | 0.001 199 | 0.078 49 | 963.73 | 1639.6 | 2603.3 | 966.78 | 1836.5 | 2803.3 | 2.5639 | 3.6863 | 6.2503 |
| 230 | 2.795 | 0.001 209 | 0.071 58 | 986.74 | 1617.2 | 2603.9 | 990.12 | 1813.8 | 2804.0 | 2.6099 | 3.6047 | 6.2146 |
| 235 | 3.060 | 0.001 219 | 0.065 37 | 1009.89 | 1594.2 | 2604.1 | 1013.62 | 1790.5 | 2804.2 | 2.6558 | 3.5233 | 6.1791 |
| 240 | 3.344 | 0.001 229 | 0.059 76 | 1033.21 | 1570.8 | 2604.0 | 1037.32 | 1766.5 | 2803.8 | 2.7015 | 3.4422 | 6.1437 |
| 245 | 3.648 | 0.001 240 | 0.054 71 | 1056.71 | 1546.7 | 2603.4 | 1061.23 | 1741.7 | 2803.0 | 2.7472 | 3.3612 | 6.1083 |
| 250 | 3.973 | 0.001 251 | 0.050 13 | 1080.39 | 1522.0 | 2602.4 | 1085.36 | 1716.2 | 2801.5 | 2.7927 | 3.2802 | 6.0730 |
| 255 | 4.319 | 0.001 263 | 0.045 98 | 1104.28 | 1496.7 | 2600.9 | 1109.73 | 1689.8 | 2799.5 | 2.8383 | 3.1992 | 6.0375 |
| 260 | 4.688 | 0.001 276 | 0.042 21 | 1128.39 | 1470.6 | 2599.0 | 1134.37 | 1662.5 | 2796.9 | 2.8838 | 3.1181 | 6.0019 |
| 265 | 5.081 | 0.001 289 | 0.038 77 | 1152.74 | 1443.9 | 2596.6 | 1159.28 | 1634.4 | 2793.6 | 2.9294 | 3.0368 | 5.9662 |
| 270 | 5.499 | 0.001 302 | 0.035 64 | 1177.36 | 1416.3 | 2593.7 | 1184.51 | 1605.2 | 2789.7 | 2.9751 | 2.9551 | 5.9301 |
| 275 | 5.942 | 0.001 317 | 0.032 79 | 1202.25 | 1387.9 | 2590.2 | 1210.07 | 1574.9 | 2785.0 | 3.0208 | 2.8730 | 5.8938 |
| 280 | 6.412 | 0.001 332 | 0.030 17 | 1227.46 | 1358.7 | 2586.1 | 1235.99 | 1543.6 | 2779.6 | 3.0668 | 2.7903 | 5.8571 |
| 285 | 6.909 | 0.001 348 | 0.027 77 | 1253.00 | 1328.4 | 2581.4 | 1262.31 | 1511.0 | 2773.3 | 3.1130 | 2.7070 | 5.8199 |
| 290 | 7.436 | 0.001 366 | 0.025 57 | 1278.92 | 1297.1 | 2576.0 | 1289.07 | 1477.1 | 2766.2 | 3.1594 | 2.6227 | 5.7821 |
| 295 | 7.993 | 0.001 384 | 0.023 54 | 1305.2 | 1264.7 | 2569.9 | 1316.3 | 1441.8 | 2758.1 | 3.2062 | 2.5375 | 5.7437 |
| 300 | 8.581 | 0.001 404 | 0.021 67 | 1332.0 | 1231.0 | 2563.0 | 1344.0 | 1404.9 | 2749.0 | 3.2534 | 2.4511 | 5.7045 |
| 305 | 9.202 | 0.001 425 | 0.019 948 | 1359.3 | 1195.9 | 2555.2 | 1372.4 | 1366.4 | 2738.7 | 3.3010 | 2.3633 | 5.6643 |
| 310 | 9.856 | 0.001 447 | 0.018 350 | 1387.1 | 1159.4 | 2546.4 | 1401.0 | 1326.0 | 2727.3 | 3.3493 | 2.2737 | 5.6230 |
| 315 | 10.547 | 0.001 472 | 0.016 867 | 1415.5 | 1121.1 | 2536.6 | 1431.0 | 1283.5 | 2714.5 | 3.3982 | 2.1821 | 5.5804 |
| 320 | 11.274 | 0.001 499 | 0.015 488 | 1444.6 | 1080.9 | 2525.5 | 1461.5 | 1238.6 | 2700.1 | 3.4480 | 2.0882 | 5.5362 |
| 330 | 12.845 | 0.001 561 | 0.012 996 | 1505.3 | 993.7 | 2498.9 | 1525.3 | 1140.6 | 2665.9 | 3.5507 | 1.8909 | 5.4417 |
| 340 | 14.586 | 0.001 638 | 0.010 797 | 1570.3 | 894.3 | 2464.6 | 1594.2 | 1027.9 | 2622.0 | 3.6594 | 1.6763 | 5.3357 |
| 350 | 16.513 | 0.001 740 | 0.008 813 | 1641.9 | 776.6 | 2418.4 | 1670.6 | 893.4 | 2563.9 | 3.7777 | 1.4335 | 5.2112 |
| 360 | 18.651 | 0.001 893 | 0.006 945 | 1725.2 | 626.3 | 2351.5 | 1760.5 | 720.5 | 2481.0 | 3.9147 | 1.1379 | 5.0526 |
| 370 | 21.03 | 0.002 213 | 0.004 925 | 1844.0 | 384.5 | 2228.5 | 1890.5 | 441.6 | 2332.1 | 4.1106 | .6865 | 4.7971 |
| 374.14 | 22.09 | 0.003 155 | 0.003 155 | 2029.6 | 0 | 2029.6 | 2099.3 | 0 | 2099.3 | 4.4298 | 0 | 4.4298 |

# Table A.1.2
## Saturated Steam: Pressure Table

| Press. kPa P | Temp. °C T | Specific Volume | | Internal Energy | | | Enthalpy | | | Entropy | | |
|---|---|---|---|---|---|---|---|---|---|---|---|---|
| | | Sat. Liquid $v_f$ | Sat. Vapor $v_g$ | Sat. Liquid $u_f$ | Evap. $u_{fg}$ | Sat. Vapor $u_g$ | Sat. Liquid $h_f$ | Evap. $h_{fg}$ | Sat. Vapor $h_g$ | Sat. Liquid $s_f$ | Evap. $s_{fg}$ | Sat. Vapor $s_g$ |
| 0.6113 | 0.01 | 0.001 000 | 206.14 | .00 | 2375.3 | 2375.3 | .01 | 2501.3 | 2501.4 | .0000 | 9.1562 | 9.1562 |
| 1.0 | 6.98 | 0.001 000 | 129.21 | 29.30 | 2355.7 | 2385.0 | 29.30 | 2484.9 | 2514.2 | .1059 | 8.8697 | 8.9756 |
| 1.5 | 13.03 | 0.001 001 | 87.98 | 54.71 | 2338.6 | 2393.3 | 54.71 | 2470.6 | 2525.3 | .1957 | 8.6322 | 8.8279 |
| 2.0 | 17.50 | 0.001 001 | 67.00 | 73.48 | 2326.0 | 2399.5 | 73.48 | 2460.0 | 2533.5 | .2607 | 8.4629 | 8.7237 |
| 2.5 | 21.08 | 0.001 002 | 54.25 | 88.48 | 2315.9 | 2404.4 | 88.49 | 2451.6 | 2540.0 | .3120 | 8.3311 | 8.6432 |
| 3.0 | 24.08 | 0.001 003 | 45.67 | 101.04 | 2307.5 | 2408.5 | 101.05 | 2444.5 | 2545.5 | .3545 | 8.2231 | 8.5776 |
| 4.0 | 28.96 | 0.001 004 | 34.80 | 121.45 | 2293.7 | 2415.2 | 121.46 | 2432.9 | 2554.4 | .4226 | 8.0520 | 8.4746 |
| 5.0 | 32.88 | 0.001 005 | 28.19 | 137.81 | 2282.7 | 2420.5 | 137.82 | 2423.7 | 2561.5 | .4764 | 7.9187 | 8.3951 |
| 7.5 | 40.29 | 0.001 008 | 19.24 | 168.78 | 2261.7 | 2430.5 | 168.79 | 2406.0 | 2574.8 | .5764 | 7.6750 | 8.2515 |
| 10 | 45.81 | 0.001 010 | 14.67 | 191.82 | 2246.1 | 2437.9 | 191.83 | 2392.8 | 2584.7 | .6493 | 7.5009 | 8.1502 |
| 15 | 53.97 | 0.001 014 | 10.02 | 225.92 | 2222.8 | 2448.7 | 225.94 | 2373.1 | 2599.1 | .7549 | 7.2536 | 8.0085 |
| 20 | 60.06 | 0.001 017 | 7.649 | 251.38 | 2205.4 | 2456.7 | 251.40 | 2358.3 | 2609.7 | .8320 | 7.0766 | 7.9085 |
| 25 | 64.97 | 0.001 020 | 6.204 | 271.90 | 2191.2 | 2463.1 | 271.93 | 2346.3 | 2618.2 | .8931 | 6.9383 | 7.8314 |
| 30 | 69.10 | 0.001 022 | 5.229 | 289.20 | 2179.2 | 2468.4 | 289.23 | 2336.1 | 2625.3 | .9439 | 6.8247 | 7.7686 |
| 40 | 75.87 | 0.001 027 | 3.993 | 317.53 | 2159.5 | 2477.0 | 317.58 | 2319.2 | 2636.8 | 1.0259 | 6.6441 | 7.6700 |
| 50 | 81.33 | 0.001 030 | 3.240 | 340.44 | 2143.4 | 2483.9 | 340.49 | 2305.4 | 2645.9 | 1.0910 | 6.5029 | 7.5939 |
| 75 | 91.78 | 0.001 037 | 2.217 | 384.31 | 2112.4 | 2496.7 | 384.39 | 2278.6 | 2663.0 | 1.2130 | 6.2434 | 7.4564 |
| MPa | | | | | | | | | | | | |
| 0.100 | 99.63 | 0.001 043 | 1.6940 | 417.36 | 2088.7 | 2506.1 | 417.46 | 2258.0 | 2675.5 | 1.3026 | 6.0568 | 7.3594 |
| 0.125 | 105.99 | 0.001 048 | 1.3749 | 444.19 | 2069.3 | 2513.5 | 444.32 | 2241.0 | 2685.4 | 1.3740 | 5.9104 | 7.2844 |
| 0.150 | 111.37 | 0.001 053 | 1.1593 | 466.94 | 2052.7 | 2519.7 | 467.11 | 2226.5 | 2693.6 | 1.4336 | 5.7897 | 7.2233 |
| 0.175 | 116.06 | 0.001 057 | 1.0036 | 486.80 | 2038.1 | 2524.9 | 486.99 | 2213.6 | 2700.6 | 1.4849 | 5.6868 | 7.1717 |
| 0.200 | 120.23 | 0.001 061 | 0.8857 | 504.49 | 2025.0 | 2529.5 | 504.70 | 2201.9 | 2706.7 | 1.5301 | 5.5970 | 7.1271 |
| 0.225 | 124.00 | 0.001 064 | 0.7933 | 520.47 | 2013.1 | 2533.6 | 520.72 | 2191.3 | 2712.1 | 1.5706 | 5.5173 | 7.0878 |

| p | t | v_f | v_g | u_f | u_fg | u_g | h_f | h_fg | h_g | s_f | s_fg | s_g |
|---|---|---|---|---|---|---|---|---|---|---|---|---|
| 0.250 | 127.44 | 0.001 067 | 0.7187 | 535.10 | 2002.1 | 2537.2 | 535.37 | 2181.5 | 2716.9 | 1.6072 | 5.4455 | 7.0527 |
| 0.275 | 130.60 | 0.001 070 | 0.6573 | 548.59 | 1991.9 | 2540.5 | 548.89 | 2172.4 | 2721.3 | 1.6408 | 5.3801 | 7.0209 |
| 0.300 | 133.55 | 0.001 073 | 0.6058 | 561.15 | 1982.4 | 2543.6 | 561.47 | 2163.8 | 2725.3 | 1.6718 | 5.3201 | 6.9919 |
| 0.325 | 136.30 | 0.001 076 | 0.5620 | 572.90 | 1973.5 | 2546.4 | 573.25 | 2155.8 | 2729.0 | 1.7006 | 5.2646 | 6.9652 |
| 0.350 | 138.88 | 0.001 079 | 0.5243 | 583.95 | 1965.0 | 2548.9 | 584.33 | 2148.1 | 2732.4 | 1.7275 | 5.2130 | 6.9405 |
| 0.375 | 141.32 | 0.001 081 | 0.4914 | 594.40 | 1956.9 | 2551.3 | 594.81 | 2140.8 | 2735.6 | 1.7528 | 5.1647 | 6.9175 |
| 0.40 | 143.63 | 0.001 084 | 0.4625 | 604.31 | 1949.3 | 2553.6 | 604.74 | 2133.8 | 2738.6 | 1.7766 | 5.1193 | 6.8959 |
| 0.45 | 147.93 | 0.001 088 | 0.4140 | 622.77 | 1934.9 | 2557.6 | 623.25 | 2120.7 | 2743.9 | 1.8207 | 5.0359 | 6.8565 |
| 0.50 | 151.86 | 0.001 093 | 0.3749 | 639.68 | 1921.6 | 2561.2 | 640.23 | 2108.5 | 2748.7 | 1.8607 | 4.9606 | 6.8213 |
| 0.55 | 155.48 | 0.001 097 | 0.3427 | 655.32 | 1909.2 | 2564.5 | 655.93 | 2097.0 | 2753.0 | 1.8973 | 4.8920 | 6.7893 |
| 0.60 | 158.85 | 0.001 101 | 0.3157 | 669.90 | 1897.5 | 2567.4 | 670.56 | 2086.3 | 2756.8 | 1.9312 | 4.8288 | 6.7600 |
| 0.65 | 162.01 | 0.001 104 | 0.2927 | 683.56 | 1886.5 | 2570.1 | 684.28 | 2076.0 | 2760.3 | 1.9627 | 4.7703 | 6.7331 |
| 0.70 | 164.97 | 0.001 108 | 0.2729 | 696.44 | 1876.1 | 2572.5 | 697.22 | 2066.3 | 2763.5 | 1.9922 | 4.7158 | 6.7080 |
| 0.75 | 167.78 | 0.001 112 | 0.2556 | 708.64 | 1866.1 | 2574.7 | 709.47 | 2057.0 | 2766.4 | 2.0200 | 4.6647 | 6.6847 |
| 0.80 | 170.43 | 0.001 115 | 0.2404 | 720.22 | 1856.6 | 2576.8 | 721.11 | 2048.0 | 2769.1 | 2.0462 | 4.6166 | 6.6628 |
| 0.85 | 172.96 | 0.001 118 | 0.2270 | 731.27 | 1847.4 | 2578.7 | 732.22 | 2039.4 | 2771.6 | 2.0710 | 4.5711 | 6.6421 |
| 0.90 | 175.38 | 0.001 121 | 0.2150 | 741.83 | 1838.6 | 2580.5 | 742.83 | 2031.1 | 2773.9 | 2.0946 | 4.5280 | 6.6226 |
| 0.95 | 177.69 | 0.001 124 | 0.2042 | 751.95 | 1830.2 | 2582.1 | 753.02 | 2023.1 | 2776.1 | 2.1172 | 4.4869 | 6.6041 |
| 1.00 | 179.91 | 0.001 127 | 0.194 44 | 761.68 | 1822.0 | 2583.6 | 762.81 | 2015.3 | 2778.1 | 2.1387 | 4.4478 | 6.5865 |
| 1.10 | 184.09 | 0.001 133 | 0.177 53 | 780.09 | 1806.3 | 2586.4 | 781.34 | 2000.4 | 2781.7 | 2.1792 | 4.3744 | 6.5536 |
| 1.20 | 187.99 | 0.001 139 | 0.163 33 | 797.29 | 1791.5 | 2588.8 | 798.65 | 1986.2 | 2784.8 | 2.2166 | 4.3067 | 6.5233 |
| 1.30 | 191.64 | 0.001 144 | 0.151 25 | 813.44 | 1777.5 | 2591.0 | 814.93 | 1972.7 | 2787.6 | 2.2515 | 4.2438 | 6.4953 |
| 1.40 | 195.07 | 0.001 149 | 0.140 84 | 828.70 | 1764.1 | 2592.8 | 830.30 | 1959.7 | 2790.0 | 2.2842 | 4.1850 | 6.4693 |
| 1.50 | 198.32 | 0.001 154 | 0.131 77 | 843.16 | 1751.3 | 2594.5 | 844.89 | 1947.3 | 2792.2 | 2.3150 | 4.1298 | 6.4448 |
| 1.75 | 205.76 | 0.001 166 | 0.113 49 | 876.46 | 1721.4 | 2597.8 | 878.50 | 1917.9 | 2796.4 | 2.3851 | 4.0044 | 6.3896 |
| 2.00 | 212.42 | 0.001 177 | 0.099 63 | 906.44 | 1693.8 | 2600.3 | 908.79 | 1890.7 | 2799.5 | 2.4474 | 3.8935 | 6.3409 |
| 2.25 | 218.45 | 0.001 187 | 0.088 75 | 933.83 | 1668.2 | 2602.0 | 936.49 | 1865.2 | 2801.7 | 2.5035 | 3.7937 | 6.2972 |
| 2.5 | 223.99 | 0.001 197 | 0.079 98 | 959.11 | 1644.0 | 2603.1 | 962.11 | 1841.0 | 2803.1 | 2.5547 | 3.7028 | 6.2575 |
| 3.0 | 233.90 | 0.001 217 | 0.066 68 | 1004.78 | 1599.3 | 2604.1 | 1008.42 | 1795.7 | 2804.2 | 2.6457 | 3.5412 | 6.1869 |

**Table A.1.2** (Continued)
Saturated Steam: Pressure Table

| Press. MPa $P$ | Temp. °C $T$ | Specific Volume | | Internal Energy | | | Enthalpy | | | Entropy | | |
|---|---|---|---|---|---|---|---|---|---|---|---|---|
| | | Sat. Liquid $v_f$ | Sat. Vapor $v_g$ | Sat. Liquid $u_f$ | Evap. $u_{fg}$ | Sat. Vapor $u_g$ | Sat. Liquid $h_f$ | Evap. $h_{fg}$ | Sat. Vapor $h_g$ | Sat. Liquid $s_f$ | Evap. $s_{fg}$ | Sat. Vapor $s_g$ |
| 3.5 | 242.60 | 0.001 235 | 0.057 07 | 1045.43 | 1558.3 | 2603.7 | 1049.75 | 1753.7 | 2803.4 | 2.7253 | 3.4000 | 6.1253 |
| 4 | 250.40 | 0.001 252 | 0.049 78 | 1082.31 | 1520.0 | 2602.3 | 1087.31 | 1714.1 | 2801.4 | 2.7964 | 3.2737 | 6.0701 |
| 5 | 263.99 | 0.001 286 | 0.039 44 | 1147.81 | 1449.3 | 2597.1 | 1154.23 | 1640.1 | 2794.3 | 2.9202 | 3.0532 | 5.9734 |
| 6 | 275.64 | 0.001 319 | 0.032 44 | 1205.44 | 1384.3 | 2589.7 | 1213.35 | 1571.0 | 2784.3 | 3.0267 | 2.8625 | 5.8892 |
| 7 | 285.88 | 0.001 351 | 0.027 37 | 1257.55 | 1323.0 | 2580.5 | 1267.00 | 1505.1 | 2772.1 | 3.1211 | 2.6922 | 5.8133 |
| 8 | 295.06 | 0.001 384 | 0.023 52 | 1305.57 | 1264.2 | 2569.8 | 1316.64 | 1441.3 | 2758.0 | 3.2068 | 2.5364 | 5.7432 |
| 9 | 303.40 | 0.001 418 | 0.020 48 | 1350.51 | 1207.3 | 2557.8 | 1363.26 | 1378.9 | 2742.1 | 3.2858 | 2.3915 | 5.6772 |
| 10 | 311.06 | 0.001 452 | 0.018 026 | 1393.04 | 1151.4 | 2544.4 | 1407.56 | 1317.1 | 2724.7 | 3.3596 | 2.2544 | 5.6141 |
| 11 | 318.15 | 0.001 489 | 0.015 987 | 1433.7 | 1096.0 | 2529.8 | 1450.1 | 1255.5 | 2705.6 | 3.4295 | 2.1233 | 5.5527 |
| 12 | 324.75 | 0.001 527 | 0.014 263 | 1473.0 | 1040.7 | 2513.7 | 1491.3 | 1193.6 | 2684.9 | 3.4962 | 1.9962 | 5.4924 |
| 13 | 330.93 | 0.001 567 | 0.012 780 | 1511.1 | 985.0 | 2496.1 | 1531.5 | 1130.7 | 2662.2 | 3.5606 | 1.8718 | 5.4323 |
| 14 | 336.75 | 0.001 611 | 0.011 485 | 1548.6 | 928.2 | 2476.8 | 1571.1 | 1066.5 | 2637.6 | 3.6232 | 1.7485 | 5.3717 |
| 15 | 342.24 | 0.001 658 | 0.010 337 | 1585.6 | 869.8 | 2455.5 | 1610.5 | 1000.0 | 2610.5 | 3.6848 | 1.6249 | 5.3098 |
| 16 | 347.44 | 0.001 711 | 0.009 306 | 1622.7 | 809.0 | 2431.7 | 1650.1 | 930.6 | 2580.6 | 3.7461 | 1.4994 | 5.2455 |
| 17 | 352.37 | 0.001 770 | 0.008 364 | 1660.2 | 744.8 | 2405.0 | 1690.3 | 856.9 | 2547.2 | 3.8079 | 1.3698 | 5.1777 |
| 18 | 357.06 | 0.001 840 | 0.007 489 | 1698.9 | 675.4 | 2374.3 | 1732.0 | 777.1 | 2509.1 | 3.8715 | 1.2329 | 5.1044 |
| 19 | 361.54 | 0.001 924 | 0.006 657 | 1739.9 | 598.1 | 2338.1 | 1776.5 | 688.0 | 2464.5 | 3.9388 | 1.0839 | 5.0228 |
| 20 | 365.81 | 0.002 036 | 0.005 834 | 1785.6 | 507.5 | 2293.0 | 1826.3 | 583.4 | 2409.7 | 4.0139 | .9130 | 4.9269 |
| 21 | 369.89 | 0.002 207 | 0.004 952 | 1842.1 | 388.5 | 2230.6 | 1888.4 | 446.2 | 2334.6 | 4.1075 | .6938 | 4.8013 |
| 22 | 373.80 | 0.002 742 | 0.003 568 | 1961.9 | 125.2 | 2087.1 | 2022.2 | 143.4 | 2165.6 | 4.3110 | .2216 | 4.5327 |
| 22.09 | 374.14 | 0.003 155 | 0.003 155 | 2029.6 | 0 | 2029.6 | 2099.3 | 0 | 2099.3 | 4.4298 | 0 | 4.4298 |

# Table A.1.3
## Superheated Vapor

| T | P = .010 MPa (45.81) v | u | h | s | P = .050 MPa (81.33) v | u | h | s | P = .10 MPa (99.63) v | u | h | s |
|---|---|---|---|---|---|---|---|---|---|---|---|---|
| Sat. | 14.674 | 2437.9 | 2584.7 | 8.1502 | 3.240 | 2483.9 | 2645.9 | 7.5939 | 1.6940 | 2506.1 | 2675.5 | 7.3594 |
| 50 | 14.869 | 2443.9 | 2592.6 | 8.1749 | | | | | | | | |
| 100 | 17.196 | 2515.5 | 2687.5 | 8.4479 | 3.418 | 2511.6 | 2682.5 | 7.6947 | 1.6958 | 2506.7 | 2676.2 | 7.3614 |
| 150 | 19.512 | 2587.9 | 2783.0 | 8.6882 | 3.889 | 2585.6 | 2780.1 | 7.9401 | 1.9364 | 2582.8 | 2776.4 | 7.6134 |
| 200 | 21.825 | 2661.3 | 2879.5 | 8.9038 | 4.356 | 2659.9 | 2877.7 | 8.1580 | 2.172 | 2658.1 | 2875.3 | 7.8343 |
| 250 | 24.136 | 2736.0 | 2977.3 | 9.1002 | 4.820 | 2735.0 | 2976.0 | 8.3556 | 2.406 | 2733.7 | 2974.3 | 8.0333 |
| 300 | 26.445 | 2812.1 | 3076.5 | 9.2813 | 5.284 | 2811.3 | 3075.5 | 8.5373 | 2.639 | 2810.4 | 3074.3 | 8.2158 |
| 400 | 31.063 | 2968.9 | 3279.6 | 9.6077 | 6.209 | 2968.5 | 3278.9 | 8.8642 | 3.103 | 2967.9 | 3278.2 | 8.5435 |
| 500 | 35.679 | 3132.3 | 3489.1 | 9.8978 | 7.134 | 3132.0 | 3488.7 | 9.1546 | 3.565 | 3131.6 | 3488.1 | 8.8342 |
| 600 | 40.295 | 3302.5 | 3705.4 | 10.1608 | 8.057 | 3302.2 | 3705.1 | 9.4178 | 4.028 | 3301.9 | 3704.7 | 9.0976 |
| 700 | 44.911 | 3479.6 | 3928.7 | 10.4028 | 8.981 | 3479.4 | 3928.5 | 9.6599 | 4.490 | 3479.2 | 3928.2 | 9.3398 |
| 800 | 49.526 | 3663.8 | 4159.0 | 10.6281 | 9.904 | 3663.6 | 4158.9 | 9.8852 | 4.952 | 3663.5 | 4158.6 | 9.5652 |
| 900 | 54.141 | 3855.0 | 4396.4 | 10.8396 | 10.828 | 3854.9 | 4396.3 | 10.0967 | 5.414 | 3854.8 | 4396.1 | 9.7767 |
| 1000 | 58.757 | 4053.0 | 4640.6 | 11.0393 | 11.751 | 4052.9 | 4640.5 | 10.2964 | 5.875 | 4052.8 | 4640.3 | 9.9764 |
| 1100 | 63.372 | 4257.5 | 4891.2 | 11.2287 | 12.674 | 4257.4 | 4891.1 | 10.4859 | 6.337 | 4257.3 | 4891.0 | 10.1659 |
| 1200 | 67.987 | 4467.9 | 5147.8 | 11.4091 | 13.597 | 4467.8 | 5147.7 | 10.6662 | 6.799 | 4467.7 | 5147.6 | 10.3463 |
| 1300 | 72.602 | 4683.7 | 5409.7 | 11.5811 | 14.521 | 4683.6 | 5409.6 | 10.8382 | 7.260 | 4683.5 | 5409.5 | 10.5183 |

| T | P = .20 MPa (120.23) v | u | h | s | P = .30 MPa (133.55) v | u | h | s | P = .40 MPa (143.63) v | u | h | s |
|---|---|---|---|---|---|---|---|---|---|---|---|---|
| Sat. | .8857 | 2529.5 | 2706.7 | 7.1272 | .6058 | 2543.6 | 2725.3 | 6.9919 | .4625 | 2553.6 | 2738.6 | 6.8959 |
| 150 | .9596 | 2576.9 | 2768.8 | 7.2795 | .6339 | 2570.8 | 2761.0 | 7.0778 | .4708 | 2564.5 | 2752.8 | 6.9299 |
| 200 | 1.0803 | 2654.4 | 2870.5 | 7.5066 | .7163 | 2650.7 | 2865.6 | 7.3115 | .5342 | 2646.8 | 2860.5 | 7.1706 |
| 250 | 1.1988 | 2731.2 | 2971.0 | 7.7086 | .7964 | 2728.7 | 2967.6 | 7.5166 | .5951 | 2726.1 | 2964.2 | 7.3789 |
| 300 | 1.3162 | 2808.6 | 3071.8 | 7.8926 | .8753 | 2806.7 | 3069.3 | 7.7022 | .6548 | 2804.8 | 3066.8 | 7.5662 |
| 400 | 1.5493 | 2966.7 | 3276.6 | 8.2218 | 1.0315 | 2965.6 | 3275.0 | 8.0330 | .7726 | 2964.4 | 3273.4 | 7.8985 |

**Table A.1.3** (Continued)
Superheated Vapor

| T | v | u | h | s | v | u | h | s | v | u | h | s |
|---|---|---|---|---|---|---|---|---|---|---|---|---|
| | P = .20 MPa (120.23) | | | | P = .30 MPa (133.55) | | | | P = .40 MPa (143.63) | | | |
| 500 | 1.7814 | 3130.8 | 3487.1 | 8.5133 | 1.1867 | 3130.0 | 3486.0 | 8.3251 | .8893 | 3129.2 | 3484.9 | 8.1913 |
| 600 | 2.013 | 3301.4 | 3704.0 | 8.7770 | 1.3414 | 3300.8 | 3703.2 | 8.5892 | 1.0055 | 3300.2 | 3702.4 | 8.4558 |
| 700 | 2.244 | 3478.8 | 3927.6 | 9.0194 | 1.4957 | 3478.4 | 3927.1 | 8.8319 | 1.1215 | 3477.9 | 3926.5 | 8.6987 |
| 800 | 2.475 | 3663.1 | 4158.2 | 9.2449 | 1.6499 | 3662.9 | 4157.8 | 9.0576 | 1.2372 | 3662.4 | 4157.3 | 8.9244 |
| 900 | 2.706 | 3854.5 | 4395.8 | 9.4566 | 1.8041 | 3854.2 | 4395.4 | 9.2692 | 1.3529 | 3853.9 | 4395.1 | 9.1362 |
| 1000 | 2.937 | 4052.5 | 4640.0 | 9.6563 | 1.9581 | 4052.3 | 4639.7 | 9.4690 | 1.4685 | 4052.0 | 4639.4 | 9.3360 |
| 1100 | 3.168 | 4257.0 | 4890.7 | 9.8458 | 2.1121 | 4256.8 | 4890.4 | 9.6585 | 1.5840 | 4256.5 | 4890.2 | 9.5256 |
| 1200 | 3.399 | 4467.5 | 5147.3 | 10.0262 | 2.2661 | 4467.2 | 5147.1 | 9.8389 | 1.6996 | 4467.0 | 5146.8 | 9.7060 |
| 1300 | 3.630 | 4683.2 | 5409.3 | 10.1982 | 2.4201 | 4683.0 | 5409.0 | 10.0110 | 1.8151 | 4682.8 | 5408.8 | 9.8780 |
| | P = .50 MPa (151.86) | | | | P = .60 MPa (158.85) | | | | P = .80 MPa (170.43) | | | |
| Sat. | .3749 | 2561.2 | 2748.7 | 6.8213 | .3157 | 2567.4 | 2756.8 | 6.7600 | .2404 | 2576.8 | 2769.1 | 6.6628 |
| 200 | .4249 | 2642.9 | 2855.4 | 7.0592 | .3520 | 2638.9 | 2850.1 | 6.9665 | .2608 | 2630.6 | 2839.3 | 6.8158 |
| 250 | .4744 | 2723.5 | 2960.7 | 7.2709 | .3938 | 2720.9 | 2957.2 | 7.1816 | .2931 | 2715.5 | 2950.0 | 7.0384 |
| 300 | .5226 | 2802.9 | 3064.2 | 7.4599 | .4344 | 2801.0 | 3061.6 | 7.3724 | .3241 | 2797.2 | 3056.5 | 7.2328 |
| 350 | .5701 | 2882.6 | 3167.7 | 7.6329 | .4742 | 2881.2 | 3165.7 | 7.5464 | .3544 | 2878.2 | 3161.7 | 7.4089 |
| 400 | .6173 | 2963.2 | 3271.9 | 7.7938 | .5137 | 2962.1 | 3270.3 | 7.7079 | .3843 | 2959.7 | 3267.1 | 7.5716 |
| 500 | .7109 | 3128.4 | 3483.9 | 8.0873 | .5920 | 3127.6 | 3482.8 | 8.0021 | .4433 | 3126.0 | 3480.6 | 7.8673 |
| 600 | .8041 | 3299.6 | 3701.7 | 8.3522 | .6697 | 3299.1 | 3700.9 | 8.2674 | .5018 | 3297.9 | 3699.4 | 8.1333 |
| 700 | .8969 | 3477.5 | 3925.9 | 8.5952 | .7472 | 3477.0 | 3925.3 | 8.5107 | .5601 | 3476.2 | 3924.2 | 8.3770 |
| 800 | .9896 | 3662.1 | 4156.9 | 8.8211 | .8245 | 3661.8 | 4156.5 | 8.7367 | .6181 | 3661.1 | 4155.6 | 8.6033 |
| 900 | 1.0822 | 3853.6 | 4394.7 | 9.0329 | .9017 | 3853.4 | 4394.4 | 8.9486 | .6761 | 3852.8 | 4393.7 | 8.8153 |
| 1000 | 1.1747 | 4051.8 | 4639.1 | 9.2328 | .9788 | 4051.5 | 4638.8 | 9.1485 | .7340 | 4051.0 | 4638.2 | 9.0153 |

(continuation rows, temperatures in °C; columns v, u, h, s for three pressures)

| T | v | u | h | s | v | u | h | s | v | u | h | s |
|---|---|---|---|---|---|---|---|---|---|---|---|---|
| 1100 | 1.2672 | 4256.3 | 4889.9 | 9.4224 | 1.0559 | 4256.1 | 4889.6 | 9.3381 | .7919 | 4255.6 | 4889.1 | 9.2050 |
| 1200 | 1.3596 | 4466.8 | 5146.6 | 9.6029 | 1.1330 | 4466.5 | 5146.3 | 9.5185 | .8497 | 4466.1 | 5145.9 | 9.3855 |
| 1300 | 1.4521 | 4682.5 | 5408.6 | 9.7749 | 1.2101 | 4682.3 | 5408.3 | 9.6906 | .9076 | 4681.8 | 5407.9 | 9.5575 |

**P = 1.00 MPa (179.91)**

| T | v | u | h | s |
|---|---|---|---|---|
| Sat. | .194 44 | 2583.6 | 2778.1 | 6.5865 |
| 200 | .2060 | 2621.9 | 2827.9 | 6.6940 |
| 250 | .2327 | 2709.9 | 2942.6 | 6.9247 |
| 300 | .2579 | 2793.2 | 3051.2 | 7.1229 |
| 350 | .2825 | 2875.2 | 3157.7 | 7.3011 |
| 400 | .3066 | 2957.3 | 3263.9 | 7.4651 |
| 500 | .3541 | 3124.4 | 3478.5 | 7.7622 |
| 600 | .4011 | 3296.8 | 3697.9 | 8.0290 |
| 700 | .4478 | 3475.3 | 3923.1 | 8.2731 |
| 800 | .4943 | 3660.4 | 4154.7 | 8.4996 |
| 900 | .5407 | 3852.2 | 4392.9 | 8.7118 |
| 1000 | .5871 | 4050.5 | 4637.6 | 8.9119 |
| 1100 | .6335 | 4255.1 | 4888.6 | 9.1017 |
| 1200 | .6798 | 4465.6 | 5145.4 | 9.2822 |
| 1300 | .7261 | 4681.3 | 5407.4 | 9.4543 |

**P = 1.20 MPa (187.99)**

| T | v | u | h | s |
|---|---|---|---|---|
| Sat. | .163 33 | 2588.8 | 2784.8 | 6.5233 |
| 200 | .169 30 | 2612.8 | 2815.9 | 6.5898 |
| 250 | .192 34 | 2704.2 | 2935.0 | 6.8294 |
| 300 | .2138 | 2789.2 | 3045.8 | 7.0317 |
| 350 | .2345 | 2872.2 | 3153.6 | 7.2121 |
| 400 | .2548 | 2954.9 | 3260.7 | 7.3774 |
| 500 | .2946 | 3122.8 | 3476.3 | 7.6759 |
| 600 | .3339 | 3295.6 | 3696.3 | 7.9435 |
| 700 | .3729 | 3474.4 | 3922.0 | 8.1881 |
| 800 | .4118 | 3659.7 | 4153.8 | 8.4148 |
| 900 | .4505 | 3851.6 | 4392.2 | 8.6272 |
| 1000 | .4892 | 4050.0 | 4637.0 | 8.8274 |
| 1100 | .5278 | 4254.6 | 4888.0 | 9.0172 |
| 1200 | .5665 | 4465.1 | 5144.9 | 9.1977 |
| 1300 | .6051 | 4680.9 | 5407.0 | 9.3698 |

**P = 1.40 MPa (195.07)**

| T | v | u | h | s |
|---|---|---|---|---|
| Sat. | .140 84 | 2592.8 | 2790.0 | 6.4693 |
| 200 | .143 02 | 2603.1 | 2803.3 | 6.4975 |
| 250 | .163 50 | 2698.3 | 2927.2 | 6.7467 |
| 300 | .182 28 | 2785.2 | 3040.4 | 6.9534 |
| 350 | .2003 | 2869.2 | 3149.5 | 7.1360 |
| 400 | .2178 | 2952.5 | 3257.5 | 7.3026 |
| 500 | .2521 | 3121.1 | 3474.1 | 7.6027 |
| 600 | .2860 | 3294.4 | 3694.8 | 7.8710 |
| 700 | .3195 | 3473.6 | 3920.8 | 8.1160 |
| 800 | .3528 | 3659.0 | 4153.0 | 8.3431 |
| 900 | .3861 | 3851.1 | 4391.5 | 8.5556 |
| 1000 | .4192 | 4049.5 | 4636.4 | 8.7559 |
| 1100 | .4524 | 4254.1 | 4887.5 | 8.9457 |
| 1200 | .4855 | 4464.7 | 5144.4 | 9.1262 |
| 1300 | .5186 | 4680.4 | 5406.5 | 9.2984 |

**P = 1.60 MPa (201.41)**

| T | v | u | h | s |
|---|---|---|---|---|
| Sat. | .123 80 | 2596.0 | 2794.0 | 6.4218 |
| 225 | .132 87 | 2644.7 | 2857.3 | 6.5518 |
| 250 | .141 84 | 2692.3 | 2919.2 | 6.6732 |
| 300 | .158 62 | 2781.1 | 3034.8 | 6.8844 |
| 350 | .174 56 | 2866.1 | 3145.4 | 7.0694 |
| 400 | .190 05 | 2950.1 | 3254.2 | 7.2374 |
| 500 | .2203 | 3119.5 | 3472.0 | 7.5390 |
| 600 | .2500 | 3293.3 | 3693.2 | 7.8080 |
| 700 | .2794 | 3472.7 | 3919.7 | 8.0535 |

**P = 1.80 MPa (207.15)**

| T | v | u | h | s |
|---|---|---|---|---|
| Sat. | .110 42 | 2598.4 | 2797.1 | 6.3794 |
| 225 | .116 73 | 2636.6 | 2846.7 | 6.4808 |
| 250 | .124 97 | 2686.0 | 2911.0 | 6.6066 |
| 300 | .140 21 | 2776.9 | 3029.2 | 6.8226 |
| 350 | .154 57 | 2863.0 | 3141.2 | 7.0100 |
| 400 | .168 47 | 2947.7 | 3250.9 | 7.1794 |
| 500 | .195 50 | 3117.9 | 3469.8 | 7.4825 |
| 600 | .2220 | 3292.1 | 3691.7 | 7.7523 |
| 700 | .2482 | 3471.8 | 3918.5 | 7.9983 |

**P = 2.00 MPa (212.42)**

| T | v | u | h | s |
|---|---|---|---|---|
| Sat. | .099 63 | 2600.3 | 2799.5 | 6.3409 |
| 225 | .103 77 | 2628.3 | 2835.8 | 6.4147 |
| 250 | .111 44 | 2679.6 | 2902.5 | 6.5453 |
| 300 | .125 47 | 2772.6 | 3023.5 | 6.7664 |
| 350 | .138 57 | 2859.8 | 3137.0 | 6.9563 |
| 400 | .151 20 | 2945.2 | 3247.6 | 7.1271 |
| 500 | .175 68 | 3116.2 | 3467.6 | 7.4317 |
| 600 | .199 60 | 3290.9 | 3690.1 | 7.7024 |
| 700 | .2232 | 3470.9 | 3917.4 | 7.9487 |

# Table A.1.3 (Continued)
## Superheated Vapor

| T | v | u | h | s | v | u | h | s | v | u | h | s |
|---|---|---|---|---|---|---|---|---|---|---|---|---|
| | P = 1.60 MPa (201.41) | | | | P = 1.80 MPa (207.15) | | | | P = 2.00 MPa (212.42) | | | |
| 800 | .3086 | 3658.3 | 4152.1 | 8.2808 | .2742 | 3657.6 | 4151.2 | 8.2258 | .2467 | 3657.0 | 4150.3 | 8.1765 |
| 900 | .3377 | 3850.5 | 4390.8 | 8.4935 | .3001 | 3849.9 | 4390.1 | 8.4386 | .2700 | 3849.3 | 4389.4 | 8.3895 |
| 1000 | .3668 | 4049.0 | 4635.8 | 8.6938 | .3260 | 4048.5 | 4635.2 | 8.6391 | .2933 | 4048.0 | 4634.6 | 8.5901 |
| 1100 | .3958 | 4253.7 | 4887.0 | 8.8837 | .3518 | 4253.2 | 4886.4 | 8.8290 | .3166 | 4252.7 | 4885.9 | 8.7800 |
| 1200 | .4248 | 4464.2 | 5143.9 | 9.0643 | .3776 | 4463.7 | 5143.4 | 9.0096 | .3398 | 4463.3 | 5142.9 | 8.9607 |
| 1300 | .4538 | 4679.9 | 5406.0 | 9.2364 | .4034 | 4679.5 | 5405.6 | 9.1818 | .3631 | 4679.0 | 5405.1 | 9.1329 |

| T | v | u | h | s | v | u | h | s | v | u | h | s |
|---|---|---|---|---|---|---|---|---|---|---|---|---|
| | P = 2.50 MPa (223.99) | | | | P = 3.00 MPa (233.90) | | | | P = 3.50 MPa (242.60) | | | |
| Sat. | .079 98 | 2603.1 | 2803.1 | 6.2575 | .066 68 | 2604.1 | 2804.2 | 6.1869 | .057 07 | 2603.7 | 2803.4 | 6.1253 |
| 225 | .080 27 | 2605.6 | 2806.3 | 6.2639 | | | | | | | | |
| 250 | .087 00 | 2662.6 | 2880.1 | 6.4085 | .070 58 | 2644.0 | 2855.8 | 6.2872 | .058 72 | 2623.7 | 2829.2 | 6.1749 |
| 300 | .098 90 | 2761.6 | 3008.8 | 6.6438 | .081 14 | 2750.1 | 2993.5 | 6.5390 | .068 42 | 2738.0 | 2977.5 | 6.4461 |
| 350 | .109 76 | 2851.9 | 3126.3 | 6.8403 | .090 53 | 2843.7 | 3115.3 | 6.7428 | .076 78 | 2835.3 | 3104.0 | 6.6579 |
| 400 | .120 10 | 2939.1 | 3239.3 | 7.0148 | .099 36 | 2932.8 | 3230.9 | 6.9212 | .084 53 | 2926.4 | 3222.3 | 6.8405 |
| 450 | .130 14 | 3025.5 | 3350.8 | 7.1746 | .107 87 | 3020.4 | 3344.0 | 7.0834 | .091 96 | 3015.3 | 3337.2 | 7.0052 |
| 500 | .139 98 | 3112.1 | 3462.1 | 7.3234 | .116 19 | 3108.0 | 3456.5 | 7.2338 | .099 18 | 3103.0 | 3450.9 | 7.1572 |
| 600 | .159 30 | 3288.0 | 3686.3 | 7.5960 | .132 43 | 3285.0 | 3682.3 | 7.5085 | .113 24 | 3282.1 | 3678.4 | 7.4339 |
| 700 | .178 32 | 3468.7 | 3914.5 | 7.8435 | .148 38 | 3466.5 | 3911.7 | 7.7571 | .126 99 | 3464.3 | 3908.8 | 7.6837 |
| 800 | .197 16 | 3655.3 | 4148.2 | 8.0720 | .164 14 | 3653.5 | 4145.9 | 7.9862 | .140 56 | 3651.8 | 4143.7 | 7.9134 |
| 900 | .215 90 | 3847.9 | 4387.9 | 8.2853 | .179 80 | 3846.5 | 4385.9 | 8.1999 | .154 02 | 3845.0 | 4384.1 | 8.1276 |
| 1000 | .2346 | 4046.7 | 4633.1 | 8.4861 | .195 41 | 4045.4 | 4631.6 | 8.4009 | .167 43 | 4044.1 | 4630.1 | 8.3288 |
| 1100 | .2532 | 4251.5 | 4884.6 | 8.6762 | .210 98 | 4250.3 | 4883.3 | 8.5912 | .180 80 | 4249.2 | 4881.9 | 8.5192 |
| 1200 | .2718 | 4462.1 | 5141.7 | 8.8569 | .226 52 | 4460.9 | 5140.5 | 8.7720 | .194 15 | 4459.8 | 5139.3 | 8.7000 |
| 1300 | .2905 | 4677.8 | 5404.0 | 9.0291 | .242 06 | 4676.6 | 5402.8 | 8.9442 | .207 49 | 4675.5 | 5401.7 | 8.8723 |

$P = 4.0$ MPa (250.40)

| T | v | u | h | s |
|---|---|---|---|---|
| Sat. | .049 78 | 2602.3 | 2801.4 | 6.0701 |
| 275 | .054 57 | 2667.9 | 2886.2 | 6.2285 |
| 300 | .058 84 | 2725.3 | 2960.7 | 6.3615 |
| 350 | .066 45 | 2826.7 | 3092.5 | 6.5821 |
| 400 | .073 41 | 2919.9 | 3213.6 | 6.7690 |
| 450 | .080 02 | 3010.2 | 3330.3 | 6.9363 |
| 500 | .086 43 | 3099.5 | 3445.3 | 7.0901 |
| 600 | .098 85 | 3279.1 | 3674.4 | 7.3688 |
| 700 | .110 95 | 3462.1 | 3905.9 | 7.6198 |
| 800 | .122 87 | 3650.0 | 4141.5 | 7.8502 |
| 900 | .134 69 | 3843.6 | 4382.3 | 8.0647 |
| 1000 | .146 45 | 4042.9 | 4628.7 | 8.2662 |
| 1100 | .158 17 | 4248.0 | 4880.6 | 8.4567 |
| 1200 | .169 87 | 4458.6 | 5138.1 | 8.6376 |
| 1300 | .181 56 | 4674.3 | 5400.5 | 8.8100 |

$P = 4.5$ MPa (257.49)

| T | v | u | h | s |
|---|---|---|---|---|
| Sat. | .044 06 | 2600.1 | 2798.3 | 6.0198 |
| 275 | .047 30 | 2650.3 | 2863.2 | 6.1401 |
| 300 | .051 35 | 2712.0 | 2943.1 | 6.2828 |
| 350 | .058 40 | 2817.8 | 3080.6 | 6.5131 |
| 400 | .064 75 | 2913.3 | 3204.7 | 6.7047 |
| 450 | .070 74 | 3005.0 | 3323.3 | 6.8746 |
| 500 | .076 51 | 3095.3 | 3439.6 | 7.0301 |
| 600 | .087 65 | 3276.0 | 3670.5 | 7.3110 |
| 700 | .098 47 | 3459.9 | 3903.0 | 7.5631 |
| 800 | .109 11 | 3648.3 | 4139.3 | 7.7942 |
| 900 | .119 65 | 3842.2 | 4380.6 | 8.0091 |
| 1000 | .130 13 | 4041.6 | 4627.2 | 8.2108 |
| 1100 | .140 56 | 4246.8 | 4879.3 | 8.4015 |
| 1200 | .150 98 | 4457.5 | 5136.9 | 8.5825 |
| 1300 | .161 39 | 4673.1 | 5399.4 | 8.7549 |

$P = 5.0$ MPa (263.99)

| T | v | u | h | s |
|---|---|---|---|---|
| Sat. | .039 44 | 2597.1 | 2794.3 | 5.9734 |
| 275 | .041 41 | 2631.3 | 2838.3 | 6.0544 |
| 300 | .045 32 | 2698.0 | 2924.5 | 6.2084 |
| 350 | .051 94 | 2808.7 | 3068.4 | 6.4493 |
| 400 | .057 81 | 2906.6 | 3195.7 | 6.6459 |
| 450 | .063 30 | 2999.7 | 3316.2 | 6.8186 |
| 500 | .068 57 | 3091.0 | 3433.8 | 6.9759 |
| 600 | .078 69 | 3273.0 | 3666.5 | 7.2589 |
| 700 | .088 49 | 3457.6 | 3900.1 | 7.5122 |
| 800 | .098 11 | 3646.6 | 4137.1 | 7.7440 |
| 900 | .107 62 | 3840.7 | 4378.8 | 7.9593 |
| 1000 | .117 07 | 4040.4 | 4625.7 | 8.1612 |
| 1100 | .126 48 | 4245.6 | 4878.0 | 8.3520 |
| 1200 | .135 87 | 4456.3 | 5135.7 | 8.5331 |
| 1300 | .145 26 | 4672.0 | 5398.2 | 8.7055 |

$P = 6.0$ MPa (275.64)

| T | v | u | h | s |
|---|---|---|---|---|
| Sat. | .032 44 | 2589.7 | 2784.3 | 5.8892 |
| 300 | .036 16 | 2667.2 | 2884.2 | 6.0674 |
| 350 | .042 23 | 2789.6 | 3043.0 | 6.3335 |
| 400 | .047 39 | 2892.9 | 3177.2 | 6.5408 |
| 450 | .052 14 | 2988.9 | 3301.8 | 6.7193 |
| 500 | .056 65 | 3082.2 | 3422.2 | 6.8803 |
| 550 | .061 01 | 3174.6 | 3540.6 | 7.0288 |
| 600 | .065 25 | 3266.9 | 3658.4 | 7.1677 |
| 700 | .073 52 | 3453.1 | 3894.2 | 7.4234 |
| 800 | .081 60 | 3643.1 | 4132.7 | 7.6566 |
| 900 | .089 58 | 3837.8 | 4375.3 | 7.8727 |
| 1000 | .097 49 | 4037.8 | 4622.7 | 8.0751 |
| 1100 | .105 36 | 4243.3 | 4875.4 | 8.2661 |

$P = 7.0$ MPa (285.88)

| T | v | u | h | s |
|---|---|---|---|---|
| Sat. | .027 37 | 2580.5 | 2772.1 | 5.8133 |
| 300 | .029 47 | 2632.2 | 2838.4 | 5.9305 |
| 350 | .035 24 | 2769.4 | 3016.0 | 6.2283 |
| 400 | .039 93 | 2878.6 | 3158.1 | 6.4478 |
| 450 | .044 16 | 2978.0 | 3287.1 | 6.6327 |
| 500 | .048 14 | 3073.4 | 3410.3 | 6.7975 |
| 550 | .051 95 | 3167.2 | 3530.9 | 6.9486 |
| 600 | .055 65 | 3260.7 | 3650.3 | 7.0894 |
| 700 | .062 83 | 3448.5 | 3888.3 | 7.3476 |
| 800 | .069 81 | 3639.5 | 4128.2 | 7.5822 |
| 900 | .076 69 | 3835.0 | 4371.8 | 7.7991 |
| 1000 | .083 50 | 4035.3 | 4619.8 | 8.0020 |
| 1100 | .090 27 | 4240.9 | 4872.8 | 8.1933 |

$P = 8.0$ MPa (295.06)

| T | v | u | h | s |
|---|---|---|---|---|
| Sat. | .023 52 | 2569.8 | 2758.0 | 5.7432 |
| 300 | .024 26 | 2590.9 | 2785.0 | 5.7906 |
| 350 | .029 95 | 2747.7 | 2987.3 | 6.1301 |
| 400 | .034 32 | 2863.8 | 3138.3 | 6.3634 |
| 450 | .038 17 | 2966.7 | 3272.0 | 6.5551 |
| 500 | .041 75 | 3064.3 | 3398.3 | 6.7240 |
| 550 | .045 16 | 3159.8 | 3521.0 | 6.8778 |
| 600 | .048 45 | 3254.4 | 3642.0 | 7.0206 |
| 700 | .054 81 | 3443.9 | 3882.4 | 7.2812 |
| 800 | .060 97 | 3636.0 | 4123.8 | 7.5173 |
| 900 | .067 02 | 3832.1 | 4368.3 | 7.7351 |
| 1000 | .073 01 | 4032.8 | 4616.9 | 7.9384 |
| 1100 | .078 96 | 4238.6 | 4870.3 | 8.1300 |

# Table A.1.3 (Continued)
## Superheated Vapor

| T | P = 6.0 MPa (275.64) v | u | h | s | P = 7.0 MPa (285.88) v | u | h | s | P = 8.0 MPa (295.06) v | u | h | s |
|---|---|---|---|---|---|---|---|---|---|---|---|---|
| 1200 | .113 21 | 4454.0 | 5133.3 | 8.4474 | .097 03 | 4451.7 | 5130.9 | 8.3747 | .084 89 | 4449.5 | 5128.5 | 8.3115 |
| 1300 | .121 06 | 4669.6 | 5396.0 | 8.6199 | .103 77 | 4667.3 | 5393.7 | 8.5473 | .090 80 | 4665.0 | 5391.5 | 8.4842 |

| T | P = 9.0 MPa (303.40) v | u | h | s | P = 10.0 MPa (311.06) v | u | h | s | P = 12.5 MPa (327.89) v | u | h | s |
|---|---|---|---|---|---|---|---|---|---|---|---|---|
| Sat. | .020 48 | 2557.8 | 2742.1 | 5.6772 | .018 026 | 2544.4 | 2724.7 | 5.6141 | .013 495 | 2505.1 | 2673.8 | 5.4624 |
| 325 | .023 27 | 2646.6 | 2856.0 | 5.8712 | .019 861 | 2610.4 | 2809.1 | 5.7568 | | | | |
| 350 | .025 80 | 2724.4 | 2956.6 | 6.0361 | .022 42 | 2699.2 | 2923.4 | 5.9443 | .016 126 | 2624.6 | 2826.2 | 5.7118 |
| 400 | .029 93 | 2848.4 | 3117.8 | 6.2854 | .026 41 | 2832.4 | 3096.5 | 6.2120 | .020 00 | 2789.3 | 3039.3 | 6.0417 |
| 450 | .033 50 | 2955.2 | 3256.6 | 6.4844 | .029 75 | 2943.4 | 3240.9 | 6.4190 | .022 99 | 2912.5 | 3199.8 | 6.2719 |
| 500 | .036 77 | 3055.2 | 3386.1 | 6.6576 | .032 79 | 3045.8 | 3373.7 | 6.5966 | .025 60 | 3021.7 | 3341.8 | 6.4618 |
| 550 | .039 87 | 3152.2 | 3511.0 | 6.8142 | .035 64 | 3144.6 | 3500.9 | 6.7561 | .028 01 | 3125.0 | 3475.2 | 6.6290 |
| 600 | .042 85 | 3248.1 | 3633.7 | 6.9589 | .038 37 | 3241.7 | 3625.3 | 6.9029 | .030 29 | 3225.4 | 3604.0 | 6.7810 |
| 650 | .045 74 | 3343.6 | 3755.3 | 7.0943 | .041 01 | 3338.2 | 3748.2 | 7.0398 | .032 48 | 3324.4 | 3730.4 | 6.9218 |
| 700 | .048 57 | 3439.3 | 3876.5 | 7.2221 | .043 58 | 3434.7 | 3870.5 | 7.1687 | .034 60 | 3422.9 | 3855.3 | 7.0536 |
| 800 | .054 09 | 3632.5 | 4119.3 | 7.4596 | .048 59 | 3628.9 | 4114.8 | 7.4077 | .038 69 | 3620.0 | 4103.6 | 7.2965 |
| 900 | .059 50 | 3829.2 | 4364.8 | 7.6783 | .053 49 | 3826.3 | 4361.2 | 7.6272 | .042 67 | 3819.1 | 4352.5 | 7.5182 |
| 1000 | .064 85 | 4030.3 | 4614.0 | 7.8821 | .058 32 | 4027.8 | 4611.0 | 7.8315 | .046 58 | 4021.6 | 4603.8 | 7.7237 |
| 1100 | .070 16 | 4236.3 | 4867.7 | 8.0740 | .063 12 | 4234.0 | 4865.1 | 8.0237 | .050 45 | 4228.2 | 4858.8 | 7.9165 |
| 1200 | .075 44 | 4447.2 | 5126.2 | 8.2556 | .067 89 | 4444.9 | 5123.8 | 8.2055 | .054 30 | 4439.3 | 5118.0 | 8.0987 |
| 1300 | .080 72 | 4662.7 | 5389.2 | 8.4284 | .072 65 | 4660.5 | 5387.0 | 8.3783 | .058 13 | 4654.8 | 5381.4 | 8.2717 |

| T (°C) | P = 15.0 MPa (342.24) $v$ | $u$ | $h$ | $s$ | P = 17.5 MPa (354.75) $v$ | $u$ | $h$ | $s$ | P = 20.0 MPa (365.81) $v$ | $u$ | $h$ | $s$ |
|---|---|---|---|---|---|---|---|---|---|---|---|---|
| Sat. | .010 337 | 2455.5 | 2610.5 | 5.3098 | .007 920 | 2390.2 | 2528.8 | 5.1419 | .005 834 | 2293.0 | 2409.7 | 4.9269 |
| 350 | .011 470 | 2520.4 | 2692.4 | 5.4421 | | | | | | | | |
| 400 | .015 649 | 2740.7 | 2975.5 | 5.8811 | .012 447 | 2685.0 | 2902.9 | 5.7213 | .009 942 | 2619.3 | 2818.1 | 5.5540 |
| 450 | .018 445 | 2879.5 | 3156.2 | 6.1404 | .015 174 | 2844.2 | 3109.7 | 6.0184 | .012 695 | 2806.2 | 3060.1 | 5.9017 |
| 500 | .020 80 | 2996.6 | 3308.6 | 6.3443 | .017 358 | 2970.3 | 3274.1 | 6.2383 | .014 768 | 2942.9 | 3238.2 | 6.1401 |
| 550 | .022 93 | 3104.7 | 3448.6 | 6.5199 | .019 288 | 3083.9 | 3421.4 | 6.4230 | .016 555 | 3062.4 | 3393.5 | 6.3348 |
| 600 | .024 91 | 3208.6 | 3582.3 | 6.6776 | .021 06 | 3191.5 | 3560.1 | 6.5866 | .018 178 | 3174.0 | 3537.6 | 6.5048 |
| 650 | .026 80 | 3310.3 | 3712.3 | 6.8224 | .022 74 | 3296.0 | 3693.9 | 6.7357 | .019 693 | 3281.4 | 3675.3 | 6.6582 |
| 700 | .028 61 | 3410.9 | 3840.1 | 6.9572 | .024 34 | 3398.7 | 3824.6 | 6.8736 | .021 13 | 3386.4 | 3809.0 | 6.7993 |
| 800 | .032 10 | 3610.9 | 4092.4 | 7.2040 | .027 38 | 3601.8 | 4081.1 | 7.1244 | .023 85 | 3592.7 | 4069.7 | 7.0544 |
| 900 | .035 46 | 3811.9 | 4343.8 | 7.4279 | .030 31 | 3804.7 | 4335.1 | 7.3507 | .026 45 | 3797.5 | 4326.4 | 7.2830 |
| 1000 | .038 75 | 4015.4 | 4596.6 | 7.6348 | .033 16 | 4009.3 | 4589.5 | 7.5589 | .028 97 | 4003.1 | 4582.5 | 7.4925 |
| 1100 | .042 00 | 4222.6 | 4852.6 | 7.8283 | .035 97 | 4216.9 | 4846.4 | 7.7531 | .031 45 | 4211.3 | 4840.2 | 7.6874 |
| 1200 | .045 23 | 4433.8 | 5112.3 | 8.0108 | .038 76 | 4428.3 | 5106.6 | 7.9360 | .033 91 | 4422.8 | 5101.0 | 7.8707 |
| 1300 | .048 45 | 4649.1 | 5376.0 | 8.1840 | .041 54 | 4643.5 | 5370.5 | 8.1093 | .036 36 | 4638.0 | 5365.1 | 8.0442 |

| T (°C) | P = 25.0 MPa $v$ | $u$ | $h$ | $s$ | P = 30.0 MPa $v$ | $u$ | $h$ | $s$ | P = 35.0 MPa $v$ | $u$ | $h$ | $s$ |
|---|---|---|---|---|---|---|---|---|---|---|---|---|
| 375 | .001 973 1 | 1798.7 | 1848.0 | 4.0320 | .001 789 2 | 1737.8 | 1791.5 | 3.9305 | .001 700 3 | 1702.9 | 1762.4 | 3.8722 |
| 400 | .006 004 | 2430.1 | 2580.2 | 5.1418 | .002 790 | 2067.4 | 2151.1 | 4.4728 | .002 100 | 1914.1 | 1987.6 | 4.2126 |
| 425 | .007 881 | 2609.2 | 2806.3 | 5.4723 | .005 303 | 2455.1 | 2614.2 | 5.1504 | .003 428 | 2253.4 | 2373.4 | 4.7747 |
| 450 | .009 162 | 2720.7 | 2949.7 | 5.6744 | .006 735 | 2619.3 | 2821.4 | 5.4424 | .004 961 | 2498.7 | 2672.4 | 5.1962 |
| 500 | .011 123 | 2884.3 | 3162.4 | 5.9592 | .008 678 | 2820.7 | 3081.1 | 5.7905 | .006 927 | 2751.9 | 2994.4 | 5.6282 |
| 550 | .012 724 | 3017.5 | 3335.6 | 6.1765 | .010 168 | 2970.3 | 3275.4 | 6.0342 | .008 345 | 2921.0 | 3213.0 | 5.9026 |
| 600 | .014 137 | 3137.9 | 3491.4 | 6.3602 | .011 446 | 3100.5 | 3443.9 | 6.2331 | .009 527 | 3062.0 | 3395.5 | 6.1179 |
| 650 | .015 433 | 3251.6 | 3637.4 | 6.5229 | .012 596 | 3221.0 | 3598.9 | 6.4058 | .010 575 | 3189.8 | 3559.9 | 6.3010 |
| 700 | .016 646 | 3361.3 | 3777.5 | 6.6707 | .013 661 | 3335.8 | 3745.6 | 6.5606 | .011 533 | 3309.8 | 3713.5 | 6.4631 |
| 800 | .018 912 | 3574.3 | 4047.1 | 6.9345 | .015 623 | 3555.5 | 4024.2 | 6.8332 | .013 278 | 3536.7 | 4001.5 | 6.7450 |
| 900 | .021 045 | 3783.0 | 4309.1 | 7.1680 | .017 448 | 3768.5 | 4291.9 | 7.0718 | .014 883 | 3754.0 | 4274.9 | 6.9886 |
| 1000 | .023 10 | 3990.9 | 4568.5 | 7.3802 | .019 196 | 3978.8 | 4554.7 | 7.2867 | .016 410 | 3966.7 | 4541.1 | 7.2064 |
| 1100 | .025 12 | 4200.2 | 4828.2 | 7.5765 | .020 903 | 4189.2 | 4816.3 | 7.4845 | .017 895 | 4178.3 | 4804.6 | 7.4057 |

# Table A.1.3 (Continued)
## Superheated Vapor

| T | v | u | h | s | v | u | h | s | v | u | h | s |
|---|---|---|---|---|---|---|---|---|---|---|---|---|
| | P = 25.0 MPa | | | | P = 30.0 MPa | | | | P = 35.0 MPa | | | |
| 1200 | .027 11 | 4412.0 | 5089.9 | 7.7605 | .022 589 | 4401.3 | 5079.0 | 7.6692 | .019 360 | 4390.7 | 5068.3 | 7.5910 |
| 1300 | .029 10 | 4626.9 | 5354.4 | 7.9342 | .024 266 | 4616.0 | 5344.0 | 7.8432 | .020 815 | 4605.1 | 5333.6 | 7.7653 |

| T | v | u | h | s | v | u | h | s | v | u | h | s |
|---|---|---|---|---|---|---|---|---|---|---|---|---|
| | P = 40.0 MPa | | | | P = 50.0 MPa | | | | P = 60.0 MPa | | | |
| 375 | .001 640 7 | 1677.1 | 1742.8 | 3.8290 | .001 559 4 | 1638.6 | 1716.6 | 3.7639 | .001 502 8 | 1609.4 | 1699.5 | 3.7141 |
| 400 | .001 907 7 | 1854.6 | 1930.9 | 4.1135 | .001 730 9 | 1788.1 | 1874.6 | 4.0031 | .001 633 5 | 1745.4 | 1843.4 | 3.9318 |
| 425 | .002 532 | 2096.9 | 2198.1 | 4.5029 | .002 007 | 1959.7 | 2060.0 | 4.2734 | .001 816 5 | 1892.7 | 2001.7 | 4.1626 |
| 450 | .003 693 | 2365.1 | 2512.8 | 4.9459 | .002 486 | 2159.6 | 2284.0 | 4.5884 | .002 085 | 2053.9 | 2179.0 | 4.4121 |
| 500 | .005 622 | 2678.4 | 2903.3 | 5.4700 | .003 892 | 2525.5 | 2720.1 | 5.1726 | .002 956 | 2390.6 | 2567.9 | 4.9321 |
| 550 | .006 984 | 2869.7 | 3149.1 | 5.7785 | .005 118 | 2763.6 | 3019.5 | 5.5485 | .003 956 | 2658.8 | 2896.2 | 5.3441 |
| 600 | .008 094 | 3022.6 | 3346.4 | 6.0114 | .006 112 | 2942.0 | 3247.6 | 5.8178 | .004 834 | 2861.1 | 3151.2 | 5.6452 |
| 650 | .009 063 | 3158.0 | 3520.6 | 6.2054 | .006 966 | 3093.5 | 3441.8 | 6.0342 | .005 595 | 3028.8 | 3364.5 | 5.8829 |
| 700 | .009 941 | 3283.6 | 3681.2 | 6.3750 | .007 727 | 3230.5 | 3616.8 | 6.2189 | .006 272 | 3177.2 | 3553.5 | 6.0824 |
| 800 | .011 523 | 3517.8 | 3978.7 | 6.6662 | .009 076 | 3479.8 | 3933.6 | 6.5290 | .007 459 | 3441.5 | 3889.1 | 6.4109 |
| 900 | .012 962 | 3739.4 | 4257.9 | 6.9150 | .010 283 | 3710.3 | 4224.4 | 6.7882 | .008 508 | 3681.0 | 4191.5 | 6.6805 |
| 1000 | .014 324 | 3954.6 | 4527.6 | 7.1356 | .011 411 | 3930.5 | 4501.1 | 7.0146 | .009 480 | 3906.4 | 4475.2 | 6.9127 |
| 1100 | .015 642 | 4167.4 | 4793.1 | 7.3364 | .012 496 | 4145.7 | 4770.5 | 7.2184 | .010 409 | 4124.1 | 4748.6 | 7.1195 |
| 1200 | .016 940 | 4380.1 | 5057.7 | 7.5224 | .013 561 | 4359.1 | 5037.2 | 7.4058 | .011 317 | 4338.2 | 5017.2 | 7.3083 |
| 1300 | .018 229 | 4594.3 | 5323.5 | 7.6969 | .014 616 | 4572.8 | 5303.6 | 7.5808 | .012 215 | 4551.4 | 5284.3 | 7.4837 |

**Table A.1.4**
Compressed Liquid

| T | P = 5 MPa (263.99) | | | | P = 10 MPa (311.06) | | | | P = 15 MPa (342.24) | | | |
|---|---|---|---|---|---|---|---|---|---|---|---|---|
| | v | u | h | s | v | u | h | s | v | u | h | s |
| Sat. | .001 285 9 | 1147.8 | 1154.2 | 2.9202 | .001 452 4 | 1393.0 | 1407.6 | 3.3596 | .001 658 1 | 1585.6 | 1610.5 | 3.6848 |
| 0 | .000 997 7 | .04 | 5.04 | .0001 | .000 995 2 | .09 | 10.04 | .0002 | .000 992 8 | .15 | 15.05 | .0004 |
| 20 | .000 999 5 | 83.65 | 88.65 | .2956 | .000 997 2 | 83.36 | 93.33 | .2945 | .000 995 0 | 83.06 | 97.99 | .2934 |
| 40 | .001 005 6 | 166.95 | 171.97 | .5705 | .001 003 4 | 166.35 | 176.38 | .5686 | .001 001 3 | 165.76 | 180.78 | .5666 |
| 60 | .001 014 9 | 250.23 | 255.30 | .8285 | .001 012 7 | 249.36 | 259.49 | .8258 | .001 010 5 | 248.51 | 263.67 | .8232 |
| 80 | .001 026 8 | 333.72 | 338.85 | 1.0720 | .001 024 5 | 332.59 | 342.83 | 1.0688 | .001 022 2 | 331.48 | 346.81 | 1.0656 |
| 100 | .001 041 0 | 417.52 | 422.72 | 1.3030 | .001 038 5 | 416.12 | 426.50 | 1.2992 | .001 036 1 | 414.74 | 430.28 | 1.2955 |
| 120 | .001 057 6 | 501.80 | 507.09 | 1.5233 | .001 054 0 | 500.08 | 510.64 | 1.5189 | .001 052 2 | 498.40 | 514.19 | 1.5145 |
| 140 | .001 076 8 | 586.76 | 592.15 | 1.7343 | .001 073 7 | 584.68 | 595.42 | 1.7292 | .001 070 7 | 582.66 | 598.72 | 1.7242 |
| 160 | .001 098 8 | 672.62 | 678.12 | 1.9375 | .001 095 3 | 670.13 | 681.08 | 1.9317 | .001 091 8 | 667.71 | 684.09 | 1.9260 |
| 180 | .001 124 0 | 759.63 | 765.25 | 2.1341 | .001 119 0 | 756.65 | 767.84 | 2.1275 | .001 115 9 | 753.76 | 770.50 | 2.1210 |
| 200 | .001 153 0 | 848.1 | 853.9 | 2.3255 | .001 148 0 | 844.5 | 856.0 | 2.3178 | .001 143 3 | 841.0 | 858.2 | 2.3104 |
| 220 | .001 186 6 | 938.4 | 944.4 | 2.5128 | .001 180 5 | 934.1 | 945.9 | 2.5039 | .001 174 8 | 929.9 | 947.5 | 2.4953 |
| 240 | .001 226 4 | 1031.4 | 1037.5 | 2.6979 | .001 218 7 | 1026.0 | 1038.1 | 2.6872 | .001 211 4 | 1020.8 | 1039.0 | 2.6771 |
| 260 | .001 274 9 | 1127.9 | 1134.3 | 2.8830 | .001 264 5 | 1121.1 | 1133.7 | 2.8699 | .001 255 0 | 1114.6 | 1133.4 | 2.8576 |
| 280 | | | | | .001 321 6 | 1220.9 | 1234.1 | 3.0548 | .001 308 4 | 1212.5 | 1232.1 | 3.0393 |
| 300 | | | | | .001 397 2 | 1328.4 | 1342.3 | 3.2469 | .001 377 0 | 1316.6 | 1337.3 | 3.2260 |
| 320 | | | | | | | | | .001 472 4 | 1431.1 | 1453.2 | 3.4247 |
| 340 | | | | | | | | | .001 631 1 | 1567.5 | 1591.9 | 3.6546 |

# Table A.1.4 (Continued)
## Compressed Liquid

| T | P = 20 MPa (365.81) | | | | P = 30 MPa | | | | P = 50 MPa | | | |
|---|---|---|---|---|---|---|---|---|---|---|---|---|
| | $v$ | $u$ | $h$ | $s$ | $v$ | $u$ | $h$ | $s$ | $v$ | $u$ | $h$ | $s$ |
| Sat. | .002 036 | 1785.6 | 1826.3 | 4.0139 | | | | | | | | |
| 0 | .000 990 4 | .19 | 20.01 | .0004 | .000 985 6 | .25 | 29.82 | .0001 | .000 976 6 | .20 | 49.03 | .0014 |
| 20 | .000 992 8 | 82.77 | 102.62 | .2923 | .000 988 6 | 82.17 | 111.84 | .2899 | .000 980 4 | 81.00 | 130.02 | .2848 |
| 40 | .000 999 2 | 165.17 | 185.16 | .5646 | .000 995 1 | 164.04 | 193.89 | .5607 | .000 987 2 | 161.86 | 211.21 | .5527 |
| 60 | .001 008 4 | 247.68 | 267.85 | .8206 | .001 004 2 | 246.06 | 276.19 | .8154 | .000 996 2 | 242.98 | 292.79 | .8052 |
| 80 | .001 019 9 | 330.40 | 350.80 | 1.0624 | .001 015 6 | 328.30 | 358.77 | 1.0561 | .001 007 3 | 324.34 | 374.70 | 1.0440 |
| 100 | .001 033 7 | 413.39 | 434.06 | 1.2917 | .001 029 0 | 410.78 | 441.66 | 1.2844 | .001 020 1 | 405.88 | 456.89 | 1.2703 |
| 120 | .001 049 6 | 496.76 | 517.76 | 1.5102 | .001 044 5 | 493.59 | 524.93 | 1.5018 | .001 034 8 | 487.65 | 539.39 | 1.4857 |
| 140 | .001 067 8 | 580.69 | 602.04 | 1.7193 | .001 062 1 | 576.88 | 608.75 | 1.7098 | .001 051 5 | 569.77 | 622.35 | 1.6915 |
| 160 | .001 088 5 | 665.35 | 687.12 | 1.9204 | .001 082 1 | 660.82 | 693.28 | 1.9096 | .001 070 3 | 652.41 | 705.92 | 1.8891 |
| 180 | .001 112 0 | 750.95 | 773.20 | 2.1147 | .001 104 7 | 745.59 | 778.73 | 2.1024 | .001 091 2 | 735.69 | 790.25 | 2.0794 |
| 200 | .001 138 8 | 837.7 | 860.5 | 2.3031 | .001 130 2 | 831.4 | 865.3 | 2.2893 | .001 114 6 | 819.7 | 875.5 | 2.2634 |
| 220 | .001 169 3 | 925.9 | 949.3 | 2.4870 | .001 159 0 | 918.3 | 953.1 | 2.4711 | .001 140 8 | 904.7 | 961.7 | 2.4419 |
| 240 | .001 204 6 | 1016.0 | 1040.0 | 2.6674 | .001 192 0 | 1006.9 | 1042.6 | 2.6490 | .001 170 2 | 990.7 | 1049.2 | 2.6158 |
| 260 | .001 246 2 | 1108.6 | 1133.5 | 2.8459 | .001 230 3 | 1097.4 | 1134.3 | 2.8243 | .001 203 4 | 1078.1 | 1138.2 | 2.7860 |
| 280 | .001 296 5 | 1204.7 | 1230.6 | 3.0248 | .001 275 5 | 1190.7 | 1229.0 | 2.9986 | .001 241 5 | 1167.2 | 1229.3 | 2.9537 |
| 300 | .001 359 6 | 1306.1 | 1333.3 | 3.2071 | .001 330 4 | 1287.9 | 1327.8 | 3.1741 | .001 286 0 | 1258.7 | 1323.0 | 3.1200 |
| 320 | .001 443 7 | 1415.7 | 1444.6 | 3.3979 | .001 399 7 | 1390.7 | 1432.7 | 3.3539 | .001 338 8 | 1353.3 | 1420.2 | 3.2868 |
| 340 | .001 568 4 | 1539.7 | 1571.0 | 3.6075 | .001 492 0 | 1501.7 | 1546.5 | 3.5426 | .001 403 2 | 1452.0 | 1522.1 | 3.4557 |
| 360 | .001 822 6 | 1702.8 | 1739.3 | 3.8772 | .001 626 5 | 1626.6 | 1675.4 | 3.7494 | .001 483 8 | 1556.0 | 1630.2 | 3.6291 |
| 380 | | | | | .001 869 1 | 1781.4 | 1837.5 | 4.0012 | .001 588 4 | 1667.2 | 1746.6 | 3.8101 |

# Table A.1.5
## Saturated Solid-Vapor

| Temp. °C $T$ | Press. kPa $P$ | Specific Volume Sat. Solid $v_i \times 10^3$ | Specific Volume Sat. Vapor $v_g$ | Internal Energy Sat. Solid $u_i$ | Internal Energy Subl. $u_{ig}$ | Internal Energy Sat. Vapor $u_g$ | Enthalpy Sat. Solid $h_i$ | Enthalpy Subl. $h_{ig}$ | Enthalpy Sat. Vapor $h_g$ | Entropy Sat. Solid $s_i$ | Entropy Subl. $s_{ig}$ | Entropy Sat. Vapor $s_g$ |
|---|---|---|---|---|---|---|---|---|---|---|---|---|
| .01 | .6113 | 1.0908 | 206.1 | −333.40 | 2708.7 | 2375.3 | −333.40 | 2834.8 | 2501.4 | −1.221 | 10.378 | 9.156 |
| 0 | .6108 | 1.0908 | 206.3 | −333.43 | 2708.8 | 2375.3 | −333.43 | 2834.8 | 2501.3 | −1.221 | 10.378 | 9.157 |
| −2 | .5176 | 1.0904 | 241.7 | −337.62 | 2710.2 | 2372.6 | −337.62 | 2835.3 | 2497.7 | −1.237 | 10.456 | 9.219 |
| −4 | .4375 | 1.0901 | 283.8 | −341.78 | 2711.6 | 2369.8 | −341.78 | 2835.7 | 2494.0 | −1.253 | 10.536 | 9.283 |
| −6 | .3689 | 1.0898 | 334.2 | −345.91 | 2712.9 | 2367.0 | −345.91 | 2836.2 | 2490.3 | −1.268 | 10.616 | 9.348 |
| −8 | .3102 | 1.0894 | 394.4 | −350.02 | 2714.2 | 2364.2 | −350.02 | 2836.6 | 2486.6 | −1.284 | 10.698 | 9.414 |
| −10 | .2602 | 1.0891 | 466.7 | −354.09 | 2715.5 | 2361.4 | −354.09 | 2837.0 | 2482.9 | −1.299 | 10.781 | 9.481 |
| −12 | .2176 | 1.0888 | 553.7 | −358.14 | 2716.8 | 2358.7 | −358.14 | 2837.3 | 2479.2 | −1.315 | 10.865 | 9.550 |
| −14 | .1815 | 1.0884 | 658.8 | −362.15 | 2718.0 | 2355.9 | −362.15 | 2837.6 | 2475.5 | −1.331 | 10.950 | 9.619 |
| −16 | .1510 | 1.0881 | 786.0 | −366.14 | 2719.2 | 2353.1 | −366.14 | 2837.9 | 2471.8 | −1.346 | 11.036 | 9.690 |
| −18 | .1252 | 1.0878 | 940.5 | −370.10 | 2720.4 | 2350.3 | −370.10 | 2838.2 | 2468.1 | −1.362 | 11.123 | 9.762 |
| −20 | .1035 | 1.0874 | 1128.6 | −374.03 | 2721.6 | 2347.5 | −374.03 | 2838.4 | 2464.3 | −1.377 | 11.212 | 9.835 |
| −22 | .0853 | 1.0871 | 1358.4 | −377.93 | 2722.7 | 2344.7 | −377.93 | 2838.6 | 2460.6 | −1.393 | 11.302 | 9.909 |
| −24 | .0701 | 1.0868 | 1640.1 | −381.80 | 2723.7 | 2342.0 | −381.80 | 2838.7 | 2456.9 | −1.408 | 11.394 | 9.985 |
| −26 | .0574 | 1.0864 | 1986.4 | −385.64 | 2724.8 | 2339.2 | −385.64 | 2838.9 | 2453.2 | −1.424 | 11.486 | 10.062 |
| −28 | .0469 | 1.0861 | 2413.7 | −389.45 | 2725.8 | 2336.4 | −389.45 | 2839.0 | 2449.5 | −1.439 | 11.580 | 10.141 |
| −30 | .0381 | 1.0858 | 2943 | −393.23 | 2726.8 | 2333.6 | −393.23 | 2839.0 | 2445.8 | −1.455 | 11.676 | 10.221 |
| −32 | .0309 | 1.0854 | 3600 | −396.98 | 2727.8 | 2330.8 | −396.98 | 2839.1 | 2442.1 | −1.471 | 11.773 | 10.303 |
| −34 | .0250 | 1.0851 | 4419 | −400.71 | 2728.7 | 2328.0 | −400.71 | 2839.1 | 2438.4 | −1.486 | 11.872 | 10.386 |
| −36 | .0201 | 1.0848 | 5444 | −404.40 | 2729.6 | 2325.2 | −404.40 | 2839.1 | 2434.7 | −1.501 | 11.972 | 10.470 |
| −38 | .0161 | 1.0844 | 6731 | −408.06 | 2730.5 | 2322.4 | −408.06 | 2839.0 | 2430.9 | −1.517 | 12.073 | 10.556 |
| −40 | .0129 | 1.0841 | 8354 | −411.70 | 2731.3 | 2319.6 | −411.70 | 2838.9 | 2427.2 | −1.532 | 12.176 | 10.644 |

# Table A.2
## Thermodynamic Properties of Ammonia[a]
# Table A.2.1
## Saturated Ammonia

| Temp. °C | Abs. Press. kPa $P$ | Specific Volume m³/kg | | | Enthalpy kJ/kg | | | Entropy kJ/kg K | | |
|---|---|---|---|---|---|---|---|---|---|---|
| | | Sat. Liquid $v_f$ | Evap. $v_{fg}$ | Sat. Vapor $v_g$ | Sat. Liquid $h_f$ | Evap. $h_{fg}$ | Sat. Vapor $h_g$ | Sat. Liquid $s_f$ | Evap. $s_{fg}$ | Sat. Vapor $s_g$ |
| −50 | 40.88 | 0.001 424 | 2.6239 | 2.6254 | −44.3 | 1416.7 | 1372.4 | −0.1942 | 6.3502 | 6.1561 |
| −48 | 45.96 | 0.001 429 | 2.3518 | 2.3533 | −35.5 | 1411.3 | 1375.8 | −0.1547 | 6.2696 | 6.1149 |
| −46 | 51.55 | 0.001 434 | 2.1126 | 2.1140 | −26.6 | 1405.8 | 1379.2 | −0.1156 | 6.1902 | 6.0746 |
| −44 | 57.69 | 0.001 439 | 1.9018 | 1.9032 | −17.8 | 1400.3 | 1382.5 | −0.0768 | 6.1120 | 6.0352 |
| −42 | 64.42 | 0.001 444 | 1.7155 | 1.7170 | −8.9 | 1394.7 | 1385.8 | −0.0382 | 6.0349 | 5.9967 |
| −40 | 71.77 | 0.001 449 | 1.5506 | 1.5521 | 0.0 | 1389.0 | 1389.0 | 0.0000 | 5.9589 | 5.9589 |
| −38 | 79.80 | 0.001 454 | 1.4043 | 1.4058 | 8.9 | 1383.3 | 1392.2 | 0.0380 | 5.8840 | 5.9220 |
| −36 | 88.54 | 0.001 460 | 1.2742 | 1.2757 | 17.8 | 1377.6 | 1395.4 | 0.0757 | 5.8101 | 5.8858 |
| −34 | 98.05 | 0.001 465 | 1.1582 | 1.1597 | 26.8 | 1371.8 | 1398.5 | 0.1132 | 5.7372 | 5.8504 |
| −32 | 108.37 | 0.001 470 | 1.0547 | 1.0562 | 35.7 | 1365.9 | 1401.6 | 0.1504 | 5.6652 | 5.8156 |
| −30 | 119.55 | 0.001 476 | 0.9621 | 0.9635 | 44.7 | 1360.0 | 1404.6 | 0.1873 | 5.5942 | 5.7815 |
| −28 | 131.64 | 0.001 481 | 0.8790 | 0.8805 | 53.6 | 1354.0 | 1407.6 | 0.2240 | 5.5241 | 5.7481 |
| −26 | 144.70 | 0.001 487 | 0.8044 | 0.8059 | 62.6 | 1347.9 | 1410.5 | 0.2605 | 5.4548 | 5.7153 |
| −24 | 158.78 | 0.001 492 | 0.7373 | 0.7388 | 71.6 | 1341.8 | 1413.4 | 0.2967 | 5.3864 | 5.6831 |
| −22 | 173.93 | 0.001 498 | 0.6768 | 0.6783 | 80.7 | 1335.6 | 1416.2 | 0.3327 | 5.3188 | 5.6515 |
| −20 | 190.22 | 0.001 504 | 0.6222 | 0.6237 | 89.7 | 1329.3 | 1419.0 | 0.3684 | 5.2520 | 5.6205 |
| −18 | 207.71 | 0.001 510 | 0.5728 | 0.5743 | 98.8 | 1322.9 | 1421.7 | 0.4040 | 5.1860 | 5.5900 |
| −16 | 226.45 | 0.001 515 | 0.5280 | 0.5296 | 107.8 | 1316.5 | 1424.4 | 0.4393 | 5.1207 | 5.5600 |
| −14 | 246.51 | 0.001 521 | 0.4874 | 0.4889 | 116.9 | 1310.0 | 1427.0 | 0.4744 | 5.0561 | 5.5305 |
| −12 | 267.95 | 0.001 528 | 0.4505 | 0.4520 | 126.0 | 1303.5 | 1429.5 | 0.5093 | 4.9922 | 5.5015 |

| Temp | P | $v_f$ | $v_g$ | | | | | $s_f$ | | $s_g$ |
|---|---|---|---|---|---|---|---|---|---|---|
| −10 | 290.85 | 0.001 534 | 0.4169 | 0.4185 | 135.2 | 1296.8 | 1432.0 | 0.5440 | 4.9290 | 5.4730 |
| −8 | 315.25 | 0.001 540 | 0.3863 | 0.3878 | 144.3 | 1290.1 | 1434.4 | 0.5785 | 4.8664 | 5.4449 |
| −6 | 341.25 | 0.001 546 | 0.3583 | 0.3599 | 153.5 | 1283.3 | 1436.8 | 0.6128 | 4.8045 | 5.4173 |
| −4 | 368.90 | 0.001 553 | 0.3328 | 0.3343 | 162.7 | 1276.4 | 1439.1 | 0.6469 | 4.7432 | 5.3901 |
| −2 | 398.27 | 0.001 559 | 0.3094 | 0.3109 | 171.9 | 1269.4 | 1441.3 | 0.6808 | 4.6825 | 5.3633 |
| 0 | 429.44 | 0.001 566 | 0.2879 | 0.2895 | 181.1 | 1262.4 | 1443.5 | 0.7145 | 4.6223 | 5.3369 |
| 2 | 462.49 | 0.001 573 | 0.2683 | 0.2698 | 190.4 | 1255.2 | 1445.6 | 0.7481 | 4.5627 | 5.3108 |
| 4 | 497.49 | 0.001 580 | 0.2502 | 0.2517 | 199.6 | 1248.0 | 1447.6 | 0.7815 | 4.5037 | 5.2852 |
| 6 | 534.51 | 0.001 587 | 0.2335 | 0.2351 | 208.9 | 1240.6 | 1449.6 | 0.8148 | 4.4451 | 5.2599 |
| 8 | 573.64 | 0.001 594 | 0.2182 | 0.2198 | 218.3 | 1233.2 | 1451.5 | 0.8479 | 4.3871 | 5.2350 |
| 10 | 614.95 | 0.001 601 | 0.2040 | 0.2056 | 227.6 | 1225.7 | 1453.3 | 0.8808 | 4.3295 | 5.2104 |
| 12 | 658.52 | 0.001 608 | 0.1910 | 0.1926 | 237.0 | 1218.1 | 1455.1 | 0.9136 | 4.2725 | 5.1861 |
| 14 | 704.44 | 0.001 616 | 0.1789 | 0.1805 | 246.4 | 1210.4 | 1456.8 | 0.9463 | 4.2159 | 5.1621 |
| 16 | 752.79 | 0.001 623 | 0.1677 | 0.1693 | 255.9 | 1202.6 | 1458.5 | 0.9788 | 4.1597 | 5.1385 |
| 18 | 803.66 | 0.001 631 | 0.1574 | 0.1590 | 265.4 | 1194.7 | 1460.0 | 1.0112 | 4.1039 | 5.1151 |
| 20 | 857.12 | 0.001 639 | 0.1477 | 0.1494 | 274.9 | 1186.7 | 1461.5 | 1.0434 | 4.0486 | 5.0920 |
| 22 | 913.27 | 0.001 647 | 0.1388 | 0.1405 | 284.4 | 1178.5 | 1462.9 | 1.0755 | 3.9937 | 5.0692 |
| 24 | 972.19 | 0.001 655 | 0.1305 | 0.1322 | 294.0 | 1170.3 | 1464.3 | 1.1075 | 3.9392 | 5.0467 |
| 26 | 1033.97 | 0.001 663 | 0.1228 | 0.1245 | 303.6 | 1162.0 | 1465.6 | 1.1394 | 3.8850 | 5.0244 |
| 28 | 1098.71 | 0.001 671 | 0.1156 | 0.1173 | 313.2 | 1153.6 | 1466.8 | 1.1711 | 3.8312 | 5.0023 |
| 30 | 1166.49 | 0.001 680 | 0.1089 | 0.1106 | 322.9 | 1145.0 | 1467.9 | 1.2028 | 3.7777 | 4.9805 |
| 32 | 1237.41 | 0.001 689 | 0.1027 | 0.1044 | 332.6 | 1136.4 | 1469.0 | 1.2343 | 3.7246 | 4.9589 |
| 34 | 1311.55 | 0.001 698 | 0.0969 | 0.0986 | 342.3 | 1127.6 | 1469.9 | 1.2656 | 3.6718 | 4.9374 |
| 36 | 1389.03 | 0.001 707 | 0.0914 | 0.0931 | 352.1 | 1118.7 | 1470.8 | 1.2969 | 3.6192 | 4.9161 |
| 38 | 1469.92 | 0.001 716 | 0.0863 | 0.0880 | 361.9 | 1109.7 | 1471.5 | 1.3281 | 3.5669 | 4.8950 |
| 40 | 1554.33 | 0.001 726 | 0.0815 | 0.0833 | 371.7 | 1100.5 | 1472.2 | 1.3591 | 3.5148 | 4.8740 |
| 42 | 1642.35 | 0.001 735 | 0.0771 | 0.0788 | 381.6 | 1091.2 | 1472.8 | 1.3901 | 3.4630 | 4.8530 |
| 44 | 1734.09 | 0.001 745 | 0.0728 | 0.0746 | 391.5 | 1081.7 | 1473.2 | 1.4209 | 3.4112 | 4.8322 |
| 46 | 1829.65 | 0.001 756 | 0.0689 | 0.0707 | 401.5 | 1072.0 | 1473.5 | 1.4518 | 3.3595 | 4.8113 |
| 48 | 1929.13 | 0.001 766 | 0.0652 | 0.0669 | 411.5 | 1062.2 | 1473.7 | 1.4826 | 3.3079 | 4.7905 |
| 50 | 2032.62 | 0.001 777 | 0.0617 | 0.0635 | 421.7 | 1052.0 | 1473.7 | 1.5135 | 3.2561 | 4.7696 |

[a] Adapted from National Bureau of Standards Circular No. 142, *Tables of Thermodynamic Properties of Ammonia.*

## Table A.2.2
## Superheated Ammonia

| Abs. Press. kPa (Sat. Temp.) °C | | Temperature, °C | | | | | | | | | | | |
|---|---|---|---|---|---|---|---|---|---|---|---|---|---|
| | | −20 | −10 | 0 | 10 | 20 | 30 | 40 | 50 | 60 | 70 | 80 | 100 |
| 50 (−46.54) | $v$ | 2.4474 | 2.5481 | 2.6482 | 2.7479 | 2.8473 | 2.9464 | 3.0453 | 3.1441 | 3.2427 | 3.3413 | 3.4397 | |
| | $h$ | 1435.8 | 1457.0 | 1478.1 | 1499.2 | 1520.4 | 1541.7 | 1563.0 | 1584.5 | 1606.1 | 1627.8 | 1649.7 | |
| | $s$ | 6.3256 | 6.4077 | 6.4865 | 6.5625 | 6.6360 | 6.7073 | 6.7766 | 6.8441 | 6.9099 | 6.9743 | 7.0372 | |
| 75 (−39.18) | $v$ | 1.6233 | 1.6915 | 1.7591 | 1.8263 | 1.8932 | 1.9597 | 2.0261 | 2.0923 | 2.1584 | 2.2244 | 2.2903 | |
| | $h$ | 1433.0 | 1454.7 | 1476.1 | 1497.5 | 1518.9 | 1540.3 | 1561.8 | 1583.4 | 1605.1 | 1626.9 | 1648.9 | |
| | $s$ | 6.1190 | 6.2028 | 6.2828 | 6.3597 | 6.4339 | 6.5058 | 6.5756 | 6.6434 | 6.7096 | 6.7742 | 6.8373 | |
| 100 (−33.61) | $v$ | 1.2110 | 1.2631 | 1.3145 | 1.3654 | 1.4160 | 1.4664 | 1.5165 | 1.5664 | 1.6163 | 1.6659 | 1.7155 | 1.8145 |
| | $h$ | 1430.1 | 1452.2 | 1474.1 | 1495.7 | 1517.3 | 1538.9 | 1560.5 | 1582.2 | 1604.1 | 1626.0 | 1648.0 | 1692.6 |
| | $s$ | 5.9695 | 6.0552 | 6.1366 | 6.2144 | 6.2894 | 6.3618 | 6.4321 | 6.5003 | 6.5668 | 6.6316 | 6.6950 | 6.8177 |
| 125 (−29.08) | $v$ | 0.9635 | 1.0059 | 1.0476 | 1.0889 | 1.1297 | 1.1703 | 1.2107 | 1.2509 | 1.2909 | 1.3309 | 1.3707 | 1.4501 |
| | $h$ | 1427.2 | 1449.8 | 1472.0 | 1493.9 | 1515.7 | 1537.5 | 1559.3 | 1581.1 | 1603.0 | 1625.0 | 1647.2 | 1691.8 |
| | $s$ | 5.8512 | 5.9389 | 6.0217 | 6.1006 | 6.1763 | 6.2494 | 6.3201 | 6.3887 | 6.4555 | 6.5206 | 6.5842 | 6.7072 |
| 150 (−25.23) | $v$ | 0.7984 | 0.8344 | 0.8697 | 0.9045 | 0.9388 | 0.9729 | 1.0068 | 1.0405 | 1.0740 | 1.1074 | 1.1408 | 1.2072 |
| | $h$ | 1424.1 | 1447.3 | 1469.8 | 1492.1 | 1514.1 | 1536.1 | 1558.0 | 1580.0 | 1602.0 | 1624.1 | 1646.3 | 1691.1 |
| | $s$ | 5.7526 | 5.8424 | 5.9266 | 6.0066 | 6.0831 | 6.1568 | 6.2280 | 6.2970 | 6.3641 | 6.4295 | 6.4933 | 6.6167 |
| 200 (−18.86) | $v$ | | 0.6199 | 0.6471 | 0.6738 | 0.7001 | 0.7261 | 0.7519 | 0.7774 | 0.8029 | 0.8282 | 0.8533 | 0.9035 |
| | $h$ | | 1442.0 | 1465.5 | 1488.4 | 1510.9 | 1533.2 | 1555.5 | 1577.7 | 1599.9 | 1622.2 | 1644.6 | 1689.6 |
| | $s$ | | 5.6863 | 5.7737 | 5.8559 | 5.9342 | 6.0091 | 6.0813 | 6.1512 | 6.2189 | 6.2849 | 6.3491 | 6.4732 |
| 250 (−13.67) | $v$ | | 0.4910 | 0.5135 | 0.5354 | 0.5568 | 0.5780 | 0.5989 | 0.6196 | 0.6401 | 0.6605 | 0.6809 | 0.7212 |
| | $h$ | | 1436.6 | 1461.0 | 1484.5 | 1507.6 | 1530.3 | 1552.9 | 1575.4 | 1597.8 | 1620.3 | 1642.8 | 1688.2 |
| | $s$ | | 5.5609 | 5.6517 | 5.7365 | 5.8165 | 5.8928 | 5.9661 | 6.0368 | 6.1052 | 6.1717 | 6.2365 | 6.3613 |
| 300 (−9.23) | $v$ | | | 0.4243 | 0.4430 | 0.4613 | 0.4792 | 0.4968 | 0.5143 | 0.5316 | 0.5488 | 0.5658 | 0.5997 |
| | $h$ | | | 1456.3 | 1480.6 | 1504.2 | 1527.4 | 1550.3 | 1573.0 | 1595.7 | 1618.4 | 1641.1 | 1686.7 |
| | $s$ | | | 5.5493 | 5.6366 | 5.7186 | 5.7963 | 5.8707 | 5.9423 | 6.0114 | 6.0785 | 6.1437 | 6.2693 |
| | $v$ | | | 0.3605 | 0.3770 | 0.3929 | 0.4086 | 0.4239 | 0.4391 | 0.4541 | 0.4689 | 0.4837 | 0.5129 |

| P (sat. temp) | | 20 | 30 | 40 | 50 | 60 | 70 | 80 | 100 | 120 | 140 | 160 | 180 |
|---|---|---|---|---|---|---|---|---|---|---|---|---|---|
| 350 (−5.35) | h | | | 1451.5 | 1476.5 | 1500.7 | 1524.4 | 1547.6 | 1570.7 | 1593.6 | 1616.5 | 1639.3 | 1685.2 |
| | s | | | 5.4600 | 5.5502 | 5.6342 | 5.7135 | 5.7890 | 5.8615 | 5.9314 | 5.9990 | 6.0647 | 6.1910 |
| 400 (−1.89) | v | | | 0.3125 | 0.3274 | 0.3417 | 0.3556 | 0.3692 | 0.3826 | 0.3959 | 0.4090 | 0.4220 | 0.4478 |
| | h | | | 1446.5 | 1472.4 | 1497.2 | 1521.3 | 1544.9 | 1568.3 | 1591.5 | 1614.5 | 1637.6 | 1683.7 |
| | s | | | 5.3803 | 5.4735 | 5.5597 | 5.6405 | 5.7173 | 5.7907 | 5.8613 | 5.9296 | 5.9957 | 6.1228 |
| 450 (1.26) | v | | | 0.2752 | 0.2887 | 0.3017 | 0.3143 | 0.3266 | 0.3387 | 0.3506 | 0.3624 | 0.3740 | 0.3971 |
| | h | | | 1441.3 | 1468.1 | 1493.6 | 1518.2 | 1542.2 | 1565.9 | 1589.3 | 1612.6 | 1635.8 | 1682.2 |
| | s | | | 5.3078 | 5.4042 | 5.4926 | 5.5752 | 5.6532 | 5.7275 | 5.7989 | 5.8678 | 5.9345 | 6.0623 |
| 500 (4.14) | v | 0.2698 | 0.2813 | 0.2926 | 0.3036 | 0.3144 | 0.3251 | 0.3357 | 0.3565 | 0.3771 | 0.3975 | | |
| | h | 1489.9 | 1515.0 | 1539.5 | 1563.4 | 1587.1 | 1610.6 | 1634.0 | 1680.7 | 1727.5 | 1774.7 | | |
| | s | 5.4314 | 5.5157 | 5.5950 | 5.6704 | 5.7425 | 5.8120 | 5.8793 | 6.0079 | 6.1301 | 6.2472 | | |
| 600 (9.29) | v | 0.2217 | 0.2317 | 0.2414 | 0.2508 | 0.2600 | 0.2691 | 0.2781 | 0.2957 | 0.3130 | 0.3302 | | |
| | h | 1482.4 | 1508.6 | 1533.8 | 1558.5 | 1582.7 | 1606.6 | 1630.4 | 1677.7 | 1724.9 | 1772.4 | | |
| | s | 5.3222 | 5.4102 | 5.4923 | 5.5697 | 5.6436 | 5.7144 | 5.7826 | 5.9129 | 6.0363 | 6.1541 | | |
| 700 (13.81) | v | 0.1874 | 0.1963 | 0.2048 | 0.2131 | 0.2212 | 0.2291 | 0.2369 | 0.2522 | 0.2672 | 0.2821 | | |
| | h | 1474.5 | 1501.9 | 1528.1 | 1553.4 | 1578.2 | 1602.6 | 1626.8 | 1674.6 | 1722.4 | 1770.2 | | |
| | s | 5.2259 | 5.3179 | 5.4029 | 5.4826 | 5.5582 | 5.6303 | 5.6997 | 5.8316 | 5.9562 | 6.0749 | | |
| 800 (17.86) | v | 0.1615 | 0.1696 | 0.1773 | 0.1848 | 0.1920 | 0.1991 | 0.2060 | 0.2196 | 0.2329 | 0.2459 | 0.2589 | |
| | h | 1466.3 | 1495.0 | 1522.2 | 1548.3 | 1573.7 | 1598.6 | 1623.1 | 1671.6 | 1719.8 | 1768.0 | 1816.4 | |
| | s | 5.1387 | 5.2351 | 5.3232 | 5.4053 | 5.4827 | 5.5562 | 5.6268 | 5.7603 | 5.8861 | 6.0057 | 6.1202 | |
| 900 (21.54) | v | | 0.1488 | 0.1559 | 0.1627 | 0.1693 | 0.1757 | 0.1820 | 0.1942 | 0.2061 | 0.2178 | 0.2294 | |
| | h | | 1488.0 | 1516.2 | 1543.0 | 1569.1 | 1594.4 | 1619.4 | 1668.5 | 1717.1 | 1765.7 | 1814.4 | |
| | s | | 5.1593 | 5.2508 | 5.3354 | 5.4147 | 5.4897 | 5.5614 | 5.6968 | 5.8237 | 5.9442 | 6.0594 | |
| 1000 (24.91) | v | | 0.1321 | 0.1388 | 0.1450 | 0.1511 | 0.1570 | 0.1627 | 0.1739 | 0.1847 | 0.1954 | 0.2058 | 0.2162 |
| | h | | 1480.6 | 1510.0 | 1537.7 | 1564.4 | 1590.3 | 1615.6 | 1665.4 | 1714.5 | 1763.4 | 1812.4 | 1861.7 |
| | s | | 5.0889 | 5.1840 | 5.2713 | 5.3525 | 5.4292 | 5.5021 | 5.6392 | 5.7674 | 5.8888 | 6.0047 | 6.1159 |
| 1200 (30.96) | v | | | 0.1129 | 0.1185 | 0.1238 | 0.1289 | 0.1338 | 0.1434 | 0.1526 | 0.1616 | 0.1705 | 0.1792 |
| | h | | | 1497.1 | 1526.6 | 1554.7 | 1581.7 | 1608.0 | 1659.2 | 1709.2 | 1758.9 | 1808.5 | 1858.2 |
| | s | | | 5.0629 | 5.1560 | 5.2416 | 5.3215 | 5.3970 | 5.5379 | 5.6687 | 5.7919 | 5.9091 | 6.0214 |

**Table A.2.2** (Continued)
Superheated Ammonia

| Abs. Press. kPa (Sat. Temp.) °C | | Temperature, °C | | | | | | | | | | | |
|---|---|---|---|---|---|---|---|---|---|---|---|---|---|
| | | 20 | 30 | 40 | 50 | 60 | 70 | 80 | 100 | 120 | 140 | 160 | 180 |
| 1400 (36.28) | v | | | 0.0944 | 0.0995 | 0.1042 | 0.1088 | 0.1132 | 0.1216 | 0.1297 | 0.1376 | 0.1452 | 0.1528 |
| | h | | | 1483.4 | 1515.1 | 1544.7 | 1573.0 | 1600.2 | 1652.8 | 1703.9 | 1754.3 | 1804.5 | 1854.7 |
| | s | | | 4.9534 | 5.0530 | 5.1434 | 5.2270 | 5.3053 | 5.4501 | 5.5836 | 5.7087 | 5.8273 | 5.9406 |
| 1600 (41.05) | v | | | | 0.0851 | 0.0895 | 0.0937 | 0.0977 | 0.1053 | 0.1125 | 0.1195 | 0.1263 | 0.1330 |
| | h | | | | 1502.9 | 1534.4 | 1564.0 | 1592.3 | 1646.4 | 1698.5 | 1749.7 | 1800.5 | 1851.2 |
| | s | | | | 4.9584 | 5.0543 | 5.1419 | 5.2232 | 5.3722 | 5.5084 | 5.6355 | 5.7555 | 5.8699 |
| 1800 (45.39) | v | | | | 0.0739 | 0.0781 | 0.0820 | 0.0856 | 0.0926 | 0.0992 | 0.1055 | 0.1116 | 0.1177 |
| | h | | | | 1490.0 | 1523.5 | 1554.6 | 1584.1 | 1639.8 | 1693.1 | 1745.1 | 1796.5 | 1847.7 |
| | s | | | | 4.8693 | 4.9715 | 5.0635 | 5.1482 | 5.3018 | 5.4409 | 5.5699 | 5.6914 | 5.8069 |
| 2000 (49.38) | v | | | | 0.0648 | 0.0688 | 0.0725 | 0.0760 | 0.0824 | 0.0885 | 0.0943 | 0.0999 | 0.1054 |
| | h | | | | 1476.1 | 1512.0 | 1544.9 | 1575.6 | 1633.2 | 1687.6 | 1740.4 | 1792.4 | 1844.1 |
| | s | | | | 4.7834 | 4.8930 | 4.9902 | 5.0786 | 5.2371 | 5.3793 | 5.5104 | 5.6333 | 5.7499 |

# Table A.3
Thermodynamic Properties of Freon-12 (Dichlorodifluoromethane)[a]

# Table A.3.1
Saturated Freon-12

| Temp. °C | Abs. Press. MPa $P$ | Specific Volume m³/kg | | | Enthalpy kJ/kg | | | Entropy kJ/kg K | | |
|---|---|---|---|---|---|---|---|---|---|---|
| | | Sat. Liquid $v_f$ | Evap. $v_{fg}$ | Sat. Vapor $v_g$ | Sat. Liquid $h_f$ | Evap. $h_{fg}$ | Sat. Vapor $h_g$ | Sat. Liquid $s_f$ | Evap. $s_{fg}$ | Sat. Vapor $s_g$ |
| −90 | 0.0028 | 0.000 608 | 4.414 937 | 4.415 545 | −43.243 | 189.618 | 146.375 | −0.2084 | 1.0352 | 0.8268 |
| −85 | 0.0042 | 0.000 612 | 3.036 704 | 3.037 316 | −38.968 | 187.608 | 148.640 | −0.1854 | 0.9970 | 0.8116 |
| −80 | 0.0062 | 0.000 617 | 2.137 728 | 2.138 345 | −34.688 | 185.612 | 150.924 | −0.1630 | 0.9609 | 0.7979 |
| −75 | 0.0088 | 0.000 622 | 1.537 030 | 1.537 651 | −30.401 | 183.625 | 153.224 | −0.1411 | 0.9266 | 0.7855 |
| −70 | 0.0123 | 0.000 627 | 1.126 654 | 1.127 280 | −26.103 | 181.640 | 155.536 | −0.1197 | 0.8940 | 0.7744 |
| −65 | 0.0168 | 0.000 632 | 0.840 534 | 0.841 166 | −21.793 | 179.651 | 157.857 | −0.0987 | 0.8630 | 0.7643 |
| −60 | 0.0226 | 0.000 637 | 0.637 274 | 0.637 910 | −17.469 | 177.653 | 160.184 | −0.0782 | 0.8334 | 0.7552 |
| −55 | 0.0300 | 0.000 642 | 0.490 358 | 0.491 000 | −13.129 | 175.641 | 162.512 | −0.0581 | 0.8051 | 0.7470 |
| −50 | 0.0391 | 0.000 648 | 0.382 457 | 0.383 105 | −8.772 | 173.611 | 164.840 | −0.0384 | 0.7779 | 0.7396 |
| −45 | 0.0504 | 0.000 654 | 0.302 029 | 0.302 682 | −4.396 | 171.558 | 167.163 | −0.0190 | 0.7519 | 0.7329 |
| −40 | 0.0642 | 0.000 659 | 0.241 251 | 0.241 910 | −0.000 | 169.479 | 169.479 | −0.0000 | 0.7269 | 0.7269 |
| −35 | 0.0807 | 0.000 666 | 0.194 732 | 0.195 398 | 4.416 | 167.368 | 171.784 | 0.0187 | 0.7027 | 0.7214 |
| −30 | 0.1004 | 0.000 672 | 0.158 703 | 0.159 375 | 8.854 | 165.222 | 174.076 | 0.0371 | 0.6795 | 0.7165 |
| −25 | 0.1237 | 0.000 679 | 0.130 487 | 0.131 166 | 13.315 | 163.037 | 176.352 | 0.0552 | 0.6570 | 0.7121 |
| −20 | 0.1509 | 0.000 685 | 0.108 162 | 0.108 847 | 17.800 | 160.810 | 178.610 | 0.0730 | 0.6352 | 0.7082 |
| −15 | 0.1826 | 0.000 693 | 0.090 326 | 0.091 018 | 22.312 | 158.534 | 180.846 | 0.0906 | 0.6141 | 0.7046 |
| −10 | 0.2191 | 0.000 700 | 0.075 946 | 0.076 646 | 26.851 | 156.207 | 183.058 | 0.1079 | 0.5936 | 0.7014 |
| −5 | 0.2610 | 0.000 708 | 0.064 255 | 0.064 963 | 31.420 | 153.823 | 185.243 | 0.1250 | 0.5736 | 0.6986 |
| 0 | 0.3086 | 0.000 716 | 0.054 673 | 0.055 389 | 36.022 | 151.376 | 187.397 | 0.1418 | 0.5542 | 0.6960 |
| 5 | 0.3626 | 0.000 724 | 0.046 761 | 0.047 485 | 40.659 | 148.859 | 189.518 | 0.1585 | 0.5351 | 0.6937 |

**Table A.3.1** (Continued)

| Temp. °C | Abs. Press. MPa $P$ | Specific Volume m³/kg | | | Enthalpy kJ/kg | | | Entropy kJ/kg K | | |
|---|---|---|---|---|---|---|---|---|---|---|
| | | Sat. Liquid $v_f$ | Evap. $v_{fg}$ | Sat. Vapor $v_g$ | Sat. Liquid $h_f$ | Evap. $h_{fg}$ | Sat. Vapor $h_g$ | Sat. Liquid $s_f$ | Evap. $s_{fg}$ | Sat. Vapor $s_g$ |
| 10 | 0.4233 | 0.000 733 | 0.040 180 | 0.040 914 | 45.337 | 146.265 | 191.602 | 0.1750 | 0.5165 | 0.6916 |
| 15 | 0.4914 | 0.000 743 | 0.034 671 | 0.035 413 | 50.058 | 143.586 | 193.644 | 0.1914 | 0.4983 | 0.6897 |
| 20 | 0.5673 | 0.000 752 | 0.030 028 | 0.030 780 | 54.828 | 140.812 | 195.641 | 0.2076 | 0.4803 | 0.6879 |
| 25 | 0.6516 | 0.000 763 | 0.026 091 | 0.026 854 | 59.653 | 137.933 | 197.586 | 0.2237 | 0.4626 | 0.6863 |
| 30 | 0.7449 | 0.000 774 | 0.022 734 | 0.023 508 | 64.539 | 134.936 | 199.475 | 0.2397 | 0.4451 | 0.6848 |
| 35 | 0.8477 | 0.000 786 | 0.019 855 | 0.020 641 | 69.494 | 131.805 | 201.299 | 0.2557 | 0.4277 | 0.6834 |
| 40 | 0.9607 | 0.000 798 | 0.017 373 | 0.018 171 | 74.527 | 128.525 | 203.051 | 0.2716 | 0.4104 | 0.6820 |
| 45 | 1.0843 | 0.000 811 | 0.015 220 | 0.016 032 | 79.647 | 125.074 | 204.722 | 0.2875 | 0.3931 | 0.6806 |
| 50 | 1.2193 | 0.000 826 | 0.013 344 | 0.014 170 | 84.868 | 121.430 | 206.298 | 0.3034 | 0.3758 | 0.6792 |
| 55 | 1.3663 | 0.000 841 | 0.011 701 | 0.012 542 | 90.201 | 117.565 | 207.766 | 0.3194 | 0.3582 | 0.6777 |
| 60 | 1.5259 | 0.000 858 | 0.010 253 | 0.011 111 | 95.665 | 113.443 | 209.109 | 0.3355 | 0.3405 | 0.6760 |
| 65 | 1.6988 | 0.000 877 | 0.008 971 | 0.009 847 | 101.279 | 109.024 | 210.303 | 0.3518 | 0.3224 | 0.6742 |
| 70 | 1.8858 | 0.000 897 | 0.007 828 | 0.008 725 | 107.067 | 104.255 | 211.321 | 0.3683 | 0.3038 | 0.6721 |
| 75 | 2.0874 | 0.000 920 | 0.006 802 | 0.007 723 | 113.058 | 99.068 | 212.126 | 0.3851 | 0.2845 | 0.6697 |
| 80 | 2.3046 | 0.000 946 | 0.005 875 | 0.006 821 | 119.291 | 93.373 | 212.665 | 0.4023 | 0.2644 | 0.6667 |
| 85 | 2.5380 | 0.000 976 | 0.005 029 | 0.006 005 | 125.818 | 87.047 | 212.865 | 0.4201 | 0.2430 | 0.6631 |
| 90 | 2.7885 | 0.001 012 | 0.004 246 | 0.005 258 | 132.708 | 79.907 | 212.614 | 0.4385 | 0.2200 | 0.6585 |
| 95 | 3.0569 | 0.001 056 | 0.003 508 | 0.004 563 | 140.068 | 71.658 | 211.726 | 0.4579 | 0.1946 | 0.6526 |
| 100 | 3.3440 | 0.001 113 | 0.002 790 | 0.003 903 | 148.076 | 61.768 | 209.843 | 0.4788 | 0.1655 | 0.6444 |

| 105 | 3.6509 | 0.001 197 | 0.002 045 | 0.003 242 | 157.085 | 49.014 | 206.099 | 0.5023 | 0.1296 | 0.6319 |
| 110 | 3.9784 | 0.001 364 | 0.001 098 | 0.002 462 | 168.059 | 28.425 | 196.484 | 0.5322 | 0.0742 | 0.6064 |
| 112 | 4.1155 | 0.001 792 | 0.000 005 | 0.001 797 | 174.920 | 0.151 | 175.071 | 0.5651 | 0.0004 | 0.5655 |

[a]Copyright 1955 and 1956, E. I. du Pont de Nemours & Company, Inc. Reprinted by permission. Adapted from English units.

# Table A.3.2
## Superheated Freon-12

| Temp. °C | 0.05 MPa | | | 0.10 MPa | | | 0.15 MPa | | |
|---|---|---|---|---|---|---|---|---|---|
| | $v$ m³/kg | $h$ kJ/kg | $s$ kJ/kg K | $v$ m³/kg | $h$ kJ/kg | $s$ kJ/kg K | $v$ m³/kg | $h$ kJ/kg | $s$ kJ/kg K |
| −20.0 | 0.341 857 | 181.042 | 0.7912 | 0.167 701 | 179.861 | 0.7401 | 0.114 716 | 184.619 | 0.7318 |
| −10.0 | 0.356 227 | 186.757 | 0.8133 | 0.175 222 | 185.707 | 0.7628 | 0.119 866 | 190.660 | 0.7543 |
| 0.0 | 0.370 508 | 192.567 | 0.8350 | 0.182 647 | 191.628 | 0.7849 | 0.124 932 | 196.762 | 0.7763 |
| 10.0 | 0.384 716 | 198.471 | 0.8562 | 0.189 994 | 197.628 | 0.8064 | 0.129 930 | 202.927 | 0.7977 |
| 20.0 | 0.398 863 | 204.469 | 0.8770 | 0.197 277 | 203.707 | 0.8275 | 0.134 873 | 209.160 | 0.8186 |
| 30.0 | 0.412 959 | 210.557 | 0.8974 | 0.204 506 | 209.866 | 0.8482 | 0.139 768 | 215.463 | 0.8390 |
| 40.0 | 0.427 012 | 216.733 | 0.9175 | 0.211 691 | 216.104 | 0.8684 | 0.144 625 | 221.835 | 0.8591 |
| 50.0 | 0.441 030 | 222.997 | 0.9372 | 0.218 839 | 222.421 | 0.8883 | 0.149 450 | 228.277 | 0.8787 |
| 60.0 | 0.455 017 | 229.344 | 0.9565 | 0.225 955 | 228.815 | 0.9078 | 0.154 247 | 234.789 | 0.8980 |
| 70.0 | 0.468 978 | 235.774 | 0.9755 | 0.233 044 | 235.285 | 0.9269 | 0.159 020 | 241.371 | 0.9169 |
| 80.0 | 0.482 917 | 242.282 | 0.9942 | 0.240 111 | 241.829 | 0.9457 | 0.163 774 | 248.020 | 0.9354 |
| 90.0 | 0.496 838 | 248.868 | 1.0126 | 0.247 159 | 248.446 | 0.9642 | | | |

| Temp. °C | 0.20 MPa | | | 0.25 MPa | | | 0.30 MPa | | |
|---|---|---|---|---|---|---|---|---|---|
| | $v$ m³/kg | $h$ kJ/kg | $s$ kJ/kg K | $v$ m³/kg | $h$ kJ/kg | $s$ kJ/kg K | $v$ m³/kg | $h$ kJ/kg | $s$ kJ/kg K |
| 0.0 | 0.088 608 | 189.669 | 0.7320 | 0.069 752 | 188.644 | 0.7139 | 0.057 150 | 187.583 | 0.6984 |
| 10.0 | 0.092 550 | 195.878 | 0.7543 | 0.073 024 | 194.969 | 0.7366 | 0.059 984 | 194.034 | 0.7216 |
| 20.0 | 0.096 418 | 202.135 | 0.7760 | 0.076 218 | 201.322 | 0.7587 | 0.062 734 | 200.490 | 0.7440 |
| 30.0 | 0.100 228 | 208.446 | 0.7972 | 0.079 350 | 207.715 | 0.7801 | 0.065 418 | 206.969 | 0.7658 |
| 40.0 | 0.103 989 | 214.814 | 0.8178 | 0.082 431 | 214.153 | 0.8010 | 0.068 049 | 213.480 | 0.7869 |
| 50.0 | 0.107 710 | 221.243 | 0.8381 | 0.085 470 | 220.642 | 0.8214 | 0.070 635 | 220.030 | 0.8075 |
| 60.0 | 0.111 397 | 227.735 | 0.8578 | 0.088 474 | 227.185 | 0.8413 | 0.073 185 | 226.627 | 0.8276 |
| 70.0 | 0.115 055 | 234.291 | 0.8772 | 0.091 449 | 233.785 | 0.8608 | 0.075 705 | 233.273 | 0.8473 |
| 80.0 | 0.118 690 | 240.910 | 0.8962 | 0.094 398 | 240.443 | 0.8800 | 0.078 200 | 239.971 | 0.8665 |

| T (°C) | | | | | | | | | |
|---|---|---|---|---|---|---|---|---|---|
| 90.0 | 0.122 304 | 247.593 | 0.9149 | 0.097 327 | 247.160 | 0.8987 | 0.080 673 | 246.723 | 0.8853 |
| 100.0 | 0.125 901 | 254.339 | 0.9332 | 0.100 238 | 253.936 | 0.9171 | 0.083 127 | 253.530 | 0.9038 |
| 110.0 | 0.129 483 | 261.147 | 0.9512 | 0.103 134 | 260.770 | 0.9352 | 0.085 566 | 260.391 | 0.9220 |
| | 0.40 MPa | | | 0.50 MPa | | | 0.60 MPa | | |
| 20.0 | 0.045 836 | 198.762 | 0.7199 | 0.035 646 | 196.935 | 0.6999 | 0.030 422 | 202.116 | 0.7063 |
| 30.0 | 0.047 971 | 205.428 | 0.7423 | 0.037 464 | 203.814 | 0.7230 | 0.031 966 | 209.154 | 0.7291 |
| 40.0 | 0.050 046 | 212.095 | 0.7639 | 0.039 214 | 210.656 | 0.7452 | 0.033 450 | 216.141 | 0.7511 |
| 50.0 | 0.052 072 | 218.779 | 0.7849 | 0.040 911 | 217.484 | 0.7667 | 0.034 887 | 223.104 | 0.7723 |
| 60.0 | 0.054 059 | 225.488 | 0.8054 | 0.042 565 | 224.315 | 0.7875 | 0.036 285 | 230.062 | 0.7929 |
| 70.0 | 0.056 014 | 232.230 | 0.8253 | 0.044 184 | 231.161 | 0.8077 | 0.037 653 | 237.027 | 0.8129 |
| 80.0 | 0.057 941 | 239.012 | 0.8448 | 0.045 774 | 238.031 | 0.8275 | 0.038 995 | 244.009 | 0.8324 |
| 90.0 | 0.059 846 | 245.837 | 0.8638 | 0.047 340 | 244.932 | 0.8467 | 0.040 316 | 251.016 | 0.8514 |
| 100.0 | 0.061 731 | 252.707 | 0.8825 | 0.048 886 | 251.869 | 0.8656 | 0.041 619 | 258.053 | 0.8700 |
| 110.0 | 0.063 600 | 259.624 | 0.9008 | 0.050 415 | 258.845 | 0.8840 | 0.042 907 | 265.124 | 0.8882 |
| 120.0 | 0.065 455 | 266.590 | 0.9187 | 0.051 929 | 265.862 | 0.9021 | 0.044 181 | 272.231 | 0.9061 |
| 130.0 | 0.067 298 | 273.605 | 0.9364 | 0.053 430 | 272.923 | 0.9198 | | | |

| T (°C) | | | | | | | | | |
|---|---|---|---|---|---|---|---|---|---|
| | 0.70 MPa | | | 0.80 MPa | | | 0.90 MPa | | |
| 40.0 | 0.026 761 | 207.580 | 0.7148 | 0.022 830 | 205.924 | 0.7016 | 0.019 744 | 204.170 | 0.6982 |
| 50.0 | 0.028 100 | 214.745 | 0.7373 | 0.024 068 | 213.290 | 0.7248 | 0.020 912 | 211.765 | 0.7131 |
| 60.0 | 0.029 387 | 221.854 | 0.7590 | 0.025 247 | 220.558 | 0.7469 | 0.022 012 | 219.212 | 0.7358 |
| 70.0 | 0.030 632 | 228.931 | 0.7799 | 0.026 380 | 227.766 | 0.7682 | 0.023 062 | 226.564 | 0.7575 |
| 80.0 | 0.031 843 | 235.997 | 0.8002 | 0.027 477 | 234.941 | 0.7888 | 0.024 072 | 233.856 | 0.7785 |
| 90.0 | 0.033 027 | 243.066 | 0.8199 | 0.028 545 | 242.101 | 0.8088 | 0.025 051 | 241.113 | 0.7987 |
| 100.0 | 0.034 189 | 250.146 | 0.8392 | 0.029 588 | 249.260 | 0.8283 | 0.026 005 | 248.355 | 0.8184 |
| 110.0 | 0.035 332 | 257.247 | 0.8579 | 0.030 612 | 256.428 | 0.8472 | 0.026 937 | 255.593 | 0.8376 |
| 120.0 | 0.036 458 | 264.374 | 0.8763 | 0.031 619 | 263.613 | 0.8657 | 0.027 851 | 262.839 | 0.8562 |
| 130.0 | 0.037 572 | 271.531 | 0.8943 | 0.032 612 | 270.820 | 0.8838 | 0.028 751 | 270.100 | 0.8745 |
| 140.0 | 0.038 673 | 278.720 | 0.9119 | 0.033 592 | 278.055 | 0.9016 | 0.029 639 | 277.381 | 0.8923 |
| 150.0 | 0.039 764 | 285.946 | 0.9292 | 0.034 563 | 285.320 | 0.9189 | 0.030 515 | 284.687 | 0.9098 |

**Table A.3.2** (Continued)
Superheated Freon-12

| Temp. °C | 1.00 MPa $v$ m³/kg | 1.00 MPa $h$ kJ/kg | 1.00 MPa $s$ kJ/kg K | 1.20 MPa $v$ m³/kg | 1.20 MPa $h$ kJ/kg | 1.20 MPa $s$ kJ/kg K | 1.40 MPa $v$ m³/kg | 1.40 MPa $h$ kJ/kg | 1.40 MPa $s$ kJ/kg K |
|---|---|---|---|---|---|---|---|---|---|
| 50.0 | 0.018 366 | 210.162 | 0.7021 | 0.014 483 | 206.661 | 0.6812 | | | |
| 60.0 | 0.019 410 | 217.810 | 0.7254 | 0.015 463 | 214.805 | 0.7060 | 0.012 579 | 211.457 | 0.6876 |
| 70.0 | 0.020 397 | 225.319 | 0.7476 | 0.016 368 | 222.687 | 0.7293 | 0.013 448 | 219.822 | 0.7123 |
| 80.0 | 0.021 341 | 232.739 | 0.7689 | 0.017 221 | 230.398 | 0.7514 | 0.014 247 | 227.891 | 0.7355 |
| 90.0 | 0.022 251 | 240.101 | 0.7895 | 0.018 032 | 237.995 | 0.7727 | 0.014 997 | 235.766 | 0.7575 |
| 100.0 | 0.023 133 | 247.430 | 0.8094 | 0.018 812 | 245.518 | 0.7931 | 0.015 710 | 243.512 | 0.7785 |
| 110.0 | 0.023 993 | 254.743 | 0.8287 | 0.019 567 | 252.993 | 0.8129 | 0.016 393 | 251.170 | 0.7988 |
| 120.0 | 0.024 835 | 262.053 | 0.8475 | 0.020 301 | 260.441 | 0.8320 | 0.017 053 | 258.770 | 0.8183 |
| 130.0 | 0.025 661 | 269.369 | 0.8659 | 0.021 018 | 267.875 | 0.8507 | 0.017 695 | 266.334 | 0.8373 |
| 140.0 | 0.026 474 | 276.699 | 0.8839 | 0.021 721 | 275.307 | 0.8689 | 0.018 321 | 273.877 | 0.8558 |
| 150.0 | 0.027 275 | 284.047 | 0.9015 | 0.022 412 | 282.745 | 0.8867 | 0.018 934 | 281.411 | 0.8738 |
| 160.0 | 0.028 068 | 291.419 | 0.9187 | 0.023 093 | 290.195 | 0.9041 | 0.019 535 | 288.946 | 0.8914 |

| Temp. °C | 1.60 MPa $v$ m³/kg | 1.60 MPa $h$ kJ/kg | 1.60 MPa $s$ kJ/kg K | 1.80 MPa $v$ m³/kg | 1.80 MPa $h$ kJ/kg | 1.80 MPa $s$ kJ/kg K | 2.00 MPa $v$ m³/kg | 2.00 MPa $h$ kJ/kg | 2.00 MPa $s$ kJ/kg K |
|---|---|---|---|---|---|---|---|---|---|
| 70.0 | 0.011 208 | 216.650 | 0.6959 | 0.009 406 | 213.049 | 0.6794 | | | |
| 80.0 | 0.011 984 | 225.177 | 0.7204 | 0.010 187 | 222.198 | 0.7057 | 0.008 704 | 218.859 | 0.6909 |
| 90.0 | 0.012 698 | 233.390 | 0.7433 | 0.010 884 | 230.835 | 0.7298 | 0.009 406 | 228.056 | 0.7166 |
| 100.0 | 0.013 366 | 241.397 | 0.7651 | 0.011 526 | 239.155 | 0.7524 | 0.010 035 | 236.760 | 0.7402 |
| 110.0 | 0.014 000 | 249.264 | 0.7859 | 0.012 126 | 247.264 | 0.7739 | 0.010 615 | 245.154 | 0.7624 |
| 120.0 | 0.014 608 | 257.035 | 0.8059 | 0.012 697 | 255.228 | 0.7944 | 0.011 159 | 253.341 | 0.7835 |
| 130.0 | 0.015 195 | 264.742 | 0.8253 | 0.013 244 | 263.094 | 0.8141 | 0.011 676 | 261.384 | 0.8037 |
| 140.0 | 0.015 765 | 272.406 | 0.8440 | 0.013 772 | 270.891 | 0.8332 | 0.012 172 | 269.327 | 0.8232 |
| 150.0 | 0.016 320 | 280.044 | 0.8623 | 0.014 284 | 278.642 | 0.8518 | 0.012 651 | 277.201 | 0.8420 |

## 3.50 MPa

| T (°C) | v | h | s |
|---|---|---|---|
| 110.0 | 0.004 324 | 222.121 | 0.6750 |
| 120.0 | 0.004 959 | 234.875 | 0.7078 |
| 130.0 | 0.005 456 | 245.661 | 0.7349 |
| 140.0 | 0.005 884 | 255.524 | 0.7591 |
| 150.0 | 0.006 270 | 264.846 | 0.7814 |
| 160.0 | 0.006 626 | 273.817 | 0.8023 |
| 170.0 | 0.006 961 | 282.545 | 0.8222 |
| 180.0 | 0.007 279 | 291.100 | 0.8413 |
| 190.0 | 0.007 584 | 299.528 | 0.8597 |
| 200.0 | 0.007 878 | 307.864 | 0.8775 |

## 3.00 MPa

| T (°C) | v | h | s |
|---|---|---|---|
| 100.0 | 0.005 231 | 220.529 | 0.6770 |
| 110.0 | 0.005 886 | 232.068 | 0.7075 |
| 120.0 | 0.006 419 | 242.208 | 0.7336 |
| 130.0 | 0.006 887 | 251.632 | 0.7573 |
| 140.0 | 0.007 313 | 260.620 | 0.7793 |
| 150.0 | 0.007 709 | 269.319 | 0.8001 |
| 160.0 | 0.008 083 | 277.817 | 0.8200 |
| 170.0 | 0.008 439 | 286.171 | 0.8391 |
| 180.0 | 0.008 782 | 294.422 | 0.8575 |
| 190.0 | 0.009 114 | 302.597 | 0.8753 |
| 200.0 | 0.009 436 | 310.718 | 0.8927 |

## 2.50 MPa

| T (°C) | v | h | s |
|---|---|---|---|
| 90.0 | 0.006 595 | 219.562 | 0.6823 |
| 100.0 | 0.007 264 | 229.852 | 0.7103 |
| 110.0 | 0.007 837 | 239.271 | 0.7352 |
| 120.0 | 0.008 351 | 248.192 | 0.7582 |
| 130.0 | 0.008 827 | 256.794 | 0.7798 |
| 140.0 | 0.009 273 | 265.180 | 0.8003 |
| 150.0 | 0.009 697 | 273.414 | 0.8200 |
| 160.0 | 0.010 104 | 281.540 | 0.8390 |
| 170.0 | 0.010 497 | 289.589 | 0.8574 |
| 180.0 | 0.010 879 | 297.583 | 0.8752 |
| 190.0 | 0.011 250 | 305.540 | 0.8926 |
| 200.0 | 0.011 614 | 313.472 | 0.9095 |

## 4.00 MPa

| T (°C) | v | h | s |
|---|---|---|---|
| 120.0 | 0.003 736 | 224.863 | 0.6771 |
| 130.0 | 0.004 325 | 238.443 | 0.7111 |
| 140.0 | 0.004 781 | 249.703 | 0.7386 |
| 150.0 | 0.005 172 | 259.904 | 0.7630 |
| 160.0 | 0.005 522 | 269.492 | 0.7854 |
| 170.0 | 0.005 845 | 278.684 | 0.8063 |
| 180.0 | 0.006 147 | 287.602 | 0.8262 |
| 190.0 | 0.006 434 | 296.326 | 0.8453 |
| 200.0 | 0.006 708 | 304.906 | 0.8636 |
| 210.0 | 0.006 972 | 313.380 | 0.8813 |
| 220.0 | 0.007 228 | 321.774 | 0.8985 |
| 230.0 | 0.007 477 | 330.108 | 0.9152 |

### Continuation rows (T = 160.0, 170.9, 180.0)

| T (°C) | v | h | s |
|---|---|---|---|
| 160.0 | 0.013 116 | 285.027 | 0.8603 |
| 170.9 | 0.013 570 | 292.822 | 0.8781 |
| 180.0 | 0.014 013 | 300.598 | 0.8955 |

| T (°C) | v | h | s |
|---|---|---|---|
| 160.0 | 0.014 784 | 286.364 | 0.8698 |
| 170.9 | 0.015 272 | 294.069 | 0.8874 |
| 180.0 | 0.015 752 | 301.767 | 0.9046 |

| T (°C) | v | h | s |
|---|---|---|---|
| 160.0 | 0.016 864 | 287.669 | 0.8801 |
| 170.9 | 0.017 398 | 295.290 | 0.8975 |
| 180.0 | 0.017 923 | 302.914 | 0.9145 |

# Table A.4
## Thermodynamic Properties of Oxygen[a]
# Table A.4.1
## Saturated Oxygen

| Temp. K | Press. MPa P | Specific Volume m³/kg | | | Enthalpy kJ/kg | | | Entropy kJ/kg K | | |
|---|---|---|---|---|---|---|---|---|---|---|
| | | Sat. Liquid $v_f$ | Evap. $v_{fg}$ | Sat. Vapor $v_g$ | Sat. Liquid $h_f$ | Evap. $h_{fg}$ | Sat. Vapor $h_g$ | Sat. Liquid $s_f$ | Evap. $s_{fg}$ | Sat. Vapor $s_g$ |
| 54.3507 | 0.00015 | 0.000 765 | 92.9658 | 92.9666 | −193.432 | 242.553 | 49.121 | 2.0938 | 4.4514 | 6.5452 |
| 60 | 0.00073 | 0.000 780 | 21.3461 | 21.3469 | −184.029 | 238.265 | 54.236 | 2.2585 | 3.9686 | 6.2271 |
| 70 | 0.00623 | 0.000 808 | 2.9085 | 2.9093 | −167.372 | 230.527 | 63.155 | 2.5151 | 3.2936 | 5.8087 |
| 80 | 0.03006 | 0.000 840 | 0.681 04 | 0.681 88 | −150.646 | 222.289 | 71.643 | 2.7382 | 2.7779 | 5.5161 |
| 90 | 0.09943 | 0.000 876 | 0.226 49 | 0.227 36 | −133.758 | 213.070 | 79.312 | 2.9364 | 2.3663 | 5.3027 |
| 100 | 0.25425 | 0.000 917 | 0.094 645 | 0.095 562 | −116.557 | 202.291 | 85.734 | 3.1161 | 2.0222 | 5.1383 |
| 110 | 0.54339 | 0.000 966 | 0.045 855 | 0.046 821 | −98.829 | 189.320 | 90.491 | 3.2823 | 1.7210 | 5.0033 |
| 120 | 1.0215 | 0.001 027 | 0.024 336 | 0.025 363 | −80.219 | 173.310 | 93.091 | 3.4401 | 1.4445 | 4.8846 |
| 130 | 1.7478 | 0.001 108 | 0.013 488 | 0.014 596 | −60.093 | 152.887 | 92.794 | 3.5948 | 1.1766 | 4.7714 |
| 140 | 2.7866 | 0.001 230 | 0.007 339 | 0.008 569 | −37.045 | 125.051 | 88.006 | 3.7567 | 0.8935 | 4.6502 |
| 150 | 4.2190 | 0.001 480 | 0.003 180 | 0.004 660 | −7.038 | 79.459 | 72.421 | 3.9498 | 0.5301 | 4.4799 |
| 154.576 | 5.0427 | 0.002 293 | 0.000 000 | 0.002 293 | 32.257 | 0.000 | 32.257 | 4.1977 | 0.0000 | 4.1977 |

[a] Adapted from L. A. Weber, *Journal of Research of the National Bureau of Standards*, **74A**: 93 (1970).

**Table A.4.2**
Superheated Oxygen

| Temp. K | 0.10 MPa v m³/kg | h kJ/kg | s kJ/kg K | 0.20 MPa v m³/kg | h kJ/kg | s kJ/kg K | 0.50 MPa v m³/kg | h kJ/kg | s kJ/kg K |
|---|---|---|---|---|---|---|---|---|---|
| 100 | 0.253 503 | 88.828 | 5.4016 | 0.125 394 | 86.864 | 5.2083 | | | |
| 125 | 0.320 717 | 112.214 | 5.6107 | 0.158 268 | 110.988 | 5.4241 | 0.060 674 | 107.093 | 5.1650 |
| 150 | 0.386 914 | 135.301 | 5.7787 | 0.192 016 | 134.440 | 5.5947 | 0.075 039 | 131.788 | 5.3448 |
| 175 | 0.452 645 | 158.255 | 5.9202 | 0.225 276 | 157.609 | 5.7376 | 0.088 842 | 155.643 | 5.4919 |
| 200 | 0.518 127 | 181.145 | 6.0427 | 0.258 282 | 180.638 | 5.8609 | 0.102 371 | 179.105 | 5.6175 |
| 225 | 0.583 465 | 204.007 | 6.1502 | 0.291 140 | 203.596 | 5.9688 | 0.115 746 | 202.359 | 5.7268 |
| 250 | 0.648 711 | 226.869 | 6.2468 | 0.323 906 | 226.529 | 6.0657 | 0.129 025 | 225.506 | 5.8246 |
| 275 | 0.713 895 | 249.769 | 6.3369 | 0.356 610 | 249.483 | 6.1560 | 0.142 242 | 248.621 | 5.9156 |
| 300 | 0.779 036 | 272.720 | 6.4140 | 0.389 271 | 272.475 | 6.2332 | 0.155 415 | 271.740 | 5.9932 |

| Temp. K | 1.00 MPa v m³/kg | h kJ/kg | s kJ/kg K | 2.00 MPa v m³/kg | h kJ/kg | s kJ/kg K | 4.00 MPa v m³/kg | h kJ/kg | s kJ/kg K |
|---|---|---|---|---|---|---|---|---|---|
| 125 | 0.027 869 | 99.653 | 4.9431 | | | | | | |
| 150 | 0.035 976 | 127.112 | 5.1433 | 0.016 270 | 116.476 | 4.9130 | 0.005 526 | 81.481 | 4.5475 |
| 175 | 0.043 341 | 152.269 | 5.2986 | 0.020 544 | 145.112 | 5.0899 | 0.009 029 | 128.618 | 4.8414 |
| 200 | 0.050 394 | 176.508 | 5.4283 | 0.024 395 | 171.150 | 5.2293 | 0.011 376 | 159.715 | 5.0080 |
| 225 | 0.057 282 | 200.280 | 5.5401 | 0.028 051 | 196.052 | 5.3464 | 0.013 444 | 187.333 | 5.1380 |
| 250 | 0.064 068 | 223.795 | 5.6394 | 0.031 597 | 220.348 | 5.4491 | 0.015 378 | 213.374 | 5.2480 |
| 275 | 0.070 790 | 247.185 | 5.7314 | 0.035 073 | 244.309 | 5.5433 | 0.017 233 | 238.560 | 5.3469 |
| 300 | 0.077 467 | 270.516 | 5.8098 | 0.038 502 | 268.076 | 5.6263 | 0.019 039 | 263.234 | 5.4300 |

675

**Table A.4.2** (Continued)
Superheated Oxygen

| Temp. K | 6.00 MPa $v$ m³/kg | $h$ kJ/kg | $s$ kJ/kg K | 8.00 MPa $v$ m³/kg | $h$ kJ/kg | $s$ kJ/kg K | 10.00 MPa $v$ m³/kg | $h$ kJ/kg | $s$ kJ/kg K |
|---|---|---|---|---|---|---|---|---|---|
| 175 | 0.005 051 | 107.496 | 4.6431 | 0.003 002 | 79.513 | 4.4384 | 0.002 020 | 52.661 | 4.2573 |
| 200 | 0.007 027 | 147.232 | 4.8565 | 0.004 864 | 133.760 | 4.7308 | 0.003 603 | 119.767 | 4.6189 |
| 225 | 0.008 589 | 178.304 | 5.0029 | 0.006 181 | 169.069 | 4.8973 | 0.004 757 | 159.686 | 4.8072 |
| 250 | 0.009 991 | 206.340 | 5.1214 | 0.007 316 | 199.317 | 5.0251 | 0.005 730 | 192.401 | 4.9455 |
| 275 | 0.011 306 | 232.848 | 5.2253 | 0.008 360 | 227.219 | 5.1344 | 0.006 606 | 221.685 | 5.0572 |
| 300 | 0.012 570 | 258.464 | 5.3116 | 0.009 351 | 253.797 | 5.2240 | 0.007 432 | 249.262 | 5.1533 |

| Temp. K | 20.00 MPa $v$ m³/kg | $h$ kJ/kg | $s$ kJ/kg K |
|---|---|---|---|
| 175 | 0.001 343 | 24.551 | 4.0086 |
| 200 | 0.001 727 | 75.318 | 4.2798 |
| 225 | 0.002 236 | 122.595 | 4.5024 |
| 250 | 0.002 755 | 163.109 | 4.6739 |
| 275 | 0.003 241 | 198.021 | 4.8069 |
| 300 | 0.003 700 | 229.655 | 4.9174 |

# Table A.5
## Thermodynamic Properties of Nitrogen[a]
## Table A.5.1
## Saturated Nitrogen

| Temp. K | Press. MPa P | Specific Volume m³/kg | | | Enthalpy kJ/kg | | | Entropy kJ/kg K | | |
|---|---|---|---|---|---|---|---|---|---|---|
| | | Sat. Liquid $v_f$ | Evap. $v_{fg}$ | Sat. Vapor $v_g$ | Sat. Liquid $h_f$ | Evap. $h_{fg}$ | Sat. Vapor $h_g$ | Sat. Liquid $s_f$ | Evap. $s_{fg}$ | Sat. Vapor $s_g$ |
| 63.143 | 0.01253 | 0.001 152 | 1.480 060 | 1.481 212 | −150.348 | 215.188 | 64.840 | 2.4310 | 3.4076 | 5.8386 |
| 65 | 0.01742 | 0.001 162 | 1.093 173 | 1.094 335 | −146.691 | 213.291 | 66.600 | 2.4845 | 3.2849 | 5.7694 |
| 70 | 0.03858 | 0.001 189 | 0.525 785 | 0.526 974 | −136.569 | 207.727 | 71.158 | 2.6345 | 2.9703 | 5.6048 |
| 75 | 0.07612 | 0.001 221 | 0.280 970 | 0.282 191 | −126.287 | 201.662 | 75.375 | 2.7755 | 2.6915 | 5.4670 |
| 77.347 | 0.101325 | 0.001 237 | 0.215 504 | 0.216 741 | −121.433 | 198.645 | 77.212 | 2.8390 | 2.5706 | 5.4096 |
| 80 | 0.1370 | 0.001 256 | 0.162 794 | 0.164 050 | −115.926 | 195.089 | 79.163 | 2.9083 | 2.4409 | 5.3492 |
| 85 | 0.2291 | 0.001 296 | 0.100 434 | 0.101 730 | −105.461 | 187.892 | 82.431 | 3.0339 | 2.2122 | 5.2461 |
| 90 | 0.3608 | 0.001 340 | 0.064 950 | 0.066 290 | −94.817 | 179.894 | 85.077 | 3.1535 | 2.0001 | 5.1536 |
| 95 | 0.5411 | 0.001 392 | 0.043 504 | 0.044 896 | −83.895 | 170.877 | 86.982 | 3.2688 | 1.7995 | 5.0683 |
| 100 | 0.7790 | 0.001 452 | 0.029 861 | 0.031 313 | −72.571 | 160.562 | 87.991 | 3.3816 | 1.6060 | 4.9876 |
| 105 | 1.0843 | 0.001 524 | 0.020 745 | 0.022 269 | −60.691 | 148.573 | 87.882 | 3.4930 | 1.4150 | 4.9080 |
| 110 | 1.4673 | 0.001 613 | 0.014 402 | 0.016 015 | −48.027 | 134.319 | 86.292 | 3.6054 | 1.2209 | 4.8263 |
| 115 | 1.9395 | 0.001 797 | 0.009 696 | 0.011 493 | −34.157 | 116.701 | 82.544 | 3.7214 | 1.0145 | 4.7359 |
| 120 | 2.5135 | 0.001 904 | 0.006 130 | 0.008 034 | −18.017 | 93.092 | 75.075 | 3.8450 | 0.7803 | 4.6253 |
| 125 | 3.2079 | 0.002 323 | 0.002 568 | 0.004 891 | +6.202 | 50.114 | 56.316 | 4.0356 | 0.3989 | 4.4345 |
| 126.1 | 3.4000 | 0.003 184 | 0.000 000 | 0.003 184 | +30.791 | 0.000 | 30.791 | 4.2269 | 0.0000 | 4.2269 |

[a] Adapted from "Thermodynamic Properties of Nitrogen Including Liquid and Vapor Phases from 63 K to 2000 K with Pressures to 10,000 Bar," R. T. Jacobsen and R. R. Stewart, *Jour. of Phys. and Chem. Ref. Data*, **2:** 757–922 (1973).

677

# Table A.5.2
## Superheated Nitrogen

| Temp. K | $v$ m³/kg | $h$ kJ/kg | $s$ kJ/kg K | $v$ m³/kg | $h$ kJ/kg | $s$ kJ/kg K | $v$ m³/kg | $h$ kJ/kg | $s$ kJ/kg K |
|---|---|---|---|---|---|---|---|---|---|
| | 0.1 MPa | | | 0.2 MPa | | | 0.5 MPa | | |
| 100 | 0.290 978 | 101.965 | 5.6944 | 0.142 475 | 100.209 | 5.4767 | 0.055 520 | 94.345 | 5.1706 |
| 125 | 0.367 217 | 128.505 | 5.9313 | 0.181 711 | 127.371 | 5.7194 | 0.073 422 | 123.824 | 5.4343 |
| 150 | 0.442 619 | 154.779 | 6.1228 | 0.220 014 | 153.962 | 5.9132 | 0.090 150 | 151.470 | 5.6361 |
| 175 | 0.517 576 | 180.935 | 6.2841 | 0.257 890 | 180.314 | 6.0760 | 0.106 394 | 178.434 | 5.8025 |
| 200 | 0.592 288 | 207.029 | 6.4234 | 0.295 531 | 206.537 | 6.2160 | 0.122 394 | 205.063 | 5.9447 |
| 225 | 0.666 552 | 233.085 | 6.5460 | 0.332 841 | 232.690 | 6.3388 | 0.138 173 | 231.459 | 6.0690 |
| 250 | 0.741 375 | 259.122 | 6.6561 | 0.370 418 | 258.796 | 6.4491 | 0.154 006 | 257.828 | 6.1801 |
| 275 | 0.815 563 | 285.144 | 6.7550 | 0.407 619 | 284.876 | 6.5485 | 0.169 642 | 284.076 | 6.2800 |
| 300 | 0.890 205 | 311.158 | 6.8457 | 0.445 047 | 310.937 | 6.6393 | 0.185 346 | 310.273 | 6.3715 |

| Temp. K | $v$ m³/kg | $h$ kJ/kg | $s$ kJ/kg K | $v$ m³/kg | $h$ kJ/kg | $s$ kJ/kg K | $v$ m³/kg | $h$ kJ/kg | $s$ kJ/kg K |
|---|---|---|---|---|---|---|---|---|---|
| | 1.0 MPa | | | 2.0 MPa | | | 4.0 MPa | | |
| 125 | 0.033 065 | 117.422 | 5.1872 | 0.014 021 | 101.489 | 4.8878 | | | |
| 150 | 0.041 884 | 147.176 | 5.4042 | 0.019 546 | 137.916 | 5.1547 | 0.008 234 | 115.716 | 4.8384 |
| 175 | 0.050 125 | 175.255 | 5.5779 | 0.024 155 | 168.709 | 5.3449 | 0.011 186 | 154.851 | 5.0804 |
| 200 | 0.058 096 | 202.596 | 5.7237 | 0.028 436 | 197.609 | 5.4992 | 0.013 648 | 187.521 | 5.2553 |
| 225 | 0.065 875 | 229.526 | 5.8502 | 0.035 697 | 225.578 | 5.6309 | 0.015 894 | 217.757 | 5.3976 |
| 250 | 0.073 634 | 256.220 | 5.9632 | 0.036 557 | 253.032 | 5.7469 | 0.018 060 | 246.793 | 5.5202 |
| 275 | 0.081 260 | 282.720 | 6.0639 | 0.040 485 | 280.132 | 5.8501 | 0.020 133 | 275.056 | 5.6277 |
| 300 | 0.088 899 | 309.173 | 6.1563 | 0.044 398 | 307.014 | 5.9436 | 0.022 178 | 302.848 | 5.7248 |

| | 6.0 MPa | | | 8.0 MPa | | | 10.0 MPa | | |
|---|---|---|---|---|---|---|---|---|---|
| 150 | 0.004 413 | 87.090 | 4.5667 | 0.002 917 | 61.903 | 4.3518 | 0.002 388 | 48.687 | 4.2287 |
| 175 | 0.006 913 | 140.183 | 4.8966 | 0.004 863 | 125.536 | 4.7470 | 0.003 750 | 112.489 | 4.6239 |
| 200 | 0.008 772 | 177.447 | 5.0961 | 0.006 390 | 167.680 | 4.9726 | 0.005 016 | 158.578 | 4.8709 |
| 225 | 0.010 396 | 210.139 | 5.2410 | 0.007 691 | 202.867 | 5.1384 | 0.006 104 | 196.079 | 5.0474 |
| 250 | 0.011 934 | 240.806 | 5.3796 | 0.008 903 | 235.141 | 5.2750 | 0.007 112 | 229.861 | 5.1900 |
| 275 | 0.013 383 | 270.222 | 5.4917 | 0.010 034 | 265.676 | 5.3910 | 0.008 046 | 261.450 | 5.3103 |
| 300 | 0.014 800 | 298.907 | 5.5916 | 0.011 133 | 295.219 | 5.4942 | 0.008 950 | 291.800 | 5.4163 |

| | 15.0 MPa | | | 20.0 MPa | | |
|---|---|---|---|---|---|---|
| 150 | 0.001 956 | 36.922 | 4.0798 | 0.001 781 | 33.637 | 3.9956 |
| 175 | 0.002 603 | 92.284 | 4.4213 | 0.002 186 | 83.453 | 4.3029 |
| 200 | 0.003 369 | 140.886 | 4.6813 | 0.002 685 | 130.291 | 4.5535 |
| 225 | 0.004 106 | 182.034 | 4.8752 | 0.003 208 | 172.307 | 4.7511 |
| 250 | 0.004 808 | 218.710 | 5.0303 | 0.003 728 | 210.456 | 4.9127 |
| 275 | 0.005 461 | 252.465 | 5.1845 | 0.004 223 | 245.640 | 5.0467 |
| 300 | 0.006 091 | 284.523 | 5.2707 | 0.004 704 | 278.942 | 5.1629 |

## Table A.6
### Thermodynamic Properties of Saturated Mercury[a]

| Press., MPa | Temp., °C | Enthalpy, kJ/kg Sat. Liquid | Enthalpy, kJ/kg Evap. | Enthalpy, kJ/kg Sat. Vapor | Entropy, kJ/kg K Sat. Liquid | Entropy, kJ/kg K Evap. | Entropy, kJ/kg K Sat. Vapor | Specific Volume Sat. Vapor, m³/kg |
|---|---|---|---|---|---|---|---|---|
| 0.000 06 | 109.2 | 15.13 | 297.20 | 312.33 | 0.0466 | 0.7774 | 0.8240 | 259.6 |
| 0.000 07 | 112.3 | 15.55 | 297.14 | 312.69 | 0.0477 | 0.7709 | 0.8186 | 224.3 |
| 0.000 08 | 115.0 | 15.93 | 297.09 | 313.02 | 0.0487 | 0.7654 | 0.8141 | 197.7 |
| 0.000 09 | 117.5 | 16.27 | 297.04 | 313.31 | 0.0496 | 0.7604 | 0.8100 | 176.8 |
| 0.000 10 | 119.7 | 16.58 | 297.00 | 313.58 | 0.0503 | 0.7560 | 0.8063 | 160.1 |
| 0.0002 | 134.9 | 18.67 | 296.71 | 315.38 | 0.0556 | 0.7271 | 0.7827 | 83.18 |
| 0.0004 | 151.5 | 20.93 | 296.40 | 317.33 | 0.0610 | 0.6981 | 0.7591 | 43.29 |
| 0.0006 | 161.8 | 22.33 | 296.21 | 318.54 | 0.0643 | 0.6811 | 0.7454 | 29.57 |
| 0.0008 | 169.4 | 23.37 | 296.06 | 319.43 | 0.0666 | 0.6690 | 0.7356 | 22.57 |
| 0.0010 | 175.5 | 24.21 | 295.95 | 320.16 | 0.0685 | 0.6596 | 0.7281 | 18.31 |
| 0.002 | 195.6 | 26.94 | 295.57 | 322.51 | 0.0744 | 0.6305 | 0.7049 | 9.570 |
| 0.004 | 217.7 | 29.92 | 295.15 | 325.07 | 0.0806 | 0.6013 | 0.6819 | 5.013 |
| 0.006 | 231.6 | 31.81 | 294.89 | 326.70 | 0.0843 | 0.5842 | 0.6685 | 3.438 |
| 0.008 | 242.0 | 33.21 | 294.70 | 327.91 | 0.0870 | 0.5721 | 0.6591 | 2.632 |
| 0.010 | 250.3 | 34.33 | 294.54 | 328.87 | 0.0892 | 0.5627 | 0.6519 | 2.140 |
| 0.02 | 278.1 | 38.05 | 294.02 | 332.07 | 0.0961 | 0.5334 | 0.6295 | 1.128 |
| 0.04 | 309.1 | 42.21 | 293.43 | 335.64 | 0.1034 | 0.5039 | 0.6073 | 0.5942 |
| 0.06 | 329.0 | 44.85 | 293.06 | 337.91 | 0.1078 | 0.4869 | 0.5947 | 0.4113 |
| 0.08 | 343.9 | 46.84 | 292.78 | 339.62 | 0.1110 | 0.4745 | 0.5855 | 0.3163 |
| 0.1 | 356.1 | 48.45 | 292.55 | 341.00 | 0.1136 | 0.4649 | 0.5785 | 0.2581 |
| 0.2 | 397.1 | 53.87 | 291.77 | 345.64 | 0.1218 | 0.4353 | 0.5571 | 0.1377 |
| 0.3 | 423.8 | 57.38 | 291.27 | 348.65 | 0.1268 | 0.4179 | 0.5447 | 0.095 51 |
| 0.4 | 444.1 | 60.03 | 290.89 | 350.92 | 0.1305 | 0.4056 | 0.5361 | 0.073 78 |
| 0.5 | 460.7 | 62.20 | 290.58 | 352.78 | 0.1334 | 0.3960 | 0.5294 | 0.060 44 |
| 0.6 | 474.9 | 64.06 | 290.31 | 354.37 | 0.1359 | 0.3881 | 0.5240 | 0.051 37 |
| 0.7 | 487.3 | 65.66 | 290.08 | 355.74 | 0.1380 | 0.3815 | 0.5195 | 0.044 79 |
| 0.8 | 498.4 | 67.11 | 289.87 | 356.98 | 0.1398 | 0.3757 | 0.5155 | 0.039 78 |
| 0.9 | 508.5 | 68.42 | 289.68 | 358.10 | 0.1415 | 0.3706 | 0.5121 | 0.035 84 |
| 1.0 | 517.8 | 69.61 | 289.50 | 359.11 | 0.1429 | 0.3660 | 0.5089 | 0.032 66 |
| 1.2 | 534.4 | 71.75 | 289.19 | 360.94 | 0.1455 | 0.3581 | 0.5036 | 0.027 81 |
| 1.4 | 549.0 | 73.63 | 288.92 | 362.55 | 0.1478 | 0.3514 | 0.4992 | 0.024 29 |
| 1.6 | 562.0 | 75.37 | 288.67 | 364.04 | 0.1498 | 0.3456 | 0.4954 | 0.021 61 |
| 1.8 | 574.0 | 76.83 | 288.45 | 365.28 | 0.1515 | 0.3405 | 0.4920 | 0.019 49 |
| 2.0 | 584.9 | 78.23 | 288.24 | 366.47 | 0.1531 | 0.3359 | 0.4890 | 0.017 78 |
| 2.2 | 595.1 | 79.54 | 288.05 | 367.59 | 0.1546 | 0.3318 | 0.4864 | 0.016 37 |
| 2.4 | 604.6 | 80.75 | 287.87 | 368.62 | 0.1559 | 0.3280 | 0.4839 | 0.015 18 |
| 2.6 | 613.5 | 81.89 | 287.70 | 369.59 | 0.1571 | 0.3245 | 0.4816 | 0.014 16 |
| 2.8 | 622.0 | 82.96 | 287.54 | 370.50 | 0.1583 | 0.3212 | 0.4795 | 0.013 29 |
| 3.0 | 630.0 | 83.97 | 287.39 | 371.36 | 0.1594 | 0.3182 | 0.4776 | 0.012 52 |
| 3.5 | 648.5 | 86.33 | 287.04 | 373.37 | 0.1619 | 0.3115 | 0.4734 | 0.010 96 |
| 4.0 | 665.1 | 88.43 | 286.73 | 375.16 | 0.1641 | 0.3056 | 0.4697 | 0.009 78 |

**Table A.6** (Continued)
Thermodynamic Properties of Saturated Mercury[a]

| Press., MPa | Temp., °C | Enthalpy, kJ/kg | | | Entropy, kJ/kg K | | | Specific Volume Sat. Vapor, m³/kg |
|---|---|---|---|---|---|---|---|---|
| | | Sat. Liquid | Evap. | Sat. Vapor | Sat. Liquid | Evap. | Sat. Vapor | |
| 4.5 | 680.3 | 90.35 | 286.44 | 376.79 | 0.1660 | 0.3004 | 0.4664 | 0.008 85 |
| 5.0 | 694.4 | 92.11 | 286.18 | 378.29 | 0.1678 | 0.2958 | 0.4636 | 0.008 09 |
| 5.5 | 707.4 | 93.76 | 285.93 | 379.69 | 0.1694 | 0.2916 | 0.4610 | 0.007 46 |
| 6.0 | 719.7 | 95.30 | 285.70 | 381.00 | 0.1709 | 0.2878 | 0.4587 | 0.006 93 |
| 6.5 | 731.3 | 96.75 | 285.48 | 382.23 | 0.1723 | 0.2842 | 0.4565 | 0.006 48 |
| 7.0 | 742.3 | 98.12 | 285.28 | 383.40 | 0.1736 | 0.2809 | 0.4545 | 0.006 09 |
| 7.5 | 752.7 | 99.42 | 285.08 | 384.50 | 0.1748 | 0.2779 | 0.4527 | 0.005 75 |

[a] Adapted from *Thermodynamic Properties of Mercury Vapor*, by Lucian A. Sheldon. ASME, 49A, 30 (1949).

## Table A.7
## Critical Constants[a]

| Substance | Formula | Molecular Weight | Temp. K | Pressure MPa | Volume m³/kmol |
|---|---|---|---|---|---|
| Ammonia | $NH_3$ | 17.03 | 405.5 | 11.28 | .0724 |
| Argon | Ar | 39.948 | 151 | 4.86 | .0749 |
| Bromine | $Br_2$ | 159.808 | 584 | 10.34 | .1355 |
| Carbon Dioxide | $CO_2$ | 44.01 | 304.2 | 7.39 | .0943 |
| Carbon Monoxide | CO | 28.011 | 133 | 3.50 | .0930 |
| Chlorine | $Cl_2$ | 70.906 | 417 | 7.71 | .1242 |
| Deuterium (Normal) | $D_2$ | 4.00 | 38.4 | 1.66 | — |
| Helium | He | 4.003 | 5.3 | 0.23 | .0578 |
| Helium³ | He | 3.00 | 3.3 | 0.12 | — |
| Hydrogen (Normal) | $H_2$ | 2.016 | 33.3 | 1.30 | .0649 |
| Krypton | Kr | 83.80 | 209.4 | 5.50 | .0924 |
| Neon | Ne | 20.183 | 44.5 | 2.73 | .0417 |
| Nitrogen | $N_2$ | 28.013 | 126.2 | 3.39 | .0899 |
| Nitrous Oxide | $N_2O$ | 44.013 | 309.7 | 7.27 | .0961 |
| Oxygen | $O_2$ | 31.999 | 154.8 | 5.08 | .0780 |
| Sulfur Dioxide | $SO_2$ | 64.063 | 430.7 | 7.88 | .1217 |
| Water | $H_2O$ | 18.015 | 647.3 | 22.09 | .0568 |
| Xenon | Xe | 131.30 | 289.8 | 5.88 | .1186 |
| Benzene | $C_6H_6$ | 78.115 | 562 | 4.92 | .2603 |
| n-Butane | $C_4H_{10}$ | 58.124 | 425.2 | 3.80 | .2547 |
| Carbon Tetrachloride | $CCl_4$ | 153.82 | 556.4 | 4.56 | .2759 |
| Chloroform | $CHCl_3$ | 119.38 | 536.6 | 5.47 | .2403 |
| Dichlorodifluoromethane | $CCl_2F_2$ | 120.91 | 384.7 | 4.01 | .2179 |
| Dichlorofluoromethane | $CHCl_2F$ | 102.92 | 451.7 | 5.17 | .1973 |
| Ethane | $C_2H_6$ | 30.070 | 305.5 | 4.88 | .1480 |
| Ethyl Alcohol | $C_2H_5OH$ | 46.07 | 516 | 6.38 | .1673 |
| Ethylene | $C_2H_4$ | 28.054 | 282.4 | 5.12 | .1242 |
| n-Hexane | $C_6H_{14}$ | 86.178 | 507.9 | 3.03 | .3677 |
| Methane | $CH_4$ | 16.043 | 191.1 | 4.64 | .0993 |
| Methyl Alcohol | $CH_3OH$ | 32.042 | 513.2 | 7.95 | .1180 |
| Methyl Chloride | $CH_3Cl$ | 50.488 | 416.3 | 6.68 | .1430 |
| Propane | $C_3H_8$ | 44.097 | 370 | 4.26 | .1998 |
| Propene | $C_3H_6$ | 42.081 | 365 | 4.62 | .1810 |
| Propyne | $C_3H_4$ | 40.065 | 401 | 5.35 | — |
| Trichlorofluoromethane | $CCl_3F$ | 137.37 | 471.2 | 4.38 | .2478 |

[a] K. A. Kobe and R. E. Lynn, Jr., *Chem. Rev.*, **52**: 117–236 (1953).

**Table A.8**
Properties of Various Ideal Gases[a]

| Gas | Chemical Formula | Molecular Weight | $R \dfrac{kJ}{kg\ K}$ | $C_{po} \dfrac{kJ}{kg\ K}$ | $C_{vo} \dfrac{kJ}{kg\ K}$ | $k$ |
|---|---|---|---|---|---|---|
| Air | — | 28.97 | 0.287 00 | 1.0035 | 0.7165 | 1.400 |
| Argon | Ar | 39.948 | 0.208 13 | 0.5203 | 0.3122 | 1.667 |
| Butane | $C_4H_{10}$ | 58.124 | 0.143 04 | 1.7164 | 1.5734 | 1.091 |
| Carbon Dioxide | $CO_2$ | 44.01 | 0.188 92 | 0.8418 | 0.6529 | 1.289 |
| Carbon Monoxide | CO | 28.01 | 0.296 83 | 1.0413 | 0.7445 | 1.400 |
| Ethane | $C_2H_6$ | 30.07 | 0.276 50 | 1.7662 | 1.4897 | 1.186 |
| Ethylene | $C_2H_4$ | 28.054 | 0.296 37 | 1.5482 | 1.2518 | 1.237 |
| Helium | He | 4.003 | 2.077 03 | 5.1926 | 3.1156 | 1.667 |
| Hydrogen | $H_2$ | 2.016 | 4.124 18 | 14.2091 | 10.0849 | 1.409 |
| Methane | $CII_4$ | 16.04 | 0.518 35 | 2.2537 | 1.7354 | 1.299 |
| Neon | Ne | 20.183 | 0.411 95 | 1.0299 | 0.6179 | 1.667 |
| Nitrogen | $N_2$ | 28.013 | 0.296 80 | 1.0416 | 0.7448 | 1.400 |
| Octane | $C_8H_{18}$ | 114.23 | 0.072 79 | 1.7113 | 1.6385 | 1.044 |
| Oxygen | $O_2$ | 31.999 | 0.259 83 | 0.9216 | 0.6618 | 1.393 |
| Propane | $C_3H_8$ | 44.097 | 0.188 55 | 1.6794 | 1.4909 | 1.126 |
| Steam | $H_2O$ | 18.015 | 0.461 52 | 1.8723 | 1.4108 | 1.327 |

[a] ($C_{po}$, $C_{vo}$, and $k$ are at 300 K).

**Table A.9**
Constant-Pressure Specific Heats of Various Ideal Gases[a]

$$\overline{C}_{po} = kJ/kmol\ K$$

$$\theta = T(Kelvin)/100$$

| Gas | | Range K | Max. Error % |
|---|---|---|---|
| $N_2$ | $\overline{C}_{po} = 39.060 - 512.79\theta^{-1.5} + 1072.7\theta^{-2} - 820.40\theta^{-3}$ | 300–3500 | 0.43 |
| $O_2$ | $\overline{C}_{po} = 37.432 + 0.020102\theta^{1.5} - 178.57\theta^{-1.5} + 236.88\theta^{-2}$ | 300–3500 | 0.30 |
| $H_2$ | $\overline{C}_{po} = 56.505 - 702.74\theta^{-0.75} + 1165.0\theta^{-1} - 560.70\theta^{-1.5}$ | 300–3500 | 0.60 |
| CO | $\overline{C}_{po} = 69.145 - 0.70463\theta^{0.75} - 200.77\theta^{-0.5} + 176.76\theta^{-0.75}$ | 300–3500 | 0.42 |
| OH | $\overline{C}_{po} = 81.546 - 59.350\theta^{0.25} + 17.329\theta^{0.75} - 4.2660\theta$ | 300–3500 | 0.43 |
| NO | $\overline{C}_{po} = 59.283 - 1.7096\theta^{0.5} - 70.613\theta^{-0.5} + 74.889\theta^{-1.5}$ | 300–3500 | 0.34 |
| $H_2O$ | $\overline{C}_{po} = 143.05 - 183.54\theta^{0.25} + 82.751\theta^{0.5} - 3.6989\theta$ | 300–3500 | 0.43 |
| $CO_2$ | $\overline{C}_{po} = -3.7357 + 30.529\theta^{0.5} - 4.1034\theta + 0.024198\theta^2$ | 300–3500 | 0.19 |

# Table A.9 (Continued)

| Gas | | Range K | Max. Error % |
|---|---|---|---|
| $NO_2$ | $\overline{C}_{po}=46.045+216.10\theta^{-0.5}-363.66\theta^{-0.75}+232.550\theta^{-2}$ | 300–3500 | 0.26 |
| $CH_4$ | $\overline{C}_{po}=-672.87+439.74\theta^{0.25}-24.875\theta^{0.75}+323.88\theta^{-0.5}$ | 300–2000 | 0.15 |
| $C_2H_4$ | $\overline{C}_{po}=-95.395+123.15\theta^{0.5}-35.641\theta^{0.75}+182.77\theta^{-3}$ | 300–2000 | 0.07 |
| $C_2H_6$ | $\overline{C}_{po}=6.895+17.26\theta-0.6402\theta^2+0.007280\theta^3$ | 300–1500 | 0.83 |
| $C_3H_8$ | $\overline{C}_{po}=-4.042+30.46\theta-1.571\theta^2+0.03171\theta^3$ | 300–1500 | 0.40 |
| $C_4H_{10}$ | $\overline{C}_{po}=3.954+37.12\theta-1.833\theta^2+0.03498\theta^3$ | 300–1500 | 0.54 |

[a] From T. C. Scott and R. E. Sonntag, Univ. of Michigan, unpublished (1971), except $C_2H_6$, $C_3H_8$, $C_4H_{10}$ from K. A. Kobe, Petroleum Refiner 28 No. 2, 113 (1949).

# Table A.10
## Thermodynamic Properties of Air at Low Pressure[a]

| $T$, K | $h$, kJ/kg | $P_r$ | $u$, kJ/kg | $v_r$ | $s°$, kJ/kg K |
|---|---|---|---|---|---|
| 100 | 99.76 | 0.029 90 | 71.06 | 2230 | 1.4143 |
| 110 | 109.77 | 0.041 71 | 78.20 | 1758.4 | 1.5098 |
| 120 | 119.79 | 0.056 52 | 85.34 | 1415.7 | 1.5971 |
| 130 | 129.81 | 0.074 74 | 92.51 | 1159.8 | 1.6773 |
| 140 | 139.84 | 0.096 81 | 99.67 | 964.2 | 1.7515 |
| 150 | 149.86 | 0.123 18 | 106.81 | 812.0 | 1.8206 |
| 160 | 159.87 | 0.154 31 | 113.95 | 691.4 | 1.8853 |
| 170 | 169.89 | 0.190 68 | 121.11 | 594.5 | 1.9461 |
| 180 | 179.92 | 0.232 79 | 128.28 | 515.6 | 2.0033 |
| 190 | 189.94 | 0.281 14 | 135.40 | 450.6 | 2.0575 |
| 200 | 199.96 | 0.3363 | 142.56 | 396.6 | 2.1088 |
| 210 | 209.97 | 0.3987 | 149.70 | 351.2 | 2.1577 |
| 220 | 219.99 | 0.4690 | 156.84 | 312.8 | 2.2043 |
| 230 | 230.01 | 0.5477 | 163.98 | 280.0 | 2.2489 |
| 240 | 240.03 | 0.6355 | 171.15 | 251.8 | 2.2915 |
| 250 | 250.05 | 0.7329 | 178.29 | 227.45 | 2.3325 |
| 260 | 260.09 | 0.8405 | 185.45 | 206.26 | 2.3717 |
| 270 | 270.12 | 0.9590 | 192.59 | 187.74 | 2.4096 |
| 280 | 280.14 | 1.0889 | 199.78 | 171.45 | 2.4461 |
| 290 | 290.17 | 1.2311 | 206.92 | 157.07 | 2.4813 |
| 300 | 300.19 | 1.3860 | 214.09 | 144.32 | 2.5153 |
| 310 | 310.24 | 1.5546 | 221.27 | 132.96 | 2.5483 |
| 320 | 320.29 | 1.7375 | 228.45 | 122.81 | 2.5802 |
| 330 | 330.34 | 1.9352 | 235.65 | 113.70 | 2.6111 |
| 340 | 340.43 | 2.149 | 242.86 | 105.51 | 2.6412 |

**Table A.10** (Continued)

| T, K | h, kJ/kg | $P_r$ | u, kJ/kg | $v_r$ | $s°$, kJ/kg K |
|---|---|---|---|---|---|
| 350 | 350.48 | 2.379 | 250.05 | 98.11 | 2.6704 |
| 360 | 360.58 | 2.626 | 257.23 | 91.40 | 2.6987 |
| 370 | 370.67 | 2.892 | 264.47 | 85.31 | 2.7264 |
| 380 | 380.77 | 3.176 | 271.72 | 79.77 | 2.7534 |
| 390 | 390.88 | 3.481 | 278.96 | 74.71 | 2.7796 |
| 400 | 400.98 | 3.806 | 286.19 | 70.07 | 2.8052 |
| 410 | 411.12 | 4.153 | 293.45 | 65.83 | 2.8302 |
| 420 | 421.26 | 4.522 | 300.73 | 61.93 | 2.8547 |
| 430 | 431.43 | 4.915 | 308.03 | 58.34 | 2.8786 |
| 440 | 441.61 | 5.332 | 315.34 | 55.02 | 2.9020 |
| 450 | 451.83 | 5.775 | 322.66 | 51.96 | 2.9249 |
| 460 | 462.01 | 6.245 | 329.99 | 49.11 | 2.9473 |
| 470 | 472.25 | 6.742 | 337.34 | 46.48 | 2.9693 |
| 480 | 482.48 | 7.268 | 344.74 | 44.04 | 2.9909 |
| 490 | 492.74 | 7.824 | 352.11 | 41.76 | 3.0120 |
| 500 | 503.02 | 8.411 | 359.53 | 39.64 | 3.0328 |
| 510 | 513.32 | 9.031 | 366.97 | 37.65 | 3.0532 |
| 520 | 523.63 | 9.684 | 374.39 | 35.80 | 3.0733 |
| 530 | 533.98 | 10.372 | 381.88 | 34.07 | 3.0930 |
| 540 | 544.35 | 11.097 | 389.40 | 32.45 | 3.1124 |
| 550 | 554.75 | 11.858 | 396.89 | 30.92 | 3.1314 |
| 560 | 565.17 | 12.659 | 404.44 | 29.50 | 3.1502 |
| 570 | 575.57 | 13.500 | 411.98 | 28.15 | 3.1686 |
| 580 | 586.04 | 14.382 | 419.56 | 26.89 | 3.1868 |
| 590 | 596.53 | 15.309 | 427.17 | 25.70 | 3.2047 |
| 600 | 607.02 | 16.278 | 434.80 | 24.58 | 3.2223 |
| 610 | 617.53 | 17.297 | 442.43 | 23.51 | 3.2397 |
| 620 | 628.07 | 18.360 | 450.13 | 22.52 | 3.2569 |
| 630 | 638.65 | 19.475 | 457.83 | 21.57 | 3.2738 |
| 640 | 649.21 | 20.64 | 465.55 | 20.674 | 3.2905 |
| 650 | 659.84 | 21.86 | 473.32 | 19.828 | 3.3069 |
| 660 | 670.47 | 23.13 | 481.06 | 19.026 | 3.3232 |
| 670 | 681.15 | 24.46 | 488.88 | 18.266 | 3.3392 |
| 680 | 691.82 | 25.85 | 496.65 | 17.543 | 3.3551 |
| 690 | 702.52 | 27.29 | 504.51 | 16.857 | 3.3707 |
| 700 | 713.27 | 28.80 | 512.37 | 16.205 | 3.3861 |
| 710 | 724.01 | 30.38 | 520.26 | 15.585 | 3.4014 |
| 720 | 734.20 | 31.92 | 527.72 | 15.027 | 3.4156 |
| 730 | 745.62 | 33.72 | 536.12 | 14.434 | 3.4314 |
| 740 | 756.44 | 35.50 | 544.05 | 13.900 | 3.4461 |
| 750 | 767.30 | 37.35 | 552.05 | 13.391 | 3.4607 |
| 760 | 778.21 | 39.27 | 560.08 | 12.905 | 3.4751 |
| 770 | 789.10 | 41.27 | 568.10 | 12.440 | 3.4894 |
| 780 | 800.03 | 43.35 | 576.15 | 11.998 | 3.5035 |
| 790 | 810.98 | 45.51 | 584.22 | 11.575 | 3.5174 |

**Table A.10** (Continued)

| T, K | h, kJ/kg | $P_r$ | u, kJ/kg | $v_r$ | $s°$, kJ/kg K |
|---|---|---|---|---|---|
| 800 | 821.94 | 47.75 | 592.34 | 11.172 | 3.5312 |
| 810 | 832.96 | 50.08 | 600.46 | 10.785 | 3.5449 |
| 820 | 843.97 | 52.49 | 608.62 | 10.416 | 3.5584 |
| 830 | 855.01 | 55.00 | 616.79 | 10.062 | 3.5718 |
| 840 | 866.09 | 57.60 | 624.97 | 9.724 | 3.5850 |
| 850 | 877.16 | 60.29 | 633.21 | 9.400 | 3.5981 |
| 860 | 888.28 | 63.09 | 641.44 | 9.090 | 3.6111 |
| 870 | 899.42 | 65.98 | 649.70 | 8.792 | 3.6240 |
| 880 | 910.56 | 68.98 | 658.00 | 8.507 | 3.6367 |
| 890 | 921.75 | 72.08 | 666.31 | 8.233 | 3.6493 |
| 900 | 932.94 | 75.29 | 674.63 | 7.971 | 3.6619 |
| 910 | 944.15 | 78.61 | 682.98 | 7.718 | 3.6743 |
| 920 | 955.38 | 82.05 | 691.33 | 7.476 | 3.6865 |
| 930 | 966.64 | 85.60 | 699.73 | 7.244 | 3.6987 |
| 940 | 977.92 | 89.28 | 708.13 | 7.020 | 3.7108 |
| 950 | 989.22 | 93.08 | 716.57 | 6.805 | 3.7227 |
| 960 | 1000.53 | 97.00 | 725.01 | 6.599 | 3.7346 |
| 970 | 1011.88 | 101.06 | 733.48 | 6.400 | 3.7463 |
| 980 | 1023.25 | 105.24 | 741.99 | 6.209 | 3.7580 |
| 990 | 1034.63 | 109.57 | 750.48 | 6.025 | 3.7695 |
| 1000 | 1046.03 | 114.03 | 759.02 | 5.847 | 3.7810 |
| 1020 | 1068.89 | 123.12 | 775.67 | 5.521 | 3.8030 |
| 1040 | 1091.85 | 133.34 | 793.35 | 5.201 | 3.8259 |
| 1060 | 1114.85 | 143.91 | 810.61 | 4.911 | 3.8478 |
| 1080 | 1137.93 | 155.15 | 827.94 | 4.641 | 3.8694 |
| 1100 | 1161.07 | 167.07 | 845.34 | 4.390 | 3.8906 |
| 1120 | 1184.28 | 179.71 | 862.85 | 4.156 | 3.9116 |
| 1140 | 1207.54 | 193.07 | 880.37 | 3.937 | 3.9322 |
| 1160 | 1230.90 | 207.24 | 897.98 | 3.732 | 3.9525 |
| 1180 | 1254.34 | 222.2 | 915.68 | 3.541 | 3.9725 |
| 1200 | 1277.79 | 238.0 | 933.40 | 3.362 | 3.9922 |
| 1220 | 1301.33 | 254.7 | 951.19 | 3.194 | 4.0117 |
| 1240 | 1324.89 | 272.3 | 969.01 | 3.037 | 4.0308 |
| 1260 | 1348.55 | 290.8 | 986.92 | 2.889 | 4.0497 |
| 1280 | 1372.25 | 310.4 | 1004.88 | 2.750 | 4.0684 |
| 1300 | 1395.97 | 330.9 | 1022.88 | 2.619 | 4.0868 |
| 1320 | 1419.77 | 352.5 | 1040.93 | 2.497 | 4.1049 |
| 1340 | 1443.61 | 375.3 | 1059.03 | 2.381 | 4.1229 |
| 1360 | 1467.50 | 399.1 | 1077.17 | 2.272 | 4.1406 |
| 1380 | 1491.43 | 424.2 | 1095.36 | 2.169 | 4.1580 |
| 1400 | 1515.41 | 450.5 | 1113.62 | 2.072 | 4.1753 |
| 1420 | 1539.44 | 478.0 | 1131.90 | 1.9808 | 4.1923 |
| 1440 | 1563.49 | 506.9 | 1150.23 | 1.8942 | 4.2092 |

## Table A.10 (Continued)

| T, K | h, kJ/kg | $P_r$ | u, kJ/kg | $v_4$ | $s°$, kJ/kg K |
|------|----------|-------|----------|-------|---------------|
| 1460 | 1587.61 | 537.1 | 1168.61 | 1.8124 | 4.2258 |
| 1480 | 1611.80 | 568.8 | 1187.03 | 1.7350 | 4.2422 |
| 1500 | 1635.99 | 601.9 | 1205.47 | 1.6617 | 4.2585 |

[a]Adapted from Table 1 in *Gas Tables*, by Joseph H. Keenan and Joseph Kaye. Copyright 1948, by Joseph H. Keenan and Joseph Kaye. Published by John Wiley & Sons, Inc., New York.

## Table A.11[a]
### Enthalpy of Formation at 25°C, Ideal Gas Enthalpy and Absolute Entropy at 0.1 MPa (1 bar) Pressure

| | Nitrogen, Diatomic ($N_2$) (9/30/65) | | Nitrogen, Monatomic (N) (3/31/61) | |
|---|---|---|---|---|
| | $(\overline{h}_f°)_{298} = 0$ kJ/kmol $M = 28.013$ | | $(\overline{h}_f°)_{298} = 472\ 646$ kJ/kmol $M = 14.007$ | |
| Temp. K | $(\overline{h}° - \overline{h}°_{298})$ kJ/kmol | $\overline{s}°$ kJ/kmol K | $(\overline{h}° - \overline{h}°_{298})$ kJ/kmol | $\overline{s}°$ kJ/kmol K |
| 0 | −8 669 | 0 | −6 197 | 0 |
| 100 | −5 770 | 159.813 | −4 117 | 130.596 |
| 200 | −2 858 | 179.988 | −2 042 | 145.006 |
| 298 | 0 | 191.611 | 0 | 153.302 |
| 300 | 54 | 191.791 | 38 | 153.432 |
| 400 | 2 971 | 200.180 | 2 117 | 159.411 |
| 500 | 5 912 | 206.740 | 4 197 | 164.051 |
| 600 | 8 891 | 212.175 | 6 276 | 167.842 |
| 700 | 11 937 | 216.866 | 8 351 | 171.047 |
| 800 | 15 046 | 221.016 | 10 431 | 173.821 |
| 900 | 18 221 | 224.757 | 12 510 | 176.268 |
| 1000 | 21 460 | 228.167 | 14 590 | 178.461 |
| 1100 | 24 757 | 231.309 | 16 669 | 180.440 |
| 1200 | 28 108 | 234.225 | 18 749 | 182.247 |
| 1300 | 31 501 | 236.941 | 20 824 | 183.803 |
| 1400 | 34 936 | 239.484 | 22 903 | 185.452 |
| 1500 | 38 405 | 241.878 | 24 983 | 186.887 |
| 1600 | 41 903 | 244.137 | 27 062 | 188.230 |
| 1700 | 45 430 | 246.275 | 29 142 | 189.490 |
| 1800 | 48 982 | 248.304 | 31 217 | 190.678 |
| 1900 | 52 551 | 250.237 | 33 296 | 191.799 |
| 2000 | 56 141 | 252.078 | 35 376 | 192.866 |

**Table A.11**[a] (Continued)
Enthalpy of Formation at 25°C, Ideal Gas Enthalpy and Absolute
Entropy at 0.1 MPa (1 bar) Pressure

| Temp. K | Nitrogen, Diatomic ($N_2$) (9/30/65) $(\bar{h}_f^\circ)_{298} = 0$ kJ/kmol $M = 28.013$ | | Nitrogen, Monatomic (N) (3/31/61) $(\bar{h}_f^\circ)_{298} = 472\ 646$ kJ/kmol $M = 14.007$ | |
|---|---|---|---|---|
| | $(\bar{h}^\circ - \bar{h}_{298}^\circ)$ kJ/kmol | $\bar{s}^\circ$ kJ/kmol K | $(\bar{h}^\circ - \bar{h}_{298}^\circ)$ kJ/kmol | $\bar{s}^\circ$ kJ/kmol K |
| 2100 | 59 748 | 253.836 | 37 455 | 193.883 |
| 2200 | 63 371 | 255.522 | 39 535 | 194.850 |
| 2300 | 67 007 | 257.137 | 41 614 | 195.774 |
| 2400 | 70 651 | 258.689 | 43 698 | 196.661 |
| 2500 | 74 312 | 260.183 | 45 777 | 197.511 |
| 2600 | 77 973 | 261.622 | 47 861 | 198.326 |
| 2700 | 81 659 | 263.011 | 49 949 | 199.113 |
| 2800 | 85 345 | 264.350 | 52 036 | 199.875 |
| 2900 | 89 036 | 265.647 | 54 124 | 200.607 |
| 3000 | 92 738 | 266.902 | 56 220 | 201.318 |
| 3200 | 100 161 | 269.295 | 60 421 | 202.674 |
| 3400 | 107 608 | 271.555 | 64 647 | 203.954 |
| 3600 | 115 081 | 273.689 | 68 906 | 205.171 |
| 3800 | 122 570 | 275.714 | 73 199 | 206.330 |
| 4000 | 130 076 | 277.638 | 77 534 | 207.443 |
| 4200 | 137 603 | 279.475 | 81 923 | 208.514 |
| 4400 | 145 143 | 281.228 | 86 370 | 209.548 |
| 4600 | 152 699 | 282.910 | 90 881 | 210.552 |
| 4800 | 160 272 | 284.521 | 95 462 | 211.527 |
| 5000 | 167 858 | 286.069 | 100 115 | 212.477 |
| 5200 | 175 456 | 287.559 | 104 847 | 213.401 |
| 5400 | 183 071 | 288.994 | 109 663 | 214.309 |
| 5600 | 190 703 | 290.383 | 114 558 | 215.201 |
| 5800 | 198 347 | 291.726 | 119 537 | 216.075 |
| 6000 | 206 008 | 293.023 | 124 600 | 216.933 |

[a] The thermochemical data in Table A.11 are calculated from the JANAF Thermochemical Tables, Thermal Research Laboratory, The Dow Chemical Company, Midland, Michigan. The date each table was issued is indicated.

**Table A.11** (Continued)
Enthalpy of Formation at 25°C, Ideal Gas Enthalpy and Absolute
Entropy at 0.1 MPa (1 bar) Pressure

| Temp. K | Oxygen, Diatomic ($O_2$) (9/30/65) | | Oxygen, Monatomic (O) (6/30/62) | |
| | $(\overline{h}_f^\circ)_{298} = 0$ kJ/kmol $M = 31.999$ | | $(\overline{h}_f^\circ)_{298} = 249\ 195$ kJ/kmol $M = 16.00$ | |
| | $(\overline{h}^\circ - \overline{h}_{298}^\circ)$ kJ/kmol | $\overline{s}^\circ$ kJ/kmol K | $(\overline{h}^\circ - \overline{h}_{298}^\circ)$ kJ/kmol | $\overline{s}^\circ$ kJ/kmol K |
|---|---|---|---|---|
| 0 | −8 682 | 0 | −6 728 | 0 |
| 100 | −5 778 | 173.306 | −4 519 | 135.947 |
| 200 | −2 866 | 193.486 | −2 188 | 152.156 |
| 298 | 0 | 205.142 | 0 | 161.060 |
| 300 | 54 | 205.322 | 42 | 161.198 |
| 400 | 3 029 | 213.874 | 2 209 | 167.432 |
| 500 | 6 088 | 220.698 | 4 343 | 172.202 |
| 600 | 9 247 | 226.455 | 6 460 | 176.063 |
| 700 | 12 502 | 231.272 | 8 569 | 179.314 |
| 800 | 15 841 | 235.924 | 10 669 | 182.118 |
| 900 | 19 246 | 239.936 | 12 770 | 184.590 |
| 1000 | 22 707 | 243.585 | 14 862 | 186.795 |
| 1100 | 26 217 | 246.928 | 16 949 | 188.787 |
| 1200 | 29 765 | 250.016 | 19 041 | 190.603 |
| 1300 | 33 351 | 252.886 | 21 125 | 192.272 |
| 1400 | 36 966 | 255.564 | 23 213 | 193.820 |
| 1500 | 40 610 | 258.078 | 25 296 | 195.260 |
| 1600 | 44 279 | 260.446 | 27 380 | 196.603 |
| 1700 | 47 970 | 262.685 | 29 464 | 197.866 |
| 1800 | 51 689 | 264.810 | 31 547 | 199.059 |
| 1900 | 55 434 | 266.835 | 33 631 | 200.184 |
| 2000 | 59 199 | 268.764 | 35 715 | 201.251 |
| 2100 | 62 986 | 270.613 | 37 798 | 202.268 |
| 2200 | 66 802 | 272.387 | 39 882 | 203.238 |
| 2300 | 70 634 | 274.090 | 41 961 | 204.163 |
| 2400 | 74 492 | 275.735 | 44 045 | 205.050 |
| 2500 | 78 375 | 277.316 | 46 133 | 205.899 |
| 2600 | 82 274 | 278.848 | 48 216 | 206.720 |
| 2700 | 86 199 | 280.329 | 50 304 | 207.506 |
| 2800 | 90 144 | 281.764 | 52 392 | 208.268 |
| 2900 | 94 111 | 283.157 | 54 484 | 209.000 |
| 3000 | 98 098 | 284.508 | 56 576 | 209.711 |
| 3200 | 106 127 | 287.098 | 60 768 | 211.063 |
| 3400 | 114 232 | 289.554 | 64 973 | 212.339 |
| 3600 | 122 399 | 291.889 | 69 191 | 213.544 |
| 3800 | 130 629 | 294.115 | 73 425 | 214.686 |

Enthalpy of Formation at 25°C, Ideal Gas Enthalpy and Absolute
Entropy at 0.1 MPa (1 bar) Pressure

| | Oxygen, Diatomic ($O_2$) (9/30/65) | | Oxygen, Monatomic (O) (6/30/62) | |
|---|---|---|---|---|
| | $(\bar{h}_f^\circ)_{298} = 0$ kJ/kmol $M = 31.999$ | | $(\bar{h}_f^\circ)_{298} = 249\ 195$ kJ/kmol $M = 16.00$ | |
| Temp. K | $(\bar{h}^\circ - \bar{h}^\circ_{298})$ kJ/kmol | $\bar{s}^\circ$ kJ/kmol K | $(\bar{h}^\circ - \bar{h}^\circ_{298})$ kJ/kmol | $\bar{s}^\circ$ kJ/kmol K |
| 4000 | 138 913 | 296.236 | 77 676 | 215.778 |
| 4200 | 147 248 | 298.270 | 81 948 | 216.820 |
| 4400 | 155 628 | 300.219 | 86 236 | 217.816 |
| 4600 | 164 046 | 302.094 | 90 546 | 218.774 |
| 4800 | 172 502 | 303.893 | 94 876 | 219.694 |
| 5000 | 180 987 | 305.621 | 99 224 | 220.585 |
| 5200 | 189 502 | 307.290 | 103 596 | 221.439 |
| 5400 | 198 037 | 308.901 | 107 985 | 222.267 |
| 5600 | 206 593 | 310.458 | 112 395 | 223.071 |
| 5800 | 215 166 | 311.964 | 116 821 | 223.849 |
| 6000 | 223 756 | 313.420 | 121 269 | 224.602 |

| | Carbon Dioxide ($CO_2$) (9/30/65) | | Carbon Monoxide (CO) (9/30/65) | |
|---|---|---|---|---|
| | $(\bar{h}_f^\circ)_{298} = -393\ 522$ kJ/kmol $M = 44.01$ | | $(\bar{h}_f^\circ)_{298} = -110\ 529$ kJ/kmol $M = 28.01$ | |
| Temp. K | $(\bar{h}^\circ - \bar{h}^\circ_{298})$ kJ/kmol | $\bar{s}^\circ$ kJ/kmol K | $(\bar{h}^\circ - \bar{h}^\circ_{298})$ kJ/kmol | $\bar{s}^\circ$ kJ/kmol K |
| 0 | −9 364 | 0 | −8 669 | 0 |
| 100 | −6 456 | 179.109 | −5 770 | 165.850 |
| 200 | −3 414 | 199.975 | −2 858 | 186.025 |
| 298 | 0 | 213.795 | 0 | 197.653 |
| 300 | 67 | 214.025 | 54 | 197.833 |
| 400 | 4 008 | 225.334 | 2 975 | 206.234 |
| 500 | 8 314 | 234.924 | 5 929 | 212.828 |
| 600 | 12 916 | 243.309 | 8 941 | 218.313 |
| 700 | 17 761 | 250.773 | 12 021 | 223.062 |
| 800 | 22 815 | 257.517 | 15 175 | 227.271 |
| 900 | 28 041 | 263.668 | 18 397 | 231.066 |
| 1000 | 33 405 | 269.325 | 21 686 | 234.531 |
| 1100 | 38 894 | 274.555 | 25 033 | 237.719 |
| 1200 | 44 484 | 279.417 | 28 426 | 240.673 |

Enthalpy of Formation at 25°C, Ideal Gas Enthalpy and Absolute
Entropy at 0.1 MPa (1 bar) Pressure

| Temp. K | Carbon Dioxide (CO$_2$) (9/30/65) $(\overline{h}_f^\circ)_{298} = -393\ 522$ kJ/kmol $M = 44.01$ | | Carbon Monoxide (CO) (9/30/65) $(\overline{h}_f^\circ)_{298} = -110\ 529$ kJ/kmol $M = 28.01$ | |
|---|---|---|---|---|
| | $(\overline{h}^\circ - \overline{h}^\circ{}_{298})$ kJ/kmol | $\overline{s}^\circ$ kJ/kmol K | $(\overline{h}^\circ - \overline{h}^\circ{}_{298})$ kJ/kmol | $\overline{s}^\circ$ kJ/kmol K |
| 1300 | 50 158 | 283.956 | 31 865 | 243.426 |
| 1400 | 55 907 | 288.216 | 35 338 | 245.999 |
| 1500 | 61 714 | 292.224 | 38 848 | 248.421 |
| 1600 | 67 580 | 296.010 | 42 384 | 250.702 |
| 1700 | 73 492 | 299.592 | 45 940 | 252.861 |
| 1800 | 79 442 | 302.993 | 49 522 | 254.907 |
| 1900 | 85 429 | 306.232 | 53 124 | 256.852 |
| 2000 | 91 450 | 309.320 | 56 739 | 258.710 |
| 2100 | 97 500 | 312.269 | 60 375 | 260.480 |
| 2200 | 103 575 | 315.098 | 64 019 | 262.174 |
| 2300 | 109 671 | 317.805 | 67 676 | 263.802 |
| 2400 | 115 788 | 320.411 | 71 346 | 265.362 |
| 2500 | 121 926 | 322.918 | 75 023 | 266.865 |
| 2600 | 128 085 | 325.332 | 78 714 | 268.312 |
| 2700 | 134 256 | 327.658 | 82 408 | 269.705 |
| 2800 | 140 444 | 329.909 | 86 115 | 271.053 |
| 2900 | 146 645 | 332.085 | 89 826 | 272.358 |
| 3000 | 152 862 | 334.193 | 93 542 | 273.618 |
| 3200 | 165 331 | 338.218 | 100 998 | 276.023 |
| 3400 | 177 849 | 342.013 | 108 479 | 278.291 |
| 3600 | 190 405 | 345.599 | 115 976 | 280.433 |
| 3800 | 202 999 | 349.005 | 123 495 | 282.467 |
| 4000 | 215 635 | 352.243 | 131 026 | 284.396 |
| 4200 | 228 304 | 355.335 | 138 578 | 286.241 |
| 4400 | 241 003 | 358.289 | 146 147 | 287.998 |
| 4600 | 253 734 | 361.122 | 153 724 | 289.684 |
| 4800 | 266 500 | 363.837 | 161 322 | 291.299 |
| 5000 | 279 295 | 366.448 | 168 929 | 292.851 |
| 5200 | 292 123 | 368.963 | 176 548 | 294.349 |
| 5400 | 304 984 | 371.389 | 184 184 | 295.789 |
| 5600 | 317 884 | 373.736 | 191 832 | 297.178 |
| 5800 | 330 821 | 376.004 | 199 489 | 298.521 |
| 6000 | 343 791 | 378.205 | 207 162 | 299.822 |

**Table A.11** (Continued)
Enthalpy of Formation at 25°C, Ideal Gas Enthalpy and Absolute
Entropy at 0.1 MPa (1 bar) Pressure

| Temp. K | Water ($H_2O$) (3/31/61) $(\bar{h}_f^\circ)_{298} = -241\ 827$ kJ/kmol $M = 18.015$ $(\bar{h}^\circ - \bar{h}_{298}^\circ)$ kJ/kmol | $\bar{s}^\circ$ kJ/kmol K | Hydroxyl (OH) (3/31/66) $(\bar{h}_f^\circ)_{298} = 39\ 463$ kJ/kmol $M = 17.007$ $(\bar{h}^\circ - \bar{h}_{298}^\circ)$ kJ/kmol | $\bar{s}^\circ$ kJ/kmol K |
|---|---|---|---|---|
| 0 | −9 904 | 0 | −9 171 | 0 |
| 100 | −6 615 | 152.390 | −6 138 | 149.587 |
| 200 | −3 280 | 175.486 | −2 975 | 171.591 |
| 298 | 0 | 188.833 | 0 | 183.703 |
| 300 | 63 | 189.038 | 54 | 183.892 |
| 400 | 3 452 | 198.783 | 3 033 | 192.465 |
| 500 | 6 920 | 206.523 | 5 991 | 199.063 |
| 600 | 10 498 | 213.037 | 8 941 | 204.443 |
| 700 | 14 184 | 218.719 | 11 903 | 209.004 |
| 800 | 17 991 | 223.803 | 14 878 | 212.979 |
| 900 | 21 924 | 228.430 | 17 887 | 216.523 |
| 1000 | 25 978 | 232.706 | 20 933 | 219.732 |
| 1100 | 30 167 | 236.694 | 24 025 | 222.677 |
| 1200 | 34 476 | 240.443 | 27 158 | 225.405 |
| 1300 | 38 903 | 243.986 | 30 342 | 227.949 |
| 1400 | 43 447 | 247.350 | 33 568 | 230.342 |
| 1500 | 48 095 | 250.560 | 36 840 | 232.598 |
| 1600 | 52 844 | 253.622 | 40 150 | 234.736 |
| 1700 | 57 685 | 256.559 | 43 501 | 236.769 |
| 1800 | 62 609 | 259.371 | 46 890 | 238.702 |
| 1900 | 67 613 | 262.078 | 50 308 | 240.551 |
| 2000 | 72 689 | 264.681 | 53 760 | 242.325 |
| 2100 | 77 831 | 267.191 | 57 241 | 244.020 |
| 2200 | 83 036 | 269.609 | 60 752 | 245.652 |
| 2300 | 88 295 | 271.948 | 64 283 | 247.225 |
| 2400 | 93 604 | 274.207 | 67 839 | 248.739 |
| 2500 | 98 964 | 276.396 | 71 417 | 250.200 |
| 2600 | 104 370 | 278.517 | 75 015 | 251.610 |
| 2700 | 109 813 | 280.571 | 78 634 | 252.974 |
| 2800 | 115 294 | 282.563 | 82 266 | 254.296 |
| 2900 | 120 813 | 284.500 | 85 918 | 255.576 |
| 3000 | 126 361 | 286.383 | 89 584 | 256.819 |
| 3200 | 137 553 | 289.994 | 96 960 | 259.199 |
| 3400 | 148 854 | 293.416 | 104 387 | 261.450 |
| 3600 | 160 247 | 296.676 | 111 859 | 263.588 |

**Table A.11** (Continued)
Enthalpy of Formation at 25°C, Ideal Gas Enthalpy and Absolute
Entropy at 0.1 MPa (1 bar) Pressure

| | Water ($H_2O$) (3/31/61) | | Hydroxyl (OH) (3/31/66) | |
| | $(\bar{h}_f^\circ)_{298}) = -241\ 827$ kJ/kmol $M = 18.015$ | | $(\bar{h}_f^\circ)_{298} = 39\ 463$ kJ/kmol $M = 17.007$ | |
| Temp. K | $(\bar{h}^\circ - \bar{h}^\circ_{298})$ kJ/kmol | $\bar{s}^\circ$ kJ/kmol K | $(\bar{h}^\circ - \bar{h}^\circ_{298})$ kJ/kmol | $\bar{s}^\circ$ kJ/kmol K |
|---|---|---|---|---|
| 3800 | 171 724 | 299.776 | 119 378 | 265.618 |
| 4000 | 183 280 | 302.742 | 126 934 | 267.559 |
| 4200 | 194 903 | 305.575 | 134 528 | 269.408 |
| 4400 | 206 585 | 308.295 | 142 156 | 271.182 |
| 4600 | 218 325 | 310.901 | 149 816 | 272.885 |
| 4800 | 230 120 | 313.412 | 157 502 | 274.521 |
| 5000 | 241 957 | 315.830 | 165 222 | 276.099 |
| 5200 | 253 839 | 318.160 | 172 967 | 277.617 |
| 5400 | 265 768 | 320.407 | 180 736 | 279.082 |
| 5600 | 277 738 | 322.587 | 188 531 | 280.500 |
| 5800 | 289 746 | 324.692 | 196 351 | 281.873 |
| 6000 | 301 796 | 326.733 | 204 192 | 283.203 |

| | Hydrogen, Diatomic ($H_2$) (3/31/61) | | Hydrogen, Monatomic (H) (9/30/65) | |
| | $(\bar{h}_f^\circ)_{298} = 0$ kJ/kmol $M = 2.016$ | | $(\bar{h}_f^\circ)_{298} = 217\ 986$ kJ/kmol $M - 1.008$ | |
| Temp. K | $(\bar{h}^\circ - \bar{h}^\circ_{298})$ kJ/kmol | $\bar{s}^\circ$ kJ/kmol K | $(\bar{h}^\circ - \bar{h}^\circ_{298})$ kJ/kmol | $\bar{s}^\circ$ kJ/kmol K |
|---|---|---|---|---|
| 0 | −8 468 | 0 | −6 197 | 0 |
| 100 | −5 293 | 102.145 | −4 117 | 92.011 |
| 200 | −2 770 | 119.437 | −2 042 | 106.417 |
| 298 | 0 | 130.684 | 0 | 114.718 |
| 300 | 54 | 130.864 | 38 | 114.847 |
| 400 | 2 958 | 139.215 | 2 117 | 120.826 |
| 500 | 5 883 | 145.738 | 4 197 | 125.466 |
| 600 | 8 812 | 151.077 | 6 276 | 129.257 |
| 700 | 11 749 | 155.608 | 8 351 | 132.458 |
| 800 | 14 703 | 159.549 | 10 431 | 135.236 |
| 900 | 17 682 | 163.060 | 12 510 | 137.684 |
| 1000 | 20 686 | 166.223 | 14 590 | 139.872 |

**Table A.11** (Continued)
Enthalpy of Formation at 25°C, Ideal Gas Enthalpy and Absolute
Entropy at 0.1 MPa (1 bar) Pressure

| | Hydrogen, Diatomic ($H_2$)<br>(3/31/61) | | Hydrogen, Monatomic (H)<br>(9/30/65) | |
| | $(\bar{h}_f^\circ)_{298} = 0$ kJ/kmol<br>$M = 2.016$ | | $(\bar{h}_f^\circ)_{298} = 217\ 986$ kJ/kmol<br>$M = 1.008$ | |
| Temp.<br>K | $(\bar{h}^\circ - \bar{h}_{298}^\circ)$<br>kJ/kmol | $\bar{s}^\circ$<br>kJ/kmol K | $(\bar{h}^\circ - \bar{h}_{298}^\circ)$<br>kJ/kmol | $\bar{s}^\circ$<br>kJ/kmol K |
|---|---|---|---|---|
| 1100 | 23 723 | 169.118 | 16 669 | 141.855 |
| 1200 | 26 794 | 171.792 | 18 749 | 143.662 |
| 1300 | 29 907 | 174.281 | 20 824 | 145.328 |
| 1400 | 33 062 | 176.620 | 22 903 | 146.867 |
| 1500 | 36 267 | 178.833 | 24 983 | 148.303 |
| 1600 | 39 522 | 180.929 | 27 062 | 149.641 |
| 1700 | 42 815 | 182.929 | 29 142 | 150.905 |
| 1800 | 46 150 | 184.833 | 31 217 | 152.093 |
| 1900 | 49 522 | 186.657 | 33 296 | 153.215 |
| 2000 | 52 932 | 188.406 | 35 376 | 154.281 |
| 2100 | 56 379 | 190.088 | 37 455 | 155.294 |
| 2200 | 59 860 | 191.707 | 39 535 | 156.265 |
| 2300 | 63 371 | 193.268 | 41 610 | 157.185 |
| 2400 | 66 915 | 194.778 | 43 689 | 158.072 |
| 2500 | 70 492 | 196.234 | 45 769 | 158.922 |
| 2600 | 74 090 | 197.649 | 47 848 | 159.737 |
| 2700 | 77 718 | 199.017 | 49 928 | 160.520 |
| 2800 | 81 370 | 200.343 | 52 007 | 161.277 |
| 2900 | 85 044 | 201.636 | 54 082 | 162.005 |
| 3000 | 88 743 | 202.887 | 56 162 | 162.708 |
| 3200 | 96 199 | 205.293 | 60 321 | 164.051 |
| 3400 | 103 738 | 207.577 | 64 475 | 165.311 |
| 3600 | 111 361 | 209.757 | 68 634 | 166.499 |
| 3800 | 119 064 | 211.841 | 72 793 | 167.624 |
| 4000 | 126 846 | 213.837 | 76 948 | 168.691 |
| 4200 | 134 700 | 215.753 | 81 107 | 169.704 |
| 4400 | 142 624 | 217.594 | 85 266 | 170.670 |
| 4600 | 150 620 | 219.372 | 89 420 | 171.595 |
| 4800 | 158 682 | 221.087 | 93 579 | 172.482 |
| 5000 | 166 808 | 222.744 | 97 734 | 173.327 |
| 5200 | 174 996 | 224.351 | 101 893 | 174.143 |
| 5400 | 183 247 | 225.907 | 106 052 | 174.930 |
| 5600 | 191 556 | 227.418 | 110 207 | 175.683 |
| 5800 | 199 924 | 228.886 | 114 365 | 176.415 |
| 6000 | 208 346 | 230.313 | 118 524 | 177.118 |

**Table A.11** (Continued)
Enthalpy of Formation at 25°C, Ideal Gas Enthalpy and Absolute
Entropy at 0.1 MPa (1 bar) Pressure

| | Nitric Oxide (NO) (6/30/63) | | Nitrogen Dioxide (NO₂) (6/30/63) | |
|---|---|---|---|---|
| | $(\bar{h}_f^\circ)_{298} = 90\ 592$ kJ/kmol $M = 30.006$ | | $(\bar{h}_f^\circ)_{298} = 33\ 723$ kJ/kmol $M = 46.005$ | |
| Temp. K | $(\bar{h}^\circ - \bar{h}^\circ_{298})$ kJ/kmol | $\bar{s}^\circ$ kJ/kmol K | $(\bar{h}^\circ - \bar{h}^\circ_{298})$ kJ/kmol | $\bar{s}^\circ$ kJ/kmol K |
| 0 | −9 192 | 0 | −10 196 | 0 |
| 100 | −6 071 | 177.034 | −6 870 | 202.431 |
| 200 | −2 950 | 198.753 | −3 502 | 225.732 |
| 298 | 0 | 210.761 | 0 | 239.953 |
| 300 | 54 | 210.950 | 67 | 240.183 |
| 400 | 3 042 | 219.535 | 3 950 | 251.321 |
| 500 | 6 058 | 226.267 | 8 150 | 260.685 |
| 600 | 9 146 | 231.890 | 12 640 | 268.865 |
| 700 | 12 309 | 236.765 | 17 368 | 276.149 |
| 800 | 15 548 | 241.091 | 22 288 | 282.714 |
| 900 | 18 857 | 244.991 | 27 359 | 288.684 |
| 1000 | 22 230 | 248.543 | 32 552 | 294.153 |
| 1100 | 25 652 | 251.806 | 37 836 | 299.190 |
| 1200 | 29 121 | 254.823 | 43 196 | 303.855 |
| 1300 | 32 627 | 257.626 | 48 618 | 308.194 |
| 1400 | 36 166 | 260.250 | 54 095 | 312.253 |
| 1500 | 39 731 | 262.710 | 59 609 | 316.056 |
| 1600 | 43 321 | 265.028 | 65 157 | 319.637 |
| 1700 | 46 932 | 267.216 | 70 739 | 323.022 |
| 1800 | 50 559 | 269.287 | 76 345 | 326.223 |
| 1900 | 54 204 | 271.258 | 81 969 | 329.265 |
| 2000 | 57 861 | 273.136 | 87 613 | 332.160 |
| 2100 | 61 530 | 274.927 | 93 274 | 334.921 |
| 2200 | 65 216 | 276.638 | 98 947 | 337.562 |
| 2300 | 68 906 | 278.279 | 104 633 | 340.089 |
| 2400 | 72 609 | 279.856 | 110 332 | 342.515 |
| 2500 | 76 320 | 281.370 | 116 039 | 344.846 |
| 2600 | 80 036 | 282.827 | 121 754 | 347.089 |
| 2700 | 83 764 | 284.232 | 127 478 | 349.248 |
| 2800 | 87 492 | 285.592 | 133 206 | 351.331 |
| 2900 | 91 232 | 286.902 | 138 942 | 353.344 |
| 3000 | 94 977 | 288.174 | 144 683 | 355.289 |
| 3200 | 102 479 | 290.592 | 156 180 | 359.000 |
| 3400 | 110 002 | 292.876 | 167 695 | 362.490 |
| 3600 | 117 545 | 295.031 | 179 222 | 365.783 |
| 3800 | 125 102 | 297.073 | 190 761 | 368.904 |

**Table A.11** (Continued)
Enthalpy of Formation at 25°C, Ideal Gas Enthalpy and Absolute
Entropy at 0.1 MPa (1 bar) Pressure

| Temp. K | Nitric Oxide (NO) (6/30/63) $(\bar{h}_f^\circ)_{298} = 90\ 592$ kJ/kmol $M = 30.006$ | | Nitrogen Dioxide (NO₂) (6/30/63) $(\bar{h}_f^\circ)_{298} = 33\ 723$ kJ/kmol $M = 46.005$ | |
|---|---|---|---|---|
| | $(\bar{h}^\circ - \bar{h}^\circ_{298})$ kJ/kmol | $\bar{s}^\circ$ kJ/kmol K | $(\bar{h}^\circ - \bar{h}^\circ_{298})$ kJ/kmol | $\bar{s}^\circ$ kJ/kmol K |
| 4000 | 132 675 | 299.014 | 202 309 | 371.866 |
| 4200 | 140 260 | 300.864 | 213 865 | 374.682 |
| 4400 | 147 863 | 302.634 | 225 430 | 377.372 |
| 4600 | 155 473 | 304.324 | 237 003 | 379.946 |
| 4800 | 163 101 | 305.947 | 248 580 | 382.410 |
| 5000 | 170 736 | 307.508 | 260 161 | 384.774 |
| 5200 | 178 381 | 309.006 | 271 751 | 387.046 |
| 5400 | 186 042 | 310.449 | 283 340 | 389.234 |
| 5600 | 193 707 | 311.847 | 294 934 | 391.343 |
| 5800 | 201 388 | 313.194 | 306 532 | 393.376 |
| 6000 | 209 074 | 314.495 | 318 130 | 395.343 |

# Table A.12
## Logarithms to the Base e of the Equilibrium Constant K

For the reaction $\nu_A A + \nu_B B \rightleftharpoons \nu_C C + \nu_D D$ the equilibrium constant $K$ is defined as

$$K = \frac{a_C^{\nu_C} a_D^{\nu_D}}{a_A^{\nu_A} a_B^{\nu_B}}$$

Base on thermodynamic data given in the JANAF Thermochemical Tables, Thermal Research Laboratory, The Dow Chemical Company, Midland, Michigan.

| Temp. K | $H_2 \rightleftharpoons 2H$ | $O_2 \rightleftharpoons 2O$ | $N_2 \rightleftharpoons 2N$ | $H_2O \rightleftharpoons H_2 + \frac{1}{2}O_2$ | $H_2O \rightleftharpoons \frac{1}{2}H_2 + OH$ | $CO_2 \rightleftharpoons CO + \frac{1}{2}O_2$ | $\frac{1}{2}N_2 + \frac{1}{2}O_2 \rightleftharpoons NO$ |
|---|---|---|---|---|---|---|---|
| 298 | −164.005 | −186.975 | −367.480 | −92.208 | −106.208 | −103.762 | −35.052 |
| 500 | −92.827 | −105.630 | −213.372 | −52.691 | −60.281 | −57.616 | −20.295 |
| 1000 | −39.803 | −45.150 | −99.127 | −23.163 | −26.034 | −23.529 | −9.388 |
| 1200 | −30.874 | −35.005 | −80.011 | −18.182 | −20.283 | −17.871 | −7.569 |
| 1400 | −24.463 | −27.742 | −66.329 | −14.609 | −16.099 | −13.842 | −6.270 |
| 1600 | −19.637 | −22.285 | −56.055 | −11.921 | −13.066 | −10.830 | −5.294 |
| 1800 | −15.866 | −18.030 | −48.051 | −9.826 | −10.657 | −8.497 | −4.536 |
| 2000 | −12.840 | −14.622 | −41.645 | −8.145 | −8.728 | −6.635 | −3.931 |
| 2200 | −10.353 | −11.827 | −36.391 | −6.768 | −7.148 | −5.120 | −3.433 |
| 2400 | −8.276 | −9.497 | −32.011 | −5.619 | −5.832 | −3.860 | −3.019 |
| 2600 | −6.517 | −7.521 | −28.304 | −4.648 | −4.719 | −2.801 | −2.671 |
| 2800 | −5.002 | −5.826 | −25.117 | −3.812 | −3.763 | −1.894 | −2.372 |
| 3000 | −3.685 | −4.357 | −22.359 | −3.086 | −2.937 | −1.111 | −2.114 |
| 3200 | −2.534 | −3.072 | −19.937 | −2.451 | −2.212 | −0.429 | −1.888 |
| 3400 | −1.516 | −1.935 | −17.800 | −1.891 | −1.576 | 0.169 | −1.690 |
| 3600 | −0.609 | −0.926 | −15.898 | −1.392 | −1.088 | 0.701 | −1.513 |
| 3800 | 0.202 | −0.019 | −14.199 | −0.945 | −0.501 | 1.176 | −1.356 |
| 4000 | 0.934 | 0.796 | −12.660 | −0.542 | −0.044 | 1.599 | −1.216 |
| 4500 | 2.486 | 2.513 | −9.414 | 0.312 | 0.920 | 2.490 | −0.921 |
| 5000 | 3.725 | 3.895 | −6.807 | 0.996 | 1.689 | 3.197 | −0.686 |
| 5500 | 4.743 | 5.023 | −4.666 | 1.560 | 2.318 | 3.771 | −0.497 |
| 6000 | 5.590 | 5.963 | −2.865 | 2.032 | 2.843 | 4.245 | −0.341 |

**Table A.13**

One-Dimensional Isentropic Compressible-Flow Functions for an Ideal Gas with Constant Specific Heat and Molecular Weight and $k = 1.4$[a]

| $M$ | $M^*$ | $\dfrac{\mathscr{A}}{\mathscr{A}^*}$ | $\dfrac{P}{P_0}$ | $\dfrac{\rho}{\rho_0}$ | $\dfrac{T}{T_0}$ |
|------|----------|----------|----------|----------|----------|
| 0 | 0 | $\infty$ | 1.000 00 | 1.000 00 | 1.000 00 |
| 0.10 | 0.109 43 | 5.8218 | 0.993 03 | 0.995 02 | 0.998 00 |
| 0.20 | 0.218 22 | 2.9635 | 0.972 50 | 0.980 27 | 0.992 06 |
| 0.30 | 0.325 72 | 2.0351 | 0.939 47 | 0.956 38 | 0.982 32 |
| 0.40 | 0.431 33 | 1.5901 | 0.895 62 | 0.924 28 | 0.968 99 |
| 0.50 | 0.534 52 | 1.3398 | 0.843 02 | 0.885 17 | 0.952 38 |
| 0.60 | 0.634 80 | 1.1882 | 0.784 00 | 0.840 45 | 0.932 84 |
| 0.70 | 0.731 79 | 1.094 37 | 0.720 92 | 0.791 58 | 0.910 75 |
| 0.80 | 0.825 14 | 1.038 23 | 0.656 02 | 0.740 00 | 0.886 52 |
| 0.90 | 0.914 60 | 1.008 86 | 0.591 26 | 0.687 04 | 0.860 58 |
| 1.00 | 1.000 00 | 1.000 00 | 0.528 28 | 0.633 94 | 0.833 33 |
| 1.10 | 1.081 24 | 1.007 93 | 0.468 35 | 0.581 69 | 0.805 15 |
| 1.20 | 1.1583 | 1.030 44 | 0.412 38 | 0.531 14 | 0.776 40 |
| 1.30 | 1.2311 | 1.066 31 | 0.360 92 | 0.482 91 | 0.747 38 |
| 1.40 | 1.2999 | 1.1149 | 0.314 24 | 0.437 42 | 0.718 39 |
| 1.50 | 1.3646 | 1.1762 | 0.272 40 | 0.394 98 | 0.689 65 |
| 1.60 | 1.4254 | 1.2502 | 0.235 27 | 0.355 73 | 0.661 38 |
| 1.70 | 1.4825 | 1.3376 | 0.202 59 | 0.319 69 | 0.633 72 |
| 1.80 | 1.5360 | 1.4390 | 0.174 04 | 0.286 82 | 0.606 80 |
| 1.90 | 1.5861 | 1.5552 | 0.149 24 | 0.256 99 | 0.580 72 |
| 2.00 | 1.6330 | 1.6875 | 0.127 80 | 0.230 05 | 0.555 56 |
| 2.10 | 1.6769 | 1.8369 | 0.109 35 | 0.205 80 | 0.531 35 |
| 2.20 | 1.7179 | 2.0050 | 0.093 52 | 0.184 05 | 0.508 13 |

**Table A.13** (Continued)
One-Dimensional Isentropic Compressible-Flow Functions for an Ideal Gas with Constant Specific Heat and Molecular Weight and $k = 1.4$[a]

| $M$ | $M^*$ | $\dfrac{\mathscr{A}}{\mathscr{A}^*}$ | $\dfrac{P}{P_0}$ | $\dfrac{\rho}{\rho_0}$ | $\dfrac{T}{T_0}$ |
|---|---|---|---|---|---|
| 2.30 | 1.7563 | 2.1931 | 0.079 97 | 0.164 58 | 0.485 91 |
| 2.40 | 1.7922 | 2.4031 | 0.068 40 | 0.147 20 | 0.464 68 |
| 2.50 | 1.8258 | 2.6367 | 0.058 53 | 0.131 69 | 0.444 44 |
| 2.60 | 1.8572 | 2.8960 | 0.050 12 | 0.117 87 | 0.425 17 |
| 2.70 | 1.8865 | 3.1830 | 0.042 95 | 0.105 57 | 0.406 84 |
| 2.80 | 1.9140 | 3.5001 | 0.036 85 | 0.094 62 | 0.389 41 |
| 2.90 | 1.9398 | 3.8498 | 0.031 65 | 0.084 89 | 0.372 86 |
| 3.00 | 1.9640 | 4.2346 | 0.027 22 | 0.076 23 | 0.357 14 |
| 3.50 | 2.0642 | 6.7896 | 0.013 11 | 0.045 23 | 0.289 86 |
| 4.00 | 2.1381 | 10.719 | 0.006 58 | 0.027 66 | 0.238 10 |
| 4.50 | 2.1936 | 16.562 | 0.003 46 | 0.017 45 | 0.198 02 |
| 5.00 | 2.2361 | 25.000 | $189(10)^{-5}$ | 0.011 34 | 0.166 67 |
| 6.00 | 2.2953 | 53.180 | $633(10)^{-6}$ | 0.005 19 | 0.121 95 |
| 7.00 | 2.3333 | 104.143 | $242(10)^{-6}$ | 0.002 61 | 0.092 59 |
| 8.00 | 2.3591 | 190.109 | $102(10)^{-6}$ | 0.001 41 | 0.072 46 |
| 9.00 | 2.3772 | 327.189 | $474(10)^{-7}$ | 0.000 815 | 0.058 14 |
| 10.00 | 2.3904 | 535.938 | $236(10)^{-7}$ | 0.000 495 | 0.047 62 |
| $\infty$ | 2.4495 | $\infty$ | 0 | 0 | 0 |

[a] Abridged from Table 30 in *Gas Tables,* by Joseph H. Keenan and Joseph Kaye. Copyright 1948, by Joseph H. Keenan and Joseph Kaye. Published by John Wiley & Sons, Inc., New York.

**Table A.14**

One-Dimensional Normal-Shock Functions for an Ideal Gas with Constant Specific Heat and Molecular Weight and $k = 1.4$[a]

| $M_x$ | $M_y$ | $\dfrac{P_y}{P_x}$ | $\dfrac{\rho_y}{\rho_x}$ | $\dfrac{T_y}{T_x}$ | $\dfrac{P_{oy}}{P_{ox}}$ | $\dfrac{P_{oy}}{P_x}$ |
|------|---------|---------|--------|--------|----------|---------|
| 1.00 | 1.000 00 | 1.0000 | 1.0000 | 1.0000 | 1.000 00 | 1.8929 |
| 1.10 | 0.911 77 | 1.2450 | 1.1691 | 1.0649 | 0.998 92 | 2.1328 |
| 1.20 | 0.842 17 | 1.5133 | 1.3416 | 1.1280 | 0.992 80 | 2.4075 |
| 1.30 | 0.785 96 | 1.8050 | 1.5157 | 1.1909 | 0.979 35 | 2.7135 |
| 1.40 | 0.739 71 | 2.1200 | 1.6896 | 1.2547 | 0.958 19 | 3.0493 |
| 1.50 | 0.701 09 | 2.4583 | 1.8621 | 1.3202 | 0.929 78 | 3.4133 |
| 1.60 | 0.668 44 | 2.8201 | 2.0317 | 1.3880 | 0.895 20 | 3.8049 |
| 1.70 | 0.640 55 | 3.2050 | 2.1977 | 1.4583 | 0.855 73 | 4.2238 |
| 1.80 | 0.616 50 | 3.6133 | 2.3592 | 1.5316 | 0.812 68 | 4.6695 |
| 1.90 | 0.595 62 | 4.0450 | 2.5157 | 1.6079 | 0.767 35 | 5.1417 |
| 2.00 | 0.577 35 | 4.5000 | 2.6666 | 1.6875 | 0.720 88 | 5.6405 |
| 2.10 | 0.561 28 | 4.9784 | 2.8119 | 1.7704 | 0.674 22 | 6.1655 |
| 2.20 | 0.547 06 | 5.4800 | 2.9512 | 1.8569 | 0.628 12 | 6.7163 |
| 2.30 | 0.534 41 | 6.0050 | 3.0846 | 1.9468 | 0.583 31 | 7.2937 |
| 2.40 | 0.523 12 | 6.5533 | 3.2119 | 2.0403 | 0.540 15 | 7.8969 |
| 2.50 | 0.512 99 | 7.1250 | 3.3333 | 2.1375 | 0.499 02 | 8.5262 |
| 2.60 | 0.503 87 | 7.7200 | 3.4489 | 2.2383 | 0.460 12 | 9.1813 |
| 2.70 | 0.495 63 | 8.3383 | 3.5590 | 2.3429 | 0.423 59 | 9.8625 |
| 2.80 | 0.488 17 | 8.9800 | 3.6635 | 2.4512 | 0.389 46 | 10.569 |
| 2.90 | 0.481 38 | 9.6450 | 3.7629 | 2.5632 | 0.357 73 | 11.302 |
| 3.00 | 0.475 19 | 10.333 | 3.8571 | 2.6790 | 0.328 34 | 12.061 |
| 4.00 | 0.434 96 | 18.500 | 4.5714 | 4.0469 | 0.138 76 | 21.068 |
| 5.00 | 0.415 23 | 29.000 | 5.0000 | 5.8000 | 0.061 72 | 32.654 |
| 10.00 | 0.387 57 | 116.50 | 5.7143 | 20.388 | 0.003 04 | 129.217 |
| ∞ | 0.377 96 | ∞ | 6.000 | ∞ | 0 | ∞ |

[a] Abridged from Table 48 in *Gas Tables*, by Joseph H. Keenan and Joseph Kaye. Copyright 1948, by Joseph H. Keenan and Joseph Kaye. Published by John Wiley & Sons, Inc., New York.

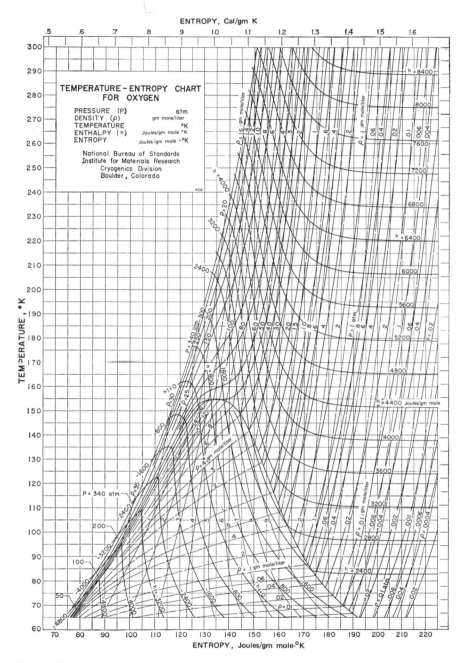

**ENTROPY, Cal/gm K**

**TEMPERATURE-ENTROPY CHART FOR OXYGEN**

PRESSURE (P)      atm
DENSITY ($\rho$)      gm mole/liter
TEMPERATURE      °K
ENTHALPY (h)      Joules/gm mole °K
ENTROPY      Joules/gm mole -°K

National Bureau of Standards
Institute for Materials Research
Cryogenics Division
Boulder, Colorado

**TEMPERATURE, °K**

**ENTROPY, Joules/gm mole-°K**

**Fig. A.2** Temperature-entropy diagram for oxygen (Reprinted by permission from ASHRAE Handbook of Fundamentals 1967).

701

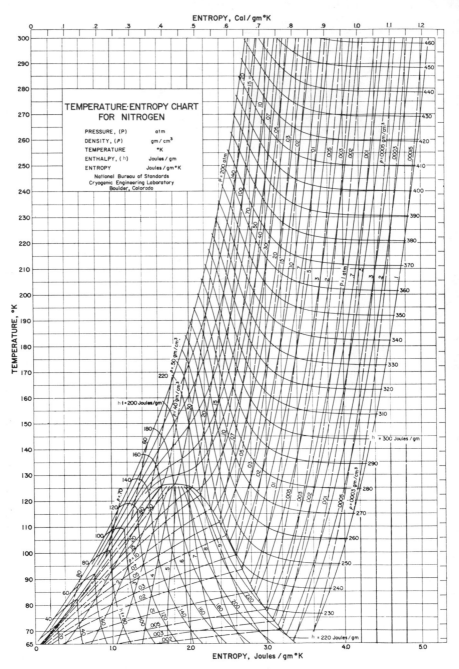

**Fig. A.3** Temperature-entropy diagram for nitrogen (Reprinted by permission from ASHRAE Handbook of Fundamentals 1967).

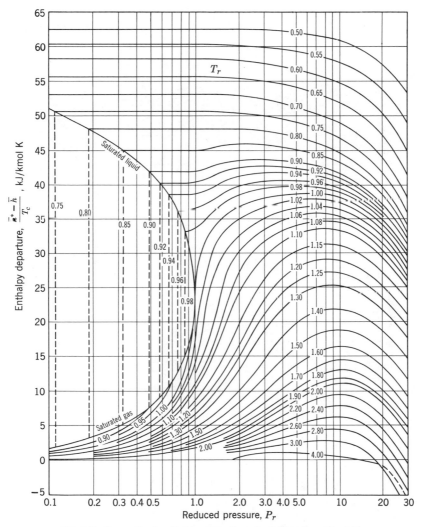

**Fig. A.6** Generalized enthalpy correction chart (Based on Fig. A.5).

**703**

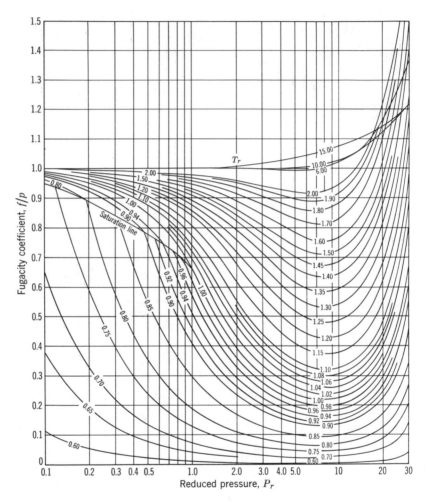

**Fig. A.7** Generalized fugacity coefficient chart (Based on Fig. A.5).

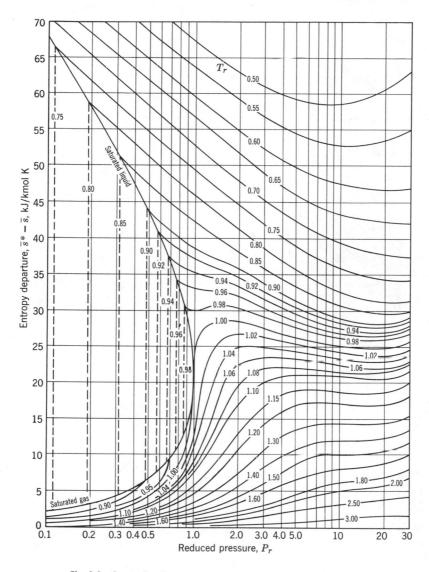

**Fig. A.8** Generalized entropy correction chart (Based on Fig. A.5).

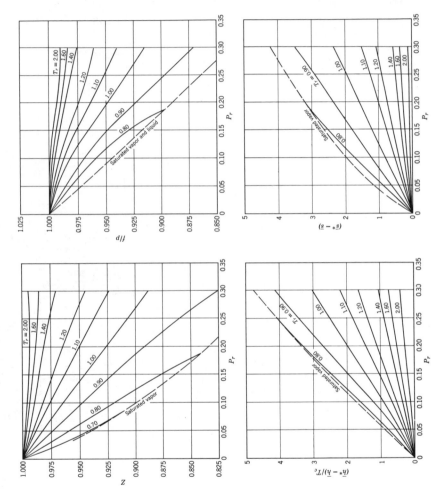

**Fig. A.9** Generalized charts at low pressure.

# Some Selected References

H. B. Callen, *Thermodynamics*, John Wiley and Sons, New York, 1960.

G. N. Hatsopoulos and J. H. Keenan, *Principles of General Thermodynamics*, John Wiley and Sons, New York, 1965.

O. A. Hougen, K. M. Watson, and R. A. Ragatz, *Chemical Process Principles, Part Two: Thermodynamics*, Second Edition, John Wiley and Sons, New York, 1959.

J. H. Keenan, *Thermodynamics*, John Wiley and Sons, New York, 1941.

J. Kestin, *A Course in Thermodynamics*, Blaisdell Publishing Co., Waltham, Mass., 1966.

E. F. Obsert, *Concepts of Thermodynamics*, McGraw-Hill Book Co., New York, 1960.

H. Reiss, *Methods of Thermodynamics*, Blaisdell Publishing Co., Waltham, Mass., 1965.

W. C. Reynolds, *Thermodynamics*, Second Edition, McGraw-Hill Book Co., New York, 1968.

A. H. Shapiro: *The Dynamics and Thermodynamics of Compressible Fluid Flow*, The Ronald Press Co., New York, 1953.

M. W. Zemansky: *Heat and Thermodynamics*, McGraw-Hill Book Co., New York, 1957.

M. W. Zemansky and H. C. Van Ness, *Basic Engineering Thermodynamics*, McGraw-Hill Book Co., New York, 1966.

# Answers to Selected Problems

| | |
|---|---|
| 2.3 | 960 N; 960 N |
| 2.6 | 2.55 kg |
| 2.9 | 2.35 kPa, 17.6 mm |
| 2.12 | 1.510 m; 1.496 m |
| 2.15 | 1.90 MPa |
| | |
| 3.3 | 0.0026 kg/minute |
| 3.9 | (a) 0.0888 m³/kg |
| | (b) 0.002 83 m³/kg |
| | (c) 0.023 08 m³/kg |
| | (d) 0.027 32 m³/kg |
| 3.15 | 33% |
| 3.18 | 0.0614 m³; 13.97 kg |
| 3.21 | 1.518 m³ |
| 3.24 | 0.067 98 m³/s; 3.3% |
| 3.27 | 0.123% |
| | |
| 4.3 | (b) 20 kJ; 50% |
| 4.6 | (a) 99.6°C |
| | (b) 159.4 kJ |
| 4.9 | (a) 423 MJ |
| | (b) 28.4 m³ |
| 4.12 | (a) 35.3 kJ |
| | (b) 5.9% |

4.15   (*a*)   247 kPa; 0.0977 m$^3$
     (*b*)   16.7 kJ
4.21   0; −265.7 kJ


5.3   (*a*)   266.7 MJ
    (*b*)   0
5.6   2111.6 MJ
5.9   115.2 hours
5.12   (*a*)   1.993 MPa; 13.033 MJ; 346.5 kJ
5.15   −13.7°C
5.18   282 kPa
5.21   (*a*)   151.6°C
    (*b*)   152.3 kJ
    (*c*)   682 kPa; 142.5°C
5.24   (*a*)   2596.3 kJ/kg
    (*b*)   520.8 kJ/kg
    (*c*)   420.9 kJ/kg
5.27   −22.9 kJ; −8.6 kJ
5.30   6.66 m/s
5.33   269.4°C; 135 mm$^2$
5.36   4.18 kW; 0.42 kW
5.39   418.3 kJ/kg
5.42   (*a*)   2673.9 kJ/kg; 97.55%
    (*b*)   18.394 MW
    (*c*)   22.489 MW
    (*d*)   3.49%
    (*e*)   126.3°C
    (*f*)   120.45 MW
    (*g*)   0.26
5.45   1890 kJ; 50.6 MJ
5.48   25°C; 3.46
5.51   95.9%
5.54   0.125 kg
5.57   7.245 kg
5.60   312°C
5.63   (*a*)   28.65 kg
    (*b*)   4024 kJ
5.66   (*a*)   0.0415
    (*b*)   0.846
5.69   (*a*)   262 kPa
    (*b*)   6.11 kJ
    (*c*)   8.52 kJ
5.72   (*a*)   137.2°C; 3.82 kg
    (*b*)   1.816 MPa; 0.112 m$^3$
    (*c*)   220 kJ

5.75    (a)  1373 kPa
        (b)  1338 kPa
        (c)  1288 kPa

6.6     6.7
6.9     1.626 kW
6.12    0.731
6.15    3 J; 0.000 33
6.18    (a)  0.91 kW
        (b)  40°C

7.3     (b)  0.9610; 0.2876
        (c)  4.6
7.6     0.843
7.9     (b)  2.789 MPa; 0.005 258 m³/kg
        (c)  −85.2 kJ; −111 kJ
7.12    700 kJ; 234.5°C
7.15    (a)  949.5 kPa; 799.8 kPa
        (b)  1.15 kg
        (c)  0.7624 kJ/K
7.18    2.0978 kJ/K
7.21    (a)  14.75 MJ
        (b)  25.742 kJ/K
7.24    (b)  136.9 K; −702.7 kJ/kg
7.27    17.86 MJ
7.30    (a)  1.2685
        (b)  0.1676 m³
        (c)  −95.3 kJ; −31.5 kJ
        (d)  0.0153 kJ/K
7.33    0.679 kW
7.36    0.882 kW
7.39    550 kPa; 324°C
7.42    (a)  16.35 MW
        (b)  4.232 MW
7.45    119.2 kW
7.48    0.0895 kJ/kg K
7.54    (a)  47.955 MJ
        (b)  −126.85 kJ/K; 150.58 kJ/K
7.57    696 m/s; 0.95
7.60    (a)  1.867; 0.9639
        (b)  3.811; 0.9982
7.63    415 kPa; −44.8°C
7.66    (a)  230.5 kPa
        (b)  0.259 kg
        (c)  1.95 kJ
        (d)  0.3203 m

7.69   108.8°C; 316°C; 0.253 kg
7.72   (a) 8.9 kg
       (b) 25.46 MJ
       (c) −1.9003 kJ/K; 19.5583 kJ/K
7.75   0; 415.5 kJ

8.3    4.3 kJ/kg
8.6    (a) 2.42 kW
       (b) 1.0 kW
8.9    −636.8 kJ/kg
8.12   (a) 13.0 kW
       (b) 78.3%; 2.61 kW
8.15   (a) 264°C
       (b) 131.7°C; 74.5
8.18   135.0 kJ/kg; 114.6 kJ/kg; 222 kJ
8.21   257.5 kJ/kg
8.24   33.37 MJ
8.27   811.6 K

9.3    28.4%; 0.1132; 34.1%; 0.1989; 36.3%, 0.2422; 37.9%; 0.2941
9.6    36.9%; 878.6 kJ/kg
9.9    (a) 42.7%
       (b) 46.1%
       (c) 47.5%
9.12   (a) 27.4%
       (b) 34.2 kJ/kg
9.15   179.19 MW; 71.54 MW
9.18   3.207; 3.347
9.21   2.818; 2.92
9.24   1.051 kJ
9.27   (b) 7.307 MPa; 3577 K; 56.6%
9.30   −197.1 kJ; −197.1 kJ; 0, 770.2 kJ;
       907.5 kJ, 907.5 kJ; 0, −770.2 kJ; 78.3%
9.33   (a) −1407 kW; 3689 kW; 32.7%
       (b) 122.6 kPa
9.36   (a) 31.21 MW
       (b) 0.359
       (c) 79.18 kg/s; 69.37 m³/s
9.39   −126.8 kJ; 423.4 kJ; 70.1%
9.42   (a) 200 kPa
       (b) 155.9 kJ; 0.641 kg/s
       (c) 750.1 K; 34.5%
9.45   285 kPa; 760 m/s
9.48   0.258

10.3   5.2 $\mu$Pa
10.15  (a) 1160 m/s
       (b) 2.23 kJ
10.18  0.400; 6.13 kg
10.30  1038 kJ/kmol
10.33  21.96 kg; 4126 kJ
10.36  292 K
10.39  (a) 15.23 kg
       (b) 2139 kJ
10.42  (a) 29.3 kJ/kg
       (b) 29.3 kJ/kg
10.45  10.54 kg
10.48  138.9 kJ
10.51  (a) −4724 kJ
       (b) 1271 kJ
10.54  4010 kJ

11.3   (a) 181 kPa; 302.3 K
       (b) 16.721 J/K
11.6   353.9 K; 6.228 J/mol K
11.9   (a) 1274 J/mol
       (b) 1274 J/mol
11.12  49.1°C; 70.44 kg
11.15  31.8°C
11.18  (a) 0.0183; 67.2%
11.21  −76 kJ; 90%; 0.0107
11.24  −5.15 kJ; 0.0123 kg; 155.9 kPa
11.27  (a) 67 kJ/kg
       (b) 0.377
11.30  (a) 500 kPa
       (b) 0.000 93 kg/s
       (c) 17.3 kW
11.33  (a) 0.1782 m³/kmol; 0.2865 m³/kmol
11.36  (a) −6.99 kJ; −5.50 kJ
       (b) −8.21 kJ; −5.40 kJ
       (c) −5.28 kJ; −5.28 kJ
11.39  (a) 4.5 MPa
       (b) −1308 kJ; 3.4 kJ/K
11.42  (a) 1.82 kg
       (b) 2.284 MPa
11.45  −3.228 kW
11.48  0.001 447 kmol; 7.7%

12.3   (a) 1.136 kg/kg
       (b) 1.411 kg/kg

12.6    2855 K

12.9    −754 031 kJ/kmol

12.12    (a)  141.6%

         (b)  0.000 431 kg/s

         (c)  0.1578 m³/s

12.15    38.66 kW; −83.3 kW

12.18    317%

12.21    1583 K

12.24    (a)  314%

         (b)  35.2°C

12.27    −8437 kJ/kg reactants

12.30    1852.7 kJ/K

12.36    −172 477 kJ; 194.46 kJ/K

12.39    1.219 V

12.42    −150 123 kJ

13.3    (a)  16.044 MPa; 88.97% $CH_4$

       (b)  16.45 MPa; 88.35% $CH_4$

13.6    449 kPa

13.9    (a)  0.003 53

       (b)  0.004 95

13.12    (a)  1.185 V

        (b)  1.109 V

13.15    (a)  3.27% N

        (b)  28.18% N

13.18    177 168 kJ

13.21    0.108 39; 32.05% $CO_2$; 11.46% CO; 56.49% $O_2$

13.24    72.8% $N_2$; 3.8% $O_2$; 11% $CO_2$; 12% $H_2O$; 0.328% NO

13.30    4.2; 1.91 MPa

13.33    (a)  9.874

        (b)  3.26% $N_2$; 9.79% $H_2$; 86.95% $NH_3$

13.36    0.6032

13.39    6.21% $CO_2$; 7.76% $H_2O$; 75.9% $N_2$; 10.07% $O_2$;

        0.052% NO; 0.001% $NO_2$

13.42    66.16% $H_2O$; 12.94% $H_2$; 5.45% $O_2$; 9.75% OH; 5.70% H

13.45    1.07% Ar; 48.40% $Ar^+$; 0.71% $Ar^{++}$; 49.82% $e^-$

14.3    703 kPa; 234°C

14.6    (b)  1345 mm²; 2598 mm²

14.9    1859 kPa

14.12    61 kPa; 294.7 K; 0.4867

14.15    55.8 kPa; 294.7 K; 0.4447

14.18    0.019 35 m²

14.21    3.735 kg/s

14.24    79.8%

14.27    (a)  293.5 kPa

        (c)  406.4 kJ/kg

# Index